BRIDGE
ENGINEERING
HANDBOOK

BRIDGE ENGINEERING
HANDBOOK

VOLUME I

EDITED BY

WAI-FAH CHEN and LIAN DUAN

CRC Press
Taylor & Francis Group
Boca Raton London New York

CRC Press is an imprint of the
Taylor & Francis Group, an **informa** business

First published 1999 by CRC Press
Taylor & Francis Group
6000 Broken Sound Parkway NW, Suite 300
Boca Raton, FL 33487-2742

Reissued 2019 by CRC Press

© 1999 by Taylor & Francis Group.
CRC Press is an imprint of Taylor & Francis Group, an Informa business

No claim to original U.S. Government works

A Library of Congress record exists under LC control number:

Publisher's Note
The publisher has gone to great lengths to ensure the quality of this reprint but points out that some imperfections in the original copies may be apparent.

Disclaimer
The publisher has made every effort to trace copyright holders and welcomes correspondence from those they have been unable to contact.

ISBN 13: 978-0-367-26344-7 (set)
ISBN 13: 978-0-367-22822-4 (hbk)
ISBN 13: 978-0-367-22825-5 (pbk)
ISBN 13: 978-0-429-27704-7 (ebk)

Visit the Taylor & Francis Web site at http://www.taylorandfrancis.com and the CRC Press Web site at http://www.crcpress.com

Foreword

Among all engineering subjects, bridge engineering is probably the most difficult on which to compose a handbook because it encompasses various fields of arts and sciences. It not only requires knowledge and experience in bridge design and construction, but often involves social, economic, and political activities. Hence, I wish to congratulate the editors and authors for having conceived this thick volume and devoted the time and energy to complete it in such short order. Not only is it the first handbook of bridge engineering as far as I know, but it contains a wealth of information not previously available to bridge engineers. It embraces almost all facets of bridge engineering except the rudimentary analyses and actual field construction of bridge structures, members, and foundations. Of course, bridge engineering is such an immense subject that engineers will always have to go beyond a handbook for additional information and guidance.

I may be somewhat biased in commenting on the background of the two editors, who both came from China, a country rich in the pioneering and design of ancient bridges and just beginning to catch up with the modern world in the science and technology of bridge engineering. It is particularly to the editors' credit to have convinced and gathered so many internationally recognized bridge engineers to contribute chapters. At the same time, younger engineers have introduced new design and construction techniques into the treatise.

This Handbook is divided into seven sections, namely:

- Fundamentals
- Superstructure Design
- Substructure Design
- Seismic Design
- Construction and Maintenance
- Special Topics
- Worldwide Practice

There are 67 chapters, beginning with bridge concepts and aesthestics, two areas only recently emphasized by bridge engineers. Some unusual features, such as rehabilitation, retrofit, and maintenance of bridges, are presented in great detail. The section devoted to seismic design includes soil-foundation-structure interaction. Another section describes and compares bridge engineering practices around the world. I am sure that these special areas will be brought up to date as the future of bridge engineering develops.

May I advise each bridge engineer to have a desk copy of this volume with which to survey and examine both the breadth and depth of bridge engineering.

T. Y. Lin
Professor Emeritus, University of California at Berkeley
Chairman, Lin Tung-Yen China, Inc.

Preface

The *Bridge Engineering Handbook* is a unique, comprehensive, and state-of-the-art reference work and resource book covering the major areas of bridge engineering with the theme "bridge to the 21st century." It has been written with practicing bridge and structural engineers in mind. The ideal readers will be M.S.-level structural and bridge engineers with a need for a single reference source to keep abreast of new developments and the state-of-the-practice, as well as to review standard practices.

The areas of bridge engineering include planning, analysis and design, construction, maintenance, and rehabilitation. To provide engineers a well-organized, user-friendly, and easy-to-follow resource, the Handbook is divided into seven sections. *Section I, Fundamentals*, presents conceptual design, aesthetics, planning, design philosophies, bridge loads, structural analysis, and modeling. *Section II, Superstructure Design*, reviews how to design various bridges made of concrete, steel, steel-concrete composites, and timbers; horizontally curved, truss, arch, cable-stayed, suspension, floating, movable, and railroad bridges; and expansion joints, deck systems, and approach slabs. *Section III, Substructure Design*, addresses the various substructure components: bearings, piers and columns, towers, abutments and retaining structures, geotechnical considerations, footings, and foundations. *Section IV, Seismic Design*, provides earthquake geotechnical and damage considerations, seismic analysis and design, seismic isolation and energy dissipation, soil–structure–foundation interactions, and seismic retrofit technology and practice. *Section V, Construction and Maintenance*, includes construction of steel and concrete bridges, substructures of major overwater bridges, construction inspections, maintenance inspection and rating, strengthening, and rehabilitation. *Section VI, Special Topics*, addresses in-depth treatments of some important topics and their recent developments in bridge engineering. *Section VII, Worldwide Practice*, provides the global picture of bridge engineering history and practice from China, Europe, Japan, and Russia to the U.S.

The Handbook stresses professional applications and practical solutions. Emphasis has been placed on ready-to-use materials, and special attention is given to rehabilitation, retrofit, and maintenance. The Handbook contains many formulas and tables that give immediate answers to questions arising from practical works. It describes the basic concepts and assumptions, omitting the derivations of formulas and theories, and covers both traditional and new, innovative practices. An overview of the structure, organization, and contents of the book can be seen by examining the table of contents presented at the beginning, while an in-depth view of a particular subject can be seen by examining the individual table of contents preceding each chapter. References at the end of each chapter can be consulted for more-detailed studies.

The chapters have been written by many internationally known authors from different countries covering bridge engineering practices, research, and development in North America, Europe, and the Pacific Rim. This Handbook may provide a glimpse of a rapidly growing trend in global economy in recent years toward international outsourcing of practice and competition in all dimensions of engineering. In general, the Handbook is aimed toward the needs of practicing engineers, but materials may be reorganized to accommodate undergraduate and graduate level bridge courses. The book may also be used as a survey of the practice of bridge engineering around the world.

The authors acknowledge with thanks the comments, suggestions, and recommendations during the development of the Handbook by Fritz Leonhardt, Professor Emeritus, Stuttgart University, Germany; Shouji Toma, Professor, Horrai-Gakuen University, Japan; Gerard F. Fox, Consulting Engineer; Jackson L. Durkee, Consulting Engineer; Michael J. Abrahams, Senior Vice President, Parsons, Brinckerhoff, Quade & Douglas, Inc.; Ben C. Gerwick, Jr., Professor Emeritus, University of California at Berkeley; Gregory F. Fenves, Professor, University of California at Berkeley; John M. Kulicki, President and Chief Engineer, Modjeski and Masters; James Chai, Senior Materials and Research Engineer, California Department of Transportation; Jinrong Wang, Senior Bridge Engineer, URS Greiner; and David W. Liu, Principal, Imbsen & Associates, Inc.

We wish to thank all the authors for their contributions and also to acknowledge at CRC Press Nora Konopka, Acquiring Editor, and Carol Whitehead and Sylvia Wood, Project Editors.

Wai-Fah Chen
Lian Duan

Editors

Wai-Fah Chen is a George E. Goodwin Distinguished Professor of Civil Engineering and Head of the Department of Structural Engineering, School of Civil Engineering at Purdue University. He received his B.S. in civil engineering from the National Cheng-Kung University, Taiwan, in 1959, M.S. in structural engineering from Lehigh University, Bethlehem, Pennsylvania in 1963, and Ph.D. in solid mechanics from Brown University, Providence, Rhode Island in 1966.

Dr. Chen's research interests cover several areas, including constitutive modeling of engineering materials, soil and concrete plasticity, structural connections, and structural stability. He is the recipient of numerous engineering awards, including the AISC T.R. Higgins Lectureship Award, the ASCE Raymond C. Reese Research Prize, and the ASCE Shortridge Hardesty Award. He was elected to the National Academy of Engineering in 1995, and was awarded an Honorary Membership in the American Society of Civil Engineers in 1997. He was most recently elected to the Academia Sinica in Taiwan.

Dr. Chen is a member of the Executive Committee of the Structural Stability Research Council, the Specification Committee of the American Institute of Steel Construction, and the editorial board of six technical journals. He has worked as a consultant for Exxon's Production and Research Division on offshore structures, for Skidmore, Owings and Merril on tall steel buildings, and for World Bank on the Chinese University Development Projects.

A widely respected author, Dr. Chen's works include *Limit Analysis and Soil Plasticity* (Elsevier, 1975), the two-volume *Theory of Beam-Columns* (McGraw-Hill, 1976–77), *Plasticity in Reinforced Concrete* (McGraw-Hill, 1982), *Plasticity for Structural Engineers* (Springer-Verlag, 1988), and *Stability Design of Steel Frames* (CRC Press, 1991). He is the editor of two book series, one in structural engineering and the other in civil engineering. He has authored or coauthored more than 500 papers in journals and conference proceedings. He is the author or coauthor of 18 books, has edited 12 books, and has contributed chapters to 28 other books. His more recent books are *Plastic Design and Second-Order Analysis of Steel Frames* (Springer-Verlag, 1994), the two-volume *Constitutive Equations for Engineering Materials* (Elsevier, 1994), *Stability Design of Semi-Rigid Frames* (Wiley-Interscience, 1995), and *LRFD Steel Design Using Advanced Analysis* (CRC Press, 1997). He is editor-in-chief of *The Civil Engineering Handbook* (CRC Press, 1995, winner of the Choice Outstanding Academic Book Award for 1996, *Choice Magazine*), and the *Handbook of Structural Engineering* (CRC Press, 1997).

Lian Duan is a Senior Bridge Engineer with the California Department of Transportation, U.S., and Professor of Structural Engineering at Taiyuan University of Technology, China.

He received his B.S. in civil engineering in 1975, M.S. in structural engineering in 1981 from Taiyuan University of Technology, and Ph.D. in structural engineering from Purdue University, West Lafayette, Indiana in 1990. Dr. Duan worked at the Northeastern China Power Design Institute from 1975 to 1978.

Dr. Duan's research interests cover areas including inelastic behavior of reinforced concrete and steel structures, structural stability and seismic bridge analysis and design. He has authored or coauthored more than 60 papers, chapters, and reports, and his research has focused on the development of unified interaction equations for steel beam-columns, flexural stiffness of reinforced concrete members, effective length factors of compression members, and design of bridge structures.

Dr. Duan is also an esteemed practicing engineer. He has designed numerous building and bridge structures. Most recently, he has been involved in the seismic retrofit design of the San Francisco-Oakland Bay Bridge West spans and made significant contributions to the project. He is coeditor of the *Structural Engineering Handbook* CRCnetBase 2000 (CRC Press, 2000).

Contributors

Michael I. Abrahams
Parsons, Brinckerhoff, Quade &
   Douglas, Inc.
New York, New York

Mohamed Akkari
California Department of
   Transportation
Sacramento, California

Fadel Alameddine
California Department of
   Transportation
Sacramento, California

Masoud Alemi
California Department of
   Transportation
Sacramento, California

S. Altman
California Department of
   Transportation
Sacramento, California

Rambabu Bavirisetty
California Department of
   Transportation
Sacramento, California

David P. Billington
Department of Civil Engineering
   and Operations Research
Princeton University
Princeton, New Jersey

Michael Blank
U.S.Army Corps of Engineers
Philadelphia, Pennsylvania

Simon A. Blank
California Department of
   Transportation
Walnut Creek, California

Michel Bruneau
Department of Civil Engineering
State University of New York
Buffalo, New York

Chun S. Cai
Florida Department of
   Transportation
Tallahassee, Florida

James Chai
California Department of
   Transportation
Sacramento, California

Hong Chen
J. Muller International, Inc.
Sacramento, California

Kang Chen
MG Engineering, Inc.
San Francisco, California

Wai-Fah Chen
School of Civil Engineering
Purdue University
West Lafayette, Indiana

Nan Deng
Bechtel Corporation
San Francisco, California

Robert J. Dexter
Department of Civil Engineering
University of Minnesota
Minneapolis, Minnesota

Ralph J. Dornsife
Washington State Department of
   Transportation
Olympia, Washington

Lian Duan
California Department of
   Transportation
Sacramento, California

Mingzhu Duan
Quincy Engineering, Inc.
Sacramento, California

Jackson Durkee
Consulting Structural Engineer
Bethlehem, Pennsylvania

Marc O. Eberhard
Department of Civil and
   Environmental Engineering
University of Washington
Seattle, Washington

Johnny Feng
J. Muller International, Inc.
Sacramento, California

Gerard F. Fox
HNTB (Ret.)
Garden City, New York

John W. Fisher
Department of Civil Engineering
Lehigh University
Bethlehem, Pennsylvania

Kenneth J. Fridley
Washington State University
Pullman, Washington

John H. Fujimoto
California Department of
   Transportation.
Sacramento, California

Mahmoud Fustok
California Department of
   Transportation
Sacramento, California

Ben C. Gerwick, Jr.
Ben C. Gerwick, Inc.
Consulting Engineers
San Francisco, California

Mahmoud Fustok
California Department of
   Transportation
Sacramento, California

Ben C. Gerwick, Jr.
Ben C. Gerwick, Inc.
Consulting Engineers
San Francisco, California

Chao Gong
ICF Kaiser Engineers
Oakland, California

Frederick Gottemoeller
Rosales Gottemoeller & Associates,
   Inc.
Columbia, Maryland

Fuat S. Guzaltan
Parsons, Brickerhoff, Quade &
   Douglas, Inc.
Princeton, New Jersey

Danjian Han
Department of Civil Engineering
South China University of
   Technology
Guangzhou, China

Ikuo Harazaki
Honshu–Shikoku Bridge Authority
Tokyo, Japan

Lars Hauge
COWI
Consulting Engineers and Planners
Lyngby, Denmark

Oscar Henriquez
Department of Civil Engineering
California State University
Long Beach, California

Susan E. Hida
California Department of
   Transportation
Sacramento, California

Dietrich L. Hommel
COWI
Consulting Engineers and Planners
Lyngby, Denmark

Ahmad M. Itani
University of Nevada
Reno, Nevada

Kevin I. Keady
California Department of
   Transportation
Sacramento, California

Michael D. Keever
California Department of
   Transportation
Sacramento, California

Sangjin Kim
Kyungpook National University
Taeg, South Korea

F. Wayne Klaiber
Department of Civil Engineering
Iowa State University
Ames, Iowa

Michael Knott
Moffatt & Nichol Engineers
Richmond, Virginia

Steven Kramer
University of Washington
Seattle, Washington

Alexander Krimotat
SC Solutions, Inc.
Santa Clara, California

John M. Kulicki
Modjeski and Masters, Inc.
Harrisburg, Pennsylvania

John Kung
California Department of
   Transportation
Sacramento, California

Farzin Lackpour
Parsons, Brickerhoff, Quade &
   Douglas, Inc.
Princeton, New Jersey

Don Lee
California Department of
   Transportation
Sacramento, California

Fritz Leonhardt
Stuttgart University
Stuttgart, Germany

Fang Li
California Department of
   Transportation
Sacramento, California

Guohao Li
Department of Bridge Engineering
Tongji University
Shanghai, People's Republic of
   China

Xila Liu
Department of Civil Engineering
Tsinghua University
Beijing, China

Luis R. Luberas
U.S.Army Corps of Engineers
Philadelphia, Pennsylvania

M. Myint Lwin
Washington State Department of
   Transportation
Olympia, Washington

Jyouru Lyang
California Department of
   Transportation
Sacramento, California

Youzhi Ma
Geomatrix Consultants, Inc.
Oakland, California

Alfred R. Mangus
California Department of
   Transportation
Sacramento, California

W. N. Marianos, Jr.
Modjeski and Masters, Inc.
Edwardsville, Illinois

Brian Maroney
California Department of
   Transportation
Sacramento, California

Serge Montens
Jean Muller International
St.-Quentin-en-Yvelines
France

Jean M. Muller
Jean M. Muller International
St.-Quentin-en-Yvelines
France

Masatsugu Nagai
Department of Civil and
  Environmental Engineering
Nagaoka University of Technology
Nagaoka, Japan

Andrzej S. Nowak
Department of Civil and
  Environmental Engineering
University of Michigan
Ann Arbor, Michigan

Atsushi Okukawa
Honshu–Shikoku Bridge Authority
Kobe, Japan

Dan Olsen
COWI
Consulting Engineers and Planners
Lyngby, Denmark

Klaus H. Ostenfeld
COWI
Consulting Engineers and Planners
Lyngby, Denmark

Joseph Penzien
International Civil Engineering
  Consultants, Inc.
Berkeley, California

Philip C. Perdikaris
Department of Civil Engineering
Case Western Reserve University
Cleveland, Ohio

Joseph M. Plecnik
Department of Civil Engineering
California State University
Long Beach, California

Oleg A. Popov
Joint Stock Company
  Giprotransmost (Tramos)
Moscow, Russia

Zolan Prucz
Modjeski and Masters, Inc.
New Orleans, Louisiana

Mark L. Reno
California Department of
  Transportation
Sacramento, California

James Roberts
California Department of
  Transportation
Sacramento, California

Norman F. Root
California Department of
  Transportation
Sacramento, California

Yusuf Saleh
California Department of
  Transportation
Sacramento, California

Thomas E. Sardo
California Department of
  Transportation
Sacramento, California

Gerard Sauvageot
J. Muller International
San Diego, California

Charles Scawthorn
EQE International
Oakland, California

Charles Seim
T. Y. Lin International
San Francisco, California

Vadim A. Seliverstov
Joint Stock Company
  Giprotransmost (Tramos)
Moscow, Russia

Li-Hong Sheng
California Department of
  Transportation
Sacramento, California

Donald F. Sorgenfrei
Modjeski and Masters, Inc.
New Orleans, Louisiana

Jim Springer
California Department of
  Transportation
Sacramento, California

Shawn Sun
California Department of
  Transportation
Sacramento, California

Shuichi Suzuki
Honshu-Shikoku Bridge Authority
Tokyo, Japan

Andrew Tan
Everest International Consultants,
  Inc.
Long Beach, California

Man-Chung Tang
T. Y. Lin International
San Francisco, California

Shouji Toma
Department of Civil Engineering
Hokkai-Gakuen University
Sapporo, Japan

M. S. Troitsky
Department of Civil Engineering
Concordia University
Montreal, Quebec
Canada

Keh-Chyuan Tsai
Department of Civil Engineering
National Taiwan University
Taipei, Taiwan
Republic of China

Keh-Chyuan Tsai
Department of Civil Engineering
National Taiwan University
Taipei, Taiwan
Republic of China

Wen-Shou Tseng
International Civil Engineering
Consultants, Inc.
Berkeley, California

Chia-Ming Uang
Department of Civil Engineering
University of California
La Jolla, California

Shigeki Unjoh
Public Works Research Institute
Tsukuba Science City, Japan

**Murugesu
Vinayagamoorthy**
California Department of
Transportation
Sacramento, California

Jinrong Wang
URS Greiner
Roseville, California

Linan Wang
California Department of
Transportation
Sacramento, California

Terry J. Wipf
Department of Civil Engineering
Iowa State University
Ames, Iowa

Zaiguang Wu
California Department of
Transportation
Sacramento, California

Rucheng Xiao
Department of Bridge Engineering
Tongji University
Shanghai, China

Yan Xiao
Department of Civil Engineering
University of Southern California
Los Angeles, California

Tetsuya Yabuki
Department of Civil Engineering

and Architecture
University of Ryukyu
Okinawa, Japan

Quansheng Yan
College of Traffic and
Communication
South China University of
Technology
Guangzhou, China

Leiming Zhang
Department of Civil Engineering
Tsinghua University
Beijing, China

Rihui Zhang
California Department of
Transportation
Sacramento, California

Ke Zhou
California Department of
Transportation
Sacramento, California

Contents

SECTION II Superstructure Design

Section I
Fundamentals

1

Conceptual Bridge Design

M. S. Troitsky
Concordia University

1.1   Introduction

Planning and designing of bridges is part art and part compromise, the most significant aspect of structural engineering. It is the manifestation of the creative capability of designers and demonstrates their imagination, innovation, and exploration [1,2]. The first question designers have to answer is what kind of structural marvel bridge design are they going to create?

The importance of conceptual analysis in bridge-designing problems cannot be emphasized strongly enough. The designer must first visualize and imagine the bridge in order to determine its fundamental function and performance.

Without question, the factors of safety and economy shape the bridge designer's thought in a very significant way. The values of technical and economic analysis are indisputable, but they do not cover the whole design process.

Bridge design is a complex engineering problem. The design process includes consideration of other important factors, such as choice of bridge system, materials, dimensions, foundations, aesthetics, and local landscape and environment. To investigate these issues and arrive at the best solution, the method of preliminary design is the subject of the discussion in this chapter.

1.2 Preliminary Design

1.2.1 Introduction

What is preliminary design? Basically, the design process of bridges consists of two major parts: (1) the preliminary design phase and (2) the final design phase. The first design phase is discussed in this section and the final design phase is discussed in Section 1.3 in more detail.

The preliminary design stage (see Tables 1.1 and 1.2)consists of a comprehensive search of current practical and analytical applications of old and new methods in structural bridge engineering. The final design stage consists of a complete treatment of a new project in all its aspects. This includes any material, steel, or concrete problems. The important argument is that with this approach a significant savings in design effort can be easily achieved, particularly in the final stage.

In order to plan and design a bridge, it is necessary first to visualize it. The fundamental creativity lies in the imagination. This is largely reflected by the designer's creativity and the designer's past experience and knowledge. Also, the designer's concept may be based on knowledge gained from comparisons of different bridge schemes.

Generally, the designer approaches the problem successively, in two steps. In preliminary design, the first and the most important part is the creation of bridge schemes. The second step is to check schemes and sketch them in a drawing. It will then be possible to determine other design needs. An examination process is then carried out for other design requirements (e.g., local conditions, span systems, construction height, profile, etc.). From an economics point of view, choice of span structure, configuration, etc. is very essential. From the cost and aesthetics prospective, the view against the local environment is important. Completing these two steps yields the desired bridge scheme that satisfies the project proposal [3].

In the preliminary design stage it is also required to find a rational scientific analysis scheme for the conceived design. Thus, an essential part of preliminary design is to select and refine various schemes in order to select the most appropriate one. This is not an easy task since there are no existing formulas and solution. It is based mainly on the designer's experience and the requirements dictated by the project.

The final stage requires a detailed study and analysis of structural behavior and stability. Economy and safety are also important aspects in bridge design, but considerable attention must be given to detailed study for the analysis, which involves the final choices of the structural system, dimensions, material, system of spans, location of foundations, wind factor, and many others.

However, the difference in preliminary schemes if all analysis is done accurately should not be substantial. Therefore, it is very important to have, from the first step, the design calculation exact and complete. The designer workload can be dramatically reduced through use of auxiliary coefficients. These coefficients can be used if the chosen scheme needs to be modified.

Design calculation is done on the basis of structural mechanics. Usually the analysis starts with the deck, stringers, and transverse beams which determine the weight of the deck. Final analysis includes a check of the main load-carrying members, determination of various loads and their effects, total weight, and analysis of bearings. Parallel to the analysis, correction of the initial construction scheme is normally carried out.

However, at the preliminary design stage it is only necessary to explain the characteristics of the alternatives. The comparison is normally based upon the weight and cost of the structure. It should also be highlighted that at this stage the weight of the structure cannot be determined with absolute precision. It is normally estimated on the basis of experimental coefficients.

As mentioned earlier, the aim of preliminary design is to compare various design schemes. This can be achieved efficiently by using computers. The designer can create a number of rational schemes and alternatives in a short period of time. A critical comparison between the various schemes should then be made. However, this is not an easy process and it is necessary to go to the next step. Various components of each scheme, such as the deck, the spans, supports, etc. should be compared with each other. It is important at this stage that the designer be able to visualize each component in the scheme, sketch it, and check its rationality, applicability, and economy. Following this, the analysis and drawings can be adjusted and corrected.

Finally, the chosen scheme should undergo a detailed design in order to establish the structure of the bridge. The analysis is applied to each component of the bridge and to the whole structure. Each part should be visualized first by the designer, sketched, analyzed, and checked for feasibility. Then it should be modified if necessary. In each case, the most beneficial alternative should be chosen. It is a very sensitive task because it is not easy to find immediate answers and the required solutions. The problem of making final choices could only be solved on the basis of general considerations and designer's particular point of view, which is undoubtedly based on personal experience and knowledge as well as professional intuition.

The sequence of analysis in detailed final design remains the same as for preliminary design except that it is more complete. The bridge structure at this stage has a physical meaning since each part has been formed and detailed on paper. Finally, the weight is estimated considering the actual volume of the bridge elements and is documented in a special form referred to as "specifications" or a list of weights. The specifications generally should be drafted at the end of the project. This sequence leads to the final stage of the project, but the process is still incomplete. The project will reach its final form only at the construction stage. For this reason, it is worth mentioning that the designer should from the beginning give serious consideration to construction problems and provide, in certain cases, complete instructions as well as methods for construction.

1.2.2   General Considerations for the Design of Bridge Schemes

Factually, the structural design scheme of the bridge presents a complex problem for the structural designer despite the presence of modern technology and advanced computer facilities. The scope of such a problem encompasses the determination of general dimensions of the structure, the span system (i.e., number and length of spans), the choice of a rational type of substructure. Also, within this scope, there is a demand to find the most advantageous solution to the problem in order to determine the maximum safety with minimum cost that is compatible with structural engineering principles. Fulfilling these demands will provide the proper solution to the technical and economic parameters, such as structure behavior, cost, safety, convenience, and external view.

Also, during the design of a bridge, crossing the river should take into consideration the cross section under the bridge that provides the required discharge of water. The opening of the bridge is measured from the level of high water as obtained at cross sections between piers, considering the configuration of the river channel, the coefficient of stream compression, and the permissible erosion of the riverbed. By changing the erosion coefficient and the cross-sectional area within the limits permitted by the standards, it is possible to obtain different acceptable dimensions of openings for the same bridge crossing. During the choice of the most expedient alternative, it is necessary also to consider that reducing the bridge opening is connected with increased cost of foundation as a result of the large depth of erosion and the need to apply more-complicated and expensive structures for stream flow. During the design of such structures as viaducts and overpasses, their total length is usually given, which may be determined by the general plan or by the landscape of the location and the relation of the cost of an embankment of great height and the bridge structure.

The design of the bridge usually starts with the development of a series of possible alternatives. By comparing alternatives, considering technical and economic parameters, we try to find the most expedient solution for the local site conditions. At the present time, the development and comparison of alternatives is the only way to find the most expedient solution. Factors influencing the choice of bridge scheme are various and their number is so great that obtaining a direct answer to what bridge scheme is most rational at a given local condition is a challenge. It is necessary to develop a few alternatives based on local conditions (geologic, hydrologic, shipping, construction, etc.) and apply the creative initiative of the designer to the choice of a structural solution. Providing structural schemes of bridge alternatives is a creative act., computers can be used to determine the

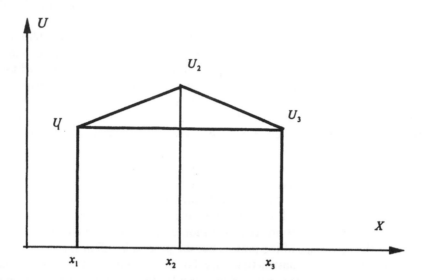

FIGURE 1.1 Quality index of the structure.

most advantageous span length and span system, to find the number of girders on the bridge having a top deck or the number of panels in the truss, and to choose the substructure. However, using computers to make a choice of rational alternatives, considering a comparison of all technical and economic parameters, is impossible. Finding an optimum alternative using different points of view often leads to different conclusions. For example, the alternative may be the most advantageous by cost, but may require great expenditure on metal or require special erection equipment, which cannot be obtained. Some alternatives may not satisfy an architectural requirement, when considering city bridges. When using computers it is still impossible to refute the conventional design method, considering all problems of specific local condition, which are practically impossible to write into a computer program.

1.2.3 Theoretical Basic Method of Preliminary Design

Methods of design cannot be invented on the basis of certain arbitrary principles. They are developed from practice. In a given theoretical study, there are enough proofs that methods of design are changing depending upon the bridge-building practice and its basic problems. Therefore, today's applied method of preliminary design is mainly determined by empirical methods.

To achieve improvement, this method is based on consistency in its exact application and explanation of its logical basis. This advanced method of preliminary design makes it possible to develop a perfect final solution for the project. It is worth mentioning at this point the importance of calculation parameters in the considered design approach.

Using mathematical models, it is possible to express (see Figure 1.1) the quality indexes U of the structure as a function of its parameters; x, y, z, i.e.,

$$U = u\ (x,\ y,\ z,\ \ldots) \tag{1.1}$$

Preliminary design provides means to determine the exact values of parameters and their quality indexes. The problem is similar to finding the limit of a function, as in calculus of variation. This analogy may be used to determine a logical basis for the method of preliminary design. It is clear that the problem of preliminary design cannot be solved in pure mathematics. The quality indexes cannot be expressed by algebraic functions. Note that the majority of parameters from one alternative to another change their size rapidly. Alternatives are shown only for consideration and to show the investigation process in order to prove the correctness of the accepted alternative.

Only in particular cases can a mathematical method be applied to find the limit. For instance, it is known that by this method it is possible to find exact dimensions of span lengths of simple-span trusses or exact heights of steel trusses those with parallel chords because of their behavior of minimal total weight of the structure.

To find the limit of the function U, it is possible to find the corresponding values of parameters x, y, z from the following equation:

$$\frac{\partial U}{\partial x} = 0, \quad \frac{\partial U}{\partial y} = 0, \quad \frac{\partial U}{\partial z} = 0 \tag{1.2}$$

These equations provide the tool to investigate the influence of each parameter as it changes the quality indexes of the structure. Leaving all other parameters constant, $\pm \Delta x$ is imposed to study the change of the value U. We can then find the value of the parameter for which ΔU changes its sign. This corresponds to the minimum of the function U.

Note that the separate parameters are interrelated. If one parameter is changed, it is necessary to modify the others. By exceeding certain limits, the span of the reinforced concrete bridge must change from a beam system to an arched system. Applying this method of preliminary design to bridges, the comparison of Eq. (1.2) leads to composition of alternatives. For each equation in Eq. (1.2), it is necessary to use a minimum of three alternatives. The first equation is formed from certain values of parameters x_1, y_1, z_1, etc. Leaving parameters y_1 and z_1 constant gives a new value of x, which is x_2 to compose a second alternative. Comparing this with the first, we establish the change of quality indexes of the bridge. If they have improved, it is necessary to change again the parameter x in the same direction, raising it to the new value x_3 to form a third alternative. Then, we compare this with the first two alternatives to determine the change of the quality indexes for the designed bridge. If, for example, they become worse, then their maximum value corresponds to x_2 (see Figure 1.1). If they improve, it is necessary to repeat the investigation for the second equation $\partial U / \partial y = 0$, and so on. All these equations must be solved simultaneously. In preliminary design, this means it is necessary to prepare many alternatives and compare them simultaneously. This process is difficult and tedious. The difficulty is increased because, unlike the purely mathematical method where the function U is given, in preliminary design the type of function is not known and should be determined.

Because of the above-mentioned difficulties, there is enough ground to assume that the first stage of the design process is based on creativity and invention.

How to build a bridge over a certain river? There are number of different answers to this question. The type of bridge can be steel or reinforced concrete, and for each case there are a number of applicable alternatives. If, for the given problem, there are several *known* solutions, there could be just as many or more undeveloped. This shows that building a bridge and creating a design are not easy tasks. Many undetermined problems face the designer. However, engineering science has proved that these difficulties could be solved in a systematic sequence, as illustrated below.

1. Equation (1.2) should be applied to the problem of structural design and solved by a method of successive approximations which follow preliminary design. This method is considered to be technically reliable and has been used in engineering successfully. In order to improve and accelerate this method (of successive approximation), it is of major importance to choose absolute precision. Experience, in a significant way, helps in making such a decision.
2. Preliminary design is generally the first approximation in the creation of a bridge project. It solves the equation for the most important parameters which have great influence on the quality indexes of the structure. Details of the structure may be investigated at a later stage.

Many solutions have been developed in practice for detailed structure. It is worth mentioning that when working with parameters, there are not many basic ones.

3. Some parameters are given and remain constant during design. Others take a limited number of values and this shortens the number of alternatives. Relations among the parameters, their correction, and the importance for the quality indexes of the bridge make it easier to carry out the methods of investigation of alternatives.

If it were possible to solve the problem by pure mathematics, then the solution would be simply to solve the equations. But we should remember that these equations (except those of the first degree) have several roots and arbitrary constants. In application to bridge design, this means, if a few equally valid alternatives are obtained, the investigation should be refined further.

The method of successive approximations should be accepted as the methodological principle because, as the process results in several alternatives for one project, we should consider only the best scientific solution. We may stop at an approximate solution, but only after we have been convinced that, in comparison with other solutions, it is the best scientific approach. This is the best way to generate designer success. The same method is applied for choosing a bridge system, as well as making a final choice for the material of bridge design, and so on.

1.2.4 Choice of Final Alternative for Reinforced-Concrete Bridges

The designer may, for instance, decide that, for a given material (say, reinforced concrete), a third alternative is chosen. Using this reasoning, the following imperfection arises: in the mathematical analogy, it was necessary to solve Eq. (1.2) simultaneously, but here each is solved separately. After determining a certain parameter, it will be kept constant and the choice for the others will follow.

There is an element of sensitivity within this method. The values of parameters are not chosen arbitrarily. The initial values are determined empirically so that their values are as exact as the real ones. The order of invention of individual parameters is also important. First, the parameters that affect the quality indexes of the structure most significantly are investigated. Then, investigation for the less important parameters follows.

Remember, the span structure implies length of span but also requires determination of the type of span structure, its form, its shape, its system, the varied types which could be uniform or unequal span structures, or the number of spans.

Within this method, the chosen alternative determines the system of span structure with minimum weight and maximum economy and safety. Now, a legitimate question may arise. Does it mean that the chosen alternative with varied type of span design (including span length, shape, form, and type) can be considered the *best* choice for the bridge project with maximum economy and safety? Although the answer may sound controversial and theocratically inconsistent, it is not. The answer is factually yes! There is a reason for that. Certain types of bridge projects require structures of various type of spans which represent economically and safely the least choice for bridge design. For refining, continue investigation by the method of successive approximation.

Note, when laying out spans for frame-beam bridges, an equal span system is often used because it provides maximum standardization of elements. However, the application of unequal span construction is also possible and in certain cases more favorable.

In following the above discussion, if an economically feasible span alternative is not satisfactory, say, does not meet the requirements of shipping regulations, or if the bridge length does not permit equal spans, then it is necessary by the method of successive approximation to find more sensitive alternative schemes with a system of spans unequal in shape, form, length, etc.

In conclusion, by changing the span system, the number of spans, or their form or shape or their combination or dimensions, it is possible to obtain a number of alternatives that will satisfy the best given local conditions at minimum cost.

For instance, changing the span system, say, by reducing the number of spans, results into a reexamination of the whole bridge design, and consequently a new bridge scheme should be drafted.

TABLE 1.1 Preliminary Design — Example 1

Design Stages	Beam System
First alternative	System of span structure, deck-type beam; bridge having three spans; construction of span structure from reinforced concrete having four main beams; supports are massive
Second alternative	Same, only two spans
Third alternative	Same, only four spans
Comparison of alternatives	The best alternative is four spans (third alternative); the first and second alternatives are canceled
Subalternatives of third alternative	1. Four-span alternative with two main beams and supports from third alternative two columns 2. Same, with prestressed concrete 3. With application of welded reinforcing frame
Comparison of alternatives	The third alternative is chosen; four-span bridges with two main reinforced concrete beams
Fourth alternative	Arch-type, three spans with four separate arches and columns above arches; supports are massive
Fifth alternative	Same, two spans
Sixth alternative	Same, four spans
Comparison of alternatives	Fourth alternative is chosen: three-span bridge with four separate arches
Subalternatives of fourth alternative	1. With two narrow arches and walls above arches 2. With box-type arches
Comparison of alternatives	Fourth alternative is chosen: three-span bridges with four separate arches

If the weight of the span structure is relieved, the span system should be modified either by decreasing the span length or span shape or form or other aspects of span structure.

It worth mentioning at this point that the significance of choosing the right alternative for the span system should not be underestimated. This is because the choice of material type for the bridge structure (e.g., monolith or prefabricated, conventional or prestressed, or reinforced concrete) is of lesser importance in cost value than the total span system, which consists of length, shape, form, number of spans, etc. Due to the relative simplicity of reinforced concrete shapes of spans and supports, calculating their volume is not a difficult process if their dimensions are given or determined from preliminary calculations.

The greatest advantage of applying theoretical methods is that the process of design is not abstract and is based on scientific analysis and quantifiable information. Therefore, during the process of choosing the best alternatives for solution, there are opportunities for eliminating imperfections for each scheme. Typical for this method is searching for the best solution through detailed investigation for each material, superstructure, bridge system, etc.

The number of alternatives obtained could be large. In the scheme of variation given in Table 1.2 it was decided to compose these alternatives with subalternatives. It takes extensive investigation, which is not always necessary. In some cases, shorter methods may be applied. In the example shown in Table 1.2, it is possible to choose the bridge material first, thus composing one alternative for steel and one for reinforced concrete systems. It is advisable to consider information from experience using an empirical approach for bridge schemes.

Example 1.1: Preliminary Design of Highway Bridge

Given
Clearance, design loads, location, bridge span, type of foundation (wells), and the material is reinforced concrete and steel.

Solution
The detailed design process is shown in Table 1.2. It can be observed that the shortened method is achieved to reduce investigation and the number of alternative projects and to increase the use of

TABLE 1.2 Preliminary Design — Example 2

Design Stages	Beam System
First alternative	Reinforced-concrete beam, three-span bridge of deck system, with four main beams; supports are massive
Second alternative	Steel beams two spans
Comparison of alternatives	The first alternative is chosen: reinforced-concrete bridge
Third alternative	Reinforced-concrete arch, two spans having four separate arches
Comparison of alternatives	After comparison of the first and third alternatives, the first alternative is chosen: beam bridge
Fourth alternative	Two spans, reinforced-concrete beam bridge
Fifth alternative	Four spans, reinforced-concrete beam bridge
Comparison of alternatives	After comparison of the first, fourth, and fifth alternatives, the fifth alternative is chosen: four-span bridge
Subalternatives of fifth alternative	1. With two main beams, monolith 2. Same, with four beams, prestressed concrete prefabricated 3. Same, with welded reinforcing frame
Comparison of alternatives	By comparison of the fifth basic and additional alternatives, fifth subalternative 2 is chosen

existing available data in practice. Thus, new investigations are unnecessary and this results in significant savings in design analysis and endeavor.

1.3 Final Design

1.3.1 Basic Trends in the Design of Bridges

In many aspects, the design of bridges is based on exact analysis and for this reason it is analogous to the solution of mathematical problems, where the results are obtained by examining the problem data and utilizing mathematical methods to arrive at a solution. This approach works well for technical and economic analyses which present very important aspects of bridge design, but it leaves out a significant part of the project.

This is because, first of all, many problems cannot be solved numerically. Second, the analysis may not correspond exactly to the actual situation. Technical analysis is valid for providing information for construction, but not significant for the solution of basic problems: choice of bridge system, choice of material, general dimensions, foundation problems, etc. These problems are solved on the basis of general considerations and the designer's judgment.

For the same problems in technical analysis or basic problems, for a bridge project there could be as many proposals as the number of the participating designers involved in engineering disputes [6]. The final choice of alternative depends to some extent on the attending participants who defend their view and support their arguments technically. It is necessary to analyze the different reasoning and determine which proposal is the most consistent with prevailing and accepted standards in the present circumstances.

The assistance of different methodological trends in bridge design is inevitable considering centuries of steady improvement and progress in bridge engineering. Progress in techniques of bridge construction depends on scientific and technological developments at each historical moment in the creation of a bridge; traditions are preserved and present views are formed.

An investigation of the history of bridges demonstrates that bridge construction has passed through several industrial stages [7]. We can separate these stages into primitive, industrial, architectural, and engineering phases. These can be subdivided still further into simpler forms and characteristics.

The influence of previous centuries on bridge design indicates that, to best understand the present trends, one must study the evolution of bridge engineering. Note that it reflects involvement of materials, the spiritual culture of the society, and the transfer of heritage. Concerning technological advances universities have had a large influence. The future engineers take from their professors the basic knowledge and new trends in design.

1.3.2 Creative Trends

In the 20th century, bridge design has undergone considerable change. With increasing demand for reinforced-concrete bridges, the need for and the creation of a new system was inevitable. The old methods had many limitations and will not be discussed here. They actually presented many obstacles for further developments in bridge engineering. It was necessary to create new specifications for reinforced-concrete structures.

The construction of highway bridges and the application of reinforced concrete presented designers with a basic problem regarding the choice of the bridge system. This created strong demand for preliminary design. This new concept required developing new methods and has put pressure on designers to look at the bridge not as a condensation of essential parts but rather as a monolithic compound unit with interrelated parts.

Because of the growing demand for reinforced-concrete and suspension bridges, the designer had large choice of materials and means to develop new bridge systems and the idea of cable-stayed bridges followed. The new century created strong demand for an analytical approach and necessitated a growing need for preliminary design with more schemes.

The acceptance that for each case there is no *one* solution but, rather, that there are several from which it is possible to choose the one most consistent with prevailing, accepted standards and most effective for the actual project leads to the basic characteristic of the second significant trend in bridge design, which will be called "creative."

Therefore, the design of each bridge is a process of finding a solution to a new problem. If there is no solution available, it must be sought. Considering the role of personal creation, this second trend may provide original new projects. Supporters of such a trend believe that creation of a bridge depends upon personal predisposition, capability, and vision. Design is considered to be a creative process that consists of a combination of structural expressions based on required knowledge and professional intuition.

1.3.3 Practical Trends

Practicability is the main consideration in this trend. The word *practical* goes hand in hand with scientific investigation using modern technology. Designers use both scientific principles and creativity for their designs only in order to solve the actual problem. In this trend, the bridge is considered as part of the highway or railway and its basic purpose is to satisfy the requirements of transportation.

The bridge should satisfy the basic requirements of safety and economic factors. The construction of the bridge should also follow the pattern of successful industrial methods.

Supporters of the creative trend considered highways and railways as areas to apply their creative capabilities and for testing their new inventions. Followers of scientific analysis investigation considered highways as large laboratories for their investigations. Adherents of practical design have borrowed their concepts from both trends, insisting that bridges must be first safe and permitted experimental structures only on secondary highways. Practical designers suggested that the structure should be standardized for industrial preparation because it could lead to faster ways of reconstruction or rehabilitation. Also, practical designers insist on use of construction techniques that require minimum maintenance and do not affect the traffic flow.

1.3.4 Basic Assumptions of Design

Methodological rules compatible with technical and applicable requirements in bridge engineering play a major role in modern progressive methods for designing bridges.

Nowadays, time is an important factor, especially in bridge construction. Progressive methods must satisfy technical swift performance as well as requirements of astute engineering economy. Such majestic structures must function effectively and, in addition, be aesthetically appealing. Bridges play the major role in the transportation system crossing rivers or other obstructions.

At different times, bridges were built for more than one purpose. The following are examples:

1. Roman bridges and those built in the Middle Ages served not only for transportation or for chariots, but also for joyful, exuberant activities for the population. These traditions were continued at later times.
2. Another trend that appeared in the Middle Ages is the construction of bridges for fortresses, castles, and towers as a protective measure against attacks by enemies. An example is the bridge at Avignon, France; also "London Tower Bridge," which was built with towers for aesthetic purposes only.
3. Another trend in the same era was to build chapels on bridges and to collect tolls to maintain them, the same old problem of upkeep (e.g., Italy, Spain, Germany).
4. During the Middle Ages and later, bridges were built to serve as dams for water mills, which were important parts of the economy in those days (e.g., Holland).
5. During the 16th and 17th centuries bridges were built as wide structures for shops and convenience in general. Good examples are London Bridge, England and Ponte Vecchio, Florence, Italy. Construction of these types of bridges was terminated toward the beginning of the 19th century.
6. In Western civilizations, bridges are sometimes built as majestic monuments to commemorate outstanding events or achievements of national importance for an important person or national hero. Examples are the monument to George Washington, the George Washington Bridge, New York City; the monument to Princess Margaret of Great Britain, The Princess Margaret Bridge, Fredericton, New Brunswick, Canada (this bridge was designed by M. S. Troitsky); the monument to the victory at the Battle of Waterloo, The Waterloo Bridge, London, England; the monument to Russian Tzar Alexander the Third, The Alexander IIIrd Bridge, Paris, France (one of the most beautiful cast-iron bridges of imperial style); the monument of the Sarajevo Association, The Gavrilo Princip Bridge, Sarajevo, Yugoslavia; the monument to Napoleon Bonaparte's victory at the battle of Austerlitz, Austerlitz Bridge, Austria.

The 19th century was characterized by industrial growth, and the use of bridges was confined to transportation as a result of the boom in building railways. Later, with Ford promoting "auto-vehicles," the building of bridges for highways became in great demand. This new trend in transport requirements put on pressure to improve safety factors as well. As a result, it is very important in modern bridge engineering to determine the carrying capacity of the bridge or the maximum value of the temporary vertical load that the bridge can bear.

Also, to avoid interruption in traffic flow, the calculations should consider the maximum number of vehicles passing in a given time. For bridges crossing navigable rivers, passing clearance must be considered. Also, similar consideration should be given to underpasses. The carrying capacity of a bridge is defined by the number of lanes, their width, and the accepted lateral clear distances of shoulders and medians required for safety considerations.

To avoid interrupted traffic flow, it is necessary for the width of the bridge to be greater than that required by the calculated carrying capacity. For example, in long bridges, it is necessary to provide an extra parking space for possible emergency cases in order to prevent a traffic jam. As a rule, the width of the roadway on the bridge is equal to the width of the highway. However, there may be deviations from this rule. For instance, although the highway may accommodate three lanes for traffic, the number of lanes on the bridge could be reduced. Also, there are examples of the reversed situation.

The condition of maximum traffic suitability and convenience is not a requirement but is preferred and attention should be paid to this issue during planning the project. Also, this issue could be considered as one of the criteria for the appraisal of the project, provided that the cost is not prohibively excessive.

The most efficient functional bridge structure is considered to be the one that embodies the most requirements of transport, with top safety factors, carrying capacity, that contains extra convenience facilities, that is most effective in labor and material, and that can be completed in a reasonable time. Since Henry Ford's time, extra pressure has been put on the transportation system, primarily on highways and railways, which has directly affected innovation in bridges. Modern-day transport is increasing in number and weight. This means bridges must be designed so that their carrying and passing capacities can accommodate heavier vehicles and larger numbers of vehicles. Designers must be resourceful and have means to overcome difficult situations effectively and to cope with the growing demands of faster and larger moving transport with the greater reserves for future growth, the longer the bridge stands without needing repair or reinforcement.

Note that by increasing the reserves for passing and carrying capacities, the cost of the bridge will increase. Determining the necessary reserve is a problem that needs to be resolved by engineering economy. The Romans did not visualize the fast development of transport and means for transportation, but concentrated their conceptual design on timelessness of the bridge structure and, for this purpose, provided great reserves for passing and carrying capacities.

The property of material is not necessarily the basic factor that defines the service time and safety of the bridge. More often, bridges are reconstructed for other reasons: too small passing and carrying capacity, insufficient clearance under the bridge, straightening of lanes or reduction of the grade.

1.3.5 Basic Requirement of the Bridge under Design

Choosing the right location is crucial for designing and planning a bridge. But above all, safety considerations that govern the technical, functional, economic, efficiencies, expeditiousness, and aesthetic requirements are very important. It is necessary for the bridge and each of its components to be safe, durable, reliable, and stable. This is usually checked by analysis using current specifications. But not all questions of durability, reliability, and stability may be answered by analysis. Therefore, in some cases it is necessary to provide special measures such as testing the performance of the structure and examining its behavior under maximum loading on the construction site.

Specifications and technical requirements should be satisfied because they guarantee the carrying capacity of the structure. From the safety point of view, all bridges designed according to the technical requirements are equal. But practically speaking, different aspects of technical requirements may be satisfied with different margins of safety.

Regarding the various bridge components, it is necessary to know that for engineering structures, the best solution should provide the appropriate material and carrying capacity.

During comparison of projects, the technical requirements should be considered. Because technical requirements may be accomplished using alternatives, consideration should always be given to additional guarantees for safety. Never compromise the safety of the passengers. Essential requirements naturally should have great importance, but they are basically satisfied by accepted clearance. Also, additional consideration must be given to issues other than elementary demands in order to make traffic flow efficiently. Note that the height of the bridge and the elevation of the roadway must be determined at an early stage, because they have influence on the traffic flow. Also, greater or smaller grades of the approaches should be designed earlier in the project. Maximum grades are defined by specifications, but for practical purposes minimum grades are the most convenient. Further, it is important to define the number of joints in the roadway that correlate to the division of the structure in separate sections.

Conditions of minimum wear of the parts of carrying construction under the influence of moving vehicles are also important to consider. Regarding the maintenance of the roadway and the bridge, it is possible to consider this as a general expense and therefore relate it to economic considerations.

Essential requirements indicate that the total cost of the bridge at all conditions should be economically rational. The overall cost of construction and bridge erection is determined in significant part by the quantity of material and the unit price. Yet, the tendency to reduce the quantity of material in order to achieve lower cost does not always lead to minimum overall cost. There are other factors that should be taken into consideration. Take, for example, steel structures: consideration should be given to quantity of steel and on top of that special attention must be given to modern industrial practices in production which in its turn may lead to conveniences in erection resulting from heavy construction with lower cost.

During comparisons of various projects, analysis of their economic criteria may reveal principles of expedience that can be applied to the project under consideration. Construction requirements are connected to economic constraints because, when the amount of material is small, the work is simple and the time required is shorter. Also, the unit price is considered as part of the economic criteria, which implies the cost of preparation and erection. All these factors affect the overall cost.

For conventional bridges to be built from a certain material, construction is carried out by established methods. Therefore, during comparison of alternatives, construction criteria are not so important. In special cases of complicated erection of bridges having large spans, or for urgent work, construction requirements are very important and may influence the choice of the bridge system and material. In these cases, it may be necessary to use a great quantity of materials, thus increasing the cost of construction and ignoring other requirements. For example, during the initial period of application, assembled reinforced-concrete constructions were more expensive than monolithic ones. However, with increased use of these constructions, the application of assembled structures is more rational and economical.

1.3.6 Aesthetic Requirements

Apart from the basic requirements of the bridge design, there are often additional demands. The first is the problem of aesthetics. Beauty should be achieved as a result of good proportions of the whole bridge and its separate parts. In spite of the tendency to build economical structures, we should not forget beauty. The importance of the architecture of the bridge should not be ignored because of economic and technical requirements. In fact, the most famous bridges are remembered by their architectural standards and magnificent structures (examples, Brooklyn Bridge, Verrazano Narrows Bridge, Golden Gate Bridge, Tower Bridge, Alexander IIIrd Bridge, Ponte Vecchio Bridge, Revelstoke Bridge, British Columbia (designed by M.S. Troitsky), Skyway Bridge, Ontario (designed by M.S. Troitsky), etc.).

There are different views regarding aesthetic practices in bridge engineering. Supporters of the rational analytical trend feel that aesthetic demands are not important and not necessary for bridges outside cities. On the other hand, designers of the creative trend consider these aesthetic values to be more important than the economic ones and equivalent to the requirements of strength and longevity.

Because of the conflicting views, this problem requires special consideration. All designers inevitably want their structure to be the most beautiful. This wish is natural and shows love and interest of the work and is necessary in order to make the designed structure head toward perfection.

During the process of design, the engineer is occupied with detailed calculations. The engineer also may be occupied with particularities and may lose sight of the complete structure. By checking the creation from an aesthetic point of view, the engineer gives attention to the wide scope and shape of the structure and has the opportunity to design details and correct if necessary. If the designer is aesthetically unsatisfied with the creation, the designer will improve it and try to find workable solutions. But the designer should always be aware of technical, economic, and safety values of the structure. Note, the architecture of bridges should not contradict either as a whole or in details the purpose of the structure. The designer's ideas should be compatible with the technical concept, surrounding conditions and environment (for example, London Tower Bridge).

It is necessary to be technically literate. Moreover, it is not enough just to design the external view of the bridge. Bridges satisfying demands and requirements of modern engineering requests and properly designed will achieve recognition and will deserve worldwide acknowledgment and credit. If designers are guided by fanciful tastes of their own, regardless of the technical concepts, they will not achieve this goal. A beautiful shape alone cannot be invested and applied to the bridge. The design should consider both the technical concepts and the structural shape.

The critical rules of proportion and the use of purely geometric shapes had, in their time, not so much an aesthetic but a technical basis. Designers based their theories on the principle of initiations and relations that they observed in nature. Historical investigation indicates that many aesthetic rules were preserved from previous centuries when they had a different basis. Even today, a bridge is considered beautiful when it has an even number of supports because it is classic and not easy to achieve with tough natural conditions. According to Palladio [7] it is clear that this rule is accepted because all birds and animals have an even number of extremities which give them better stability. Freeing themselves from prejudice and carrying out independent investigations to find the shape corresponding to the contents should lead designers toward the development of the theory of true aesthetics in bridge engineering. History has shown how the shapes of bridges were changed depending upon the general development of the cultural and economic life of a nation. For this reason, the problem of aesthetics in bridge engineering should be viewed in a historical perspective. A designer should be able to judge the bridge by considering its external view and scheme of construction.

Followers of the historic direction renounced such investigations and by this changed their principles and were more attracted to the design of bridges. However, a joint venture by engineers and architects is not always useful for solving a problem of bridge design. Nowadays, architects specialize in the construction of buildings which is reflected in their aesthetic taste. Although architectural rules and views may be correct for buildings, they may not be applicable to bridges. For example, when designing a building, architects usually use steel construction as a frame for the building which requires certain covering. For the bridge designer, steel construction is a force polygon that clearly demonstrates the transfer of forces. For an attractive external view for the bridge, detailed design and proper accomplishment of the construction are important. The external view may be spoiled by careless work. The technical concepts of structure and the architectural shape should not be separate, but should satisfy the local conditions and cover a wide scope of requirements. By understanding the validity of recognizing special aesthetic criteria a proper alternative can be selected. The final choice of alternative is the solution of some technical problem in correspondence with the basic purpose of the bridge as part of the roadway.

If the bridge is not considered a monument commemorating an outstanding event or an outstanding historic figure or a significant happening in the world, but serves only for traffic for a certain period of time, then it is not necessary to design this bridge as a highly aesthetic creation. We may be satisfied by more modest wishes with regard to its external view. Practice indicates that designers may create, and actually have created, attractive bridges even when they were governed only by the technical and economic requirements during the design process.

A bridge that is properly designed from the technical and economic point of view cannot contradict the basic rules of architecture. The general basis of architecture consists of the idea that masses of material should be distributed expediently. The properties of the material should be used correspondingly, and the whole structure should correspond to its purpose.

Generally, economic considerations of bridge design are the same as those stated above. An economic design is achieved by (1) the expedient distribution of material, choice of the most economical system, cross sections of the members, and considering working conditions; and (2) the use of proper material (members in tension use steel, members in compression use concrete).

Therefore, economic expediency and architectural conception are determined by the same criteria. From this, it is impossible to contrast aesthetic criteria with technical and economic aspects.

For example, it is advisable to reject a beam bridge for an arch in the case when the first by all other properties is better, or to prefer a single-span bridge to the more expedient two spans. Also, it is possible to say that the choice of alternative, considering technical and economic criteria, should not deviate from the proper way to achieve the aesthetic aims.

Finally, the bridge will only be perfect in an aesthetic sense, when its system as a whole and its separate members are chosen not on the basis of personal taste of the designer, but considering technical and economic expedience.

All other proofs that are often applied by the authors of separate projects to defend unsuccessful technical and economic alternatives should be rejected. All these proofs are based on the unstable and changeable bases of personal opinion. Such proofs are only declarations of personal impressions and tend not to prove anything but only to convince people by the use of feeble verbal arguments.

Many definitions are expressed using varied terminology synonymous in meaning, but with drastically different shades in the positive and negative sense. For example, regarding the structure of the bridge, when the deck is at the bottom chord the defender may say that this structure is "expressive," "easily seen," or "stands out with a beautiful shape on the sky." The opponent, however, may object and say that this structure "obstructs view," "hangs on the observer," etc. By the skillful use of such terminology, it is possible to convince the inexperienced that a beautifully presented perspective is not as worthy of praise as a less successful project.

1.3.7 Requirement for Scientific Research

The second additional requirement sometimes asked of bridges under design is called the scientific research or "innovation." This requires that the bridge contain a new achievement due to scientific research or a new invention.

The design of a bridge always contains something new. Even if the project is worked using old examples and applying typical projects, the designer uses new contributions along with the known. Therefore, there is always a certain degree of novelty. A good designer or engineer should not only be familiar with previous designs but should also be updated with modern scientific research and benefit from that by using advanced technical sciences in the design as the project changes.

It is natural for the designer to search for novelty; yet new solutions should be born only from the tendency to reach the best solution by starting from the existing conditions at the project. Therefore, the "novelty" requirement cannot run contrary; they should complement each other.

The history of the evolution of bridge engineering is the progression from simple to more complex, and it was achieved gradually and unevenly. Some periods were distinguished by invention and the appearance of new shapes, systems, and types of bridges; other periods were characterized by mastering and perfecting existing systems and the development of scientific research work. For example, at the end of the 19th century, a great step forward was made in the area of stone bridges. Perhaps the most significant achievement in the modern era was the appearance of the cast-iron arch and iron-suspension systems with different members of trusses and large spans. All these novelties resulted from the impact of growing industry and transportation.

The first 40 or 50 years of the 19th century were spent creating the iron beam bridges, and the second part of the century was devoted to developing expedient systems and improving the construction. Significant periods in later history were devoted to the development of reinforced-concrete bridges. The initial period of trials and creation of the construction was 1880 to 1890, and the period of mastery was 1900 to 1910. However, it is necessary to note that with the general development of science and technology, the role of scientific research is increasingly racing together with novelty, rationalization, and invention. It is obvious that the necessity for novelty results from the general economic conditions and sociocultural requirements.

The attitude toward novelty in bridge engineering has been modified. Adherents of the rational, analytical direction preferred to hold on to some classical models, considering that the search for new shapes should be related only to scientific research work of creative direction, however, tending toward the new and original by ignoring any old pattern. A realistic approach to a new idea should be based on understanding that novelty is not an aim in itself and that the new idea should be a solid ground for improvement.

It is necessary to consider the criterion of novelty because it sometimes appears as an independent factor during appraisal of projects and choice of alternatives. Because novelty is not an aim in itself, it should not be a special criterion, forcing a preference for new construction irrespective of its quality. When by basic conditions the new idea is better and there is no doubt regarding its quality, then it should be adopted and should replace the old. In the opposite case, it should be refused.

Not every novelty leads to progress in bridge engineering. If the novelty is sound, it may be developed to such a degree that it would lead to a new method, but if it is not better than the old method or not yet developed, its development at a later stage may be helped by abstaining from early application. Early application leads to lowering the quality of bridges and may compromise new ideas before they reach full appreciation and are fully evaluated.

The criterion of novelty may be considered independent only in separate cases when economy requires the introduction of a new type of construction. An example is the introduction of prefabricated reinforced-concrete construction. At the present time it is expedient to use prefabricated reinforced-concrete construction, but initially it was more expensive than conventional construction. The criterion of novelty then was contrary to other criteria. It had to be solved for each case, especially when the novelty was not an aim in itself, but was required for economic and commercial demands.

One reason for introducing new construction techniques is related to the necessity of experimental and practical checking of the scientific research work, which is certainly necessary.

Regarding bridges on main highways, however, it is not advisable to subject them to experiment, because their basic designation is to serve transportation. Only separate experimental structures and special controls are permitted. However, in each case, the problems of special scientific research and structural experimentation should be performed at a scientific institution.

1.3.8 Basic Parameters of the Bridge

The quality of the structure is evaluated considering different criteria: technical, functional, economic, construction, and, in addition, the material of the system and the geometric dimensions of the bridge. All these criteria are temporary parameters defining the quality of the structure.

The problem of design generally consists of the way to find the values of these parameters that will correspond to a better quality of the structure. It is necessary to consider first, in detail, basic factors influencing the quality of the structure. All the parameters interact, but their influence on the quality of the structure is different. Their influence on each other is different: one may depend little on another; another may greatly influence the other. For example, basic parameters for material may not influence basic parameters of foundation and so on.

During preliminary design, the determination of basic parameters interacts and has major influence in making decisions about the location of the bridge, the span, the material, the type of foundation, the system of the bridge, the length of separate spans, the type of superstructure, and the type of supports.

The location of the bridge usually does not much depend on other parameters, but does have an impact on them. For small bridges, the location is defined by the intersection of the highway with the river, ravine, etc. For medium to large bridges, it is possible to compare a number of alternatives, such as the basic value of the highway and the cost of approaches and highway installations. The cost of the bridge itself plays a deciding role because its span at all alternatives is usually an unchangeable constant. For this reason, during selection of bridge location it is possible to propose an often-used bridge type without detailed study. However, there are two exceptions to this general rule. First, if the river is not used for shipping and has sandbanks, then at the location of largest curvature the span of the bridge obtained is smaller, but the depth of the water here is greater. Therefore, foundations are complicated and the installation of pile supports may be impossible. On the sandbanks where the span is increased, but the water depth is shallower, it is possible to build a simple viaduct-type bridge supported by the piles. If the bridge is proposed to be built from timber, its location should be chosen over sandbank. Therefore, during choice of crossing, it is necessary to consider both alternative types of bridges.

The second exception is the design of viaducts across mountain ravines. In this case, the change of the crossing has substantial impact on the choice of the span of the bridge and it is reflected in its cost. It is true that the type of the bridge for the first comparison may be left unchanged (e.g., reinforced concrete arch type, etc.), but it may be designed for all alternatives because the cost of the viaduct will have impact on the choice of location of the crossing.

The above exceptions do not occur often and should be considered separately; for this reason the location of the crossing may be chosen before preliminary design and must be made by the investigators with designers' efforts only in order to check the correctness of the choice. The size of the bridge opening is defined by hydraulic and hydroanalytic investigations and is assumed for the design. It some cases, however, during the design process it is possible to change the span. The size of opening, as shown above, depends on the crossing location. It also depends on the type and depth of the foundation. At greater depths, greater washout is permitted, with corresponding diminishing of the opening. At shallow foundations the reverse could occur.

In principle, two opposite solutions may exist:

1. Build bridge supports as safe against washing, squeeze the river by flow-directed dikes, and obtain a minimum opening.
2. Not squeeze the river, cross the whole river during flood, and thus the concern that the supports will wash out will no longer be a problem.

The first solution is used as a rule for rivers on the plain and can be justified economically and technically. Only for a timber bridge is it expedient to cover the whole flood area by the approach viaducts. Here the size of the opening depends upon the bridge material. The second solution may often be expedient for mountain rivers in which the main channel is often changing and threatens to wash out the flood embankment.

Generally, the size of the opening may change a little depending on the type of foundation. If the type of foundation as a whole is determined by the local conditions (e.g., by using caissons or wells), then the size of the opening for all alternatives remains unchanged. Choice of material is the most substantial problem during preliminary design and depends not only on the designer's point of view but also on other conditions that must be considered before preparing the project. Each material has its own area of application and the problem of material choice arises when these areas intersect.

Timber bridges are usually used as temporary structures. Spans greater than 160 ft often present difficulties. For permanent bridges, the choice is usually between reinforced concrete and steel structures. The following are some recommendations concerning the material selection for the bridge:

1. For spans ranging between 65 to 100 ft reinforced-concrete beam-type bridges are mainly used and steel is considered for overpasses and underpasses.
2. For spans ranging between 330 and 500 ft, steel bridges are often preferred.
3. For spans ranging between 650 and 800 ft, it is expedient to use steel bridges.

Therefore, the choice between reinforced concrete and steel bridges is generally for spans ranging between 65 and 330 ft.

The type of foundation for the bridge is determined mainly by the geologic investigation of ground in the riverbanks and in the main channel, and also by the depth and behavior of the water.

Relatively, the type of foundation influences the superstructure, size of separate spans, and type of supports. Foundations built at the present time may be divided into two basic groups:

1. Piler foundation in which timber pilers are used for shallow foundations and reinforced concrete and steel piles are used for deep foundations.
2. Massive, shallow foundation (between others or piles) and deep foundations (caissons and wells). It is obvious that for large spans it is necessary to use a massive foundation.

Shallow pile foundations are possible for viaduct bridges having small spans. Regarding the bridge system, it should be emphasized that pile foundations almost define the beam system and arches. Suspension bridges require a massive foundation and supports, but there might be other alternatives. During design, the following parameters remain constant or are slightly modified for the bridge system:

1. Size of spans (unequal or uniform);
2. Span system;
3. Type of supports.

1.3.9 Bridge System

The bridge system (i.e., beam, arch, suspension) is integrally related to the chosen material. Beam systems are mostly used for small and medium spans. An arch system is mainly used for large spans and a suspension system is used for long spans.

When using reinforced-concrete bridges, the following should be taken into consideration:

1. For spans up to 130 ft, a beam system is recommended.
2. For spans ranging between 130 and 200 ft, either a beam or arch system can be used.
3. For longer spans, an arch system is recommended.

For steel bridges, beam systems are mainly used. The arch system is expedient to use for spans longer than 160 ft. All the above span lengths are approximate and can be used as preliminary guidance in the early stage of the investigation in order to determine the appropriate system to use. The bridge system depends also on other parameters. It is impossible to investigate all other parameters without assuming the material type for the structure and the bridge system in the early stage of the investigation.

1.3.10 Size of Separate System

The size of the separate system greatly influences the cost of bridges. Determination of the span system involves a number of basic problems that need to be solved during the preliminary design.

For beam bridges having steel trusses, a known rule exists. The cost of the main truss with bracing per span should equal the cost of one pier with foundation. For all other cases, the length of span depends upon the type of foundation and pier.

Similarly, the system of the span construction has influence on the system of the span. With arch bridges, the cost of support is generally greater than that of beam type. For this reason (all things being equal), the span of the arch bridge should be greater than a beam bridge. The exceptions are high viaducts having rising high arches which are more economical. The limits of changes to span length are governed by clearances for ships and typical uses of span structures. The clearances for ships regulate the minimum size of the span. Usually the span is greater than the most economical length. For this reason, during crossings of navigable rivers the size of the span at the main channel in most cases is predetermined. It is necessary to change only side and approach spans. When choosing approach spans, it is necessary to consider typical projects because the use of typical construction is more rational and useful.

From this it follows that the length of spans is not arbitrary. They are chosen from defined conditions. The span length is closely connected with the system of span structure. Therefore, it is necessary at the early stage of the project to assume the proper system of span structure, noting that the choice of the system significantly determines the bridge system.

1.3.11 Type of Span Construction

The type of span construction is closely related to the bridge system. After assuming a bridge system, the span structure should be determined. There might be some problems related to the type of structure (e.g., solid or truss type for steel, monolith or prefabricated for reinforced concrete), the number of main girders, the basic dimensions, etc. Detailed study for each case is needed. Many problems common to particular cases can be investigated earlier, during the preparation of typical projects.

The use of typical projects substantially helps the individual design. For example, in the majority of medium-span bridges typical projects may be used. The use of typical projects simplifies fabrication of the structure, reduces the time necessary for design and construction, and makes the structure more economical to execute. However, the immediate use of typical projects should not be considered as a rule. They should be considered as a first solution, which in many cases can be improved. Each project has different circumstances, and typical projects do not provide solutions to all possible design problems. In some projects there might be some local conditions that need to be dealt with and were not addressed in previous projects. This problem is especially recognized in the design and construction of long-span bridges. Examples of already built bridges may provide a rational starting point. Together with this experience in the design and building of bridges, it is possible to establish some useful relations such as the ratio of truss height to span to the number and length of panels, etc.

The design of bridge structures starts with the critical study and the use of existing bridges to prepare the first alternative of the structure and continues during the investigation to separate parameters to prepare the next alternatives.

1.3.12 Type of Support

Supports can be divided into two groups: columns and massive supports. The second group is used in the presence of large floating ice and arch-type span structures. Column-type supports are most expedient with small-beam structures.

1.4 Remarks and Conclusions

A proper design method should meet two basic criteria:

1. First, the design method should be based on scientific engineering research and analysis. From comprehensive research, design derives logical conclusions.
2. Design methods should be achieved by practice and previous experience in the design and construction of bridges. Also, modifications should always be performed to improve the design. This is largely reflected by the designer's creative capability, sense of invention, and innovation.

Therefore, the integrated part of preliminary design is a comprehensive search of scientific, practical findings and analysis.

References

1. Waddell, J. A. L., *Bridge Engineering*, Vol. 1, John Wiley & Sons, New York, 1916, 267–280.
2. Mitropolskii, N. M., *Methodology of Bridges Design*, scientific-technical edition, Avtotransportni Literatury, Moscow, 1958, 215–242 [in Russian].

3. Polivanov, N. I., *Design and Calculation of the Reinforced Concrete and Metal Highway Bridges,* Transport, Moscow, 1970, 5–36 [in Russian].
4. Steinman, D. B. and Watson. S. R., *Bridges and Their Builders,* Dover, New York, 1957, 378–391.
5. DeMare, E., *Your Book of Bridges,* Faber and Faber, London, 1963, 11–28.
6. Vitruvius, *The Ten Books on Architecture,* translated by M. H. Morgan, Dover, New York, 1960, 13–16.
7. Palladio, A., *The Four Books on Architecture,* Stroiizdat, Moscow, 1952 [in Russian].
8. Holgate, A., *The Art in Structural Design,* Clarendon Press, Oxford, 1986, 1–6, 24–30, 187–195.
9. Francis, A. J., *Introducing Structures,* Pergamon Press, New York, 1980, 221–260.

2

Aesthetics — Basics*

Fritz Leonhardt
Stuttgart University, Germany

2.1 Introduction

Aesthetics falls within the scope of philosophy, physiology, and psychology. How then, you may ask, can I as an engineer presume to express an opinion on aesthetics, an opinion which will seem to experts to be that of a layman. Nevertheless, I am going to try.

For over 50 years I have been concerned with, and have read a great deal about, questions concerning the aesthetic design of building projects and judgment of the aesthetic qualities of works in areas of the performing arts. I have been disappointed by all but a few philosophical treatises on aesthetics. I find the mental acrobatics of many philosophers — whether, for example, existence is the existence of existing — difficult to follow. Philosophy is the love of Truth, but truth is elusive and hard to pin down. Books by great building masters are full of observations and considerations from which we can learn in the same way that we study modern natural scientists.

My ideas on aesthetics are based largely on my own observations, the results of years of questioning — why do we find this beautiful or that ugly? — and on innumerable discussions with architects who also were not content with the slogans and "isms" of the times, but tried to think critically and logically.

---

*Much of the material of this chapter was taken from Leonhardt, F., *Bridges — Aesthetics and Design*, Chapter 2: The basics of aesthetics, OVA, Stuttgart, Germany, 1984, with permission.

The question of aesthetics cannot be understood purely by critical reasoning. It reaches to emotion, where logic and rationality lose their precision. Undaunted, I will personally address these questions, so pertinent to all of us, as rationally as possible. I will confine myself to the aesthetics of building works, of man-made objects, although from time to time a glance at the beauty of nature as created by God may help us reinforce our findings.

I would beg you to pardon the deficiencies that have arisen because of my outside position as a layman. This work is intended to encourage people to study questions of aesthetics using the methods of the natural scientist (observation, experiment, analysis, hypothesis, theory) and to restore the respect and value which it enjoyed in many cultures.

2.2 The Terms

The Greek word *aisthetike* means the science of sensory perception and very early on was attributed to the perception of the beautiful. Here we will define it as follows:

Aesthetics: The science or study of the quality of beauty an object possesses, and communicates to our perceptions through our senses (expression and impression according to Klages [1]).

Aesthetic: In relation to the qualities of beauty or its effects; aesthetic is not immediately beautiful but includes the possibility of nonbeauty or ugliness. Aesthetic is not limited to *forms*, but includes surroundings, light, shadows, and color.

2.3 Do Objects Have Aesthetic Qualities?

Two different opinions were expressed in old philosophical studies of aesthetics:

1. Beauty is not a quality of the objects themselves, but exists only in the imagination of the observer and is dependent on the observer's experience [2]. Smith said in his "Plea for Aesthetics" [3], "Aesthetic value is not an inborn quality of things, but something lent by the mind of the observer, an interpretation by understanding and feeling." But how can we interpret what does not exist? Some philosophers went so far as questioning the existence of objects at all, saying they are only vibrating atoms, and everything we perceive is subjective and only pictured by our sensory organs. This begs the question, then, is it possible to picture the forms and colors of objects on film using a camera? These machines definitely have no human sensory organs.

2. The second school of thought maintains that objects have qualities of beauty. Kant [4] in his *Critique of Pure Reason* said, "Beauty is what is generally and without definition, pleasing." It is not immediately clear what is meant by "without definition," perhaps without explaining and grasping the qualities of beauty consciously. What is "generally pleasing" must mean that the majority of observers "like" it. Paul [5] expressed similar thoughts in his *Vorschule der Aesthetik* and remarked that Kant's constraint "without definition" is unnecessary. Thomas Aquinas (1225–1274) simply said, "A thing is beautiful if it pleases when observed. Beauty consists of completeness, in suitable proportions, and in the luster of colors." At another time, Kant said that objects may arouse pleasure independent of their purpose or usefulness. He discussed "disinterested pleasure," a pleasure free from any interest in objects: "When perceiving beauty, I have no interest in the existence of the object." This emphasizes the subjective aspect of aesthetic perception, but nonetheless bases the origin of beauty in the object.

Is one right? Most would side with Kant and grant that all objects have aesthetic qualities, whether we perceive them or not. Aesthetic value is transmitted by the object as a message or simulation and its power to ourselves depends on how well we are tuned for reception. This example drawn from modern technology should be seen only as an aid to understanding. If a person is receptive to transmissions of beauty, it then depends very largely on how sensitive and developed are the person's senses for aesthetic messages, whether the person has any feeling for quality at all. We will look at this question more closely in Section 2.4.

On the other hand, Schmitz, in his *Neue Phänomenologie* [6], sees in this simple approach "one of the worst original sins in the theory of cognition." ...This *physiologism* limits the information for human perception to messages that reach the sensory organs and the brain in the form of physical signals and are therefore metaphysically raised to consciousness in a strangely transformed shape." We must see the relationships between the object and circumstances, associations, and situations. More important is the situation and observer's background and experience. The observer is "affectively influenced," [6] i.e., the effect depends on the health of the observer's senses, on the observer's mood, on the observer's mental condition; the observer will have different perceptions when sad or happy. The observer's background experience arouses concepts and facts for which the observer is prepared subconsciously or which are suggested by the situation. Such "protensions" [6] influence the effects of the object perceived, and include prejudices which are held by most people and which are often a strong and permanent hindrance to objective cognition and judgment. However, none of this phenomenology denies the existence of the aesthetic qualities of objects.

Aesthetic quality is not limited to any particular fixed value by the characteristics of the object, but varies within a range of values dependent on a variety of characteristics of the observer. Judgment occurs in a process of communication. Bahrdt [7], the sociologist, said, "As a rule aesthetic judgment takes place in a context of social situations in which the observers are currently operating. The observers may be a group, a public audience, or individuals who may be part of a community or public. The situation can arise at work together, during leisure time, or during a secluded break from the rush of daily life. In each of these different situations the observer has a different perspective and interpretation, and thus a different aesthetic experience [impression]."

Aesthetic characteristics are expressed not only by form, color, light, and shadow of the object, but by the immediate surroundings of the object and thus are dependent on object environment. This fact is well known to photographers who can make an object appear much more beautiful by careful choice of light and backdrop. Often a photograph of a work of art radiates a stronger aesthetic message than the object itself (if badly exhibited) in a gallery. With buildings, the effect is very dependent on the weather, position of the sun, and on the foreground and background. It remains undisputed that there is an infinite number and variety of objects (which all normal healthy human beings find beautiful). Nature's beauty is a most powerful source of health for humans, giving credence to the suggestion that we have an inborn aesthetic sense.

The existence of aesthetic qualities in buildings is clearly demonstrated by the fact that there are many buildings, groups of buildings, or civic areas which are so beautifully designed that they have been admired by multitudes of people for centuries, and which today, despite our artless, materialistic attitudes to life, are still visited by thousands and still radiate vital power. We speak of classical beauty. All cultures have such works, and people go to great lengths to preserve and protect them; substantial assistance has come from all over the world to help preserve Venice, whose enchanting beauty is so varied and persuasive.

We can also give negative evidence for the existence of aesthetic qualities in objects in our man-made environment. Think of the ugliness of city slums, or depressing monotonous apartment blocks, or huge blocky concrete structures. These products of the "brutalist" school have provoked waves of protest. This affront to our senses prompted the Swiss architect Rolf Keller to write his widely read book *Bauen als Umweltzerstörung* [8].

All these observations and experiences point to the conclusion that objects have aesthetic qualities. We must now look at the question of how humans receive and process these aesthetic messages.

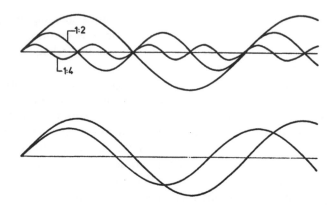

FIGURE 2.1   Wave diagrams for consonant and dissonant tones.

2.4   How Do Humans Perceive Aesthetic Values?

Humans as the receivers of aesthetic messages use all of their senses: they see with their eyes, hear with their ears, feel by touch, and perceive temperature and radiation by sensors distributed in the body, sensors for which there is no one name. Our sensory organs receive different waveforms, wavelengths, and intensities. We read shapes by light rays, whose wavelengths give us information about the colors of objects at the same time. The wavelength of visible light ranges from 400 μm (violet) to 700 μm (red) (1 μm = 1 millionth of 1 mm). Our ears can hear frequencies from about 2 to 20,000 Hz.

The signals received are transmitted to the brain and there the aesthetic reaction occurs — satisfaction, pleasure, enjoyment, disapproval, or disgust. In modern Gestalt psychology, Arnheim [9] explained the processes of the brain as the creation of electrochemical charge fields which are topologically similar to the observed object. If such a field is in equilibrium, the observer feels aesthetic satisfaction, in other cases the observer may feel discomfort or even pain. Much research needs to be done to verify such explanations of brain functions, but they do seem plausible. However, for most of us we do not need to know brain functions exactly.

During the course of evolution, which we assume to have taken many millions of years, the eye and ear have developed into refined sensory organs with varied reactions to different kinds of waveforms. Special tone sequences can stimulate so much pleasure that we like to hear them — they are consonant or in harmony with one another. If, however, the waveforms have no common nodes (Figure 2.1) the result is dissonance or beats, which can be painful to our ear. Dissonances are often used in music to create excitement or tension.

The positive or negative effects are a result not only of the charge fields in the brain, but the anatomy of our ear, a complex structure of drum oscular bones, spiral cochlea, and basilar membrane. Whether we find tones pleasant or uncomfortable would seem to be physiological and thus genetically conditioned. There are naturally individual differences in the sense of hearing, differences which occur in all areas and in all forms of plant and animal life.

There are also pleasant and painful messages for the eye. The effects are partly dependent on the condition of the eye, as, for example, when we emerge from a dark room into light. Color effects of a physiological nature were described in much detail by Goethe in his color theory [10]. In the following, we will discuss the effects of physical colors on the rested, healthy eye, and will not address color effects caused by the refraction or reflection of light.

Some bright chemical colors cause painful reactions, but most colors occurring naturally seem pleasant or beautiful. Again, the cause lies in waves. The monotonous waves of pure spectral colors have a weak effect. The eye reacts more favorably to superimposed waves or to the interaction of two separate colors, especially complementary colors.

We feel that such combinations of complementary colors are harmonious, and speak of "color harmony." Great painters have given us many examples of color harmony, such as the blue and yellow in the coat of Leonardo da Vinci's *Madonna of the Grotto*.

We all know that colors can have different psychological effects: red spurs aggression; green and blue have a calming effect. There are whole books devoted to color psychology and its influence on human moods and attitudes.

We can assume that the eye's aesthetic judgment is also physiologically and genetically controlled, and that harmonic waveforms are perceived as more pleasant than dissonant ones. Our eyes sense not only color but can form images of the three-dimensional, spatial characteristics of objects, which is vital for judging the aesthetic effects of buildings. We react primarily to proportions of objects, to the relationships between width and length and between width and height, or between these dimensions and depth in space. The objects can have unbroken surfaces or be articulated. Illumination gives rise to an interplay of light and shadow, whose proportions are also important.

Here the question of whether there are genetic reasons for perceiving certain proportions as beautiful or whether upbringing, education, or habit play a role cannot be answered as easily as for those of acoustic tone and color. Let us first look at the role proportions play.

2.5   The Cultural Role of Proportions

Proportions exist not only between geometric lengths, but between the frequencies of musical tones and colors. An interplay between harmonic proportions in music, color, and geometric dimensions was discovered very early, and has preoccupied the thinkers of many different cultural eras.

Pythagoras of Samos, a Greek philosopher (571–497 B.C.) noted that proportion between small whole numbers (1:2, 2:3, 3:4, or 4:3, and 3:2) has a pleasing effect for tones and lengths. He demonstrated this with the monochord, a stretched string whose length he divided into equal sections, comparing the tones generated by the portions of the string at either side of an intermediate support or with the open tone [11–13].

In music these harmonic or consonant tone intervals are well known, for example,

String Length	Frequencies	
1:2	2:1	Octave
2:3	3:2	Fifth
3:4	4:3	Fourth
4:5	5:4	Major third

The more the harmonies of two tones agree, the better their consonance; the nodes of the harmonies are congruent with the nodes of the basic tones. Later, different tone scales were developed to appeal to our feelings in a different way depending on the degree of consonance of the intervals; think of major and minor keys with their different emotional effects.

A correspondence between harmonic proportions in music and good geometric proportions in architecture was suggested and studied at an early stage. In Greek temples many proportions corresponding with Pythagoras's musical intervals can be identified. Kayser [14] has recorded these relationships for the Poseidon temple of Paestum.

H. Kayser (1891–1964) dedicated his working life to researching the "harmony of the World." For him, the heart of the Pythagorean approach is the coupling of the tone of the monochord string with the lengths of the string sections, which relates the qualitative (tone perception) to the quantitative (dimension). The monochord may be compared with a guitar. If you pull the string of a guitar, it gives a tone; the height of the tone (quality) depends on the length (dimension = quantity) and the tension of the string. Kayser considered the qualitative factor (tones) as judgment by emotional feeling. It is from this coupling of tone and dimension, of perception and logic, of feeling and knowledge, that the emotional sense for the proportions of buildings originates — the tones of buildings, if you will.

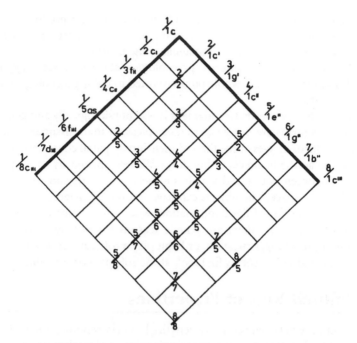

FIGURE 2.2 Giorgio numerical analogy in Λ-shape.

Kayser also had shown that Pythagorean harmonies can be traced back to older cultures such as Egyptian, Babylonian, and Chinese, and that knowledge of harmonic proportions in music and building are about 3000 years old. Kayser's research has been continued by R. Haasse at the Kayser Institute for Harmonic Research at the Vienna College of Music and Performing Arts.

Let us return to our historical survey. In his famous 10 books *De Architectura,* Marcus Vitruvius Pollio (84–14 B.C.) noted the Grecian relationships between music and architecture and based his theories of proportion on them.

Wittkower [12] mentions an interesting text by the monk Francesco Giorgio of Venice. Writing in 1535 on the design of the Church of S. Francesco della Vigna in Venice (shortened extract):

> To build a church with correct, harmonic proportions, I would make the width of the nave nine double paces, which is the square of three, the most perfect and holy number. The length of the nave should be twenty-seven, three times nine, that is an octave and a fifth. ... We have held it necessary to follow this order, whose master and author is God himself, the great master builder. ... Whoever should dare to break these rules, he would create a deformity, he would blaspheme against the laws of Nature."

So strictly were the laws of harmony, God's harmony, obeyed.

In his book *Harmonia,* Francesco Giorgio represented his mystic number analogies in the form of the Greek letter Λ. Thimus [15] revised this "Lambdoma" for contemporary readers (Figure 2.2).

"Rediscovered" for curing the ills of today's architecture, Andea di Piero da Padova — known to us as *Palladio* [16], was a dedicated disciple of harmonic proportions. He wrote, "The pure proportions of tones are harmonious for the ear, the corresponding harmonies of spatial dimensions are harmonious for the eye. Such harmonies give us feelings of delight, but no-one knows why — except he who studies the causes of things."

Palladio's buildings and designs prove that beautiful structures can be created using these harmonic proportions when they are applied by a sensitive master. Palladio also studied proportions in spatial perspective, where the dimensions are continuously reduced along the line of vision. He confirmed the view already stated by Brunelleschi (1377–1446) that objective laws of harmony also apply to perspective space.

Even before Palladio, Leon Batista Alberti (1404–1472), had written about the proportions of buildings, Pythagoras had said:

> The numbers which thrill our ear with the harmony of tones are entirely the same as those which delight our eye and understanding. ... [We] shall thus take all our rules for harmonic relationships from the musicians who know these numbers well, and from those particular things in which Nature shows herself so excellent and perfect.

We can see how completely classical architecture, particularly during the Renaissance, was ruled by harmonic proportions. In the Gothic age master builders kept their canon of numbers secret. Not until a few years ago did the book *Die Geheimnisse der Kathedrale von Chartres* (The Secrets of Chartres Cathedral) by the Frenchman L. Charpentier appear [13], in which he deciphered the proportions of this famous work. It reads like an exciting novel. The proportions correspond with the first Gregorian scale, based on *re* with the main tones of *re-fa-la*. Relationships to the course of the sun and the stars are demonstrated.

Ancient philosophers spent much of their time attempting to prove that God's sun, moon, stars, and planets obeyed these harmonic laws. In his work *Harmonice Mundi* Johannes Kepler (1571–1630) showed that there are a great number of musical harmonies. He discovered his third planetary law by means of harmonic deliberations, the so-called octavoperations. Some spoke of "the music of the spheres" (Boethius, Musica mundana).

Villard de Honnecourt, the 13th-century cathedral builder from Picardy, gave us an interesting illustration of harmonic canon for division based on the upper tone series 1–½–⅓–¼, etc. For Gothic cathedrals he started with a rectangle of 2:1. This Villard diagram (Figure 2.3) [13, 17] was probably used for the design of the Bern cathedral. Whole-number proportions of the fourth and third series can be seen in the articulation of the tower of Ulm Cathedral. A Villard diagram can be drawn for a square, and it then, for example, fits the cross section of the earlier basilica of St. Peter's Cathedral in Rome.

When speaking of proportion, many think of the golden mean, but this does not form a series of whole-number relationships and does not play the important role in architecture which is often ascribed to it. This proportion results from the division of a length $a + b$ where $b < a$ so that

$$\frac{b}{a} = \frac{a}{a+b} \tag{2.1}$$

This is the case if

$$a = \frac{\sqrt{5} + 1}{2} b = 1.618b \tag{2.2}$$

the reciprocal value is $b = 0.618a$, which is close to the value of the minor sixth at ⅝ = 0.625 or ⁸⁄₅ = 1.6. The golden mean is a result of the convergence of the Fibonacci series, which is based on the proportion of $a{:}b$, $b{:}(a + b)$, etc.:

$$
\begin{array}{llll}
a{:}b & = & 1{:}2 & = 0.500 = \text{octave} \\
b{:}(a + b) = & & 2{:}3 & = 0.667 = \text{fifth} \\
& & 3{:}5 & = 0.600 = \text{major sixth} \\
& & 5{:}8 & = 0.625 = \text{minor sixth} \\
& & 8{:}13 & = 0.615 \\
& & 13{:}21 & = 0.619 \\
& & 21{:}34 & = 0.618 = \text{Golden Mean}
\end{array}
$$

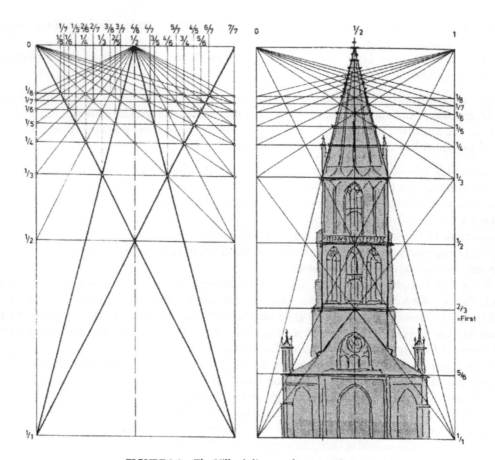

FIGURE 2.3   The Villard diagram for rectangle 2:1.

This numerical value is interesting in that:

$$\frac{1.618}{1.618-1} = \frac{1.618}{0.618} = 2.618$$

and

$$2.168 \ (6/5) = 3.1416 = \pi$$

The golden mean thus provided the key to squaring the circle, as can be found in Chartres Cathedral. It can be constructed by dividing the circle into five (Figure 2.4).

The Fibonacci series is also used to construct a logarithmic spiral, which occurs in nature in snail and ammonite shells, and which is considered particularly beautiful for ornaments. Le Corbusier (1887–1965) used the golden mean to construct his "Modulor" based on an assumed body height of 1.829 m but the Modulor is in itself not a guarantee of harmony.

An interesting proportion is $a: b = 1: \sqrt{3} = 1: 1.73$. It is close to the golden mean but for technical applications has the important characteristic that the angles to the diagonals are 30° or 60° (equilateral triangle) and the length of the diagonal is $2a$ or $2b$ (Figure 2.5). A grid with sides in the ratio of $1: \sqrt{3}$ was patented on July 8, 1976 by Johann Klocker of Strasslach. He used this grid to design carpets, which were awarded prizes for their harmonious appearance.

During the last 50 years architects have largely discarded the use of harmonic proportions. The result has been a lack of aesthetic quality in many buildings where the architect did not choose

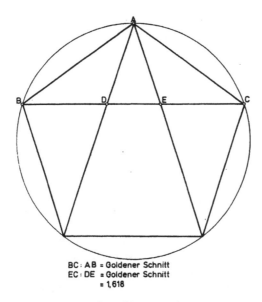

BC : AB = Goldener Schnitt
EC : DE = Goldener Schnitt
= 1,618

FIGURE 2.4 The golden mean in a pentagon.

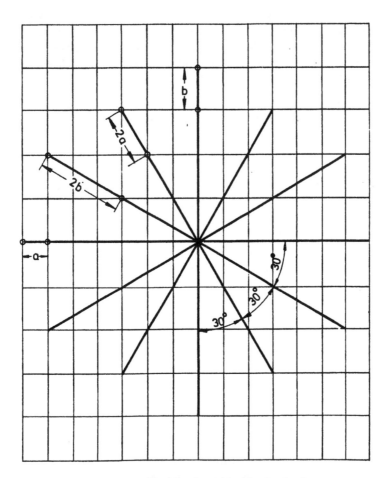

FIGURE 2.5 The Kloecker grid with $a{:}b = 1{:}\sqrt{3}$.

good proportions intuitively as a result of his artistic sensitivity. There were exceptions, as always. The Swiss architect Andre M. Studer [18] and the Finn Aulis Blomstdt consciously built "harmonically." One result of the wave of nostalgia of the 1970s is a return in many places to such aesthetics. Kayser in Reference [14] and P. Jesberg in the *Deutsche Bauzeitschrift* DBZ 9/1977 gave a full description of harmonic proportions.

2.6 How Do We Perceive Geometric Proportions?

In music we can assert plausibly that a feeling and sense for harmonic tone series is controlled genetically and physiologically through the inborn characteristics of the ear. What about the proportions of lengths, dimensions of objects, and volumes? Helmcke [19], of the Technical University of Berlin, wholeheartedly supported the idea of a genetic basis for the aesthetic perception of proportions and he argued as follows:

> During the evolution of animals and Man the choice of partner has undoubtedly always played an important role. Since ancient times men have chosen women as partners, who in their eyes were the most beautiful and well proportioned and equally women have chosen men as partners the strongest and most well-built in their eyes. Through natural selection [Darwin] during the evolution of a species this must have led to the evolution of aesthetic perception and feeling and resulted in the development in Man of a genetically coded aesthetic ideal for human partners, passed on from generation to generation. We fall in love more easily with a beautiful partner; love at first sight is directed mostly by an instinctive feeling for beauty, and not by logic. Nobody who knows Man and his history will doubt that there is an inherited human ideal of beauty. Every culture has demonstrated its ideal of human beauty, and if we study the famous sculptures of Greek artists we recognize that the European ideal of beauty in female and male bodies has not changed in the last 3000 years.

For the Greeks the erotic character of the beauty of the human body played a dominant role. At the Symposium of Xenophon (ca. 390 B.C.) Socrates made a speech in praise of Eros. According to Grassi [20], the term *beautiful* is used preferentially for the human body.

The Spanish engineer Eduardo Torroja (1899–1961), whose structures were widely recognized for their beauty, wrote in his book *The Logic of Form* [21] that "truly the most perfect and attractive work of Nature is woman." Helmcke said that "Man's aesthetic feeling, while perceiving certain proportions of a body, developed parallel to the evolution of Man himself and is programmed genetically in our cells as a hereditary trigger mechanism."

According to this the proportions of a beautiful human body would be the basis of our hereditary sense of beauty. This view is too narrow because thousands of other natural objects radiate beauty, but let us continue to study "Man" for the time being.

Fortunately, all humans differ in their hereditary, attributes, and appearance, although generally only slightly. This means that our canons of beauty cannot be tied to strictly specific geometric forms and their proportions. There must be a certain range of scatter. This range covers the differences in the ideals of beauty held by different races. It ensures that during the search for a partner each individual's ideal will differ, keeping the competition for available partners within reasonable bounds.

We can also explain this distribution physiologically. Our eyes have to work much harder than our ears. The messages received by the eye span a range about a thousand times wider than the scale of tones to be processed by the ear. This means that with colors and geometric proportions harmony and disharmony are not so sharply defined as with musical tones. The eye can be deceived more easily and is not as quickly offended or aggravated as the ear, which reacts sensitively to the smallest dissonance.

More evidence for a hereditary sense of beauty is provided by the fact that even during their first year, children express pleasure at beautiful things and are offended, even to the point of weeping, by ugly objects. How children's eyes sparkle when they see a pretty flower.

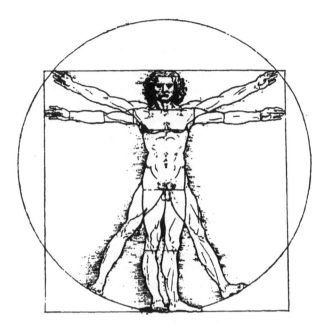

FIGURE 2.6   Image of man in circle and square according to Leonardo da Vinci.

Evidence against the idea that we have a hereditary sense of beauty is suggested by the fact that people argue so much about what is beautiful or ugly, demonstrating a great deal of insecurity in the judgment of aesthetic qualities. We will give this further thought in Section 2.7.

Our ability to differentiate between good and bad using our senses of taste and smell has also developed genetically and with certain variations is the same for most people [22]. With this background of genetic development it is understandable that the proportions of those human bodies considered beautiful have been studied throughout the ages. A Greek sculptor Polyklet of Kikyon (465–420 B.C.) defined the following proportions:

two handbreadths	= height of the face and height of the breast, distance breast to navel, navel to end of trunk
three handbreadths	= height of skull, length of foot
four handbreadths	= distance shoulder to elbow, elbow to fingertips
six handbreadths	= ear to navel, navel to knee, length of trunk, length of thigh

Plyklet based his "canon for the ideal figure" on these relationships. These studies had the greatest influence on art during the age of humanism, for example, through the *Vier Büecher von menschlicher Proportione* 1528 by Albrecht Dürer (1471–1528).

Vitruvius also dealt with the human body in his books *De Architectura* and used the handbreadth as a unit of measure. Leonardo da Vinci followed Vitruvius's theories when drawing his image of man inscribed in a square (Figure 2.6). Leonardo's friend, the mathematician Luca Pacioli (ca. 1445–1514) began his work *De Divina Proportione,* 1508, with the words:

Let us first speak of the proportions of Man because all measures and their relationships are derived from the human body and here are to be found all numerical relationships, through which God reveals the innermost secrets of Nature. Once the ancients had studied the correct proportions of the human body they proportioned all their works, particularly the temples, accordingly. (Quoted by Wittkower [12])

The human body with outstretched arms and legs inscribed in a square and circle became a favorite emblem for humanistically oriented artists right up to Le Corbusier and Ernst Neufert. Let us close this section with a quotation from one of Helmcke's [19] works:

> The intellectual prowess of earlier cultures is revealed to us whenever their artists, architects, and patrons succeeded in incorporating, consciously or unconsciously, our hereditary, genetically programmed canon of proportions in their works; in achieving this they come close to our genetically controlled search for satisfaction of our sense of aesthetics. It reveals the spiritual pauperism of today's artists, architects and patrons when, despite good historical examples and despite advances in the natural sciences and the humanities they do not know of these simple biologically, anatomically based relationships or are too ungifted to perceive, understand, and realize them. Those who deprecate our search for the formal canons of our aesthetic feelings as a foolish and thus unnecessary pastime must expect to have their opinion ascribed to arrogant ignorance and to the lack of a sure instinctive sense of beauty, and already ethnologically known as a sign of decadence due to domestication.
>
> The only criticism which, in my [Helmcke's] opinion can be leveled at the thousands of years' old search for universally valid canons of form, lies in the assumption that these canons shall consist of fixed proportions and shall thus be valid for all mankind. ...
>
> What is needed is experience of and insight into the range of scatter of proportional relations and insight into the limits within our hereditary aesthetic sense reacts positively, and beyond which it reacts negatively.

2.7 Perception of Beauty in the Subconscious

We are not generally aware of how strongly our world of feelings, our degree of well-being, comfort, disquiet, or rejection is dependent on impressions from our surroundings. Neurologists know that parts of our brain are capable of reacting to external stimuli without reference to the conscious mind and of processing extensive amounts of information. This takes place in the limbic system of the primitive structures of the midbrain and the brain stem. For all those activities of the subconscious which deal with the processing of aesthetic messages, Smith [3] used the phrase "limbic aesthetics" and dedicated a whole chapter of his very readable book to them.

Our subconscious sense of beauty is almost always active, whether we are at home, in the city marketplace, in a church, in a beautiful landscape, or in the desert. Our surroundings affect us through their aesthetic characteristics even if our conscious thoughts are occupied with entirely different matters and impressions.

Smith wrote of the sensory appetite of these primitive parts of the brain for pleasant surroundings, for the magic of the city, and for the beauty of nature. The limbic system reacts to an oversupply of stimuli with rejection or anxiety.

Symbolic values connected with certain parts of our environment also act on the subconscious. The home, the church, school, garden, etc. have always possessed symbolic values created by learning and experience. These are related mostly to basic human situations and cause emotional reactions, without ever reaching the conscious level.

This perception of beauty at the subconscious level plays a particularly strong role in city dwellers. Their basic feeling of well-being is doubtless influenced by the aesthetic qualities of their environment in this way. This has social consequences (see Section 2.10) and underlines our responsibility to care about the beauty of the environment.

2.8 Aesthetic Judgment and Taste

When two observers are not agreed in their judgment of a work of art, the discussion is all too often ended with the old proverb, "De gustibus non disputandum est." We like to use a little Latin to show our classical education, which, as we know, is supposed to include an understanding of art. This "there's no accounting for taste" is an idle avoidance tactic, serving only to show that the speaker has never really made a serious effort to study aesthetics and thus has educational deficiencies in the realm of assessing works of art.

Of course, taste is subject to continual change, which in turn depends on current ideals, fashions, and is dependent on historical and cultural background. The popular taste in any given period of time or even the taste of single individuals is never a reliable measure of aesthetic qualities.

On the other hand, genetic studies have shown that we have a certain basic hereditary sense of beauty. Smith [3] said that this aesthetic perception has developed into one of the highest capabilities of our central nervous system and is a source of deep satisfaction and joy.

The judgment of aesthetic characteristics is largely dependent on feelings which are derived from our sensory perceptions. Beauty, then, despite some theories (Bense, Maser) cannot be rationally measured. When looking at the nature of feelings we must admit the fact that despite all our research and science, we know very little about humanity or about ourselves. We can, however, call upon observations and experiences which are helpful.

We repeatedly experience that the majority of people agree that a certain landscape, great painting, or building is beautiful. When entering a room, for example, in an old church, or while wandering through a street, feelings are aroused which are pleasant, comfortable, even elevating, if we sense a radiation of beauty. If we enter a slum area, we feel revulsion or alarm, as we perceive the disorder and decay. We can be more or less aware of these feelings, depending on how strongly our thoughts are occupied elsewhere. Sensitivities and abilities to sense beauty naturally differ from person to person, as is true of our other talents. This sensitivity is influenced by impressions from our environment, by experience, by relationships with our companions at home, at school, and with our friends. Two people judging the qualities of beauty of an object are likely to give different opinions.

Beautiful surroundings arouse feelings of delight in almost all people, but an ugly, dirty environment causes discomfort. Only the degree of discomfort will differ. In our everyday life such feelings often occur only at a subconscious level and often their cause is only perceived after subsequent reflection.

We can develop a clear capacity for judging aesthetic qualities only when we study the message emanated by an object consciously and ask ourselves whether or not we like a building or a room. Next, we must ask ourselves why. Why do I like this and not that? Only by frequent analysis, evaluation, and consideration of consciously perceived aesthetic values can we develop that capacity of judgment which we commonly call taste — taste about which we must argue, so that we can strengthen and refine it. Taste, then, demands self-education, which can be cultivated by critical discussion with others or by guidance from those more experienced. Good judgment of aesthetic values requires a broad education. It can be compared to an art and requires skill, and like art it takes not only talent, but a lot of work.

We need not be afraid that such analysis will weaken our creative skills; in fact, the opposite is true: the goal of analysis is the discovery of the truth through creative thinking [23]. People have different talents and inclinations since they grow up in different circles with different cultural backgrounds and therefore their tastes will always differ. In any given culture, however, there is a certain polarity on the judgment of beauty. Psychologists call this agreement "normal behavior," "a normal reaction of the majority." This again corresponds with Kant's view that beauty is what is generally thought to be beautiful by the majority of people.

Beauty cannot be strictly proved, however; so we must be tolerant in questions of taste and must give freedom to what is generally felt to be beautiful and what ugly. That there is a generally recognized concept of beauty is proved by the consistent judgment of the classical works of art of all great cultures, visited year after year by thousands of people. Think of the popularity of exhibitions of great historic art today. It is history that has the last word on the judgment of aesthetic values, long after fashions have faded.

Fashions: Artistic creation will never be entirely free from fashion. The drive to create something new is the hallmark of creative beings. If the new becomes popular, it is soon copied, and so fashions are born. They are born of the ambition and vanity of humans and please both. The desire to impress often plays a role. Up to a certain point, fashions are necessary; in certain new directions true art may develop through the fashionable, acquiring stability through a maturing process and enduring beyond the original fashion. Often, such new developments are rejected, because we are strongly influenced by the familiar, by what we are used to seeing, and only later realize the value of the new. Again, history pronounces a balanced judgment.

Confusion is often caused in our sense of judgment by modern artists who deliberately represent ugliness in order to mirror the warped mental state of our industrial society. Some of this work has no real quality, but is nonetheless acclaimed as modern art. The majority dares not question this for fear of rejection, slander, and peer pressure.

Although some works that consciously display ugliness or repulsiveness may well be art, we must seriously question the sanity and honesty of the patrons of primitive smearings, tangles of scrap iron, or old baby baths covered in Elastoplast strips (J. Beuys) when such efforts are exhibited as works of art. Happily, the courage to reject clearly such affronts and to put them in their place is on the increase. We only need to read Claus Borgeest's book, *Das Kunsturteil*, [24], in which he wrote, "the belief in such 'art' is a modern form of self-inflicted immaturity, whose price is the self-deprivation of reason, man's supreme attribute."

In any case, it would be wrong to describe as beautiful works, those haunted by ugliness, even if they have the quality of art. The artist intends to provoke and to encourage deliberation. However, the educational effects of such artistic creations are questionable, because we usually avoid their repeated study. Painters and sculptors, however, should be free to paint and sculpt as hatefully and repulsively as they wish — we do not have to look at their works. It is an entirely different case with buildings; they are not a private affair, but a public one. It follows that the designer has responsibility to the rest of humankind and a duty to produce beautiful buildings so that the designer does not give offence. Rightly, the ancient Greeks forbade public showings of ugliness, because their effects are largely negative.

We seldom find anyone who will hang ugly works of art in his or her home. It is beyond a doubt that in the long term we feel comfortable only in beautiful surroundings and that beauty is a significant requirement for the well-being of our soul; this is much more important for people's happiness than we today care to admit.

2.9 Characteristics of Aesthetic Qualities Lead to Guidelines for Designing

The search for explanations, the analysis of aesthetic values, are bound to lead to useful results, at least for man-made buildings and structures. We will now try to subject matters of feelings, emotions, to the clear light of recognition and understanding.

If we do this, we can certainly find answers to the question, "Why is this beautiful and this ugly?" For recognized masterpieces of architecture generally considered beautiful, there have been answers since ancient times, many of which are given in the quoted literature on proportions. Such buildings reveal certain characteristics of quality and from these we can deduce guidelines for design, such as certain proportions, symmetry, rhythm, repeats, contrasts, and similar factors, The master schools of old had such rules or guidelines, such as those of Vitruvius and Palladio. Today, these rules are surely valid and must be rediscovered for the sake of future architecture. They can prove a valuable aid in the design of building structures and at the very least contribute toward avoiding gross design errors.

Many architects and engineers reject rules, but in their statements about buildings we still find references to harmony, proportion, rhythm, dominance, function, etc. Torroja [21] rejected rules, but he said "the enjoyment and conscious understanding of aesthetic pleasure will without doubt be much greater if, through a knowledge of the rules of harmony, we can enjoy all the refinements and perfections of the building in question." Rules of harmony are based on rules of proportion, and somehow the striving for individual artistic freedom prevents us from recognizing relationships often imposed upon us by ethics.

Let us then attempt to formulate such characteristics, rules, or guidelines as they apply to building structures, particularly bridges.

2.9.1   Fulfillment of Purpose–Function

Buildings or bridge structures are erected for a purpose. The first requirement is that the buildings and bridges be designed to optimally suit this purpose. To meet the specific purpose, a bridge may have different structural types: arches, beams, or suspensions. The structure should reveal itself in a pure, clear form and impart a feeling of stability. We must seek simplicity here. The form of the basic structure must also correspond to the materials used. Brick and wood dictate different forms from those for steel or concrete. We speak of form justified by the material, or of "logic of form" [21]. This reminds us of the architect Sullivan's rule "form follows function" which became an often misunderstood maxim for building design. The function of a building is not only that it stand up. One must fulfill all the various requirements of the people that inhabit the building. These include hygiene, comfort, shelter from weather, beauty, even cosiness. The fulfillment of the functional requirements of buildings includes favorable thermal, climatic, acoustic, and aesthetic qualities. Sullivan undoubtedly intends us to interpret his rule in this sense. For buildings the functional requirements are very complex, but in engineering structures, functions besides load-carrying capacity must be fulfilled, such as adequate protection against weather, limitation of deformation and oscillation, among others, and all these factors affect design. Quality and beauty must be united, and quality takes first priority!

2.9.2   Proportion

An important characteristic necessary to achieve beauty of a building is good, harmonious proportions, in three-dimensional space. Good proportions must exist between the relative sizes of the various parts of a building, between its height, width, and breadth, between masses and voids, closed surfaces and openings, between the light and dark caused by sunlight and shadow. These proportions should convey an impression of balance. Tassios [25] preferred "expressive proportions" which emphasize the desired character of a building (see Section 2.9.8).

For structures it is not sufficient that their design is "statically correct." A ponderous beam can be as structurally correct as a slender beam, but it expresses something totally different. Not only are the proportions of the geometric dimensions of individual parts of the building important, but also those of the masses of the structure. In a bridge, for instance, these relationships may be between the suspended superstructure and the supporting columns, between the depth and the span of the beam, or between the height, length, and width of the openings. Harmony is also achieved by the repetition of the same proportions in the entire structure or in its various parts. This is particularly true in buildings.

Sometimes contrasting proportion can be a suitable element. The detailed discussion is referred to Chapter 4 of my book [26], which shows what good proportions can mean for bridges.

2.9.3   Order

A third important rule is the principle of order in the lines and edges of a building, an order achieved by limiting the directions of these lines and edges to only a few in space.

Too many directions of edges, struts, and the like create disquiet, confuse the observer, and arouse disagreeable emotions. Nature offers us many examples of how order can lead to beauty; just think of the enchanting shapes of snow crystals and of many flowers [27, 28]. Good order must be observed between the proportions occurring in a building; for instance, rectangles of 0.8:1 should not be placed next to slim rectangles of 1:3. Symmetry is a well-tried element of order whenever the functional requirements allow symmetry without constraint.

We can include the repetition of equal elements under the rule of order. Repetition provides rhythm, which creates satisfaction. Too many repetitions, on the other hand, lead to monotony, which we encounter in the modular architecture of many high-rise buildings. Where too many repetitions occur, they should be interrupted by other design elements.

The selection of one girder system throughout the structure provides an element of good order. Interrupting a series of arches with a beam gives rise to aesthetic design problems. Under the principle of order for bridges we may include the desire to avoid unnecessary accessories. The design should be so refined that we can neither remove nor add any element without disturbing the harmony of the whole.

2.9.4 Refining the Form

In many cases, bodies formed by parallel straight lines appear stiff and static, producing uncomfortable optical illusions. Tall bridge piers or towers with parallel sides appear from below to be wider at the top than at the bottom, which would be unnatural. Nor does this uniform thickness conform to our concept of functionality, because the forces decrease with increasing height. For this reason, the Egyptians and Greeks gave the columns of their temples a very slight taper, which in many cases is actually curved. Towers are built tapered or stepped. On high towers and bridge piers, a parabolic taper looks better than a straight taper.

The spans of a viaduct crossing a valley should become smaller on the slopes, and even the depth of the girders or edge fascia can be adjusted to the varying spans. Long beams of which the bottom edge is exactly horizontal look as if they are sagging, and so we give them a slight camber.

We must also check the appearance of the design from all possible vantage points of the future observer. Often the pure elevation on the drawing board is entirely satisfactory, but in skew angle views of unpleasant overlapping are found. We must also consider the effects of light and shadow. A wide cantilever deck slab can throw bridge girders into shadow and make them appear light, whereas similar shadows break the expressive character of an arch. Models are strongly recommended for checking a design from all possible viewpoints.

These refinements of form are based on long experience and must be studied with models from case to case.

2.9.5 Integration into the Environment

As the next rule, we recognize the need to integrate a structure or a building into its environment, landscape, or cityscape, particularly where its dimensional relationships and scale are concerned. In this respect many mistakes have been made during the past decades by placing massive concrete blocks in the heart of old city areas. Many factories and supermarkets also show this lack of sensitive integration. Sometimes long-span bridges with deep, heavy beams spoil lovely valley landscapes or towns with old houses lining the riverbank.

The dimensions of buildings must also be related to the human scale. We feel uneasy and uncomfortable moving between gigantic high-rise buildings. Heavy, brutal forms are often deliberately chosen by architects working with prefabricated concrete elements, but they are simply offensive. It is precisely their lack of scale and proportion that has led to the revolt against the brutality of this kind of architecture.

2.9.6 Surface Texture

When integrating a building with its surroundings, a major role is played by the *choice of materials*, the *texture of the surfaces*, and particularly by *color*. How beautiful and vital a natural stone wall can appear if we choose the right stone. By contrast, how repulsive are many concrete facades; not only do they have a dull gray color from the beginning, but they weather badly, producing an ugly patina and appear dirty after only a few years. Rough surfaces are suitable for piers and abutments; smooth surfaces work well on fascia-beams, girders, and slender columns. As a rule, surfaces should be matte and not glossy.

2.9.7 Color

Color plays a significant role in the overall aesthetic effect. Many researchers have studied the psychological effects of color. Here, too, ancient rules of harmonious color composition apply, but today successful harmonious color schemes are rare. Often, we find the fatal urge for sensation, for startling aggressive effects, which can be satisfied all too easily with the use of dissonant colors, especially with modern synthetic pop — or shocking — colors. We can find, however, many examples of harmonious coloring, generally in town renovation programs. Bavaria has provided several examples where good taste has prevailed.

2.9.8 Character

A building and bridge should have character; it should have a certain deliberate effect on people. The nature of this desired effect depends on the purpose, the situation, the type of society, and on sociological relationships and intentions. Monarchies and dictatorships try to intimidate by creating monumental buildings, which make people feel small and weak. We can hope this belongs to the past. Only large banks and companies still make attempts to impress their customers with monumentalism. Churches should lead inward to peace of mind or convey a sense of release and joy of life as in the Baroque or Rococo. Simple dwellings should radiate safety, shelter, comfort, and warmth. Beautiful houses can stimulate happiness.

Buildings of the last few decades express an air of austere objectivity, monotony, coldness, confinement, and, in cities, confusion, restlessness, and lack of composition; there is too much individuality and egoism. All this dulls people's senses and saddens them.

We seem to have forgotten that people also want to meet with joy in their man-made environment. Modern buildings seem to lack entirely the qualities of cheerfulness, buoyancy, charm, and relaxation. We should once again become familiar with design features that radiate cheerfulness without lapsing into Baroque profusion.

2.9.9 Complexity — Stimulation by Variety

Smith [3] postulated a "second aesthetic order," suggested by findings made by biologists and psychologists [29]. According to this, beauty can be enhanced by the tension between variety and similarity, between complexity and order. Baumgarten expressed this as early as 1750, "Abundance and variety should be combined with clarity. Beauty offers a twofold reward: a feeling of well being both from the perception of newness, originality and variation as well as from coherence, simplicity, and clarity." Leibniz in 1714 demanded for the achievement of perfection as much variety as possible, but with the greatest possible order.

Berlyne [30] considered the sequence of tension and relaxation to be a significant characteristic of aesthetic experience. Venturi [31], a rebel against the "rasteritis" (modular disease) architecture of Mies van der Rohe, said, "A departure from order — but with artistic sensitivity — can create pleasant poetic tension."

A certain amount of excitement caused by a surprising object is experienced as pleasant if neighboring objects within the order ease the release of tension. If variety dominates our orientation, reflex is overtaxed and feelings ranging from distaste to rejection are aroused. Disorder is not beautiful.

This complexity doubtless requires artistic skill to be successful. It can be used well in bridge design if, for instance, in a long, multispan bridge the main span is accented by a variation in the girder form. The interplay of complexity and order is important in architecture, particularly in city planning. Palladio was one of the first to extend the classical understanding of harmony by means of the complexity of architectural elements and ornamentation.

2.9.10 Incorporating Nature

We will always find the highest degree of beauty in nature, in plants, flowers, animals, crystals, and throughout the universe in such a variety of forms and colors that awe and admiration make it extremely difficult to begin an analysis. As we explore deeper into the realm of beauty we also find in nature rules and order, but there are always exceptions. It must also remain possible to incorporate such exceptions in the masterpieces of art made by creative humans [28].

The beauty of nature is a rich source for the needs of the soul, and for humans' psychic well-being. All of us know how nature can heal the effects of sorrow and grief. Walk through beautiful countryside — it often works wonders. As human beings we need a direct relationship with nature, because we are a part of her and for thousands of years have been formed by her.

This understanding of the beneficial effects of natural beauty should lead us to insist that nature again be given more room in our man-made environment. This is already happening in many of our cities, but we must introduce many more green areas and groups of trees. Here we must mention the valuable work of Seifert [32] during the building of the first autobahns in Germany.

2.9.11 Closing Remarks on the Rules

We must not assume that the simple application of these rules will in itself lead to beautiful buildings or bridges. The designer must still possess imagination, intuition, and a sense for both form and beauty. Some are born with these gifts, but they must be practiced and perfected. The act of designing must always begin with individual freedom, which in any case will be restricted by all the functional requirements, by the limits of the site, and not least by building regulations that are usually too strict.

The rules, however, provide us with a better point of departure and help us with the critical appraisal of our design, particularly at the model stage, thus making us aware of design errors.

The artistically gifted may be able to produce masterpieces of beauty intuitively without reference to any rules and without rational procedures. However, the many functional requirements imposed on today's buildings and structures demand that our work must include a significant degree of conscious, rational, and methodical reasoning.

2.10 Aesthetics and Ethics

Aesthetics and ethics are in a sense related; by ethics we mean our moral responsibility to humanity and nature. Ethics also infers humility and modesty, virtues which we find lacking in many designers of the last few decades and which have been replaced by a tendency toward the spectacular, the sensational, and the gigantic in design. Due to exaggerated ambition and vanity and spurred by the desire to impress, unnecessary superlatives of fashions were created, lacking true qualities of beauty. Most of these works lack the characteristics needed to satisfy the requirements of the users of these buildings.

As a responsibility, ethics requires a full consideration of all functional requirements. In our man-made environment we must emphasize the categories of quality and beauty. In his *Acht Todsünden der Menschheit*, Loreanz [33] once said that "the senses of aesthetics and ethics are apparently very closely related, so that the aesthetic quality of the environment must directly affect Man's ethical behavior." He said further, "The beauty of Nature and the beauty of the man-made cultural environment are apparently both necessary to maintain Man's mental and psychic health. Total blindness of the soul for all that is beautiful is a mental disease that is rapidly spreading today and which we must take seriously because it makes us insensitive to the ethically obnoxious."

In one of his last important works, in *To Have or to Be* [34] Erich Fromm also said that the category of "goodness" must be an important prerequisite for the category "beauty," if beauty is to be an enduring value. Fromm goes so far as to say that "the physical survival of mankind is dependent on a radical spiritual change in Man." The demand for aesthetics is only a part of the general demand for changes in the development of "Man." These changes have been called for at least in part and at intervals by humanism, but their full realization in turn demands a new kind of humanism, as well expressed in the appeal by Peccei [35].

2.11 Summary

In order to reach a good capacity of judging aesthetic qualities of buildings or bridge structures, it is necessary to go deep into our human capacities of perception and feelings. The views of many authors who treated aesthetics may help to come to some understanding, which shall help us to design with good aesthetic quality.

References

1. Klages, L., *Grundlagen der Wissenschaft vom Ansdruck*, 9, Auflage, Bonn, Germany, 1970.
2. Hume, H., *On the Standard of Taste*, London, U.K., 1882.
3. Smith, P. F, *Architektur und Aesthetik*, Stuttgart, Germany, 1981. Original: *Architecture and the Human Dimension*, London, 1979.
4. Kant, I., *Kritik der Urteilskraft*, Reklam, Stuttgart, Germany, 1995.
5. Paul, J., *Vorschule der Aesthetik*, München, Germany, 1974.
6. Schmitz, H., *Neue Phänomenologie*, Bonn, Germany, 1980.
7. Bahrdt, H. P., *Vortrag Hannover*, Stiftung FVS, Hamburg, Germany, 1979.
8. Keller, R., *Bauen als Umweltzerstörung*, Zürich, Switzerland, 1973.
9. Arnheim, R., *Art and Visual Perception*, Berkeley, CA, 1954. German: *Kunst und Sehen*, Berlin, Germany, 1978.
10. Goethe, J. W., *Farbenlehre*, Stuttgart, Germany, 1979.
11. Szabo, I., *Anfänge der griechischen Mathematik*, Berlin, Germany, 1969.
12. Wittlkower, R., *Architectural Principles in the Age of Humanism*, London, 1952. German: *Grundlagen der Architektur im Zeitalter des Humanismus*, München, Germany, 1969.
13. Charpentier, L., *Die Geheimnisse der Kathedrale von Chartres*, Köln, Germany, 1974.
14. Kayser, H., *Paestum*, Heidelberg, Germany, 1958.
15. Thimus, A. von, *Die Harmonische Symbolik des Altertums*, Köln, Germany, 1968–1976.
16. Puppi, L., *Andrea Palladio*, Stuttgart, Germany, 1977.
17. Strübin, M., Das Villard Diagramm, *Schw. Bauz.*, 1947, 527.
18. Studer, A. M., Architektur, Zahlen und Werte, *Dtsch Bauz.*, 9, Stuttgart, Germany, 1965.
19. Helmcke, J. G., *Ist das Empfinden von aesthetisch schoenen Formen angeboren oder anerzogen*, Heft 3 des SFB 64 der Universität Stuttgart, Germany, 1976, 59; see also *Grenzen menschlicher Anpassung*, IL 14, Universität Stuttgart, Germany, 1975.
20. Grassi, E., *Die Theorie des Schönen in der Antike*, Köln, Germnay, 1980.
21. Torroja, E., *Logik der Form*, München, Germany, 1961.
22. Tellenbach, H., *Geschmack und Atmosphaere*, Salzburg, Austria, 1968.
23. Grimm, C. T., Rationalized esthetics in civil engineering, *J. Struct. Div.*, ASCE, 1975.
24. Borgeest, C., *Das Kunsturteil*, Frankfurt, Germany, 1979. Also: *Das sogenannte Schöne*, Frankfurt, Germany, 1977.
25. Tassios, T. P., Relativity and optimization of aesthetic rules for structures, *IABSE Congr. Rep.*, Zürich, Switzerland, 1980.
26. Leonhartdt, F., *Bridges — Aesthetics and Design*, DVA, Stuttgart, Germany, 1984.

27. Heydemann, B., Auswirkungen des angeborenen Schönheitssinnes bei Mensch und Tier, *Nat., Horst Sterns Umweltmag.*, 0, 1980.

28. Kayser, H., *Harmonia Plantarum*, Basel, Switzerland. Also: H. Akroasis, *Die Lehre von der Harmonik der Welt*, Stuttgart, Germany, 1976.

29. Humphrey, N., The illusion of beauty, *Perception Bd.*, 2, 1973.

30. Berlyne, D. E., *Aesthetics and Psycho-Biology*, New York, 1971.

31. Venturi, R., *Complexity and Contradiction in Architecture*, New York, 1966.

32. Seifert, A., *Ein Leben für die Landschaft*, Düsseldorf-Köln, Germany, 1962.

33. Lorenz, K., *Acht Todsünden der zivilisierten Menschheit*, München, Germany, 1973.

34. Fromm, E., *Haben oder Sein*, Stuttgart, Germany, 1976.

35. Peccei, A., *Die Zukunft in unserer Hand*, München, Germany, 1981.

3

Bridge Aesthetics — Structural Art

David P. Billington
Princeton University

Frederick Gottemoeller
Rosales Gottemoeller Associates, Inc.

3.1 Introduction

In recent years it has become apparent that the real problems of bridge design include more than the structural or construction issues relating to the spanning of a gap. The public often expresses concern over the appearance of bridges, having recognized that a bridge's visual impact on its community is lasting and must receive serious consideration.

The public knows that civilization forms around civil works: for water, transportation, and shelter. The quality of public life depends, therefore, on the quality of such civil works as aqueducts, bridges, towers, terminals, and meeting halls: their efficiency of design, their economy of construction, and the visual appearance of their completed forms. At their best, these civil works function reliably, cost the public as little as possible, and, when sensitively designed, become works of art.

Thus, engineers all over the world are being forced to address the issues of aesthetics. Engineers cannot avoid aesthetic issues by taking care of the structural elements and leaving the visual quality to someone else. It is the shapes and sizes of the structural components themselves that dominate the appearance of the bridge, not the details, color, or surfaces. Since they control the shapes and sizes of the structural components, engineers must acknowledge the fact that they are ultimately responsible for the appearance of their structures. Engineers are used to dealing with issues of performance, efficiency, and cost. Now, they must also be prepared to deal with issues of appearance.

FIGURE 3.1 Thomas Telford's Craigellachie Bridge.

3.2 The Engineer's Aesthetic and Structural Art

"Aesthetics" is a mysterious subject to most engineers, not lending itself to the engineer's usual tools of analysis. It is a topic rarely taught in engineering schools. Many contemporary engineers are not aware that a long line of engineers have made aesthetics an explicit element in their work, beginning with the British engineer Thomas Telford. In 1812, Telford defined structural art as the personal expression of structure within the disciplines of efficiency and economy. Efficiency here meant reliable performance with minimum materials, and economy implied the construction with competitive costs and restricted maintenance expenses. Within these bounds, structural artists find the means to choose forms and details that express their own vision, as Telford did in his Craigellachie Bridge (Figure 3.1). The arch is shaped to be an efficient structural form in cast iron, while his diamond pattern of spandrel bars, at a location in the bridge where structural considerations permit many options, is clearly chosen with an eye to its appearance.

Those engineers who were most conscious of the centrality of aesthetics for structure have also been regarded as the best in a purely technical sense. Starting with Thomas Telford (1757–1834), we can identify Gustave Eiffel (1832–1923) and John Roebling (1806–1869) as the undisputed leaders in their fields during the 19th century. They designed the largest and most technically challenging structures, and they were leaders of their professions. Telford was the first president of the first formal engineering society, the Institution of Civil Engineers, and remained president for 14 years until his death. Eiffel directed his own design–construction–fabrication company and created the longest spanning arches and the highest tower; Roebling founded his large scale wire rope manufacturing organization while building the world's longest spanning bridges (Figure 3.2).

In reinforced concrete, Robert Maillart (1872–1940) was the major structural artist of the early 20th century. First in his 1905 Tavanasa Bridge, and later with the 1930 Salginatobel (Figure 3.3) and 1936 Vessy designs, he imagined a new form for three-hinged arches that included his own invention of the hollow box in reinforced concrete. The Swiss engineer Christian Menn (1927–) has demonstrated how a deep understanding of arches, prestressing, and cable-stayed forms can lead to structures worthy of exhibition in art museums. Especially noteworthy are the 1964 Reichenau Arch, the 1974 Felsenau prestressed cantilever, and the 1980 concrete cable-stayed Ganter Bridge. Meanwhile, German engineer

FIGURE 3.2 John Roebling's Brooklyn Bridge.

Jorg Schlaich has developed new ideas for light structures often using cables, characterized by a series of elegant footbridges in and around Stuttgart (Figure 3.4).

The engineers' aesthetic results from the conscious choice of form by engineers who seek the expression of structure. It is neither the unconscious result of the search for economy nor the product of supposedly optimizing calculations. Many of the best structural engineers have recognized the possibility for structural engineering to be an art form parallel to but independent of architecture. These people have, over the past two centuries, defined a new tradition, structural art, which we take here to be the ideal for an engineer's aesthetic.

Although structural art is emphatically modern, it cannot be labeled as just another movement in modern art. For one thing, its forms and its ideals have changed little since they were first expressed by Thomas Telford. It is not accidental that these ideals emerged in societies that were struggling with the consequences not only of industrial revolutions but also of democratic ones. The tradition of structural art is a democratic one.

In our own age the works of structural art provide evidence that the common life flourishes best when the goals of freedom and discipline are held in balance. The disciplines of structural art are efficiency and economy, and its freedom lies in the potential it offers the individual designer for the expression of a personal style motivated by the conscious aesthetic search for engineering elegance. These are the three leading ideals of structural art — efficiency, economy, and elegance.

FIGURE 3.3   Robert Maillert's Salginotobel Bridge.

FIGURE 3.4   One of Jorg Schlaich's footbridges.

3.3   The Three Dimensions of Structure

Its first dimension is a scientific one. Each working structure or machine must perform in accordance with the laws of nature. In this sense, then, technology becomes part of the natural world. Methods of analysis useful to scientists for explaining natural phenomena are often useful to engineers for describing the behavior of their artificial creations. It is this similarity of method that helps to feed the fallacy that engineering is applied science. But scientists seek to discover preexisting form and explain its behavior by inventing formulas, whereas engineers want to invent forms, using preexisting formulas to check their designs. Because the forms studied by scientists are so different from those of engineers, the methods of analysis will differ; yet, because both sets of forms exist in the natural world, both must obey the same natural laws. This scientific dimension is measured by efficiency.

Technological forms live also in the social world. Their forms are shaped by the patterns of politics and economics as well as by the laws of nature. The second dimension of structure is a social one. In the past or in primitive places of the present, completed structures and machines might, in their most elementary forms, be merely the products of a single person; in the civilized modern world, however, these technological forms, although at their best designed by one person, are the products of a society. The public must support them, either through public taxation or through private commerce. Economy measures the social dimension of structure.

Technological objects visually dominate our industrial, urban landscape. They are among the most powerful symbols of the modern age. Structures and machines define our environment. The locomotive of the 19th century has given way to the automobile and airplane of the 20th. Large-scale complexes that include structures and machines become major public issues. Power plants, weapons systems, refineries, river works — all have come to symbolize the promises and problems of industrial civilization.

The Golden Gate, the George Washington, and the Verrazano Bridges carry on the traditions set by the Brooklyn Bridge. The Chicago Hancock and Sears Towers, and the New York Woolworth, Empire State, and World Trade Center Towers all bring the promise of the Eiffel Tower into the utility of city office and apartment buildings. The Astrodome, the Kingdome, and the Superdome carry into the late 20th century the vision of huge permanently covered meeting spaces first dramatized by the 1851 Crystal Palace in London and the 1889 Gallery of Machines in Paris.

Nearly every American knows something about these immense 20th-century structures, and modern cities repeatedly publicize themselves by visual reference to these works. As Montgomery Schuyler, the first American critic of structures, wrote in the 19th century for the opening of the Brooklyn Bridge, "It so happens that the work which is likely to be our most durable monument, and to convey some knowledge of us to the most remote posterity, is a work of bare utility; not a shrine, not a fortress, not a palace but a bridge. This is in itself characteristic of our time."[1].

So it is that the third dimension of technology is symbolic, and it is, of course, this dimension that opens up the possibility for the new engineering to be structural art. Although there can be no measure for a symbolic dimension, we recognize a symbol by its elegance and its expressive power. Thus, the Sunshine Skyway (Figure 3.5) has become a symbol of both Florida's Tampa Bay area and the best of late-20th-century technology.

There are three types of designers who work with forms in space: the engineer, the architect, and the sculptor. In making a form, each designer must consider the three dimensions or criteria we have discussed. The first, or scientific criterion, essentially comes down to making structures with a minimum of materials and yet with enough resistance to loads and environment so that they will last. This efficiency–endurance analysis is arbitrated by the concern for safety. The second, or social criterion, comprises mainly analyses of costs as compared with the usefulness of the forms by society. Such cost–benefit analyses are set in the context of politics. Finally, the third criterion, the symbolic, consists of studies in appearance, along with a consideration of how elegance can be achieved within the constraints set by the scientific and social criteria. This is the aesthetic/ethical basis upon which the individual designer builds his or her work.

For the structural designer the scientific criterion is primary (as is the social criterion for the architect and the symbolic criterion for the sculptor). Yet the structural designer must balance the primary criterion with the other two. It is true that all structural art springs from the central ideal of artificial forms controlling natural forces. Structural forms will, however, never get built if they do not gain some social acceptance. The will of the designer is never enough. Finally, the designer must think aesthetically for structural form to become structural art. All of the leading artists of

FIGURE 3.5   The Sunshine Skyway.

structure thought about the appearance of their designs. These engineers consciously made aesthetic choices to arrive at their final designs. Their writings about aesthetics show that they did not base design only on the scientific and social criteria of efficiency and economy. Within those two constraints, they found the freedom to invent form. It was precisely the austere discipline of minimizing materials and costs that gave them the license to create new images that could be built and endure.

3.4   Structure and Architecture

The modern world tends to classify towers, stadiums, and even bridges as architecture, creating an important, but subtle, fallacy. Even the word is a problem, because *architect* comes from the Greek word meaning chief technician. But, beginning with the Industrial Revolution, structure has become an art form separate from architecture. The visible forms of the Eiffel Tower, Seattle's Kingdome, and the Brooklyn Bridge result directly from technological ideas and from the experience and imagination of individual structural engineers. Sometimes, the engineers have worked with architects just as with mechanical or electrical engineers, but the forms have come from structural engineering ideas.

Structural designers give form to objects that are of relatively large scale and of single use, and these designers see forms as the means of controlling the forces of nature to be resisted. Architectural designers, on the other hand, give form to objects that are of relatively small scale and of complex human use, and these designers see forms as the means of controlling the spaces to be used by people. The prototypical engineering form — the public bridge — requires no architect. The prototypical architectural form — the private house — requires no engineer. Structural engineers and architects learn from each other and sometimes collaborate fruitfully, especially when, as with tall buildings, large scale goes together with complex use. But the two types of designers act predominately in different spheres.

The works of structural art have sprung from the imagination of engineers who have, for the most part, come from a new type of school — the polytechnical school, unheard of prior to the late 18th century. Engineers organized new professional societies, worked with new materials, and stimulated political thinkers to devise new images of future society. Their schools developed curricula that decidedly cut whatever bond had previously existed between those who made architectural forms and those who began to make — out of industrialized metal and later from reinforced concrete — the new engineering forms by which we everywhere recognize the modern world. For these forms the ideas inherited from the masonry world of antiquity no longer applied; new ideas were essential in order to build with the new materials. But as these new ideas broke so radically with conventional taste, they were rejected by the cultural establishment.

This is, of course, a classic problem in the history of art: new forms often offend the academics. In this case, it was beaux arts against structural arts. The skeletal metal of the 19th century offended most architects and cultural leaders. New buildings and city bridges suffered from valiant attempts to cover up or contort their structure into some reflection of stone form. In the 20th century, the use of reinforced concrete led to similar attempts. Although some people were able to see the potential for lightness and energy, most architects tried gamely to make concrete look like stone or, later on, like the emerging abstractions of modern art. There was a deep sense that engineering alone was insufficient.

The conservative, plodding, hip-booted technicians might be, as the architect Le Corbusier said, "healthy and virile, active and useful, balanced and happy in their work, but only the architect, by his arrangement of forms, realizes an order which is pure creation of his spirit ... it is then that we experience the sense of beauty." The belief that the happy engineer, like the noble savage, gives us useful things but only the architect can make them into art is one that ignores the centrality of aesthetics to the structural artist. In towers, bridges, free-spanning roofs, and many types of industrial buildings, aesthetic considerations provide important criteria for the engineer's design. The best of such engineering works are examples of structural art, made by engineers, and they have appeared with enough frequency to justify the identification of structural art as a mature tradition with a unique character. One of the most recent manifestations is Christian Menn's Sunniberg Bridge (Figure 3.6).

3.5   Application to Everyday Design

Many of today's engineers see themselves as a type of applied scientist, analyzing preexisting structural forms that have been established by others. Seeing oneself as an applied scientist is an unfortunate state of mind for a design engineer. It eliminates the imaginative half of the design process and forfeits the opportunity for the integration of form and structural requirements that can result in structural art. Design must start with the selection of a structural form. It is a decision that can be made well only by the engineer because it must be based on a knowledge of structural forms and how they control forces and movements.

In the case of most everyday bridges the selection of form is based largely on precedents and standards established by the bridge-building agency. For example, the form of a highway overpass may be predetermined by the client agency to be a welded plate girder bridge because that is what the agency prefers or what local steel fabricators are used to or even because the steel industry is a dominant political force in the state. In other cases, the form may be established by an architect or urban designer for reasons outside structural requirements. Thus the form is set without any serious consideration of whether or not that is in fact the best form for that particular site.

Creative form determination consists not of applying free visual imagination alone nor in applying rigorous scientific analysis alone, but of applying both together, at the same time. The art starts with a vision of what might be. The development of that vision is the key. Many engineers call the development of the vision conceptual engineering. It is the most important part of design. It is the stage at which all plausible forms are examined. The examination must include, to a rough level of precision, the whole range of considerations; performance, cost, and appearance. All that follows, including the aesthetic impression the bridge makes, will depend on the quality of the form selected. This stage is often ignored or foreclosed, based on precedents, standards, preconceived ideas or prior experience that may or may not apply.

FIGURE 3.6   Christian Menn's Sunniberg Bridge.

The reasons often given for shortchanging this stage include, "Everybody knows that [steel plate girders, precast concrete girders, cast-in-place concrete] are the most economical structure for this location," or "We always build [steel plate girders, precast concrete girders, cast-in-place concrete] in this state," or "Let's use the same design as we did for [any bridge] last year."

At this point someone will protest that other considerations (costs, the preferences of the local contracting industry, etc.) will indeed differentiate and determine the form. Too often these reasons are based on unexamined assumptions, such as, "The local contracting industry will not adjust to a different form," or "Cost differentials from [a past project] still apply," or "The client will never consider a different idea." Or the belief is based on a misleading analysis of costs which relies too much on assumed unit costs. Or that belief may be simply habit — either the engineer's or the client's — often expressed in the phrase, "We've always done it that way." Accepting these assumptions and beliefs places an unfortunate and unnecessary limitation on the quality of the resulting bridge for, by definition, improvements must come from the realm of ideas not tried before.

As Captain James B. Eads put it in the preliminary report on his great bridge over the Mississippi River at St. Louis:

Must we admit that because a thing has never been done, it never can be, when our knowledge and judgment assure us that it is entirely practicable?[2].

FIGURE 3.7   MD 18 over U.S. 50.

FIGURE 3.8   Another possibility for MD 18 over U.S. 50.

The engineer's first job is to question all such determinations, assumptions and beliefs. From that questioning will come the open mind that is necessary to develop a vision of what each structure can be at its best.

Unless such questioning is the starting point it is unlikely that the most promising ideas will ever appear. No design will occur. Instead, there will be a premature assumption of the bridge form, and the engineer will move immediately into the analysis of the assumed form. That is why so many engineers mistake analysis for design. Design is more correctly the selection of the form in the first place, which most engineers have not been permitted to do. Design is by far the more important of the two activities.

Engineers also focus on analysis in the belief that the form (shape and dimensions) will be determined by the forces as calculated in the analysis. But, in fact, there are a large number of forms that can be shown by the analysis to work equally well. It is the engineer's option to choose among them, and in so doing to determine the forces by means of the form, not the other way around.

Take the simple example of a two-span continuous girder bridge, using an existing structure, MD 18 over U.S. 50 (Figure 3.7). Here the engineer has a wide range of possibilities such as a girder with parallel flanges, or with various haunches having a wide range of proportions (Figure 3.8). The moments will depend on the stiffness at each point, which in turn will depend on the presence or absence of a haunch and its shape (Figure 3.9). The engineer's choice of shape and dimensions will determine the moments at each point along the girder. The forces will follow the choice of form. Within limits, the engineer can direct the forces.

Let's examine which form the engineer should choose. All can support the required load. Depending on the specifics of the local contracting industry, many of them will be essentially equal in cost. All would perform equally well and all are comparable in cost, leaving the engineer a decision that can only be made on aesthetic grounds. Why not pick the one the engineer believes looks best?

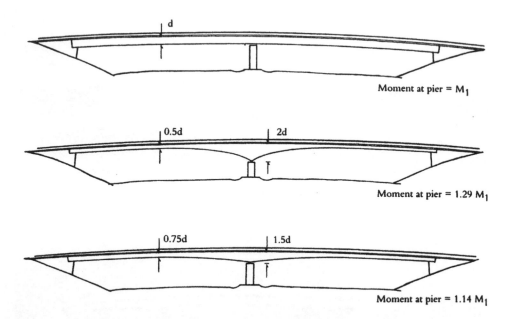

FIGURE 3.9   Forces determined by the engineer's choice of form.

That, in a nutshell, is the process that all of the great engineers have followed. Maillart's work, as one example, shows that the engineer cannot choose form as freely as a sculptor, but the engineer is not restricted to the discovery of preexisting forms as the scientist is. The engineer invents form, and Maillart's career shows that such invention has both a visual and a scientific basis. When either is denied, engineering design ceases. For Maillart, the dimensions were not to be determined by the calculations, and even the results of the calculations could be changed (by adjusting the form) because a designer rather than an analyst is at work. Analysis and calculation are the servants of design. Design, analysis, and must work together. In the words of Spanish engineer Eduardo Torroja,

> The imagination alone cannot reach such [elegant] designs unaided by reason, nor can a process of deduction, advancing by successive cycles of refinement, be so logical and determinate as to lead inevitably to them.[3]

The engineering challenge is not just to find the least costly solution. The engineering challenge is to bring forth elegance from utility: We should not be content with bridges that only move vehicles and people. They should move our spirits as well.

3.6   The Role of Case Studies in Bridge Design

Bridge design, even of highway overpasses, often involves standard problems but always in different situations. Case studies can help in the design of these standard problems by showing models and points of comparison for a large number of bridges without implying that each such bridge be mere imitation.

The primary goals of a case study are to look carefully at all major aspects of the completed bridge, to understand the reasons for each design decision, and to discuss alternatives, all to the end of improving future designs. Such cases help to define more general ideas or principles. Case studies are well recognized by engineers when designing for acceptable performance and low cost; they can be useful when considering appearance as well.

A common organization of these studies will help identify standard problems and make comparisons easier. First comes an overall evaluation of the bridge as a justification for studying it. Is it a good example that can be better? Is it a model of near perfection? Is it a bad example to be avoided?, Second comes a description of the complete bridge, which is divided into parts roughly coinciding with easily identifiable costs and including modifications to each part as suggested improvements. In this major description section there is an order to the parts that implies a priority for the structural engineer: concept and form of the entire structure, superstructure, supports, deck, color, and landscaping.

1. The *concept and form* of the completed bridge goes together with a summary of the bridge performance history (including maintenance) and of its construction cost, usually given per square meter of bridge. Required clearances, foundation conditions, hydraulic requirements, traffic issues, and other general requirements would be covered here.
2. The *superstructure* here includes primarily the main horizontal spanning members such as continuous girders, arches, trusses, etc. In continuous steel girder bridges, the cost is primarily identified with the fabricated steel cost. Modification in design by haunching, changing span lengths, or making girders continuous with columns would be discussed including their influence on cost.
3. The *pier supports* are most frequently columns or frames either in the median or outside the shoulders, or at both places in highway overpasses. These are normally highly visible elements and can have many possible forms. Different designs for the relationship among steel girder, bearings, and columns can make major improvements in appearance without detriment to cost or performance.
4. The *abutment supports* are also highly visible parts of the bridge, which include bearings, cantilever walls, cheek walls, and wing walls.
5. The *deck* includes the concrete slab or orthotropic steel deck, overhangs, railings, parapets, and provisions for drainage, all of which have an influence on performance as well as on the appearance either when seen in profile or from beneath the bridge.
6. The *color* is especially significant for steel structures that are painted, and *texture* can be important for concrete surfaces of piers, abutments, and deck.
7. The *landscaping/guardrail* includes plantings and other features that can have important visual consequences to the design.   .

The order of these parts is significant because it focuses attention on the engineering design. The performance of a weak structural concept cannot be saved by good deck details. An ugly form cannot be salvaged by color or landscaping. The first four parts are structural, the fifth is in part structural, whereas the last two, while essential for the bridge engineer to consider, involve primarily nonstructural ideas.

Third, the case study can give a critique of the concept and form by comparison with other similar bridges or bridge designs for similar conditions, including those with very different forms, as a stimulus to design imagination.

Fourth and finally, the case should conclude with some discussion of the relationship of this study to a theory of bridge design. Clearly, any such study must be based upon a set of ideas about design which often implicitly bias the writer who should make these ideas explicit. This conclusion should show how the present study illustrates a theory and even at times forces a modification of it. General ideas form only out of specific examples.

3.7   Case Study in Colorado: Buckley Road over I-76

Colorado's Buckley Road over I-76 (Figure 3.10) offers the application of an innovative form to prestressed concrete girders in order to achieve longer than normal spans, with a visually unique result. It is therefore a worthy subject for a case study.

FIGURE 3.10   Buckley Road over I-76.

3.7.1   Description of the Bridge

In *concept* this is a three-span continuous beam bridge with a 47° skew made of precast prestressed girders set onto cast-in-place concrete piers and abutments.

The *superstructure* consists of seven girders spaced approximately 3 m apart and each made up of five precast prestressed concrete segments (Figure 3.11). The main span is approximately 56 m and each side span is approximately 50 m. Segments one and five are 37.8 m long and behave essentially as simply supported beams between the abutments and the cantilever segments two and four. These latter cantilever 12 m into the side spans and 15 m into the main span. Segment three is 25.6 m long and behaves approximately as a simply supported beam within the main span. There are 0.15 m spacings between segments for closure pours. The cantilever segments have a linear haunch of 0.6 m from the girder depth of 1.8 m for the other three segments, which are Colorado BT72 girders.

The two *piers* each consist of a pier cap beam 1.2 m wide by 1.8 m deep and 28 m long supported by three walls each 1.2 m wide, about 7.6 m high, and 6 m long at their tops tapering to 3 m long at the footings. The cap beam extends about 1.5 m beyond the centerline of the exterior girder or about 1.1 m beyond the edge of that girder's bottom flange. The pier next to the railroad has a crash wall built into the three tapered walls.

The *abutments* are shallow concrete beams 0.9 m wide supported on piles and carrying the precast girders. The *deck* is a series of precast pretensioned concrete panels made composite with the precast girders. Bounded by Jersey barriers, the deck is 20.4 m in width and overhangs the exterior girders by 0.96 m or slightly over half the depth of the unhaunched girder segments.

3.7.2   Critique of the Bridge

The *concept* of a fully precast superstructure, a three-span continuous beam, and cast-in-place piers has led to an economical structure and fits well the site conditions of crossing both I-76 and the double-track railroad. Other reasonable concepts include a two-span bridge and a three-span bridge with the cantilever segments two and four cast in place with the piers (as illustrated by the Stewart

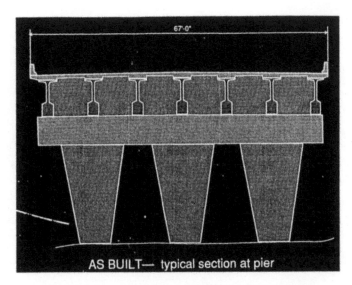

67'-0"

AS BUILT— typical section at pier

FIGURE 3.11 Typical section of Buckley Road as built.

Park Bridge in Oregon). This critique will confine itself to the present concept, but a comparison of this bridge and the Oregon one will follow. In each case, the ideas of structural art will form the basis for a critique.

The *superstructure* represents an unusual use of a precast bulb T girder whose bulb is extended vertically to create a haunch at the two interior supports. The profile view expresses the increased forces at the interior supports and the construction photos (with temporary walkway) show the lightness achieved by an overhang that is about the same dimension as the girder depth.

The *haunches* would be more effective visually were they deeper and the segments one, three, and five correspondingly shallower. For example, with the Colorado C68 girder, the depth would decrease to 1.7 m and a haunch of 2.6 m would more strikingly express the flow of forces. At the same time, the girder spacing would be reduced to 2.9 m to permit an overhang of 1.4 m.

Another solution would be to retain the Colorado BT72 girders, increase the haunch to 3 m, and reduce the number of girders from seven to six, thus again increasing the overhang. If the six girders were spaced 3.35 m on centers, then the overhang would be 1.7 m or nearly the depth of the BT72 girders.

The *piers* are visually prominent and look heavy. They also have a formal shaping which does not clearly express the structure. Specifically, the horizontal lines of the hammerhead beam separate it from the supporting walls and the 1.8-m depth of that beam is far greater than needed to carry the girder loads over the 2.4-m span between the wide supporting walls below. Since these piers are relatively short compared with the long spans of the girders, their massive appearance is accentuated by the lack of structural expression. It is clear from beneath the bridge that the 6-m-wide walls can easily be made to support all the girders directly without any hammerhead beam (Figure 3.12). The walls will therefore be higher and, if carefully shaped, will form a striking integration with the deck girders. The cast-in-place diaphragm can then be structurally integrated with the walls and the girders to form a cross frame for live loads.

The *abutments* can be improved by eliminating the wall that hides the girder ends and bearings. Along with the lighter-looking girders, this structural expression at the abutments will increase the already striking appearance of the bridge profile.

The *deck* overhang, by being increased, lends lightness to the girders. Otherwise, the system used is good and avoids the staining that can arise when metal slab forms are left in place.

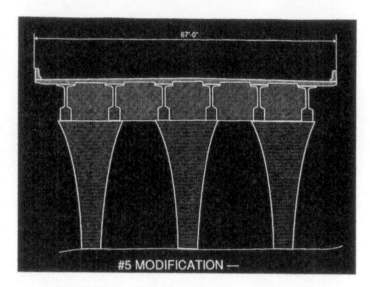

FIGURE 3.12   Possible modification to Buckley Road.

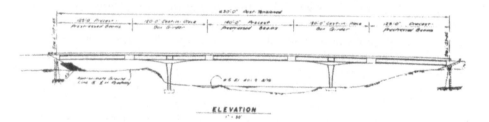

FIGURE 3.13   Elevation of Louis Pierce's Stewart Park Bridge.

3.7.3 The Stewart Park Bridge

The *concept* for this 1978 bridge (Figure 3.13) designed by Louis Pierce is the same as for Buckley Road except for the cantilever segments two and four which are cast-in-place prestressed concrete hollow boxes. Because the spans (56.4, 79.2, 56.4 m) are longer than those for Buckley Road, segments one, three, and five are each made of two separate precast pieces.

The *superstructure* and the *piers* are thus integrated into one form rather than separated into two forms as at Buckley Road. The boxes are haunched from the 2.4 m of the constant section segments to 3.65 m at the two interior supports for a ratio of 1.55. But the boxes are 2.4 m deep along their exterior faces and haunch laterally to 3.65 m over a distance of 2.5 m. Just as at Buckley Road, the *deck* overhang is too short, about 0.8 m for a girder depth of 2.4 m.

The shape of the two piers are walls 7.6 m wide, 2.3 m thick at the top, tapering to 5.8 m wide and 1.4 m thick at the base. The total height is 12.7 m above the footing but only about 7.6 m above the ground line. This shaping of piers, having about the same height as those of Buckley Road, gives an impression of lightness missing from the latter structure (Figure 3.14).

3.7.4 Summary

The Buckley Road bridge represents a good design. A similar concept can be improved in future designs by relatively small changes in the superstructure through stronger haunching and wider deck overhangs and by major changes in the pier form. The use of cast-in-place cantilever sections offers increased possibilities for elegant forms and closer integration of superstructure with piers.

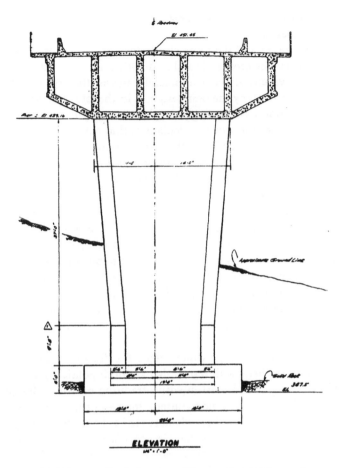

FIGURE 3.14 Typical section of Stewart Park Bridge at pier.

This case study gives an example of how a good bridge can provide an excellent basis for further study and improvement.

3.8 Achieving Structural Art in Modern Society: Computer Analysis and Design Competitions

Most people would agree that the ideals of structural art coincide with those of an urban society: conservation of natural resources, minimization of public expenditures, and the creation of a more visually appealing environment. As the history of structural art shows, some engineers have already turned these ideals into realities. But these are isolated cases. How might they become the rule instead of the exception? We can address this question historically, by identifying the central ideas that have been associated with great structural art. These ideas reflect each of the three dimensions: the scientific, social, and symbolic.

The leading scientific idea might be stated as that of reducing analysis. In structural art, this idea has coexisted with the opposite tendency to overemphasize analysis, which today is typified by the heavy use of the computer for structural calculations. One striking example comes from the design of thin concrete vaults — thin shell roofs. Here, the major advances between 1955 and 1980 — a time of intense analytic developments — were achieved, not by performing complex analyses using computers, but rather by reducing analysis to very simple ideas based on observed physical behavior. Roof vaults characterize this advance and they carry forward the central scientific idea in structural art: the analyst of the form, being also the creator of the form, is free to change shapes so that analytic complexity disappears.

The form controls the forces and the more clearly that designers can visualize those forces the surer they are of their forms. The great early and mid-20th-century structural artists such as Robert Maillart and Pier Luigi Nervi have all written forcefully against the urge to complicate analysis. We see the same arguments put forth by the best designers in the late 20th century. When the form is well chosen, its analysis becomes astoundingly simple. The computer, of course, has become more and more useful as a time saver for routine calculations that come after the design is set. It is also increasingly valuable in aiding the designer through computer graphics. But like any machine, although it can reduce human labor, it cannot substitute for human creativity.

Turning to the social dimension, a leading idea that has come out of structural art is the effectiveness of public design competitions. Design quality arises from the stimulus of competing designs for the same project rather than from complex regulations imposed upon a single designer. The progress of modern bridge design illustrates the benefit and meaning of alternative designs. Many alternative designs have been prepared pursuant to design competitions, which bring the public into the process in a positive way. It is not enough for the public merely to protest the building of ugly, expensive designs. A positive activity is essential, and that can only come about when the public sees the alternative designs that are possible for a project. Thus, governments can ensure better designs by relinquishing some of their control over who designs and on what forms are chosen, and by giving some of this control to an informed jury which includes representatives of the lay public.

Although there is little tradition in the United States for design competitions in bridges, such a tradition is firmly rooted elsewhere, with results that are both politically and aesthetically spectacular. Switzerland has the longest and most intensive tradition of bridge design competitions, and it is no coincidence that, by nearly common consent, the two greatest bridge designers of the first half of the 20th century were Swiss: Robert Maillart (1872–1940), who designed in concrete, and Othmar Amman (1879–1966), designer of the George Washington and Verrazano Bridges, who designed in steel. That Switzerland, one sixth the size of Colorado, and with fewer people than New York City, could achieve such world prominence is due to the centrality of economics and aesthetics for both their engineering teachers and their practicing designers, a centrality which is encouraged by design competitions.

Maillart's concrete arches in Switzerland were often the least expensive proposals in design competitions, and they were later to provide the main focus for the first art museum exhibition ever devoted exclusively to the work of one engineer: the New York City Museum of Modern Art's 1947 exhibition on Maillart's structures. Amman has been similarly honored. His centennial was celebrated by symposia both in Boston and in New York and by an exhibition held in Switzerland. Both Maillart and Amman wrote articulately on the appearance as well as on the economy of bridges. They are prime examples of structural artists.

This Swiss bridge tradition continues today with a large number of striking new bridges in concrete that follow Maillart in principle if not in imitative detail. The most impressive post-World War II works are those of Christian Menn, whose long-span arches and cantilevers extend the new technique of prestressing to its limits, as Maillart's three-hinged and deck-stiffened arches did earlier with reinforced concrete.

Design competitions stimulated these engineers and also educated the general public. To be effective, uch competitions must be accepted by political authorities, judged by engineers and informed lay members whose opinions will be debated in the public press, and controlled by carefully drawn rules.

It is false images of engineering that keep us from insisting on following our normal instinct for open competition. The American politics of public works falsely compares the engineering designer either with a medical doctor or with a building contractor.

Supporters of the first comparison argue that you would never hold a competition to decide who will repair your heart; rather, you would choose professionals on the basis of reputation and then leave them alone to do the skilled work for which they are trained. However, there is a key difference between hearts and bridges. For most people, there is only one heart which will do the job. Picking a "best" heart is not a consideration. On the other hand, for a given bridge site, there are many bridge designs that will solve the problem. The more minds that are put to the problem, the more likely that an outstanding design will emerge. After all, the ultimate goal is to pick the best bridge, not the best bridge designer.

Furthermore, developing the engineer's imagination creates a valuable asset for society. That imagination needs more chances to exercise than there are chances to build, and it is stimulated by competition. However, frustrating it may be to lose a competition, the activity is healthy and maturing, especially when even the losers are compensated financially for their time, as they often are in Switzerland.

For proponents of the second false comparison, design competitions are to be run just as building competitions in which the lowest bid for design cost gets the design contract. In American public structures, design and construction are legally distinct activities. The cost of design is normally 5% or less of the cost of construction. Therefore, a brilliant engineer might spend more preparing a design which, as can often happen, will cost the owners substantially less overall. By the same token, an engineer who cuts the design fee to get the job may have to make a more conservative design which could easily cost the owner more in overall costs. Hence, large amounts of potential savings to the public are lost by a foolish policy of saving a little during the first stage of a project.

In one type of Swiss design competition, a small number of designers are invited to compete, some of their costs are covered, and they get additional prize funds in the order recommended by the jury. The winner usually gets the commission for the detailed design. Only several such competitions a year are needed to stimulate the entire profession and to show the general public the numerous possibilities available as good solutions to any one problem. This method of design award opens up the political process to local people far more than does the cumbersome and largely negative one of protest, legal action, and negation of building that so dominates public action in late-20th-century America.

The state of Maryland is leading the way in the United States. In 1988, Maryland held a design competition for a new structure over the Severn River adjoining the U.S. Naval Academy in Annapolis. The competition was patterned on the Swiss practice. The results of the competition resolved an acrimonious community controversy. The winning structure, by Thomas Jenkins (Figure 3.15), was recognized by the American Institute of Steel Construction as the outstanding medium-span structure constructed in 1995–96. In 1998, Maryland, together with the state of Virginia, the District of Columbia, and the Federal Highway Administration, conducted a competition to select the design of the new Woodrow Wilson Bridge over the Potomac River at Washington, D.C. The winning design (Figure 3.16) was prepared by a team led by the Parsons Transportation Group.

Properly defined design competitions reveal truths about society that are otherwise difficult to define. The resulting designs, therefore, became unique symbols of their time and place. This brings us to the third leading idea that has been associated with great structural art — the idea that its materials and forms possess a particular symbolic significance. Perceptive painters, poets, and writers have recognized in structural art a new type of symbol — first in metal and then in concrete — which fits mysteriously closely both to the engineering possibilities and to the possibilities inherent in democracy. The thinness and openness of the Eiffel Tower, Brooklyn Bridge, and Maillart's arches, as well as the stark contrast between their forms and their surroundings, have a deep affinity to both the political traditions and era in which they arose. They symbolize the artificial rather than the natural, the democratic rather than the autocratic, and the transparent rather than the impenetrable. Their forms reflect directly the inner springs of creativity emerging from contemporary industrial societies.

FIGURE 3.15   Thomas Jenkins's U.S. Naval Academy Bridge over the Severn River.

These forms imply a democratic rather than an autocratic life. When structure and form are one, the result is a lightness, even fragility, which closely parallels the essence of a free and open society. The workings of a democratic government are transparent, conducted in full public view, and although a democracy may be far from perfect, its form and its actual workings (its structure) are inseparable. Furthermore, the public must continually inspect its handiwork: constant maintenance and periodic renewal are essential to its exposed structure. Politicians do not have life tenure; they must be inspected, chastised, and purified from time to time, and replaced when found corrupt or inept. So it is with the works of structural art. They, too, are subject to the weathering and fatigue of open use. They remind us that our institutions belong to us and not to some elite. If we let them deteriorate, as we flagrantly have in our older cities and transportation networks, then that outward sign betokens an inner corruption of the common life in a free democratic society.

3.9   The Engineer's Goal

The ideal bridge is structurally straightforward and elegant. It should provide safe passage and visual delight for drivers, pedestrians, and people living or working nearby. Society holds engineers responsible for the quality of their work, including its appearance. For the same reason engineers would not build a bridge that is unsafe, they should not build one that is ugly. Bridge designers must consider visual quality as fundamental a criterion in their work as performance, cost, and safety.

There are no fast rules or generic formulas conducive to outstanding visual quality in bridge design. Each bridge is unique and should be studied individually, always taking into consideration all the issues, constraints, and opportunities of its particular setting or environment. Nevertheless, by observing other bridges, using case studies and design guidelines, engineers can learn what makes bridges visually outstanding and develop their abilities to make their own bridges attractive. They can achieve outstanding visual quality in bridge design while maintaining structural integrity and meeting their budgets.

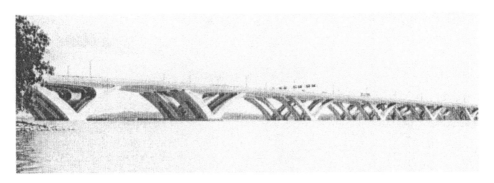

FIGURE 3.16   The competition-winning design for the new Woodrow Wilson Bridge over the Potomac River at Washington, D.C.

References

1. Schuyler, M., The bridge as a monument, *Harpers Weekly* 27, 326, 1883; reprinted in *American Architecture and Other Writings*, W. H. Jordy and R. Coe, Eds., Atheneum, New York, 1964, 164.
2. Eads, J.B., Report of the Engineer-in-Chief of the Illinois and St. Louis Bridge Company, St. Louis, Missouri Democrat Book and Job Printing House, 1868. Reprinted in *Engineers of Dreams*, H. Petroski, Alfred A. Knopf, New York, 1995, 54.
3. Torroja, E., *The Structures of Eduardo Torroja, an Autobiography of Engineering Accomplishment*, F.W. Dodge, New York, 1958, 7.

Bibliography

Much of the discussion in this chapter is contained in a more complete form in the following four books by the authors:

David P. Billington, *Robert Maillart and the Art of Reinforced Concrete*, MIT Press, Cambridge, MA, 1990.
David P. Billington, *Robert Maillart's Bridges — The Art of Engineering*, Princeton University Press, Princeton, NJ, 1979.
David P. Billington, *The Tower and the Bridge — The New Art of Structural Engineering*, Basic Books, New York, 1983.
Frederick Gottemoeller, *Bridgescape, the Art of Designing Bridges*, John Wiley & Sons, New York, 1998.

The following should be part of the reference library of any engineer interested in bridge aesthetics (not in any particular order):

Fritz Leonhardt, *Brucken*, MIT Press, Cambridge, MA, 1982.
Max Bill, *Robert Maillart, Bridges & Construction*, 1st ed. 1949, Praeger, Westport, CT, 1969.
Steward C. Watson and M. K. Hurd, *Esthetics in Concrete Bridge Design*, American Concrete Institute, Detroit, 1990.
Hans Jochen Oster et al., *Fussgangerbrucken, Jorg Schlaich und Rudolph Bergermann Katalog zur Ausstellung an der ETH Zurich*, Institut für Baustatik und Konstruktion, Zürich, Switzerland, 1994.
Martin P. Burke, Jr. and the General Structures Committee, *Bridge Aesthetics around the World*, Transportation Research Board, Washington, D.C., 1991.

4

Planning of
Major Fixed Links

Klaus H. Ostenfeld
COWI, Denmark

Dietrich L. Hommel
COWI, Denmark

Dan Olsen
COWI, Denmark

Lars Hauge
COWI, Denmark

4.1 Introduction

Characteristics of Fixed Links

Within the infrastructure of land transportation, fixed links are defined as permanent structures across large stretches of water allowing for uninterrupted passage of highway and/or railway traffic with adequate safety, efficiency, and comfort.

Traffic services are often provided by ferries before a fixed link is established. Normally, a fixed link offers shorter traveling times and higher traffic capacities than the ferry services. The establishment of a fixed link may therefore have a strong positive impact on the industrial and economic development of the areas to be served by the link. This together with an increased reliability in connection with climatic conditions are the major reasons for considering the implementation of a fixed link.

The waters to be passed by the links are often navigable; the link structures may present obstacles to the vessel traffic and are thus subject to the risk of impact from vessels. If the vessel traffic is important, the link traffic may be better separated from the crossing vessel traffic for general traffic safety. The water flow is often influenced by the link structures and this may affect the environment both near and far from the site. Furthermore, the water stretches and areas to be passed are part of beautiful territories forming important habitats for wildlife fauna and flora. The protection and preservation of the environment will therefore often be a major issue in the political discussions prior to the establishment of the links. These aspects have to be realized and considered in the very beginning of the planning process.

TABLE 4.1 Major Fixed Links Opened Since 1988

Name of Link	Total Length and Types of Structures	Status Early 1998	Traffic Mode
Confederation Bridge, Canada	12.9 km, high-level concrete box girder bridge	Open to traffic 1997	Highway traffic
Vasco da Gama Bridge, Portugal	12.3 km, viaducts and high-level cable-stayed bridge	Open to traffic 1998	Highway traffic
Second Severn Bridge, Great Britain	5.1 km, viaducts and high-level cable-stayed bridge	Open to traffic 1996	Highway traffic
Honshu–Shikoku Connection, Japan			
• Kojima–Sakaide Route	37.3 km, a o high-level suspension bridges	Open to traffic 1988	Highway and railway
• Kobe–Naruto Route	89.6 km, a o high-level suspension bridges	South part open 1998	Highway traffic
• Onomichi–Imabari Route	59.4 km, a o high-level suspension bridges	Under construction	Highway traffic
Lantau Fixed Crossing, Hong Kong	3.4 km High-level suspension bridge	Open to traffic 1997	Highway and railway
• Tsing Ma Bridge	High level cable stayed bridge	Open to traffic 1997	Highway and railway
• Kap Shui Mun Bridge			
Boca Tigris Bridge, China	4.6 km, high-level suspension bridge	Open to traffic 1997	Highway traffic
Great Belt link, Denmark	17.5 km		
• West Bridge	6.6 km, low-level concrete box girder bridge	Open to traffic 1997	Highway and railway
• East Bridge	6.8 km, high-level suspension bridge	Open to traffic 1998	Highway traffic
• East Railway Tunnel	8.0 km, bored tunnel, two tubes	Open to traffic 1997	Railway traffic
Øresund link, Sweden–Denmark	16 km, immersed tunnel, artificial island, high-level cable-stayed bridge, viaducts	Under construction	Highway and railway
Rion–Antirion Bridge, Golf of Corinthe, Greece	2.9 km, high-level cable-stayed bridge, viaducts	Construction started in 1998	Highway traffic
Channel Tunnel, Great Britain–France	50.5 km, bored tunnel, three tubes	Open to traffic 1994	Railway with car and lorry shuttle
Trans-Tokyo Bay Crossing, Japan	15.1 km, bored tunnel, artificial islands, high- and low-level steel box girder bridges	Open to traffic 1997	Highway traffic

Generally, the term *fixed link* is associated with highway and/or railway sections of considerable length and a fixed link may comprise a combination of different civil engineering structures such as tunnels, artificial islands, causeways, and different types of bridges. Selected examples of major fixed links opened or are under construction since 1988 are listed in Table 4.1.

Planning Activities for Major Fixed Links

Major fixed links represent important investments for the society and may have considerable influence on the development potential of the areas they serve.

The political discussions about the decision to design a fixed link may be extended over decades or even centuries. In this period planning activities on a society level are necessary to demonstrate the need for the fixed link and to determine positive and negative effects of the implementation. These early planning considerations are outside the scope of this chapter, but the outcome of the early planning activities may highly influence the tasks in the later planning phases after the final decision is made.

In the early planning phases, basic principles and criteria are dealt with, such as

- Ownership and financing
- Approximate location
- Expected service lifetime
- Necessary traffic capacity
- Considerations for other forms of traffic like vessel traffic and air traffic
- Principles for environmental evaluation
- Risk policy
- International conventions

Section 4.2 will explain the later planning phases by describing major steps in project development with emphasis on the consideration of all relevant aspects. The focus will be on the technical and civil engineering aspects of bridges as fixed links, but most of the methods and principles described can be applied to other types of link structures. In the case of complex fixed link arrangements (comprising more than one type of structure), some of these structures may be alternative solutions. Several combined solutions are therefore studied and for each combination it is normally necessary to perform the planning for the entire link as a whole.

The elements of the project basis for a major fixed link are further detailed in Section 4.3, and examples of major fixed links recently built, under construction, or in the planning stage in the Scandinavian area are described in Section 4.4.

The chapter does not treat aesthetic and environmental issues individually, but assumes that all alternatives are evaluated according to the same principles. Public approval processes are beyond the aim of this chapter; readers are referred to References [1–3].

Fixed links are unique in size and cost, and the political environment differs from project to project. It is thus not possible to provide a recipe for planning major fixed links. The present chapter describes some of the elements, which the authors believe are important in the complex, multidisciplinary planning process of all fixed links.

Many important fixed links still remain to be planned and built. One of the more spectacular ones is the Gibraltar link between Africa and Europe. Figure 4.1 shows an artist's impression of the bridge pylons for the planned Gibraltar link. Examples of other future links are the Messina Strait crossing in Italy, the Mallaca Strait crossing between Malaysia and Indonesia, and the Río de la Plata Bridge connecting Argentina and Uruguay.

4.2 Project Development

4.2.1 Initial Studies

The first step in project development consists of a review of all information relevant to the link and includes an investigation of the most likely and feasible technical solutions for the structures.

The transportation mode, highway and railway traffic, and the amount of traffic is determined based on a traffic estimate. The prognosis of traffic is often associated with considerable uncertainty since fixed links will not only satisfy the existing demands but may also create new demands due to the increased quality of the transport. For railway traffic, it has to be decided whether a railway line will accommodate one or two tracks. Similarly, the highway traffic can either be transported on shuttle trains or the bridge can be accommodated with a carriageway designed to a variety of standards, the main characteristics being the number of lanes.

The decision on the expected traffic demands and the associated traffic solution models is often based on a mix of technical, economic, socioeconomic, and political parameters. The decision may be confirmed at later stages of the planning when more information is available.

FIGURE 4.1 Artist's impression of 465-m-high pylon on 300 m water for planned Gibraltar link with 3,500 m spans. (Courtesy of Dissing +Weitling, Architects, Denmark.)

A *fixed link concept study* will review alignment possibilities and define an appropriate corridor for further studies. It will consider the onshore interchanges for the anticipated traffic modes and identify potential conflict areas. It describes all feasible arrangements for the structures from coast to coast, and reviews the requirement for special onshore structures. Finally, the study defines the concepts to be investigated in greater depth in subsequent phases.

An *environmental condition study* aims to identify potential effects the structures may have on the environment and to review the legal environmental framework. It also identifies important conflict areas and describes the project study area. It will review the available information on the marine and onshore environment and define the need for additional investigations.

A *technical site condition study* will address the geological, the foundation, the navigation, the climatic, and the hydraulic conditions. It will review the topographic situation and define additional studies or investigations for the following project phase.

A *preliminary design basis study* will review the statutory requirements, codes, and standards and identify the need for relevant safety and durability requirements.

Finally, a *preliminary costing basis study* will define the cost estimation technique and provide first preliminary cost estimates.

Considering the results of these studies a comprehensive investigation program for the next project step — the conceptual study — will be defined.

4.2.2 Conceptual Study

The conceptual study is an iterative process, where all the aspects likely to influence the project should be considered, weighted, and clarified to achieve the most suitable solution for the intended purpose and location. These aspects are cost, construction, structural, navigational, environmental, aesthetic, risk, geological, vessel collision, wind, and earthquake.

After an interim selection of various alternatives, conceptual studies are undertaken for each selected alternative solution. Preliminary site investigations like subsoil investigations in the defined alignment corridor, wind, earthquake, and vessel traffic investigations will be carried out simultaneously with the conceptual studies.

The conceptual study comprises development of a project basis, including

- Defining functional requirements
- Reviewing and defining the navigational aspects
- Establishing risk policy and procedures for risk management
- Specifying design basis including structure-specific requirements
- Developing the costing basis

Each of the selected solutions will be developed in a conceptual design and described through drawings and descriptions. The conceptual design will comply with the project basis and further consider:

- Preliminary site investigations
- Structural aspects
- Architectural aspects
- Environmental aspects
- Mechanical and electrical installations and utilities
- Definitions and constraints for operation and maintenance
- Cost aspects
- Major construction stages

Practically, it is not possible to satisfy all the above requirements, but effort should be made to achieve a balanced solution. The conceptual study phase is concluded by a comparison analysis with predeterminant weighting of parameters, which provides the technical ranking of all alternative solutions.

4.2.3   Project Selection and Procurement Strategy

Project Selection

By using the results from the technical ranking of the solutions, the basis for a project selection has to be established by the owner organization. This requires information from other investigations carried out in parallel with the technical studies that cover:

- Environmental impact assessment including hydraulic studies;
- Traffic demand studies including possible tariff structures;
- Layout, cost, and requirements for connections to the existing network outside the study areas;
- Definition of the project implementation and the tendering procedure.

The information obtained from the above studies may be used as input into a cost–benefit model of the anticipated solutions. These final results usually provide the basis to make a decision that will best consider local and global political viewpoints. Public hearings may be necessary in addition to the investigations. The result of this process is the selection of the solution of choice.

Procurement Strategy

The optimum procurement method should ensure that the work and activities are distributed on and executed by the most qualified party (owner, consultant, contractor) at all phases, to meet the required quality level, at the lowest overall cost. The procurement strategy should clarify tendering procedures with commercial and legal regulations for the region. In the following, three main contracting concepts in the definition of the procurement strategy for a fixed link project are presented.

Contracting Concepts

The three main concepts are as follows:

- *Separate Design and Construction (SDC)* is a concept in which the construction contract documents are prepared by the owner, often assisted by an engineering consulting team, and the construction is performed by a contractor.
- *Design-Build (DB)* is a concept in which both engineering design and construction responsibilities are assigned to a single entity, most often the contractor.
- *Design-Built-Operate and Transfer (BOT)* is a concept in which the financing, design, construction, and operation are assigned to a concessionaire. After an agreed number of years of operation, the link is transferred to the owner. The BOT is not described further because it uses the same design and construction procedure as the DB.
- The *SDC* concept requires that the owner and the consultant participate actively during all phases to influence and control the quality and performance capability of the completed facility. The main differences between the various forms of the SDC are the degree of detailing at the tender stage and whether alternatives will be permitted. Completing the detailed design prior to inviting tenders is good if the strategy is to obtain lump-sum bids in full compliance with the owner's conditions.

Tendering based on a partial design — often 60 to 70% — represents a compromise between initial design costs and definitions of the owner's requirements to serve as a reference for alternative tenders. This procurement strategy has been applied for large construction works from the 1970s. Advantages are that the early start of the construction work can be achieved while completing the design work and that innovative ideas may be developed between the owner, contractor and the consultant, and incorporated in the design. The procedure usually allows contractors to submit alternatives in which case the tender design serves the important purpose of outlining the required quality standards. A disadvantage, however, is the risk for later claims due to the fact that the final design is made after awarding the construction contract. The more aggressive contracting environment and the development of international tender rules have made it desirable to procure on a completely fixed basis.

The *DB* concept assigns a high degree of autonomy to the contractor, and, as a consequence, the owner's's direct influence on the quality and performance of the completed facility is reduced. To ensure that the contractor delivers a project that meets the expectations of the owner, it is necessary to specify these in the tender documents. Aesthetic, functional, maintenance, durability, and other technical standards and requirements should be defined. Also legal, environmental, financial, time, interface, and other more or less transparent constraints to the contractor's freedom of performance should be described in the tender documents in order to ensure comparable solutions and prices. Substantial requirements to the contractor's quality assurance system are essential in combination with close follow-up by the owner.

Tenders for major bridges may be difficult to evaluate if they are based on substandard and marginal designs or on radical and unusual designs. The owner then has the dilemma of either rejecting a low tender or accepting it and paying high additional costs for subsequent upgrading.

Contract Packaging

The total bridge project can be divided into reasonable contract parts:

- Vertical separation (e.g., main bridge, approach bridges, viaducts, and interchanges);
- Horizontal separation (e.g., substructure and superstructure);
- Disciplinary separation (e.g., concrete and steel works).

The application of these general principles depends on the specific situation of each project. Furthermore, the achievement of the intended quality level, together with contract sizes allowing for competitive bidding, should be considered in the final choice. Definition and control of interfaces between the different contractors is an important task for the owner's organization.

4.2.4 Tender Design

The main purpose of a tender design or a bid design is to describe the complexity of the structure and to determine the construction quantities, allowing the contractors to prepare a bid for the construction work. The goal for a tender design is as low cost as possible within the given framework. This is normally identical to the lowest quantities and/or the most suitable method. It is essential that the project basis be updated and completed prior to the commencement of the tender design. This will minimize the risk of contract disputes.

It is vital that a common understanding between consultant and owner is achieved. Assumptions regarding the physical conditions of the site are important, especially subsoil, wind, and earthquake conditions. Awareness that these factors might have a significant impact on the design and thereby on the quantities and complexity is important. The subsoil conditions for the most important structures should always be determined prior to the tender design to minimize the uncertainty.

Determination of the quantities is also necessary. For instance, if splice lengths in the reinforcement are included, if holes or cutouts in the structure are included, what material strengths are assumed. There must be stipulated an estimate of the expected variation of quantities (global or local quantities).

The structures in the tender design shall be constructible. In an SDC contract, the tender design should be based on safe and well-established production and erection procedures. In the case of DB, the tender design is carried out in close cooperation between the, contractor and the consultant. This assures that the design accommodates the contractor's methods and the available equipment.

The tender design is often carried out within a short period of time. It should focus on elements with large cost impact and on elements with large uncertainties in order to arrive as closely as possible at the actual quantities and to describe the complexity of the structure efficiently from a costing point of view. A tender design comprises layout drawings of the main structural elements, detailed drawings of typical details with a high degree of repetition, typical reinforcement arrangement, and material distribution.

Aesthetics are normally treated during the tender design. It is important that extreme event loads such as vessel collision, train derailment, cable rupture, earthquake, and ice impact should be considered in the tender design phase as they often govern the design. Durability, operation, and maintenance aspects should be considered in the tender design. Experience from operation and maintenance of similar bridges allows a proper service life design to be carried out. It is at the early design stages that the construction methods should be chosen which have a significant effect on further operation and maintenance costs.

It is not unusual that a tender design is prepared for more than one solution to obtain the optimal solution. It could, for instance, be two solutions with different materials (concrete and steel) as for the Storebælt East Bridge. It could also be two solutions with traffic arranged differently (one level or two levels) as for the Øresund link. Different structural layouts as cable stayed and suspension bridge could be relevant to investigate under certain conditions. After the designs are prepared to a certain level, a selection can be carried out based on a preliminary pricing, and one or more solutions are brought all the way to tender.

Tender documents to follow the drawings should be prepared. The tender documents comprise bill of quantities, special specifications, and the like.

4.2.5   Tender Evaluation

The objective of tender evaluation is to select the overall most advantageous tender including capitalized owner's risk and cost for operation and maintenance. A basis should be established via a rating system where all tenders become directly comparable. The rating system is predefined by the owner, and should be part of the tender documents.

The tender evaluation activities can be split up in phases:

1. Preparation
2. Compilation and checking of tenders
3. Evaluation of tenders
4. Preparation for contract negotiations
5. Negotiation and award of contract

The *preparation* phase covers activities up to the receipt of tenders. The main activities are as follows:

- Establish the owner's risk for each of the tendered projects, using the owner's cost estimate;
- Define tender opening procedures and tender opening committee;
- Quantify the differences in present value due to function, operation, maintenance, and owner's risk for each of the tendered projects, using the owner's cost estimate.

After receipt of tenders, a summary report, which collects the information supplied in the different tenders into a single summarizing document and presents a recommendation of tenders for detailed review, as a result of *compilation and checking of tenders*, should be prepared. Typical activities are as follows:

- Check completeness of compliance of all tenders, including arithmetical correctness and errors or omissions;
- Identify possible qualifications and reservations;
- Identify parts of tenders where clarification is needed, or more detailed examination required;
- Prepare a preliminary list of questions for clarification by the tenderers;
- Review compliance with requirements for alternative designs;
- Upgrade alternative tender design and pricing to the design basis requirements for tender design.

The *evaluation of tenders* comprises the following:

- Provide initial questionnaires for tender clarification to tenderers, arrange clarification meetings, and request tenderers' written clarification answers;
- Adjust tender prices to a comparable basis taking account of revised quantities due to modified tender design effects of combined tenders, alternatives, options, reservations, and differences in present value;
- Appraise the financial components of the tenders;
- Assess owner's risk;
- Review technical issues of alternatives and their effect on interfaces;
- Review the proposed tender time schedule;
- Evaluate proposed subcontractors, suppliers, consultants, testing institutes, etc.
- Review method statements and similar information;
- Establish list of total project cost.

The assessment of owner's risk concerns exceeding budgets and time limits. An evaluation of the split of financial consequences between contractor and owner should be carried out.

Preparation for contract negotiations should be performed, allowing all aspects for the actual project type to be taken into account. Typical activities are as follows:

- Modify tender design to take current status of the project development into account to establish an accurate contract basis;
- Modify tender design to accommodate alternatives;
- Coordinate with the third parties regarding contractual interfaces;
- Coordinate with interfacing authorities;
- Establish strategies and recommendations for contract negotiations.

The probable extent and nature of the negotiations will become apparent from the tender evaluation. Typical activities during *negotiations and award of contract* are as follows:

- Prepare draft contract documents;
- Clarify technical, financial, and legal matters;
- Finalize contract documents.

4.2.6 Detailed Design

The detailed design is either carried out before (SDC contracts) or after signing of the construction contact (DB contracts). In the case the detailed design is carried out in parallel with the construction work, the completion of the detailed design should be planned and coordinated with the execution. A detailed planning of the design work is required when the parts of the structure, typically the foundation structures, need to be designed and constructed before the completion of the design of the entire structure. Design of temporary works is normally conducted in-house by the contractor, whereas the design of the permanent works is carried out by the consultant.

The purpose of the detailed design is to prepare drawings for construction in accordance with various requirements and specifications. Detailed design drawings define all measures and material qualities for the structure. Shop drawings for steel works are generally prepared by the steel fabricator. Detailed reinforcement arrangements and bar schedules are either prepared by the contractor or the consultant. It is important that the consultant prescribes the tolerance requirements of the design.

The detailed design should consider the serviceability limit state (deflection and comfort), the ultimate limit state (strength and stability), and the extreme event limit state (collapse of the structure). To ensure the adequacy of the design, substantial analyses, including three-dimensional global finite-element analyses, local finite-element analyses, and nonlinear analyses both in geometry and materials, should be carried out. Dynamic calculations, typically response spectrum analyses, are usually performed to determine the response from wind. The dynamic amplifications of traffic loads and cable rupture are determined by a time-history analysis, which is also frequently used for vessel collision and earthquake analyses.

For large cable-supported bridges, wind tunnel testing is conducted as part of the detailed design. Preliminary wind tunnel testing is often carried out in the tender design phase to investigate the aerodynamic stability of the structure. Other tests, such as scour protection and fatigue tests can be carried out to ensure design satisfactions. Detailed subsoil investigations for all foundation locations are carried out prior to, or in parallel with, the detailed design.

The operation and maintenance (O&M) objectives should be implemented in the detailed design in a way which:

- Gives an overall cost-effective operation and maintenance;
- Causes a minimum of traffic restrictions due to O&M works;
- Provides optimal personnel safety;
- Protects the environment;
- Allows for an easy documentation of maintenance needs and results.

In addition, the contractor should provide a forecast schedule for the replacement of major equipment during the lifetime of the bridge.

4.2.7 Follow-Up during Construction

During the construction period the consultant monitors the construction work to verify that it is performed in accordance with the intentions of the design. This design follow-up, or general supervision, is an activity which is carried out in cooperation with (and within the framework of) the owner's supervision organization.

The general supervision activities include review of the contractor's quality assurance manuals, method statements, work procedures, work instructions, and design of temporary structures, as well as proper inspections on the construction site during important construction activities. The quality of workmanship and materials is verified by spot-checking the contractor's quality control documentation.

When the work results in mistakes or nonconformances, the general supervision team evaluates the contractor's proposals for rectification or evaluates whether or not the structural element in question can be used as built, without any modifications. The general supervision team also evaluates proposals for changes to the design submitted by the contractor and issues recommendations on approval of such proposals.

The duties of the general supervision team also include preparation of technical supervision plans, which are manuals used by the supervision organization as a basis for the technical supervision of the construction work. These manuals should be based on inputs from the consultants and experienced engineers to avoid mistakes during the construction work.

The general supervision team monitors the performance of the supervision organization and receives feedback on experience gained by the supervision organization, as in some cases it may be found necessary and advantageous to adjust the design of the project to suit the contractor's actual performance.

The general supervision team provides advice on the necessity for expert assistance, special testing of materials, and special investigations. The general supervision team evaluates the results of such activities and issues recommendations to the owner. Special testing institutes are often involved in the third-party controls which normally are performed as spot checks only. Examples are nondestructive testing of welds, mechanical and chemical analyses of steel materials, and testing of concrete constituents such as cement, aggregates, and admixtures at official laboratories.

The general supervision team assists the supervision organization with the final inspection of the works prior to the contractor's handing over of the works. The general supervision team assists the consultant with the preparation of operation and maintenance manuals and procedures for inspections and maintenance during the operation phase. Some of these instructions are based on detailed manuals prepared by the contractor's suppliers. This can apply to bearings, expansion joints, electrical installations, or special equipment such as dehumidification systems or buffers. Preparation of these manuals by the suppliers is part of their contractual obligations, and the manuals should be prepared in the required language of the country where the project is situated.

4.3 Project Basis

4.3.1 Introduction

The project basis is all the information and requirements that are decisive in the planning and design of a fixed link. The project basis is developed simultaneously with the early design activities, and it is important to have the owner's main requirements defined as early as possible, and to be precise about what types of link solutions are to be included.

4.3.2 Geometric Requirements

Most geometric requirements for the fixed links stem from the operational requirements of traffic and all the important installations. However, geometric requirements may also be necessary to mitigate accidents and to provide the needed space for safety and emergency situations. Geometric considerations should be addressed in the risk analyses.

4.3.3 Structural Requirements

Design Basis

A main purpose of a design basis is to provide a set of requirements to ensure an adequate structural layout, safety, and performance of the load-bearing structures and installations for the intended use.

Structural Design Codes

The structures must resist load effects from self-weight and a variety of external loads and environmental phenomena (climate and degradation effects). To obtain an adequately uniform level of structural safety, the statistical nature of the generating phenomena as well as the structural capacity should be considered. A rational approach is to adapt probabilistic methods, but these are generally inefficient for standard design situations, and consequently it is recommended that a format as used in codes of practice be applied. These codes are calibrated to achieve a uniform level of structural safety for ordinary loading situations, and probabilistic methods can subsequently be used to calibrate the safety factors for loads and/or design situations that are not covered by the codes of practice.

The safety level — expressed as formal probability of failure or exceeding of limit state — is of the order 10^{-6} to 10^{-7}/year for ultimate limit states for important structures in major links.

4.3.4 Environmental Requirements

Fixed links crossing environmentally sensitive water stretches need to be developed with due attention to environmental requirements. Environmental strategies should be directed toward modification of the structural design to reduce any impact and to consider compensation or mitigation for unavoidable impacts. Guidelines for environmental considerations in the structural layout and detailing and in the construction planning are developed by a consultant, and these should typically address the following areas:

- Geometry of structures affecting the hydraulic situation;
- Space occupied by bridge structures, ramps and depot areas;
- Amount and character of excavated soils;
- Amount of external resources (raw materials winning);
- Methodology of earth works (dredging and related spill).

Consequences of the environmental requirements should be considered in the various project phases. Typical examples for possible improvements are selection of spans as large as possible or reasonable, shaping of the underwater part of foundations to reduce their blocking effect, orientation of structures parallel to the prevailing current direction, minimizing and streamlining of protection structures, reduction of embankment length, optimal layout of depot areas close to the shorelines, and reuse of excavated material.

The process should be started at the very early planning stages and continued until the link is completed and the impact on the environment should be monitored and assessed.

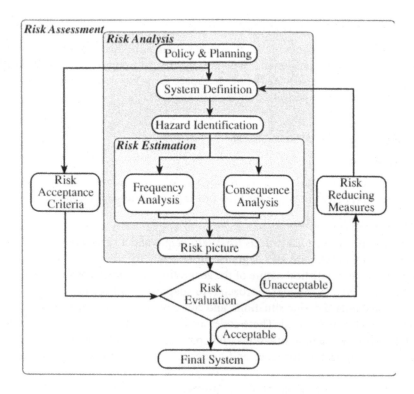

FIGURE 4.2   Risk management components.

4.3.5   Risk Requirements

Types of Risk

Risk studies and risk management have gained a widespread application within the planning, design, and construction of fixed links. Risks are inherent in major transportation links, and therefore it is important for the owner and society that risks are identified and included in the project basis together with the technical and economic aspects. Risks are often studied separately according to the consequences of concern:

- Economic risk (rate of interest, inflation, exceeding of budget, changing traffic patterns);
- Operational risk (accidents, loss of lives, impact to environment, disruption of the traffic, loss of assets, loss of income);
- Construction risks (failure to meet time schedule or quality standards, unexpected ground conditions, accidents).

Economic risks in the project may be important for decisions on whether to initiate the project at all. The construction risk may have important implications on the selection of the structural concept and construction methods.

Risk Management Framework

The main risk management components are shown in Figure 4.2. The risk policy is formulated by the owner in few words: "The safety of the transportation link must be comparable with the safety for the same length of similar traffic on land."

The risk acceptance criteria are an engineering formulation of the risk policy in terms of upper limits of risk. The risk policy also specifies the types of risk selected to be considered, typically user fatalities and financial loss. In some cases, other risks are specifically studied, e.g., risk of traffic disruption and risk of environmental damage, but these risks may conveniently be converted into financial losses.

The risk analysis consists of a systematic hazard identification and an estimation of the two components of the risk, the likelihood and the consequence. Finally, the risk is evaluated against the acceptance criteria. If the risk is found unacceptable, risk-reducing measures are required. It is recommended to develop and maintain an accurate accounting system for the risks and to plan to update the risk assessment in pace with the project development.

In the following, three common risk evaluation methods are discussed: fixed limits, cost efficiency, and ALARP, i.e., as low as reasonably practicable.

Fixed limits is the classical form of acceptance criteria. Fixed limits are also known from legislation and it may easily be determined whether a determined risk is acceptable or not. On the other hand, the determination of limits, which can ensure an optimal risk level, may be difficult.

With a pure *cost efficiency* consideration, an upper limit is not defined, but all cost-efficient risk-reducing measures are introduced. For this cost–benefit consideration it is necessary to establish direct quantification of the consequences in units comparable to costs.

The *ALARP* method applies a cost–benefit consideration in which it, however, is stated that the risk shall be reduced until the cost of the reduction measures is in disproportion with the risk-reducing effect. This will result in a lower risk level than the pure cost efficiency. In ALARP a constraint of the acceptable risk is further introduced as an upper limit beyond which the risk is unconditionally unacceptable.

Often it is claimed that society regards one accident with 100 fatalities as worse than 100 accidents each with one fatality. Such an attitude toward risk aversion can be introduced in the risk policy and the risk acceptance criteria. The aversion against large accidents can also be modeled with aversion factors that are multiplied on the consequences of accidents with many fatalities; the more fatalities, the higher the factor. The sensitivity of the evaluations of risk should be considered by the representation of the uncertainty of the information in the models.

Risk Studies in Different Project Phases

The general result of the risk management is a documentation of the risk level, basis for decisions, and basis for risk communication. The specific aims and purposes for risk management depend on the phase of the project. Here some few examples of the purposes of risk management in conceptual study, tender design, detailed design, and operation, are given.

During initial studies, the risk should be crudely analyzed using more qualitative assessments of the risks. A risk management framework should be defined early in the design process. In the beginning of the project, some investigations should be initiated in order to establish a basis for the more-detailed work in later phases; for example, vessel traffic observations should be performed to provide the basis for the estimation of vessel impact probability. In later phases detailed special studies on single probabilities or consequences may be undertaken.

In the conceptual study the most important activities are to identify all relevant events, focus on events with significant risk contributions and risks with potential impact on geometry (safety, rescue, span width). Extreme event loads are established based on the risk studies. In the tender design phase, the main risks are examined in more detail, in particular risks with potential impact on the project basis. In the detailed design phase, the final documentation of the risk level should be established and modifications to the operational procedures should be made.

4.3.6 Aesthetic Requirements

The final structures and components of a fixed link are a result of a careful aesthetic appraisal and design of all the constituent elements. The purpose of aesthetic requirements is to obtain an optimal technical and sculptural form of individual elements and to obtain an overall aesthetic quality and visual consistency between the elements and the setting. Although difficult, it is recommended to establish guidelines for aesthetic questions.

4.3.7 Navigation Conditions

The shipping routes and the proposed arrangement for a major bridge across navigable waters may be such that both substructure and superstructure could be exposed to vessel collisions. General examples of consequences of vessel collisions are as follows:

- Fatalities and injuries to users of the bridge and to crew and vessel passengers;
- Pollution of the environment, in the case of an accidental release of the hazardous cargo;
- Damage or total loss of bridge;
- Damage or total loss of vessels;
- Economic loss in connection with prolonged traffic disruption of the bridge link.

A bridge design that is able to withstand worst-case vessel impact loads on other piers than the navigation piers is normally not cost-effective. Furthermore, such a deterministic approach does not reduce the risk to the environment and to the vessels. Therefore, a probabilistic approach addressing the main risks in a systematic and comprehensive way is recommended. This approach should include studies of safe navigation conditions, vessel collision risk analysis and vessel collision design criteria, as outlined below.

Navigation risks should be addressed as early as possible in the planning phases. The general approach outlined here is in accordance with the IABSE Green Ship Collision Book [4] and the AASTHO Guide Specification [5]. The approach has been applied in the development of the three major fixed links discussed in Section 4.4.

Safe Navigation Conditions

Good navigation conditions are a prerequisite for the safe passage of the bridge such that vessel collisions with the bridge will not occur under normal conditions, but only as a result of navigation error or technical failure on-board during approach.

The proposed bridge concept should be analyzed in relation to the characteristics of the vessel traffic. The main aspects to be considered are as follows:

- Preliminary design of bridge;
- Definition of navigation routes and navigation patterns;
- Data on weather conditions, currents, and visibility;
- Distribution of vessel movements with respect to type and size;
- Information on rules and practice for navigation, including use of pilots and tugs;
- Records of vessel accidents in the vicinity of the bridge;
- Analysis of local factors influencing the navigation conditions;
- Identification of special hazards from barges, long tows and other special vessels;
- Future navigation channel arrangements;
- Forecast of future vessel traffic and navigation conditions to the relevant study period;
- Identification of largest safe vessel and tow and of preventive measures for ensuring full control with larger passing vessels.

Vessel Collision Analysis

An analysis should be used to support the selection of design criteria for vessel impact. Frequencies of collisions and frequencies of bridge collapse should be estimated for each bridge element exposed to vessel collision. Relevant types of hazards to the bridge should be identified and modeled, hazards from ordinary vessel traffic which is laterally too far out of the ordinary route, hazards from vessels failing to turn properly at a bend near the bridge, and from vessels sailing on more or less random courses.

The frequencies of collapse depend on the design criteria for vessel impact. The overall design principle is that the design vessels are selected such that the estimated bridge collapse frequency fulfills an acceptance criterion.

Vessel Collision Design Criteria

Design criteria for vessel impact should be developed. This includes selection of design vessels for the various bridge elements which can be hit. It also includes estimates of sizes of impact loads and rules for application of the loads. Both bow collisions and sideways collisions should be considered. Design capacities of the exposed girders against impact from a deck house shall be specified.

The vessel impact loads are preferably expressed as load indentation curves applicable for dynamic analysis of bridge response. Rules for application of the loads should be proposed. It is proposed that impact loads will be estimated on the basis of general formulas described in Ref. [4].

4.3.8 Wind Conditions

Bridges exposed to the actions of wind should be designed to be consistent with the type of bridge structure, the overall wind climate at the site, and the reliability of site-specific wind data. Wind effects on traffic could also be an important issue to be considered.

Susceptibility of Bridge Structures to Wind

Winds generally introduce time-variant actions on all bridge structures. The susceptibility of a given bridge to the actions of wind depends on a number of structural properties such as overall stiffness, mass, and shape of deck structure and support conditions.

Cable-supported bridges and long-span beam structures are often relatively light and flexible structures in which case wind actions may yield significant contributions to structural loading as well as influence user comfort. Site-specific wind data are desirable for the design. Engineering codes and standards often provide useful information on mean wind properties, whereas codification of turbulence properties are rare. Guidelines for turbulence properties for generic types of terrain (sea, open farmland, moderately built-up areas) may be found in specialized literature. If the bridge is located in complex hilly/mountainous terrain or in the proximity of large structures (buildings, bridges, dams), it is advisable to carry out field investigations of the wind climate at the bridge site. Important wind effects from isolated obstacles located near the planned bridge may often be investigated by means of wind tunnel model testing.

In general, it is recommended that aerodynamic design studies be included in the designs process. Traditionally, aerodynamic design have relied extensively on wind tunnel testing for screening and evaluation of design alternatives. Today, computational fluid dynamics methods are becoming increasingly popular due to speed and efficiency as compared with experimental methods.

Wind Climate Data

The properties of turbulence in the atmospheric boundary layer change with latitude, season and topography of the site, but must be known with a certain accuracy in order to design a bridge to a desired level of safety. The following wind climatic data should be available for a particular site for design of wind-sensitive bridge structures:
Mean wind:

- Maximum of the 10-min average wind speed corresponding to the design lifetime of the structure;
- Vertical wind speed profile;
- Maximum short-duration wind speed (3-s gust wind speed).

Turbulence properties (along-wind, cross-wind lateral, and cross-wind vertical):

- Intensity;
- Spectral distribution;
- Spatial coherence.

The magnitude of the mean wind governs the steady-state wind load to be carried by the bridge structure and is determinant for the development of aeroelastic instability phenomena. The turbulence properties govern the narrowband random oscillatory buffeting response of the structure, which is similar to the sway of trees and bushes in storm winds.

4.3.9 Earthquake Conditions

Structures should be able to resist regional seismic loads in a robust manner, avoiding loss of human lives and major damages, except for the very rare but large earthquake. The design methods should be consistent with the level of seismicity and the amount of available reliable information. Available codes and standards typically do not cover important lifeline structures such as a fixed link, but they may be used for inspiration for the development of a design basis.

4.3.10 Ice Conditions

The geographic location of a bridge site indicates whether or not ice loads are of concern for that structure. The ice loads may be defined as live loads or extreme event loads (exceptional environmental loads which are not included in live loads).

From recent studies carried out for the Great Belt link, the following main experience was obtained:

- Ice loads have a high dynamic component very likely to lock in the resonance frequencies of the bridge structure;
- Bearing capacity of the soil is dependent on number and type of load cycles, so dynamic soil testing is needed;
- Damping in soil and change of stiffness cause important reductions in the dynamic response;
- If possible, the piers should be given an inclined surface at the water level;
- High ductility of the structure should be achieved.

4.3.11 Costing Basis

The cost estimate is often decisive for the decision on undertaking the construction of the link, for selection of solution models, and for the selection of concepts for tendering. The estimate may also be important for decisions of detailed design items on the bridge.

Cost Uncertainty Estimation

To define the cost uncertainty it may be helpful to divide it into two conceptions which may overlap: (1) the uncertainties of the basis and input in the estimation (mainly on cost and time) and (2) risk of unwanted events. The uncertainty is in principle defined for each single item in the cost estimate. The risks can in principle be taken from a construction risk analysis.

Cost Estimates at Different Project Phases

Cost estimates are made in different phases of the planning of a project. In the first considerations of a project, the aim is to investigate whether the cost of the project is of a realistic magnitude and whether it is worthwhile to continue with conceptual studies. Later, the cost estimates are used to compare solution concepts, and to evaluate designs and design modification until the final cost estimate before the tender is used to evaluate the overall profitability of the project and to compare with the received bids. Different degrees of detailing of the estimates are needed in these stages. In the early phases an "overall unit cost" approach may be the only realistic method for estimating a price, whereas in the later phases it is necessary to have a detailed breakdown of the cost items and the associated risks and uncertainties.

Life Cycle Costs

The life cycle cost is an integration of the entire cost for a bridge from the first planning to the final demolition. The life cycle cost is normally expressed as a present value figure. Hence, the interest rate used is very important as it is a weighting of future expenses against initial expenses.

It shall initially be defined how the lifetime costs are to be considered. For example, disturbance of the traffic resulting in waiting time for the users can be regarded as an operational cost to society whereas it is only a cost for the owner if it influences the users' behavior so that income will be less.

Important contributors to the life cycle cost for bridges are as follows:

- The total construction cost, including costs for the owner's organization;
- Future modifications or expansion of the bridge;
- Risks and major repairs;
- Income from the operation of the bridge;
- Demolition costs.

Comparison Analysis

In the development of a project numerous situations are encountered in which comparisons and rankings must be made as bases for decisions. The decisions may be of different nature, conditions may be developing, and the decision maker may change. The comparisons should be based on a planning and management tool which can rationalize, support, and document the decision making.

A framework for the description of the solutions can be established and maintained. This framework may be modified to suit the purpose of the different situations. It is likely that factual information can be reused in a later phase.

The main components of the comparison can be as follows:

- Establishment of decision alternatives;
- Criteria for evaluation of the alternatives;
- Quantitative assessment of impacts of the various alternatives utilizing an evaluation grid;
- Preference patterns for one or more decision makers with associated importance of criteria;
- Assessment of uncertainties.

The decision maker must define the comparison method using a combination of technical, environmental and financial criteria. Quantitative assessment of all criteria is performed. Decision-making theories from economics and mathematical tools are used.

Establishment of Alternatives and Their Characteristics

All decision alternatives should be identified. After a brief evaluation, the most obviously nonconforming alternatives may be excluded from the study. In complex cases a continued process of detailing of analysis and reduction of number of alternatives may be pursued. The selection of parameters for which it is most appropriate to make more-detailed analyses can be made on the basis of a sensitivity analysis of the parameters with respect to the utility value.

Risks may be regarded as uncertain events with adverse consequences. Of particular interest are the different risk pictures of the alternatives. These risk pictures should be quantified by the use of preferences so that they can be part of the comparison.

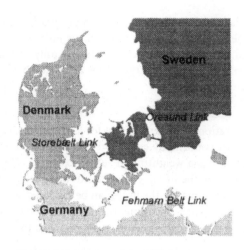

FIGURE 4.3   Denmark and neighboring countries.

Comparisons at Different Project Stages

After the initial identification of all possible solution models, the purpose of the first comparison may be to reduce the number of solution models to be investigated in the later phases. The solution models may here be alternative design concepts. In this first ranking the detail of the analysis should be adequate to determine the least attractive solutions with an appropriate certainty. This will in most cases imply that a relatively crude model can be used at this stage. A partly qualitative assessment of some of the parameters, based on an experienced professional's judgment, can be used.

At a later phase decisions should be made on which models to select for tender design, and later in the tender evaluation, which tenderer to award the contract. In these comparisons the basis and the input should be more well established, as the comparison here should be able to select the single best solution with sufficient certainty.

Weighting the criteria is necessary. Although a strictly rational weighting and conversion of these criteria directly into terms of financial units may not be possible, it is often sufficient if the weighting and selection process are shown to the tenderers before the tender.

An example of the comparison and selection process can be the following, which is performed in stages. Each stage consists of an evaluation and shortlisting of the tenders eliminating the low ranked tenders. At each stage the tenderers not on the shortlist are informed about the weak points and they are given the opportunity of changing their tender within a short deadline. At the last stage the remaining tenderers are requested to state their final offer improving on the technical quality and financial aspects raised by the owner during negotiations. Then the owner can select the financially most advantageous tender.

4.4   Recent Examples of Fixed Links

4.4.1   Introduction

Since the 1980s three major fixed links have been designed or planned in Denmark and neighboring countries, Figure 4.3. A combined tunnel and bridge link for railway and highway traffic has been constructed across the 18-km-wide Storebælt, a 16-km tunnel and bridge link for railway and highway traffic between Denmark and Sweden will be inaugurated in year 2000, and the conceptual study has been completed (1998) for a fixed link across the 19-km-wide Fehmarn Belt between Denmark and Germany.

4.4.2 The Storebælt Link

Over the years, more or less realistic projects for a fixed link across the Storebælt have been presented. At 18-km-wide, the belt is part of the inland sea area and divides Denmark's population and economy into nearly equal halves.

The belt is divided into two channels, east and west, by the small island, Sprogø, which has been as an obvious stepping-stone, an integral part of all plans for fixed link projects. The international vessel traffic between the Baltic Sea and the North Sea navigates the eastern channel, whereas the western channel is a national waterway. To bridge the eastern channel has therefore always been the main challenge of the project.

The first tender design for a combined railway and highway bridge across the eastern channel was prepared in 1977–78. However, only 1½ month short of issuing tender documents and call for bids, the progress of the project was temporarily stopped by the government. This was in August 1978. Several state-of-the-art investigations such as vessel impact, fatigue, and wind loads were carried out for two selected navigation spans: a 780-m main span cable-stayed bridge and a 1416-m main span suspension bridge, both designed for a heavy duty double-track railway and a six-lane highway.

The construction of the fixed link was again politically agreed upon on June 12, 1986, and the main principles for the link were set out. It should consist of a low-level bridge for combined railway and highway traffic, the West Bridge, across the western channel; whereas the eastern channel should be crossed by a bored or an immersed tunnel for the railway, the East Tunnel, and a high-level bridge for the highway traffic, the East Bridge.

A company, A/S Storebælt, was established January 23, 1987 and registered as a limited company with the Danish State as sole shareholder. The purpose of the company was to plan, design, implement, and operate the fixed link. The project is financed by government-guaranteed commercial loans to be paid back via user tolls. A/S Storebæltsforbindelsen has published a series of reports on the link structures, see Reference [6–8].

The East Bridge

Project Development
In 1987 conceptual design was carried out for the East Bridge. The main objectives were to develop a global optimization with regard to the following:

- Alignment, profile, and navigation clearance;
- Position of main navigation channel;
- Navigation span solutions, based on robust and proven design and construction technology;
- Constructable and cost-competitive solutions for the approach spans, focusing on repetitive industrialized production methods onshore;
- Master time schedule;
- Master budget.

In 1989–90 pretender studies, tender design, and tender documents were prepared. During the pretender phase, comparative studies of four alternative main bridge concepts were carried out to evaluate thoroughly the technical, financial, and environmental effects of the range of main spans:

Cable-stayed bridge	916 m main span
Cable-stayed bridge	1204 m main span
Suspension bridge	1448 m main span
Suspension bridge	1688 m main span

Navigation risk studies found only the 1688-m main span adequate to cross the existing navigation route without affecting the navigation conditions negatively. This during tender design was reduced to 1624 m, which together with a relocated navigation route, proved to be sufficient and was selected for tender and construction.

The pylons were tendered in both steel and concrete. For the approach span superstructure, 124-m-long concrete spans and 168-m-long steel spans as well as composite steel/concrete concepts were developed. Although an equally competitive economy was found, it was decided to limit the tender designs to concrete and steel spans. The East Bridge was tendered as SDC.

The tender documents were subdivided into four packages to be priced by the contractors: superstructure and substructure inclusive pylons for the suspension bridge (2) and superstructures and substructures for the approach spans. The tender documents were released to prequalified contractors and consortia in June 1990.

In December 1990 the tenders were received. Eight consortia submitted 32 tenders inclusive smaller alternatives and four major alternatives to the basic tender design.

In October 1991, construction contracts were signed with two international consortia; a German, Dutch, and Danish joint venture for the substructures, inclusive of concrete pylons, and an Italian contractor for an alternative superstructure tender where high-strength steel was applied to a more or less unchanged basic cross section, thereby increasing the span length for the approach spans from 168 m to 193 m.

The suspension span is designed with a main cable sag corresponding to ⅑ of span length. The steel bridge girder is suspended from 800-mm-diameter main cables in hangers each 24 m.

The girder is continuous over the full cable-supported length of 2.7 km between the two anchor blocks. The traditional expansion joints at the tower positions are thus avoided. Expansions joints are arranged in four positions only, at the anchor blocks and at the abutments of the approach spans.

The concrete pylons rise 254 m above sea level. They are founded on caissons placed directly on crushed stone beds.

The anchor blocks must resist cable forces of 600,000 tonnes. They are founded on caissons placed on wedge-shaped foundation bases suitable for large horizontal loading. An anchor block caisson covers an area of 6100 m².

The caissons for the pylons, the anchor blocks, and the approach spans as well as for the approach span pier shafts have been constructed at a prefabrication site established by the contractor 30 nautical miles from the bridge site. The larger caissons were cast in two dry docks, and the smaller caissons and the pier shafts for the approach spans on a quay area, established for this purpose. A pylon caisson weighed 32,000 tonnes and an anchor block caisson 36,000 tonnes when they were towed from the dry dock by tug boats to their final position in the bridge alignment.

Both the suspension bridge girder and the approach span girders are designed as closed steel boxes and constructed of few basic elements: flat panels with trough stiffeners and transverse bulkhead trusses. The two approach bridges, 2530 and 1538 m, respectively, are continuous from the abutments to the anchor blocks. The suspension bridge girder is 31.0 m wide and 4.0 m deep; the girder for the approach spans is 6.7 m deep. They are fabricated in sections, starting in Italy. In Portugal, on their way by barge to Denmark, a major preassembly yard was established for girder sections to be assembled, before they were finally joined to full-span girders in Denmark. The East Bridge (Figure 4.4), was inaugurated by the Danish Queen on June 14, 1998 and the link was opened to highway traffic.

Project Basis
The project basis was throughout its development reviewed by international panels of experts.

Structural Requirements
Danish codes, standards, rules, and regulations were applied wherever applicable and supplemented with specific additional criteria and requirements, regarding various extreme event loads.

Environmental
The environmental design criteria required that the construction should be executed with no effect on the water flow through the belt. This was achieved by dredging, short ramps, long spans, and hydraulic shaped piers and pylons. The blocking effect to be compensated for was only about 0.5% of the total flow in the belt.

FIGURE 4.4 Storebælt East Bridge.

Risk
Risk acceptance criteria were established early and a series of risk analyses regarding train accidents, fire and explosion, ice loads, and vessel collision were carried out to ensure adequate and consistent safety level for the entire link. The acceptance criteria required that the probability of disruption of a duration of more than 1 month should not exceed a specified level, and that the risk level for fatalities for crossing should be comparable to the risk for a similar length of traffic on land. The analyses were followed up by risk management through the subsequent phases to ensure that the objectives were met.

Navigation
With 18,000 vessel passages each year through the eastern channel, important considerations were given for navigation. Comprehensive vessel simulations and collision analysis studies were performed, leading to an improved knowledge about safe navigation conditions and also to a set of probabilistically based criteria for the required impact resistance of the bridge piers and girders. Vessel impact has been the governing load criterion for all the bridge piers. A vessel traffic service (VTS) system was established mainly for prevention of collision accidents to the low West Bridge.

Wind
The local wind climate at Storebælt was investigated by measurements from a 70-m-high tower on Sprogø. For the East Bridge, aerodynamic investigations were carried out on 16 different highway girder box section configurations in a wind tunnel. The testing determined the critical wind speed for flutter for the selected girder shape to be 74 m/s which was safely above the design critical wind speed of 60 m/s. For the detailed design an aeroelastic full bridge model of 1:200 scale was tested under simulated turbulent wind conditions.

The West Bridge
The 6.6-km West Bridge (Figure 4.5) was tendered in three alternative types of superstructure; a double-deck composite girder, triple independent concrete girders side by side, and a single steel box girder. All three bridge alternatives shared a common gravity-founded sand-filled caisson substructure, topped by pier shafts of varying layout.

FIGURE 4.5   Storebælt West Bridge.

Tender documents were issued to six prequalified consortia in April 1988, and 13 offers on the tender solutions as well as three major alternatives and nine smaller alternatives were received from five groups. Tender evaluation resulted in selecting an alternative design: two haunched concrete box girders with a typical span length of 110.4 m, reduced to 81.75 m at the abutments and the expansion joints. The total length was subdivided into six continuous girders, requiring seven expansion joints.

It was originally intended to tender the West Bridge as an SDC, but as an alternative design was selected, the contract ended up being similar to a DB contract.

Altogether, 324 elements, comprising 62 caissons, 124 pier shafts, and 138 girders, have been cast in five production lines at a reclaimed area close to the bridge site. All the elements were cast, moved by sliding, stored on piled production lines, and later discharged without use of heavy gantry cranes. The maximum weight of an element was 7400 tonnes. The further transportation and installation was carried out by *Svanen,* a large purpose-built catamaran crane vessel.

By this concept, which was originally presented in the tender design, but further developed in the contractor's design, the entire prefabrication system was optimized in regard to resources, quality, and time. The bridge was handed over on January 26, 1994.

The East Tunnel

Two immersed tunnel solutions as well as a bored tunnel were considered for the 9 km wide eastern channel. After tender, the bored tunnel was selected for financial and environmental reasons.

The tunnel consists of two 7.7-m-internal-diameter tubes, each 7412 m long and 25 m apart. At the deepest point, the rails are 75 m below sea level.

Four purpose-built tunnel boring machines of the earth-balance pressure type have bored the tunnels, launched from each end of both tubes. The tunnel tubes are connected at about 250 m intervals by 4.5-m-diameter cross passages which provide safe evacuation of passengers and are the location for all electrical equipment. About 250 m of reinforced concrete cut-and-cover tunnels are built at each end of the bored tubes. The tunnel is lined with precast concrete segmental rings, bolted together with synthetic rubber gaskets. Altogether, 62,000 segments have been produced. A number of protective measures has been taken to ensure a 100-year service life design.

FIGURE 4.6   Storebælt East Tunnel.

On April 7, 1995, the final tunnel lining segment was installed. Thus, the construction of the tunnel tubes was completed, almost 5 years after work commenced. Railway systems were installed and in June 1997 the railway connection (Figure 4.6) was opened to traffic and changes in the traffic pattern between East and West Denmark started.

4.4.3   The Øresund Link

The 16-km fixed link for combined railway and highway traffic between Denmark and Sweden consists of three major projects: a 3.7-km immersed tunnel, a 7.8-km bridge, and an artificial island which connects the tunnel and the bridge.

The tunnel contains a four-lane highway and two railway tracks. The different traffic routes are separated by walls, and a service tunnel will be placed between the highway's two directions. The tunnel will be about 40 m wide and 8 m high. The 20 reinforced concrete tunnel elements, 175 m long and weighing 50,000 tonnes, are being prefabricated at the Danish side, and towed to the alignment.

The owner organization of the Øresund link is Øresundskonsortiet, established as a consortium agreement between the Danish company A/S Øresundsforbindelsen and the Swedish company Svensk-Danska Broförbindelsen on January 27, 1992. The two parties own 50% each of the consortium. The purpose of the consortium is to own, plan, design, finance, construct and operate the fixed link across Øresund.

The project is financed by commercial loans, guaranteed jointly and severally between the Danish and the Swedish governments. The highway part will be paid by user tolls, whereas the railway companies of the two countries will pay fixed installments per year. The revenue also has to cover the construction work expenses for the Danish and Swedish land-based connections.

Prequalified consultants were asked in February 1993 to prepare a conceptual design, as part of a proposal to become the in-house consultant for the owner. Two consultants were selected to prepare tender documents for the tunnel, the artificial island and the bridge, respectively.

The Øresund Bridge

In July 1994, the Øresundkonsortiet prequalified a number of contractors to build the bridge on a design and construct basis.

FIGURE 4.7   Øresund Main Bridge.

The bridge was tendered in three parts; the approach bridge from Sweden, the high-level bridge with a 490-m main span and a vertical clearance of 57 m, and the approach spans toward Denmark. Two solutions for the bridge were suggested: primarily, a two-level concept with the carriageway on the top deck and the two-track railway on the lower deck; secondarily a one-level bridge. Both concepts were based on cable-stayed main bridges.

Five consortia were prequalified to participate in the competition for the high-level bridge, and six consortia for the approach bridges. In June 1995, the bids for the Øresund Bridge were delivered. The two-level concept was selected as the financially most favorable solution. In November 1995, the contract for the entire bridge was awarded to a Swedish–German–Danish consortium.

The 7.8-km bridge includes a 1090-m cable-stayed bridge (Figure 4.7) with a main span of 490 m. The 3013 and 3739 m approach bridges have spans of 140 m. The entire superstructure is a composite structure with steel truss girders between the four-lane highway on the upper concrete deck and the dual-track railway on the lower deck.

Fabrication of the steel trusses and casting of the concrete deck of the approach bridges are carried out in Spain. The complete 140-m-long girder sections, weighing up to 7000 tonnes, are tugged on flat barges to the bridge site and lifted into position on the piers. Steel trusses for the cable-stayed bridge are fabricated in Sweden and transported to the casting yard close to the bridge site, where the concrete decks are cast.

On the cable-stayed bridge the girder will also be erected in 140-m sections on temporary supports before being suspended by the stays. This method is unusual for a cable-stayed bridge, but it is attractive because of the availability of the heavy-lift vessel *Svanen,* and it reduces the construction time and limits vessel traffic disturbance. (*Svanen* was, as mentioned earlier, purpose-built for the Storebælt West Bridge. After its service there *Svanen* crossed the Atlantic to be upgraded and used for the erection works at the Confederation Bridge in Canada. Back again in Europe *Svanen* performs an important job at Øresund). During the construction period two VTS systems have been in operation, the Drogden VTS on the Danish side, and the Flint VTS on the Swedish side. The main tasks for the VTS systems are to provide vessels with necessary information in order to ensure safe navigation and avoid dangerous situations in the vicinity of the working areas. The VTS systems have proved their usefulness on several occasions.

The cable system consists of two vertical cable planes with parallel stays, the so-called harp-shaped cable system. In combination with the flexural rigid truss girder and an efficient pier support in the side spans, a high stiffness is achieved.

The module of the truss remains 20 m both in the approach and in the main spans. This results in stay cable forces of up to 16,000 tonnes which is beyond the range of most suppliers of prefabricated cables. Four prefabricated strands in a square configuration have therefore been adopted for each stay cable.

The concrete pylons are 203.5 m high and founded on limestone. Caissons, prefabricated on the Swedish side of Øresund, are placed in 15 m water depth, and the cast-in-place pylon shafts are progressing. Artificial islands will be established around the pylons and nearby piers to protect against vessel impact. All caissons, piers, and pier shafts are being prefabricated onshore to be assembled offshore. The bridge is scheduled to be opened for traffic in year 2000.

Project Basis

General Requirements

The Eurocode system was selected to constitute the normative basis for the project. Project application documents (PADs) have been prepared as companion documents to each of the Eurocodes. The PADs perform the same function as the national application documents (NADs) developed by the member countries implementing the Eurocodes.

The partial safety and load combination factors are determined by reliability calibration. The target reliability index of $\beta = 4.7$, specified by the owner, corresponds to high safety class as commonly used for important structures in the Nordic countries.

In addition to the Eurocodes and the PADs, general design requirements were specified by the owner to cover special features of a large civil work. This is in line with what is normally done on similar projects. The general design requirements cover the following areas:

- Functional and aesthetic requirements as alignment, gradients, cross sections, and clearance profiles;
- Civil and structural loads, load combinations, and partial safety coefficients; methods of structural analysis and design;
- Soil mechanics requirements to foundation design and construction, including soil strength and deformation parameters;
- Mechanical and electrical requirements to tunnel and bridge installations, including systems for supervision, control and data acquisition (SCADA), power distribution, traffic control, communication.

Risk

LHRisk acceptance criteria were developed such that the individual user risk for crossing the link would be equal to the average risk on a highway and railway on land of similar length and traffic intensity. In addition, the societal risk aspects concerning accidents with larger numbers of fatalities were controlled as well.

The ALARP-principle — as presented in Section 4.3 — was applied to reduce consequences from risks within a cost–benefit approach. Especially the disruption risks were controlled in this way. Risk-reducing measures were studied to reduce the frequency and consequences of hazardous events. The analyses carried out addressed main events due to fire, explosion, toxic releases, vessel collision and grounding, flooding, aircraft crash, and train derailment.

Navigation

Øresund is being used by local vessel traffic and vessels in transit up to a certain limit set by the water depth in the channels Drogden and Flinterännan. The Drogden channel near the Danish coast will be crossed by the immersed tunnel and only requirements regarding accidental vessel impact to tunnel structures have been specified. The Flinterännan near the Swedish coast is being crossed by the bridge, and the navigation route will be improved for safety reasons. Design criteria against vessel impact have been specifically developed on a probabilistic basis, and main piers will be protected by artificial islands.

4.4.4 The Fehmarn Belt Crossing

In 1995, the Danish and German Ministry of Transport invited eight consulting consortia to tender for the preliminary investigations for a fixed link across the 19-km-wide Fehmarn Belt.

Two Danish/German consortia were selected; one to carry out the geological and the subsoil investigations, and the other to investigate technical solution models, the environmental impact, and to carry out the day-to-day coordination of all the investigations.

In the first phase, seven different technical solutions were investigated, and in the second phase five recommended solutions were the basis for a concept study:

- A bored railway tunnel with shuttle services;
- An immersed railway tunnel with shuttle services;
- A combined highway and railway bridge;
- A combined highway and railway bored tunnel;
- A combined highway and railway immersed tunnel.

With a set of more detailed and refined functional requirements, various concepts for each of the five solution models have been studied in more detail than in the first phase. This concept study was finalized in early July 1997 with the submission of an interim report. The conceptual design started in December 1997 and is planned to last 7 months. To provide an adequate basis for a vessel collision study and the associated part of the risk analysis, vessel traffic observations are carried out by the German Navy. In parallel, the environmental investigations are continued, whereas the geological and the subsoil investigations are concluded.

The results of the study will constitute the basis for public discussions and political decisions whether or not to establish a fixed link, and also which solution model should be preferred.

References

1. The Danish Transport Council, Facts about Fehmarn Belt, Report 95-02, February 1995.
2. The Danish Transport Council, Fehmarn Belt. Issues of Accountability, Report 95-03, May 1995.
3. The Interaction between Major Engineering Structures and the Marine Environment, Report from IABSE Colloquium, Nyborg, Denmark, 1991.
4. Ole Damgaard Larsen: Ship Collision with Bridges, IABSE Structural Engineering Documents, 1993.
5. Guide Specification and Commentary for Vessel Collision Design of Highway Bridges, Vol. 1, Final Report, AASHTO, 1991.
6. The Storebælt Publications: East Tunnel, A/S Storebæltsforbindelsen, København, Denmark, 1997.
7. The Storebælt Publications: West Bridge, A/S Storebæltsforbindelsen, København, Denmark, 1998.
8. The Storebælt Publications: East Bridge, A/S Storebæltsforbindelsen, København, Denmark, 1998.

5
Design Philosophies for Highway Bridges

John M. Kulicki
Modjeski and Masters, Inc.

5.1 Introduction

Several bridge design specifications will be referred to repeatedly herein. In order to simplify the references, the "Standard Specifications" means the *AASHTO Standard Specifications for Highway Bridges* [1], and the sixteenth edition will be referenced unless otherwise stated. The "LRFD Specifications" means the *AASHTO LRFD Bridge Design Specifications* [2], and the first edition will be referenced, unless otherwise stated. This latter document was developed in the period 1988 to 1993 when statistically based probability methods were available, and which became the basis of quantifying safety. Because this is a more modern philosophy than either the load factor design method or the allowable stress design method, both of which are available in the Standard Specifications, and neither of which have a mathematical basis for establishing safety, much of the chapter will deal primarily with the LRFD Specifications.

There are many issues that make up a design philosophy — for example, the expected service life of a structure, the degree to which future maintenance should be assumed to preserve the original resistance of the structure or should be assumed to be relatively nonexistent, the ways brittle behavior can be avoided, how much redundancy and ductility are needed, the degree to which analysis is expected to represent accurately the force effects actually experienced by the structure, the extent to which loads are thought to be understood and predictable, the degree to which the designers' intent will be upheld by vigorous material-testing requirements and thorough inspection during construction, the balance between the need for high precision during construction in terms of alignment and positioning compared with allowing for misalignment and compensating for it in the design, and, perhaps most fundamentally, the basis for establishing safety in the design specifications. It is this last issue, the way that specifications seek to establish safety, that is dealt with in this chapter.

5.2 Limit States

All comprehensive design specifications are written to establish an acceptable level of safety. There are many methods of attempting to provide safety and the method inherent in many modern bridge design specifications, including the LRFD Specifications, the Ontario Highway Bridge Design Code [3], and the Canadian Highway Bridge Design Code [4], is probability-based reliability analysis. The method for treating safety issues in modern specifications is the establishment of "limit states" to define groups of events or circumstances that could cause a structure to be unserviceable for its original intent.

The LRFD Specifications are written in a probability-based limit state format requiring examination of some, or all, of the four limit states defined below for each design component of a bridge.

- The *service limit state* deals with restrictions on stress, deformation, and crack width under regular service conditions. These provisions are intended to ensure the bridge performs acceptably during its design life.

- The *fatigue and fracture limit state* deals with restrictions on stress range under regular service conditions reflecting the number of expected stress range excursions. These provisions are intended to limit crack growth under repetitive loads to prevent fracture during the design life of the bridge.

- The *strength limit state* is intended to ensure that strength and stability, both local and global, are provided to resist the statistically significant load combinations that a bridge will experience in its design life. Extensive distress and structural damage may occur under strength limit state conditions, but overall structural integrity is expected to be maintained.

- The *extreme event limit state* is intended to ensure the structural survival of a bridge during a major earthquake, or when collided by a vessel, vehicle, or ice flow, or where the foundation is subject to the scour that would accompany a flood of extreme recurrence, usually considered to be 500 years. These provisions deal with circumstances considered to be unique occurrences whose return period is significantly greater than the design life of the bridge. The joint probability of these events is extremely low, and, therefore, they are specified to be applied separately. Under these extreme conditions, the structure is expected to undergo considerable inelastic deformation by which locked-in force effects due to temperature effects, creep, shrinkage, and settlement will be relieved.

5.3 Philosophy of Safety

5.3.1 Introduction

A review of the philosophy used in a variety of specifications resulted in three possibilities, allowable stress design (ASD), load factor design (LFD), and reliability-based design, a particular application of which is referred to as load and resistance factor design (LRFD). These philosophies are discussed below.

5.3.2 Allowable Stress Design

ASD is based on the premise that one or more factors of safety can be established based primarily on experience and judgment which will assure the safety of a bridge component over its design life; for example, this design philosophy for a member resisting moments is characterized by design criteria such as

$$\Sigma M/S \le F_y/1.82 \tag{5.1}$$

where

ΣM = sum of applied moments
F_y = specified yield stress
S = elastic section modules

The constant 1.82 is the factor of safety.

The "allowable stress" is assumed to be an indicator of the resistance and is compared with the results of stress analysis of loads discussed below. Allowable stresses are determined by dividing the elastic stress at the onset of some assumed undesirable response, e.g., yielding of steel or aluminum, crushing of concrete, loss of stability, by a safety factor. In some circumstances, the allowable stresses were increased on the basis that more representative measures of resistance, usually based on inelastic methods, indicated that some behaviors are stronger than others. For example, the ratio of fully yielded cross-sectional resistance (no consideration of loss of stability) to elastic resistance based on first yield is about 1.12 to 1.15 for most rolled shapes bent about their major axis. For a rolled shape bent about its minor axis, this ratio is 1.5 for all practical purposes. This increased plastic strength inherent in weak axis bending was recognized by increasing the basic allowable stress for this illustration from 0.55 F_y to 0.60 F_y and retaining the elastic calculation of stress.

The specified loads are the working basis for stress analysis. Individual loads, particularly environmental loads, such as wind forces or earthquake forces, may be selected based on some committee-determined recurrence interval. Design events are specified through the use of load combinations discussed in Section 5.4.1.4. This philosophy treats each load in a given load combination on the structure as equal from the viewpoint of statistical variability. A "commonsense" approach may be taken to recognize that some combinations of loading are less likely to occur than others; e.g., a load combination involving a 160 km/h wind, dead load, full shrinkage, and temperature may be thought to be far less likely than a load combination involving the dead load and the full design live load. For example, in ASD the former load combination is permitted to produce a stress equal to four thirds of the latter. There is no consideration of the probability of both a higher-than-expected load and a lower-than-expected strength occurring at the same time and place. There is little or no direct relationship between the ASD procedure and the actual resistance of many components in bridges, or to the probability of events actually occurring.

These drawbacks notwithstanding, ASD has produced bridges which, for the most part, have served very well. Given that this is the historical basis for bridge design in the United States, it is important to proceed to other, more robust design philosophies of safety with a clear understanding of the type of safety currently inherent in the system.

5.3.3 Load Factor Design

In LFD a preliminary effort was made to recognize that the live load, in particular, was more highly variable than the dead load. This thought is embodied in the concept of using a different multiplier on dead and live load; e.g., a design criteria can be expressed as

$$1.30 M_D + 2.17 \left(M_{L+I} \right) \leq \phi M_u \tag{5.2}$$

where

M_D = moment from dead loads
M_{L+I} = moment from live load and impact
M_u = resistance
ϕ = a strength reduction factor

Resistance is usually based on attainment of either loss of stability of a component or the attainment of inelastic cross-sectional strength. Continuing the rolled beam example cited above, the distinction between weak axis and strong axis bending would not need to be identified because the cross-sectional resistance is the product of yield strength and plastic section modulus in both cases. In some cases, the resistance is reduced by a "strength reduction factor," which is based on the possibility that a component may be undersized, the material may be understrength, or the method of calculation may be more or less accurate than typical. In some cases, these factors have been based on statistical analysis of resistance itself. The joint probability of higher-than-expected loads and less-than-expected resistance occurring at the same time and place is not considered.

In the Standard Specifications, the same loads are used for ASD and LFD. In the case of LFD, the loads are multiplied by factors greater than unity and added to other factored loads to produce load combinations for design purposes. These combinations will be discussed further in Section 5.4.3.1.

The drawback to load factor design as seen from the viewpoint of probabilistic design is that the load factors and resistance factors were not calibrated on a basis that takes into account the statistical variability of design parameters in nature. In fact, the factors for steel girder bridges were established for one correlation at a simple span of 40 ft (12.2 m). At that span, both load factor design and service load design are intended to give the same basic structure. For shorter spans, load factor design is intended to result in slightly more capacity, whereas, for spans over 40 feet, it is intended to result in slightly less capacity with the difference increasing with span length. The development of this one point calibration for steel structures is given by Vincent in 1969 [5].

5.3.4 Probability- and Reliability-Based Design

Probability-based design seeks to take into account directly the statistical mean resistance, the statistical mean loads, the nominal or notional value of resistance, the nominal or notional value of the loads, and the dispersion of resistance and loads as measured by either the standard deviation or the coefficient of variation, i.e., the standard deviation divided by the mean. This process can be used directly to compute probability of failure for a given set of loads, statistical data, and the designer's estimate of the nominal resistance of the component being designed. Thus, it is possible to vary the designer's estimated resistance to achieve a criterion which might be expressed in terms, such as the component (or system) must have a probability of failure of less than 0.0001, or whatever variable is acceptable to society. Design based on probability of failure is used in numerous engineering disciplines, but its application to bridge engineering has been relatively small. The AASHTO "Guide Specification and Commentary for Vessel Collision Design of Highway Bridges" [6] is one of the few codifications of probability of failure in U.S. bridge design.

Alternatively, the probabilistic methods can be used to develop a quantity known as the "reliability index" which is somewhat, but not directly, relatable to the probability of failure. Using a reliability-based code in the purest sense, the designer is asked to calculate the value of the reliability index provided by his or her design and then compare that to a code-specified minimum value. Through a process of calibrating load and resistance factors to reliability indexes in simulated trial designs, it is possible to develop a set of load and resistance factors, so that the design process looks very much like the existing LFD methodology. The concept of the reliability index and a process for reverse-engineering load and resistance factors is discussed in Section 5.3.5.

In the case of the LRFD Specifications, some loads and resistances have been modernized as compared with the Standard Specifications. In many cases, the resistances are very similar. Most of the load and resistance factors have been calculated using a statistically based probability method which considers the joint probability of extreme loads and extreme resistance. In the parlance of the LRFD Specifications, "extreme" encompasses both maximum and minimum events.

5.3.5 The Probabilistic Basis of the LRFD Specifications

5.3.5.1 Introduction to Reliability as a Basis of Design Philosophy

A consideration of probability-based reliability theory can be simplified considerably by initially considering that natural phenomena can be represented mathematically as normal random variables,

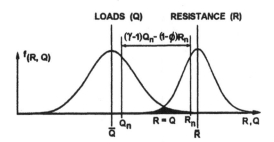

FIGURE 5.1 Separation of loads and resistance. (*Source:* Kulicki, J.M., et al., NH, Course 13061, Federal Highway Administration, Washington, D.C., 1994.)

as indicated by the well-known bell-shaped curve. This assumption leads to closed-form solutions for areas under parts of this curve, as given in many mathematical handbooks and programmed into many hand calculators.

Accepting the notion that both load and resistance are normal random variables, we can plot the bell-shaped curve corresponding to each of them in a combined presentation dealing with distribution as the vertical axis against the value of load, Q, or resistance, R, as shown in Figure 5.1 from Kulicki et al. [7]. The mean value of load, $\overline{Q}$, and the mean value of resistance, $\overline{R}$, are also shown. For both the load and the resistance, a second value somewhat offset from the mean value, which is the "nominal" value, or the number that designers calculate the load or the resistance to be, is also shown. The ratio of the mean value divided by the nominal value is called the "bias." The objective of a design philosophy based on reliability theory, or probability theory, is to separate the distribution of resistance from the distribution of load, such that the area of overlap, i.e., the area where load is greater than resistance, is tolerably small. In the particular case of the LRFD formulation of a probability-based specification, load factors and resistance factors are developed together in a way that forces the relationship between the resistance and load to be such that the area of overlap in Figure 5.1 is less than or equal to the value that a code-writing body accepts. Note in Figure 5.1 that it is the nominal load and the nominal resistance, not the mean values, which are factored.

A conceptual distribution of the difference between resistance and loads, combining the individual curves discussed above, is shown in Figure 5.2. It now becomes convenient to define the mean value of resistance minus load as some number of standard deviations, $\beta\sigma$, from the origin. The variable β is called the "reliability index" and σ is the standard deviation of the quantity $R - Q$. The problem with this presentation is that the variation of the quantity $R - Q$ is not explicitly known. Much is already known about the variation of loads by themselves or resistances by themselves, but the difference between these has not yet been quantified. However, from the probability theory, it is known that if load and resistance are both normal and random variables, then the standard deviation of the difference is

$$\sigma_{(R-Q)} = \sqrt{\sigma_R^2 + \sigma_Q^2} \qquad (5.3)$$

Given the standard deviation, and considering Figure 5.2 and the mathematical rule that the mean of the sum or difference of normal random variables is the sum or difference of their individual means, we can now define the reliability index, β, as

$$\beta = \frac{\overline{R} - \overline{Q}}{\sqrt{\sigma_R^2 + \sigma_Q^2}} \qquad (5.4)$$

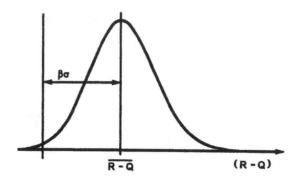

FIGURE 5.2 Definition of reliability index, β. (*Source:* Kulicki, J.M., et al., NH, Course 13061, Federal Highway Administration, Washington, D.C., 1994.)

Comparable closed-form equations can also be established for other distributions of data, e.g., log-normal distribution. A "trial-and-error" process is used for solving for β when the variable in question does not fit one of the already existing closed-form solutions.

The process of calibrating load and resistance factors starts with Eq. (5.4) and the basic design relationship; the factored resistance must be greater than or equal to the sum of the factored loads:

$$\phi R = Q = \Sigma \gamma_i x_i \qquad (5.5)$$

Solving for the average value of resistance yields:

$$\overline{R} = \overline{Q} + \beta \sqrt{\sigma_R^2 + \sigma_Q^2} = \lambda R = \frac{1}{\phi} \lambda \Sigma \gamma_i x_i \qquad (5.6)$$

By using the definition of bias, indicated by the symbol λ, Eq. (5.6) leads to the second equality in Eq. (5.6). A straightforward solution for the resistance factor, φ, is

$$\phi = \frac{\lambda \Sigma \gamma_i x_i}{\overline{Q} + \beta \sqrt{\sigma_R^2 + \sigma_Q^2}} \qquad (5.7)$$

Unfortunately, Eq. (5.7) contains three unknowns, i.e., the resistance factor, φ, the reliability index, β, and the load factors, γ.

The acceptable value of the reliability index, β, must be chosen by a code-writing body. While not explicitly correct, we can conceive of β as an indicator of the fraction of times that a design criterion will be met or exceeded during the design life, analogous to using standard deviation as an indication of the total amount of population included or not included by a normal distribution curve. Utilizing this analogy, a β of 2.0 corresponds to approximately 97.3% of the values being included under the bell-shaped curve, or 2.7 of 100 values not included. When β is increased to 3.5, for example, now only two values in approximately 10,000 are not included.

It is more technically correct to consider the reliability index to be a comparative indicator. One group of bridges having a reliability index that is greater than a second group of bridges also has more safety. Thus, this can be a way of comparing a new group of bridges designed by some new process to a database of existing bridges designed by either ASD or LFD. This is, perhaps, the most correct and most effective use of the reliability index. It is this use which formed the basis for determining the target, or code specified, reliability index, and the load and resistance factors in the LRFD Specifications, as will be discussed in the next two sections.

The probability-based LRFD for bridge design may be seen as a logical extension of the current LFD procedure. ASD does not recognize that various loads are more variable than others. The introduction of the load factor design methodology brought with it the major philosophical change of recognizing that some loads are more accurately represented than others. The conversion to probability-based LRFD methodology could be thought of as a mechanism to select the load and resistance factors more systematically and rationally than was done with the information available when load factor design was introduced.

5.3.5.2 Calibration of Load and Resistance Factors

Assuming that a code-writing body has established a target value reliability index β, usually denoted β_T, Eq. (5.7) still indicates that both the load and resistance factors must be found. One way to deal with this problem is to select the load factors and then calculate the resistance factors. This process has been used by several code-writing authorities [2–4]. The steps in the process follow:

- Factored loads can be defined as the average value of load, plus some number of standard deviation of the load, as shown as the first part of Eq. (5.6) below.

$$\gamma_i x_i = \bar{x}_i + n\sigma_i = \bar{x}_i + nV_i\bar{x}_i \qquad (5.8)$$

Defining the "variance," V_i, as equal to the standard deviation divided by the average value leads to the second half of Eq. (5.8). By utilizing the concept of bias one more time, Eq. (5.6) can now be condensed into Eq. (5.9).

$$\gamma_i = \lambda(1 + nV_i) \qquad (5.9)$$

Thus, it can be seen that load factors can be written in terms of the bias and the variance. This gives rise to the philosophical concept that load factors can be defined so that all loads have the same probability of being exceeded during the design life. This is not to say that the load factors are identical, just that the probability of the loads being exceeded is the same.

- By using Eq. (5.7) for a given set of load factors, the value of the resistance factor can be assumed for various types of structural members and for various load components, e.g., shear, moment, etc. on the various structural components. Computer simulations of a representative body of structural members can be done, yielding a large number of values for the reliability index.
- Reliability indexes are compared with the target reliability index. If close clustering results, a suitable combination of load and resistance factors has been obtained.
- If close clustering does not result, a new trial set of load factors can be used and the process repeated until the reliability indexes do cluster around, and acceptably close to, the target reliability index.
- The resulting load and resistance factors taken together will yield reliability indexes close to the target value selected by the code-writing body as acceptable.

The outline above assumes that suitable load factors are assumed. If the process of varying the resistance factors and calculating the reliability indexes does not converge to a suitable narrowly grouped set of reliability indexes, then the load factor assumptions must be revised. In fact, several sets of proposed load factors may have to be investigated to determine their effect on the clustering of reliability indexes.

The process described above is very general. To understand how it is used to develop data for a specific situation, the rest of this section will illustrate the application to calibration of the load and resistance factors for the LRFD Specifications. The basic steps were as follows:

- Develop a database of sample current bridges.
- Extract load effects by percentage of total load.
- Develop a simulation bridge set for calculation purposes.
- Estimate the reliability indexes implicit in current designs.
- Revise loads-per-component to be consistent with the LRFD Specifications.
- Assume load factors.
- Vary resistance factors until suitable reliability indexes result.

Approximately 200 representative bridges were selected from various regions of the United States by requesting sample bridge plans from various states. The selection was based on structural type, material, and geographic location to represent a full range of materials and design practices as they vary around the country. Anticipated future trends should also be considered. In the particular case of the LRFD Specifications, this was done by sending questionnaires to various departments of transportation asking them to identify the types of bridges they are expecting to design in the near future.

For each of the bridges in the database, the load indicated by the contract drawings was subdivided by the following characteristic components:

- The dead load due to the weight of factory-made components;
- The dead load of cast-in-place components;
- The dead load due to asphaltic wearing surfaces where applicable;
- The dead weight due to miscellaneous items;
- The live load due to the HS20 loading;
- The dynamic load allowance or impact prescribed in the 1989 AASHTO Specifications.

Full tabulations for all these loads for the full set of bridges in the database are presented in Nowak [8].

Statistically projected live load and the notional values of live load force effects were calculated. Resistance was calculated in terms of moment and shear capacity for each structure according to the prevailing requirements, in this case the AASHTO Standard Specifications for load factor design.

Based on the relative amounts of the loads identified in the preceding section for each of the combination of span and spacing and type of construction indicated by the database, a simulated set of 175 bridges was developed, comprising the following:

- In all; 25 noncomposite steel girder bridge simulations for bending moments and shear with spans of 9, 18, 27, 36, and 60 m and, for each of those spans, spacings of 1.2, 1.8, 2.4, 3.0, and 3.6 m;
- Representative composite steel girder bridges for bending moments and shear having the same parameters as those identified above;
- Representative reinforced concrete T-beam bridges for bending moments and shear having spans of 9, 18, 27, and 39 m, with spacings of 1.2, 1.8, 2.4, and 3.6 m in each span group;
- Representative prestressed concrete I-beam bridges for moments and shear having the same span and spacing parameters as those used for the steel bridges.

Full tabulations of these bridges and their representative amounts of the various loads are presented in Nowak [8].

The reliability indexes were calculated for each simulated and each actual bridge for both shear and moment. The range of reliability indexes which resulted from this phase of the calibration process is presented in Figure 5.3 from Kulicki et al. [7]. It can be seen that a wide range of values was obtained using the current specifications, but this was anticipated based on previous calibration work done for the Ontario Highway Bridge Design Code (OHBDC) [9].

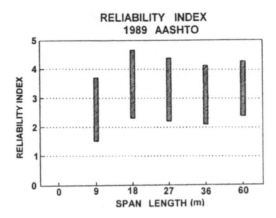

FIGURE 5.3 Reliability indexes inherent in the 1989 AASHTO Standard Specifications. (*Source:* Kulicki, J.M., et al., NH, Course 13061, Federal Highway Administration, Washington, D.C., 1994.)

TABLE 5.1 Parameters of Bridge Load Components

Load Component	Bias Factor	Coefficient of Variation	Load Factor		
			$n = 1.5$	$n = 2.0$	$n = 2.5$
Dead load, shop built	1.03	0.08	1.15	1.20	1.24
Dead load, field built	1.05	0.10	1.20	1.25	1.30
Dead load, asphalt and utilities	1.00	0.25	1.375	1.50	1.65
Live load (with impact)	1.10–1.20	0.18	1.40–1.50	1.50–1.60	1.60–1.70

Source: Nowak, A.S., Report UMCE 92-25, University of Michigan, Ann Arbor, 1993. With permission.

These calculated reliability indexes, as well as past calibration of other specifications, serve as a basis for selection of the target reliability index, β_T. A target reliability index of 3.5 was selected for the OHBDC and is under consideration for other reliability-based specifications. A consideration of the data shown in Figure 5.3 indicates that a β of 3.5 is representative of past LFD practice. Hence, this value was selected as a target for the calibration of the LRFD Specifications.

5.3.5.3 Load and Resistance Factors

The parameters of bridge load components and various sets of load factors, corresponding to different values of the parameter n in Eq. (5.9) are summarized in Table 5.1 from Nowak [8].

Recommended values of load factors correspond to $n = 2$. For simplicity of the designer, one factor is specified for shop-built and field-built components, $\gamma = 1.25$. For D_3, weight of asphalt and utilities, $\gamma = 1.50$. For live load and impact, the value of load factor corresponding to n = 2 is $\gamma = 1.60$. However, a more conservative value of $\gamma = 1.75$ is utilized in the LRFD Specifications.

The acceptance criterion in the selection of resistance factors is how close the calculated reliability indexes are to the target value of the reliability index, β_T. Various sets of resistance factors, ϕ, are considered. Resistance factors used in the code are rounded off to the nearest 0.05.

Calculations were performed using the load components for each of the 175 simulated bridges using the range of resistance factors shown in Table 5.3. For a given resistance factor, material, span, and girder spacing, the reliability index is computed. Values of β were calculated for live-load factors, $\gamma = 1.75$. For comparison, the results are also shown for live-load factor, $\gamma = 1.60$. The calculations are performed for the resistance factors, ϕ, listed in Table 5.2 from Nowak [8].

Reliability indexes were recalculated for each of the 175 simulated cases and each of the actual bridges from which the simulated bridges were produced. The range of values obtained using the new load and resistance factors is indicated in Figure 5.4.

TABLE 5.2   Considered Resistance Factors

Material	Limit State	Resistance Factors, ϕ	
		Lower	Upper
Noncomposite steel	Moment	0.95	1.00
	Shear	0.95	1.00
Composite steel	Moment	0.95	1.00
	Shear	0.95	1.00
Reinforced concrete	Moment	0.85	0.90
	Shear	0.90	0.90
Prestressed concrete	Moment	0.95	1.00
	Shear	0.90	0.95

Source: Nowak, A.S., Report UMCE 92-25, University of Michigan, Ann Arbor, 1993. With permission.

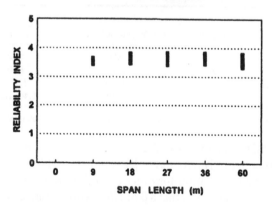

FIGURE 5.4   Reliability indexes inherent in LRFD Specifications. (*Source:* Kulicki, J.M., et al., NH, Course 13061, Federal Highway Administration, Washington, D.C., 1994.)

Figure 5.4 from Kulicki et al. [7] shows that the new calibrated load and resistance factors and new load models and load distribution techniques work together to produce very narrowly clustered reliability indexes. This was the objective of developing the new factors. Correspondence to a reliability index of 3.5 is something which can now be altered by AASHTO. The target reliability index could be raised or lowered as may be advisable in the future and the factors can be recalculated accordingly. This ability to adjust the design parameters in a coordinated manner is one of the strengths of a probabilistically based reliability design.

5.4   Design Objectives

5.4.1   Safety

5.4.1.1   Introduction

Public safety is the primary responsibility of the design engineer. All other aspects of design, including serviceability, maintainability, economics, and aesthetics are secondary to the requirement for safety. This does not mean that other objectives are not important, but safety is paramount.

5.4.1.2 The Equation of Sufficiency

In design specifications the issue of safety is usually codified by an application of the general statement the design resistances must be greater than, or equal to, the design load effects. In ASD, Eq. (5.1) can be generalized as

$$\Sigma Q_i \le R_E/\text{FS} \tag{5.10}$$

where
Q_i = a load
R_E = elastic resistance
FS = factor of safety

In LFD, Eq. (5.2) can be generalized as

$$\Sigma \gamma_i Q_i \le \phi R \tag{5.11}$$

where
γ_i = a load factor
Q_i = a load
R = resistance
ϕ = a strength reduction factor

In LRFD, Eq. (5.2) can be generalized as

$$\Sigma \eta_i y_i Q_i \le \phi R_n = R_r \tag{5.12}$$

where
η_i = $\eta_D \eta_R \eta_I$; $\eta_i = \eta_D \eta_R \eta_I \ge 0.95$ for loads for which a maximum value of γ_i is appropriate and $\eta_i = 1/(\eta_i \eta_D \eta_R) \le 1.0$ for loads for which a minimum value of γ_i is appropriate
γ_i = load factor: a statistically based multiplier on force effects
ϕ = resistance factor: a statistically based multiplier applied to nominal resistance
η_i = load modifier
η_D = a factor relating to ductility
η_R = a factor relating to redundancy
η_I = a factor relating to operational importance
Q_i = nominal force effect: a deformation, stress, or stress resultant
R_n = nominal resistance: based on the dimensions as shown on the plans and on permissible stresses, deformations, or specified strength of materials
R_r = factored resistance: ϕR_n

Eq. (5.12) is applied to each designed component and connection as appropriate for each limit state under consideration.

5.4.1.3 Special Requirements of the LRFD Specifications

Comparison of the equation of sufficiency as it was written above for ASD, LFD, and LRFD shows that, as the design philosophy evolved through these three stages, more aspects of the component under design and its relation to its environment and its function to society must be expressly considered. This is not to say that a designer using ASD necessarily considers less than a designer using LFD or LRFD. The specification provisions are the minimum requirements, and prudent designers often consider additional aspects. However, as specifications mature and become more reflective of the real world, additional criteria are often needed to assure adequate safety which may have been provided, albeit nonuniformly, by simpler provisions. Therefore, it is not surprising to find that the LRFD Specifications require explicit consideration of ductility, redundancy, and operational importance in Eq. (5.12), while the Standard Specifications does not.

Ductility, redundancy, and operational importance are significant aspects affecting the margin of safety of bridges. While the first two directly relate to the physical behavior, the last concerns the consequences of the bridge being out of service. The grouping of these aspects is, therefore, arbitrary; however, it constitutes a first effort of codification. In the absence of more precise information, each effect, except that for fatigue and fracture, is estimated as ±5%, accumulated geometrically, a clearly subjective approach. With time, improved quantification of ductility, redundancy, and operational importance, and their interaction, may be attained.

Ductility

The response of structural components or connections beyond the elastic limit can be characterized by either brittle or ductile behavior. Brittle behavior is undesirable because it implies the sudden loss of load-carrying capacity immediately when the elastic limit is exceeded. Ductile behavior is characterized by significant inelastic deformations before any loss of load-carrying capacity occurs. Ductile behavior provides warning of structural failure by large inelastic deformations. Under cyclic loading, large reversed cycles of inelastic deformation dissipate energy and have a beneficial effect on structure response.

If, by means of confinement or other measures, a structural component or connection made of brittle materials can sustain inelastic deformations without significant loss of load-carrying capacity, this component can be considered ductile. Such ductile performance should be verified by experimental testing.

Behavior that is ductile in a static context, but that is not ductile during dynamic response, should also be avoided. Examples of this behavior are shear and bond failures in concrete members and loss of composite action in flexural members.

The ductility capacity of structural components or connections may either be established by full- or large-scale experimental testing, or with analytical models that are based on realistic material behavior. The ductility capacity for a structural system may be determined by integrating local deformations over the entire structural system.

Given proper controls on the innate ductility of basic materials, proper proportioning and detailing of a structural system are the key consideration in ensuring the development of significant, visible, inelastic deformations, prior to failure, at the strength and extreme event limit states.

For the fatigue and fracture limit state for fracture-critical members and for the strength limit state for all members:

η_D ≥ 1.05 for nonductile components and connections,
= 1.00 for conventional designs and details complying with these specifications
≥ 0.95 for components and connections for which additional ductility-enhancing measures have been specified beyond those required by these specifications

For all other limit states:

η_D = 1.00

Redundancy

Redundancy is usually defined by stating the opposite, e.g., a nonredundant structure is one in which the loss of a component results in collapse or a nonredundant component is one whose loss results in complete or partial collapse. Multiple load path structures should be used, unless there are compelling reasons to the contrary. The LRFD Specifications require additional resistance in order to reduce probability of loss of nonredundant component and to provide additional resistance to accommodate load redistribution.

For the strength limit state:

η_R ≥ 1.05 for nonredundant members
= 1.00 for conventional levels of redundancy
≥ 0.95 for exceptional levels of redundancy

For all other limit states:

$\eta_R = 1.00$

The factors currently specified were based solely on judgment and were included to require more explicit consideration of redundancy. Research is under way by Ghosn and Moses [10] to provide more rational requirements based on reliability indexes thought to be acceptable in damaged bridges which must remain in service for a period of about 2 years. The "reverse engineering" concept is being applied to develop values similar in intent to η_R.

Operational Importance
The concept of operational importance is applied to the strength and extreme event limit states. The owner may declare a bridge, or any structural component or connection, thereof, to be of operational importance. Such classification should be based on social/survival and/or security/defense requirements. If a bridge is deemed of operational importance, η_I is taken as ≥1.05. Otherwise, η_I is taken as 1.0 for typical bridges and may be reduced to 0.95 for relatively less important bridges.

5.4.1.4 Design Load Combinations in ASD, LFD, and LRFD
The following permanent and transient loads and forces are considered in the ASD and LFD using the Standard Specifications, and in LRFD using the LRFD Specifications.

The load factors for various loads, making up a design load combination, are indicated in Table 5.4 and Table 5.5 for LRFD and Table 5.6 for ASD and LFD. In the case of the LRFD Specifications, all of the load combinations are related to the appropriate limit state. Any, or all, of the four limit states may be required in the design of any particular component and those which are the minimum necessary for consideration are indicated in the specifications where appropriate. Thus, a design might involve any load combination in Table 5.4.

In the case of ASD or LFD, there is no direct relationship between the load combinations specified in Table 5.6 and limit states, as the design requirements in the Standard Specifications are not organized in that manner. A design by ASD uses those combinations in Table 5.5 indicated for the allowable stress design method as appropriate for the component under consideration. The load combinations indicated for LFD are not used in conjunction with allowable stress design. The opposite is true for LFD.

The application of the load combinations in Table 5.6 for ASD and LFD has been available to bridge designers for decades and is relatively well understood. Numerous textbooks have dealt with these subjects. For this reason, the remainder of this section will deal primarily with the relatively newer LRFD Specifications.

All relevant subsets of the load combinations in Table 5.4 should be investigated. The factors should be selected to produce the total factored extreme force effect. For each load combination, both positive and negative extremes should be investigated. In load combinations where one force effect decreases the effect of another, the minimum value should be applied to load reducing the force effect. For each load combination, every load that is indicated, including all significant effects due to distortion, should be multiplied by the appropriate load factor.

It can be seen in Table 5.4 that some of the load combinations have a choice of two load factors. The larger of the two values for load factors shown for TU, TG, CR, SH, and SE are to be used when calculating deformations; the smaller value should be used when calculating all other force

TABLE 5.3 Load Designations

Name of Load	LRFD Designation	Standard of Specification Designation
	Permanent Loads	
Downdrag	DD	
Dead load of structural components attachments	DC	D
Dead load of wearing surfaces and utilities	DW	D
Dead load of earth fill	EF	D
Horizontal earth pressure	EH	E
Earth surcharge load	ES	E
Vertical earth pressure	EV	D
	Transient Loads	
Vehicular braking force	BR	LF
Vehicular centrifugal force	CE	CF
Creep	CR	R
Vehicular collision force	CT	—
Vessel collision force	CV	—
Earthquake	EQ	EQ
Friction	FR	—
Ice load	IC	ICE
Vehicular dynamic load allowance	IM	I
Vehicular live load	LL	L
Live–load surcharge	LS	L
Pedestrian live load	PL	L
Settlement	SE	—
Shrinkage	SH	S
Temperature gradient	TG	—
Uniform temperature	TU	T
Water load and stream pressure	WA	SF
Wind on live load	WL	WL
Wind load on structure	WS	W

TABLE 5.4 Load Combinations and Load Factors in LRFD

Limit State Load Combinations	DC DD DW EH EV ES	LL IM CE BR PL LS	WA	WS	WL	FR	TU CR SH	TG	SE	Use One of These at a Time			
										EQ	IC	CT	CV
Strength I	γ_p	1.75	1.00	—	—	1.00	0.50/1.20	γ_{TG}	γ_{SE}	—	—	—	—
Strength II	γ_p	1.35	1.00	—	—	1.00	0.50/1.20	γ_{TG}	γ_{SE}	—	—	—	—
Strength III	γ_p	—	1.00	1.40	—	1.00	0.50/1.20	γ_{TG}	γ_{SE}	—	—	—	—
Strength IV													
EH, EV, ES, DW	γ_p	—	1.00	—	—	1.00	0.50/1.20	—	—	—	—	—	—
DC only	1.5									—	—	—	—
Strength V	γ_p	1.35	1.00	0.40	0.40	1.00	0.50/1.20	γ_{TG}	γ_{SE}	—	—	—	—
Extreme Event I	γ_p	γ_{EQ}	1.00	—	—	1.00	—	—	—	1.00	—	—	—
Extreme Event II	γ_p	0.50	1.00	—	—	1.00	—	—	—	—	1.00	1.00	1.00
Service I	1.00	1.00	1.00	0.30	0.30	1.00	1.00/1.20	γ_{TG}	γ_{SE}	—	—	—	—
Service II	1.00	1.30	1.00	—	—	1.00	1.00/1.20	—	—	—	—	—	—
Service III	1.00	0.80	1.00	—	—	1.00	1.00/1.20	γ_{TG}	γ_{SE}	—	—	—	—
Fatigue LL, IM and CE only	—	0.75	—	—	—	—	—	—	—	—	—	—	—

TABLE 5.5   Load Factors for Permanent Loads, γ_p in LRFD

Type of Load	Load Factor	
	Maximum	Minimum
DC: Component and attachments	1.25	0.90
DD: Downdrag	1.80	0.45
DW: Wearing surfaces and utilities	1.50	0.65
EH: Horizontal earth pressure		
• Active	1.50	0.90
• At rest	1.35	0.90
EV: Vertical earth pressure		
• Overall stability	1.35	N/A
• Retaining structure	1.35	1.00
• Rigid buried structure	1.30	0.90
• Rigid frames	1.35	0.90
• Flexible buried structures other than metal box culverts	1.95	0.90
• Flexible metal box culverts	1.50	0.90
ES: Earth surcharge	1.50	0.75

effects. Where movements are calculated for the sizing of expansion dams, the design of bearing, or similar situations where consideration of unexpectedly large movements is advisable, the larger factor should be used. When considering the effect of these loads on forces that are compatibility generated, the lower factor may be used. This latter use requires structural insight.

Consideration of the variability of loads in nature indicates that loads may be either larger or smaller than the nominal load used in the design specifications. While the concept of variability of permanent loads receives little coverage in ASD, it is codified expressly in LFD. Note that in Table 5.6 the LFD load combinations contain a dead load modifier, indicated as β_E or β_D. These β terms are not to be confused with the reliability index, heretofore referred to as β. The purpose of the modifying factors β_E and β_D is to account for conditions where it is inadvisable to consider either that all of the dead load exists all of the time or that the dead load may be less than the nominal values indicated in the specifications. Thus, for example, the use of the β_D factor 0.75 when checking members for minimum axial load maximum moment means when designing columns and those fixtures which abut the columns, such as footings, it is necessary to evaluate not just the maximum bending moment and the maximum axial load, based on assuming that all the elements of a load combination are thought to obtain their maximum values, but also a load combination in which it is assumed that the dead load is lighter than the nominal load. In the case where the majority of the axial load comes from the dead load and the majority of the bending moment comes from lateral load or live load, this modified combination will tend to produce a maximum eccentricity and hence could control the design of columns and footings.

The specified values of β_E are given below:

β_E 1.00 for vertical and lateral loads on all other structures

β_E 1.3 for lateral earth pressure for retaining walls and rigid frames, excluding rigid culverts; for lateral at-rest earth pressures, $\beta_E = 1.15$

β_E 0.5 for lateral earth pressure when checking positive moments in rigid frames; this complies with Section 3.20

β_E 1.0 for vertical earth pressure

β_D 0.75 when checking member for minimum axial load and maximum moment or maximum eccentricity — for column design

β_D 1.0 when checking member for maximum axial load and minimum moment — for column design

β_D 1.0 for flexural and tension members

β_E 1.0 for rigid culverts

β_E 1.5 for flexible culverts

TABLE 5.6　Table of Coefficients γ and β in ASD and LFD

	Group	γ	D	(L+I)$_n$	(L+I)$_p$	CF	E	B	SF	W	WL	LF	R + S + T	EQ	Ice	%
		1	2	3	3A	4	5	6	7	8	9	10	11	12	13	14
									β Factors							
SERVICE LOAD	I	1.0	1	1	0	1	β$_E$	B	1	0	0	0	0	0	0	100
	IA	1.0	1	2	0	0	0	1	0	0	0	0	0	0	0	150
	IB	1.0	1	0	1	1	β$_E$	0	1	0	0	0	0	0	0	b
	II	1.0	1	0	0	0	1	1	1	1	0	0	0	0	0	125
	III	1.0	1	1	0	1	β$_E$	1	1	0.3	1	1	0	0	0	125
	IV	1.0	1	1	0	1	β$_E$	1	1	0	0	0	1	0	0	125
	V	1.0	1	0	0	0	1	1	1	1	0	0	1	0	0	140
	VI	1.0	1	1	0	1	β$_E$	1	1	0.3	1	1	1	0	0	140
	VII	1.0	1	0	0	0	1	1	1	0	0	0	0	1	0	133
	VIII	1.0	1	1	0	1	1	1	1	0	0	0	0	0	1	140
	IX	1.0	1	0	0	0	1	1	1	1	0	0	0	0	1	150
	X	1.0	1	1	0	0	β$_E$	0	0	0	0	0	0	0	0	100
LOAD FACTOR DESIGN	I	1.3	β$_D$	1.67ᵃ	0	1.0	β$_E$	1	1	0	0	0	0	0	0	
	IA	1.3	β$_D$	2.20	0	0	0	0	0	0	0	0	0	0	0	
	IB	1.3	β$_D$	0	1	1.0	β$_E$	1	1	0	0	0	0	0	0	NOT APPLICABLE
	II	1.3	β$_D$	0	0	0	β$_E$	1	1	1	0	0	0	0	0	
	III	1.3	β$_D$	1	0	1	β$_E$	1	1	0.3	1	1	0	0	0	
	IV	1.3	β$_D$	1	0	1	β$_E$	1	1	0	0	0	1	0	0	
	V	1.25	β$_D$	0	0	0	β$_E$	1	1	1	0	0	1	0	0	
	VI	1.25	β$_D$	1	0	1	β$_E$	1	1	0.3	1	1	1	0	0	
	VII	1.3	β$_D$	0	0	0	β$_E$	1	0	0	0	0	0	1	0	
	VIII	1.3	β$_D$	1	0	1	β$_E$	1	1	0	0	0	0	0	1	
	IX	1.20	β$_D$	0	0	0	β$_E$	1	1	1	0	0	0	0	1	
	X	1.30	1	1.67	0	0	β$_E$	0	0	0	0	0	0	0	0	

- (L + I)$_n$ = Live load plus impact for AASHTO Highway H or HS loading.
- (L + I)$_p$ = Live load plus impact consistent with the overload criteria of the operation agency.
- % (col. 14) = percentage of basic unit stress.
- No increase in allowable unit stresses shall be permitted for members or connections carrying wind loads only.

ᵃ 1.25 may be used for design of outside roadway beam when combination of sidewalk live load, and traffic live load plus impact governs the design, but the capacity of the section should not be less than required for highway traffic live load only, using a β factor of 1.67. 1.00 may be used for design of deck slab with combination of loads as described in Article 3.24.2.2.

ᵇ Percentage $= \dfrac{\text{Maximum Unit Stress (Operating Rating)}}{\text{Allowable Basic Unit Stress}} \times 100$.

The LRFD Specifications recognize the variability of permanent loads by providing both maximum and minimum load factors for the permanent loads, as indicated in Table 5.5. For permanent force effects, the load factor that produces the more critical combination should be selected from Table 5.4. In the application of permanent loads, force effects for each of the specified six load types should be computed separately. Assuming variation of one type of load by span, length, or component within a bridge is not necessary. For each force effect, both extreme combinations may need to be investigated by applying either the high or the low load factor, as appropriate. The algebraic sums of these products are the total force effects for which the bridge and its components should be designed. This reinforces the traditional method of selecting load combinations to obtain realistic extreme effects.

When the permanent load increases the stability or load-carrying capacity of a component or bridge, the minimum value of the load factor for that permanent load should also be investigated. Uplift, which is treated as a separate load case in past editions of the AASHTO Standard Specifications for Highway Bridges, becomes a Strength I load combination. For example, when the dead-load reaction is positive and live load can cause a negative reaction, the load combination for maximum uplift force would be 0.9DC + 0.65DW + 1.75(LL+IM). If both reactions were negative, the load combination would be 1.25DC + 1.50DW + 1.75(LL+IM).

The load combinations for various limit states shown in Table 5.4 are described below.

Strength I   Basic load combination relating to the normal vehicular use of the bridge without wind.

Strength II   Load combination relating to the use of the bridge by permit vehicles without wind. If a permit vehicle is traveling unescorted, or if control is not provided by the escorts, the other lanes may be assumed to be occupied by the vehicular live load herein specified. For bridges longer than the permit vehicle, addition of the lane load, preceding and following the permit load in its lane, should be considered.

Strength III   Load combination relating to the bridge exposed to maximum wind velocity which prevents the presence of significant live load on the bridge.

Strength IV   Load combination relating to very high ratios of dead load to live load force effect. This calibration process had been carried out for a large number of bridges with spans not exceeding 60 m. Spot checks had also been made on a few bridges up to 180 m spans. For the primary components of large bridges, the ratio of dead and live load force effects is rather high and could result in a set of resistance factors different from those found acceptable for small- and medium-span bridges. It is believed to be more practical to investigate one more load case, rather than requiring the use of two sets of resistance factors with the load factors provided in Strength I, depending on other permanent loads present. This Load Combination IV is expected to govern when the ratio of dead load to live load force effect exceeds about 7.0.

Strength V   Load combination relating to normal vehicular use of the bridge with wind of 90 km/h velocity.

Extreme Event I   Load combination relating to earthquake. The designer-supplied live-load factor signifies a low probability of the presence of maximum vehicular live load at the time when the earthquake occurs. In ASD and LFD the live load is ignored when designing for earthquake.

Extreme Event II   Load combination relating to reduced live load in combination with a major ice event, or a vessel collision, or a vehicular impact.

Service I   Load combination relating to the normal operational use of the bridge with 90 km/h wind. All loads are taken at their nominal values and extreme load conditions are excluded. This combination is also used for checking deflection of certain buried structures and for the investigation of slope stability.

Service II   Load combination whose objective is to prevent yielding of steel structures due to vehicular live load, approximately halfway between that used for Service I and Strength I limit state, for which case the effect of wind is of no significance. This load combination corresponds to the overload provision for steel structures in past editions of the AASHTO Standard Specifications for the Design of Highway Bridges.

Service III   Load combination relating only to prestressed concrete structures with the primary objective of crack control. The addition of this load combination followed a series of trial designs done by 14 states and several industry groups during 1991 and early 1992. Trial designs for prestressed concrete elements indicated significantly more prestressing would be needed to support the loads specified in the proposed specifications. There is no nationwide physical evidence that these vehicles used to develop the notional live loads have caused detrimental cracking in existing prestressed concrete components. The statistical significance

Fatigue

of the 0.80 factor on live load is that the event is expected to occur about once a year for bridges with two design lanes, less often for bridges with more than two design lanes, and about once a day for the bridges with a single design lane. Fatigue and fracture load combination relating to gravitational vehicular live load and dynamic response, consequently BR and PL need not be considered. The load factor reflects a load level which has been found to be representative of the truck population, with respect to large number of return cycles.

5.4.2 Serviceability

The LRFD Specification treats serviceability from the view points of durability, inspectibility, maintainability, rideability, deformation control, and future widening.

Contract documents should call for high-quality materials and require that those materials that are subject to deterioration from moisture content and/or salt attack be protected. Inspectibility is to be assured through adequate means for permitting inspectors to view all parts of the structure which have structural or maintenance significance. The provisions related to inspectibility are relatively short, but as all departments of transportation have begun to realize, bridge inspection can be very expensive and is a recurring cost due to the need for biennial inspections. Therefore, the cost of providing walkways and other access means and adequate room for people and inspection equipment to be moved about on the structure is usually a good investment.

Maintainability is treated in the specification in a manner similar to durability; there is a list of desirable attributes to be considered.

The subject of live-load deflections and other deformations remains a very difficult issue. On the one hand, there is very little direct correlation between live-load deflection and premature deterioration of bridges. There is much speculation that "excessive" live-load deflection contributes to premature deck deterioration, but, to date (late 1997), no causative relationship has been statistically established.

Rider comfort is often advanced as a basis for deflection control. Studies in human response to motion have shown that it is not the magnitude of the motion, but rather the acceleration that most people perceive, especially in moving vehicles. Many people have experienced the sensation of being on a bridge and feeling a definite movement, especially when traffic is stopped. This movement is often related to the movement of floor systems, which are really quite small in magnitude, but noticeable nonetheless. There being no direct correlation between magnitude (not acceleration) of movement and discomfort has not prevented the design profession from finding comfort in controlling the gross stiffness of bridges through a deflection limit. As a compromise between the need for establishing comfort levels and the lack of compelling evidence that deflection was a cause of structural distress, the deflection criteria, other than those pertaining to relative deflections of ribs of orthotropic decks and components of some wood decks, were written as voluntary provisions to be activated by those states that so chose. Deflection limits, stated as span divided by some number, were established for most cases, and additional provisions of absolute relative displacement between planks and panels of wooden decks and ribs of orthotropic decks were also added. Similarly, optional criteria were established for a span-to-depth ratio for guidance primarily in starting preliminary designs, but also as a mechanism for checking when a given design deviated significantly from past successful practice.

5.4.3 Constructibility

Several new provisions were included in the LRFD Specification related to:

- The need to design bridges so that they can be fabricated and built without undue difficulty and with control over locked-in construction force effects;

- The need to document one feasible method of construction in the contract documents, unless the type of construction is self-evident; and
- A clear indication of the need to provide strengthening and/or temporary bracing or support during erection, but not requiring the complete design thereof.

References

1. American Association of State Highway and Transportation Officials, *Standard Specifications for Highway Bridges,* 16th ed., AASHTO, Washington, D.C., 1996.
2. American Association of State Highway and Transportation Officials, *Load Resistance Factor Design,* AASHTO, Washington, D.C., 1996.
3. Ontario Ministry of Transportation and Communications, *Ontario Highway Bridge Design Code,* OMTC, Toronto, Ontario, Canada, 1994.
4. Canadian Standards Association, *Canadian Highway Bridge Design Code,* Canadian Standards Association, Rexdale, Ontario, Canada, 1998.
5. Vincent, G. S., Load factor design of steel highway bridges, *AISI Bull.,* 15, March, 1969.
6. American Association of State Highway and Transportation Officials, *Guide Specification and Commentary for Vessel Collision Design of Highway Bridges,* Vol. I: Final Report, AASHTO, Washington, D.C., February 1991.
7. Kulicki, J. M., Mertz, D. R., and Wassef, W. G., LRFD Design of Highway Bridges, NHI Course 13061, Federal Highway Administration, Washington, D.C., 1994.
8. Nowak, A. S., Calibration of LRFD Bridge Design Code, Department of Civil and Environmental Engineering Report UMCE 92-25, University of Michigan, Ann Arbor, 1993.
9. Nowak, A. S. and Lind, N. C., Practical bridge code calibration, *ASCE J. Struct. Div.,* 105 (ST12), 2497–2510, 1979.
10. Ghosn, M. and Moses, F., Redundancy in Highway Bridge Superstructures, Draft Report to NCHRP, February 1997.

6
Highway Bridge Loads and Load Distribution

Susan E. Hida
*California Department
of Transportation*

6.1 Introduction

This chapter deals with highway bridge loads and load distribution as specified in the AASHTO Load and Resistance Factor Design (LRFD) Specifications [1]. Stream flow, ice loads, vessel collision loads, loads for barrier design, loads for anchored and mechanically stabilized walls, seismic forces, and loads due to soil–structure interaction will be addressed in subsequent chapters. Load combinations are discussed in Chapter 5.

When proceeding from one component to another in bridge design, the controlling load and the controlling factored load combination will change. For example, permit vehicles, factored and combined for one load group, may control girder design for bending in one location. The standard design vehicular live load, factored and combined for a different load group, may control girder design for shear in another location. Still other loads, such as those due to seismic events, may control column and footing design.

Note that in this chapter, superstructure refers to the deck, beams or truss elements, and any other appurtenances above the bridge soffit. Substructure refers to those components that support loads from the superstructure and transfer load to the ground, such as bent caps, columns, pier walls, footings, piles, pile extensions, and caissons. Longitudinal refers to the axis parallel to the direction of traffic. Transverse refers to the axis perpendicular to the longitudinal axis.

6.2 Permanent Loads

The LRFD Specification refers to the weights of the following as "permanent loads":

- The structure
- Formwork which becomes part of the structure
- Utility ducts or casings and contents
- Signs
- Concrete barriers
- Wearing surface and/or potential deck overlay(s)
- Other elements deemed permanent loads by the design engineer and owner
- Earth pressure, earth surcharge, and downdrag

The permanent load is distributed to the girders by assigning to each all loads from superstructure elements within half the distance to the adjacent girder. This includes the dead load of the girder itself and the soffit, in the case of box girder structures. The dead loads due to concrete barrier, sidewalks and curbs, and sound walls, however, may be equally distributed to all girders.

6.3 Vehicular Live Loads

The design vehicular live load was replaced in 1993 because of heavier truck configurations on the road today, and because a statistically representative, notional load was needed to achieve a "consistent level of safety." The notional load that was found to best represent "exclusion vehicles," i.e., trucks with loading configurations greater than allowed but routinely granted permits by agency bridge rating personnel, was adopted by AASHTO and named "Highway Load '93" or HL93. The mean and standard deviation of truck traffic was determined and used in the calibration of the load factors for HL93. It is notional in that it does not represent any specific vehicle [2].

The distribution of loads per the LRFD Specification is more complex than in the Standard Specifications for Highway Bridge Design [3]. This change is warranted because of the complexity in bridges today, increased knowledge of load paths, and technology available to be more rational in performing design calculations. The end result will be more appropriately designed structures.

6.3.1 Design Vehicular Live Load

The AASHTO "design vehicular live load," HL93, is a combination of a "design truck" or "design tandem" and a "design lane." The design truck is the former Highway Semitrailer 20-ton design truck (HS20-44) adopted by AASHO (now AASHTO) in 1944 and used in the previous Standard Specification. Similarly, the design lane is the HS20 lane loading from the AASHTO Standard Specifications. A shorter, but heavier, design tandem is new to AASHTO and is combined with the design lane if a worse condition is created than with the design truck. Superstructures with very short spans, especially those less than 12 m in length, are often controlled by the tandem combination.

The AASHTO design truck is shown in Figure 6.1. The variable axle spacing between the 145 kN loads is adjusted to create a critical condition for the design of each location in the structure. In the transverse direction, the design truck is 3 m wide and may be placed anywhere in the standard 3.6-m-wide lane. The wheel load, however, may not be positioned any closer than 0.6 m from the lane line, or 0.3 m from the face of curb, barrier, or railing.

The AASHTO design tandem consists of two 110-kN axles spaced at 1.2 m on center. The AASHTO design lane loading is equal to 9.3 N/mm and emulates a caravan of trucks. Similar to the truck loading, the lane load is spread over a 3-m-wide area in the standard 3.6-m lane. The lane loading is not interrupted except when creating an extreme force effect such as in "patch" loading of alternate spans. Only the axles contributing to the extreme being sought are loaded.

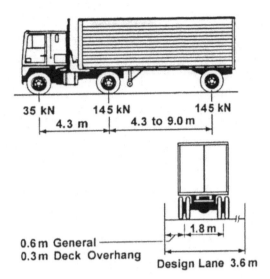

FIGURE 6.1 AASHTO-LRFD design truck. (AASHTO LRFD Bridge Design Specifications 2nd. ed., American Association of State Highway and Transportation Officials. Washington, D.C., 1998. With permission.)

When checking an extreme reaction at an interior pier or negative moment between points of contraflexure in the superstructure, two design trucks with a 4.3-m spacing between the 145-kN axles are to be placed on the bridge with a minimum of 15 m between the rear axle of the first truck and the lead axle of the second truck. Only 90% of the truck and lane load is used. This procedure differs from the Standard Specification which used shear and moment riders.

6.3.2 Permit Vehicles

Most U.S. states have developed their own "Permit Design Vehicle" to account for vehicles routinely granted permission to travel a given route, despite force effects greater than those due the design truck, i.e., the old HS20 loading. California uses anywhere from a 5- to 13-axle design vehicle as shown in Figure 6.2 [4]. Some states use an HS25 design truck, the configuration being identical to the HS20 but axle loads 25% greater.

The permit vehicular live load is combined with other loads in the Strength Limit State II as discussed in Chapter 5. Early editions of the AASHTO Specifications expect the design permit vehicle to be preceded and proceeded by a lane load. Furthermore, adjacent lanes may be loaded with the new HL93 load, unless restricted by escort vehicles.

6.3.3 Fatigue Loads

For fatigue loading, the LRFD Specification uses the design truck alone with a constant axle spacing of 9 m. The load is placed to produce extreme force effects. In lieu of more exact information, the frequency of the fatigue load for a single lane may be determined by multiplying the average daily truck traffic by p, where p is 1.00 in the case of one lane available to trucks, 0.85 in the case of two lanes available to trucks, and 0.80 in the case of three or more lanes available to trucks. If the average daily truck traffic is not known, 20% of the average daily traffic may be used on rural interstate bridges, 15% for other rural and urban interstate bridges, and 10% for bridges in urban areas.

6.3.4 Load Distribution for Superstructure Design

Figure 6.3 summarizes load distribution for design of longitudinal superstructure elements. Load distribution tables and the "lever rule" are approximate methods and intended for most designs.

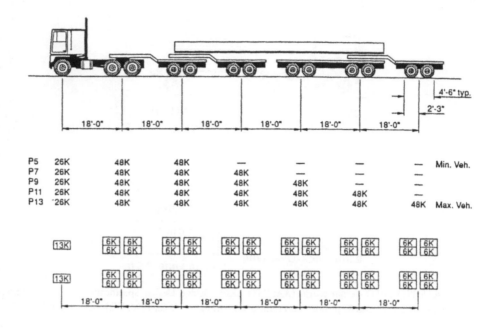

P5	26K	48K	48K	—	—	—	—	Min. Veh.
P7	26K	48K	48K	48K	—	—	—
P9	26K	48K	48K	48K	48K	—	—
P11	26K	48K	48K	48K	48K	48K	—
P13	26K	48K	48K	48K	48K	48K	48K	Max. Veh.

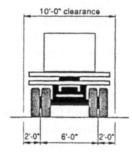

FIGURE 6.2 Caltrans permit truck. (AASHTO LRFD Bridge Design Specifications 2nd. ed., American Association of State Highway and Transportation Officials. Washington, D.C., 1998. With permission.)

The lever rule considers the slab between two girders to be simply supported. The reaction is determined by summing the reactions from the slabs on either side of the beam under consideration. "Refined analysis" refers to a three-dimensional consideration of the loads and is to be used on more complex structures. In other words, classical force and displacement, finite difference, finite element, folded plate, finite strip, grillage analogy, series/harmonic, or yield line methods are required to obtain load effects for superstructure design.

Note that, by definition of the vehicular design live load, no more than one truck can be in one lane simultaneously, except as previously described to generate maximum reactions or negative moments. After forces have been determined from the longitudinal load distribution and the longitudinal members have been designed, the designer may commence load distribution in the transverse direction for deck and substructure design.

6.3.4.1 Decks

Decks may be designed for vehicular live loads using empirical methods or by distributing loads on to "effective strip widths" and analyzing the strips as continuous or simply supported beams.

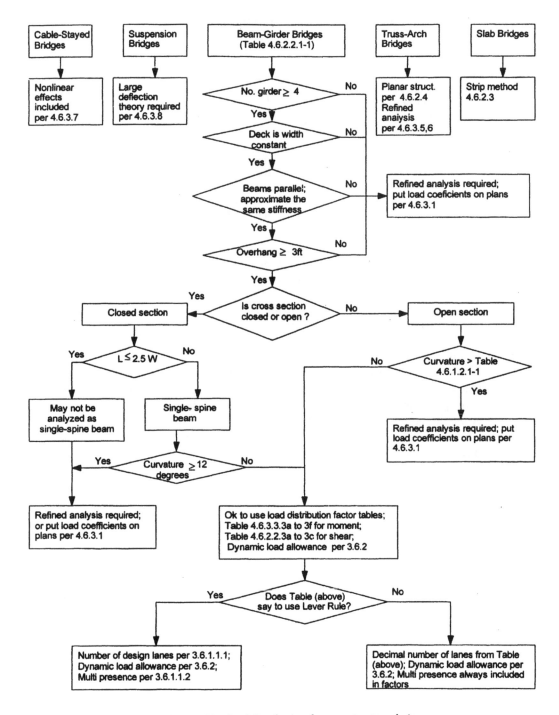

FIGURE 6.3 Live-load distribution for superstructure design.

Empirical methods rely on transfer of forces by arching of the concrete and shifting of the neutral axis. Loading is discussed in Chapter 24, Bridge Decks and Approach Slabs.

6.3.4.2 Beam–Slab Bridges

Approximate methods for load distribution on beam–slab bridges are appropriate for the types of cross sections shown in Table 4.6.2.2.1-1 of the AASHTO LRFD Specification. Load distribution

TABLE 6.1　Reduction of Load Distribution Factors for Moment in Longitudinal Beams on Skewed Supports

Type of Superstructure	Applicable Cross Section from Table 4.6.2.2.1-1	Any Number of Design Lanes Loaded	Range of Applicability
Concrete deck, filled grid, or partially filled grid on steel or concrete beams, concrete T-beams, or double T-sections	a, e, k i, j, if sufficiently connected to act as a unit	$1 - c_1(\tan\theta)^{1.5}$ $c_1 = 0.25\left(\dfrac{K_g}{Lt_s^3}\right)^{0.25}\left(\dfrac{S}{L}\right)^{0.25}$ if $\theta < 30°$, then $c_1 = 0$ if $\theta > 60°$, use $\theta = 60°$	$30° \leq \theta \leq 60°$ $1100 \leq S \leq 4900$ $6000 \leq L \leq 73{,}000$ $N_b \geq 4$
Concrete deck on concrete spread box beams, concrete box beams, and double T-sections used in multibeam decks	b, c, f, g	$1.05 - 0.25\tan\theta \leq 1.0$ if $\theta > 60°$, use $\theta = 60°$	$0 \leq \theta \leq 60°$

Source: AASHTO LRFD Bridge Design Specifications, 2nd. ed., American Association of State Highway and Transportation Officials. Washington, D.C., 1998. With permission.

factors, generated from expressions found in AASHTO LRFD Tables 4.6.2.2.2a–f and 4.6.2.2.3a–c, result in a decimal number of lanes and are used for girder design. Three-dimensional effects are accounted for. These expressions are a function of beam area, beam width, beam depth, overhang width, polar moment of inertia, St. Venant's torsional constant, stiffness, beam span, number of beams, number of cells, beam spacing, depth of deck, and deck width. Verification was done using detailed bridge deck analysis, simpler grillage analyses, and a data set of approximately 200 bridges of varying type, geometry, and span length. Limitations on girder spacing, span length, and span depth reflect the limitations of this data set.

The load distribution factors for moment and shear at the obtuse corner are multiplied by skew factors as shown in Tables 6.1 and 6.2, respectively.

6.3.4.3　Slab-Type Bridges

Cast-in-place concrete slabs or voided slabs, stressed wood decks, and glued/spiked wood panels with spreader beams are designed for an equivalent width of longitudinal strip per lane for both shear and moment. That width, E (mm), is determined from the formula:

$$E = 250 + 0.42\sqrt{L_1 W_1} \tag{6.1}$$

when one lane is loaded, and

$$E = 2100 + 0.12\sqrt{L_1 W_1} \leq W/N_L \tag{6.2}$$

when more than one lane is loaded. L_1 is the lesser of the actual span or 18,000 mm, W_1 is the lesser of the edge-to-edge width of bridge and 18,000 mm in the case of single-lane loading, and 18,000 mm in the case of multilane loading, and N_L is the numbr of design lanes.

6.3.5　Load Distribution for Substructure Design

Bridge substructure includes bent caps, columns, pier walls, pile caps, spread footings, caissons, and piles. These components are designed by placing one or more design vehicular live loads on the traveled way as previously described for maximum reaction and negative bending moment, not exceeding the maximum number of vehicular lanes permitted on the bridge. This maximum may be determined by dividing the width of the traveled way by the standard lane width (3.6 m), and "rounding down," i.e., disregarding any fractional lanes. Note that (1) the traveled way need not be

TABLE 6.2   Correction Factors for Load Distribution Factors for Support Shear of the Obtuse Corner

Type of Superstructue	Applicable Cross Section from Table 4.6.2.2.1-1	Correction Factor	Range of Applicability
Concrete deck, filled grid, or partially filled grid on steel or concrete beams, concrete T-beams or double T-sections	a, e, k i, j, if sufficiently connected to act as a unit	— $1.0 + 2.0 \left(\dfrac{L t_s^3}{K_g} \right)^{0.3} \tan\theta$	$0° \leq \theta \leq 60°$ $1100 \leq S \leq 4900$ $6000 \leq L \leq 73{,}000$ $N_b \geq 4$
Multicell concrete box beams, box sections	d	$1.0 + \left[0.25 + \dfrac{L}{70d} \right] \tan\theta$	$0° \leq \theta \leq 60°$ $1800 \leq S \leq 4000$ $6000 \leq L \leq 73000$ $900 \leq d \leq 2700$ $N_b \geq 3$
Concrete deck on spread concrete box beams	b, c	$1.0 + \dfrac{\sqrt{Ld}}{6S} \tan\theta$	$0° \leq \theta \leq 60°$ $1800 \leq S \leq 3500$ $6000 \leq L \leq 43{,}000$ $450 \leq d \leq 1700$ $N_b \geq 3$
Concrete box beams used in multibeam decks	f, g	$1.0 + \dfrac{L\sqrt{\tan\theta}}{90d}$	$0° \leq \theta \leq 60°$ $6000 \leq L \leq 37{,}000$ $430 \leq d \leq 1500$ $900 \leq b \leq 1500$ $5 \leq N_b \leq 20$

Source: AASHTO LRFD Bridge Design Specifications, 2nd. ed., American Association of State Highway and Transportation Officials. Washington, D.C., 1998. With permission.

measured from the edge of deck if curbs or traffic barriers will restrict the traveled way for the life of the structure and (2) the fractional number of lanes determined using the previously mentioned load distribution charts for girder design is not used for substructure design.

Figure 6.4 shows selected load configurations for substructure elements. A critical load configuration may result from not using the maximum number of lanes permissible. For example, Figure 6.4a shows a load configuration that may generate the critical loads for bent cap design and Figure 6.4b shows a load configuration that may generate the critical bending moment for column design. Figure 6.4c shows a load configuration that may generate the critical compressive load for design of the piles. Other load configurations will be needed to complete design of a bridge footing. Note that girder locations are often ignored in determination of substructure design moments and shears: loads are assumed to be transferred directly to the structural support, disregarding load transfer through girders in the case of beam–slab bridges. Adjustments are made to account for the likelihood of fully loaded vehicles occurring side-by-side simultaneously. This "multiple presence factor" is discussed in the next section.

In the case of rigid frame structures, bending moments in the longitudinal direction will also be needed to complete column (or pier wall) as well as foundation designs. Load configurations which generate these three cases must be checked:

1. Maximum/minimum axial load with associated transverse and longitudinal moments;
2. Maximum/minimum transverse moment with associated axial load and longitudinal moment;
3. Maximum/minimum longitudinal moment with associated axial load and transverse moment.

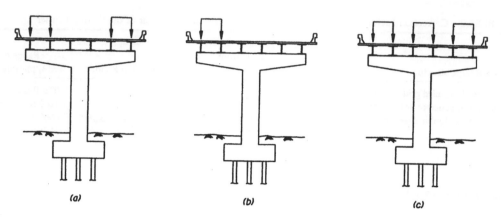

FIGURE 6.4   Various load configurations for substructure design.

TABLE 6.3   Multiple Presence Factors

Number of Loaded Lanes	Multiple Presence Factors m
1	1.20
2	1.00
3	0.85
>3	0.65

Source: AASHTO LRFD Bridge Design Specifications, 2nd. ed., American Association of State Highway and Transportation Officials. Washington, D.C., 1998. With permission.

If a permit vehicle is also being designed for, then these three cases must also be checked for the load combination associated with Strength Limit State II (discussed in Chapter 5).

6.3.6   Multiple Presence of Live-Load Lanes

Multiple presence factors modify the vehicular live loads for the probability that vehicular live loads occur together in a fully loaded state. The factors are shown in Table 6.3.

These factors should be applied prior to analysis or design only when using the lever rule or doing three-dimensional modeling or working with substructures. Sidewalks greater than 600 mm can be treated as a fully loaded lane. If a two-dimensional girder line analysis is being done and distribution factors are being used for a beam-and-slab type of bridge, multiple presence factors are not used because the load distribution factors already consider three-dimensional effects. For the fatigue limit state, the multiple presence factors are also not used.

6.3.7   Dynamic Load Allowance

Vehicular live loads are assigned a "dynamic load allowance" load factor of 1.75 at deck joints, 1.15 for all other components in the fatigue and fracture limit state, and 1.33 for all other components and limit states. This factor accounts for hammering when riding surface discontinuities exist, and long undulations when settlement or resonant excitation occurs. If a component such as a footing is completely below grade or a component such as a retaining wall is not subject to vertical reactions from the superstructure, this increase is not taken. Wood bridges or any wood component is factored at a lower level, i.e., 1.375 for deck joints, 1.075 for fatigue, and 1.165 typical, because of the energy-absorbing characteristic of wood. Likewise, buried structures such as culverts are subject to the dynamic load allowance but are a function of depth of cover, D_E (mm):

$$IM = 40(1.0 - 4.1 \times 10^{-4} D_E) \geq 0\% \qquad (6.3)$$

6.3.8 Horizontal Loads Due to Vehicular Traffic

Substructure design of vertical elements requires that horizontal effects of vehicular live loads be designed for. Centrifugal forces and braking effects are applied horizontally at a distance 1.80 m above the roadway surface. The centrifugal force is determined by multiplying the design truck or design tandem — alone — by the following factor:

$$C = \frac{4v^2}{3gR} \qquad (6.4)$$

Highway design speed, v, is in m/s; gravitational acceleration, g, is 9.807 m/s²; and radius of curvature in traffic lane, R, is in m. Likewise, the braking force is determined by multiplying the design truck or design tandem from all lanes likely to be unidirectional in the future, by 0.25. In this case, the lane load is not used because braking effects would be damped out on a fully loaded lane.

6.4 Pedestrian Loads

Live loads also include pedestrians and bicycles. The LRFD Specification calls for a 3.6×10^{-3} MPa load simultaneous with highway loads on sidewalks wider than 0.6 m. "Pedestrian- or bicycle-only" bridges are to be designed for 4.1×10^{-3} MPa. If the pedestrian- or bicycle-only bridge is required to carry maintenance or emergency vehicles, these vehicles are designed for, omitting the dynamic load allowance. Loads due to these vehicles are infrequent and factoring up for dynamic loads is inappropriate.

TABLE 6.4 Base Wind Pressures, P_B, corresponding to $V_B = 160$ km/h

Structural Component	Windward Load, MPa	Leeward Load, MPa
Trusses, columns, and arches	0.0024	0.0012
Beams	0.0024	NA
Large flat surfaces	0.0019	NA

Source: AASHTO LRFD Bridge Design Specifications, 2nd. ed., American Association of State Highway and Transportation Officials. Washington, D.C., 1998. With permission.

6.5 Wind Loads

The LRFD Specification provides wind loads as a function of base design wind velocity, V_B equal to 100 mph; and base pressures, P_B, corresponding to wind speed V_B. Values for P_B are listed in Table 6.4. The design wind pressure, P_D, is then calculated as

$$P_D = P_B \left(\frac{V_{DZ}}{V_B} \right)^2 = P_B \frac{V_{DZ}^2}{25,600} \qquad (6.5)$$

where V_{DZ} is the design wind velocity at design elevation Z in km/h. V_{DZ} is a function of the friction velocity, V_0 (km/h), multiplied by the ratio of the actual wind velocity to the base wind velocity both at 10 m above grade, and the natural logarithm of the ratio of height to a meteorological constant length for given surface conditions:

TABLE 6.5 Values of V_o and Z_o for Various Upstream Surface Conditions

Condition	Open Country	Suburban	City
V_o (km/h)	13.2	15.2	19.4
Z_o (mm)	70	300	800

Source: AASHTO LRFD Bridge Design Specifications, 2nd. ed., American Association of State Highway and Transportation Officials. Washington, D.C., 1998. With permission.

TABLE 6.6 Temperature Ranges, °C

Climate	Steel or Aluminum	Concrete	Wood
Moderate	−18 to 50	−12 to 27	−12 to 24
Cold	−35 to 50	−18 to 27	−18 to 24

Source: AASHTO LRFD Bridge Design Specifications, 2nd. ed., American Association of State Highway and Transportation Officials. Washington, D.C., 1998. With permission.

$$V_{DZ} = 2.5V_0\left(\frac{V_{10}}{V_B}\right)\ln\left(\frac{Z}{Z_0}\right) \tag{6.6}$$

Values for V_0 and Z_0 are shown in Table 6.5.

The resultant design pressure is then applied to the surface area of the superstructure as seen in elevation. Solid-type traffic barriers and sound walls are considered as part of the loading surface. If the product of the resultant design pressure and applicable loading surface depth is less than a lineal load of 4.4 N/mm on the windward chord, or 2.2 N/mm on the leeward chord, minimum loads of 4.4 and 2.2 N/mm, respectively, are designed for.

Wind loads are combined with other loads in Strength Limit States III and V, and Service Limit State I, as defined in Chapter 5. Wind forces due to the additional surface area from trucks is accounted for by applying a 1.46 N/mm load 1800 mm above the bridge deck.

Wind loads for substructure design are of two types: loads applied to the substructure and those applied to the superstructure and transmitted to the substructure. Loads applied to the superstructure are as previously described. A base wind pressure of 1.9×10^{-3} MPa force is applied directly to the substructure, and is resolved into components (perpendicular to the front and end elevations) when the structure is skewed.

In absence of live loads, an upward load of 9.6×10^{-4} MPa is multiplied by the width of the superstructure and applied at the windward quarter point simultaneously with the horizontal wind loads applied perpendicular to the length of the bridge. This uplift load may create a worst condition for substructure design when seismic loads are not of concern.

6.6 Effects Due to Superimposed Deformations

Elements of a structure may change size or position due to settlement, shrinkage, creep, or temperature. Changes in geometry cause additional stresses which are of particular concern at connections. Determining effects from foundation settlement are a matter of structural analysis. Effects due to shrinkage and creep are material dependent and the reader is referred to design chapters elsewhere

TABLE 6.7 Basis of Temperature Gradients

Zone	Concrete		50 mm Asphalt		100 mm Asphalt	
	T_1 (°C)	T_2 (°C)	T_1 (°C)	T_2 (°C)	T_1 (°C)	T_2 (°C)
1	30	7.8	24	7.8	17	5
2	25	6.7	20	6.7	14	5.5
3	23	6	18	6	13	6
4	21	5	16	5	12	6

Source: AASHTO LRFD Bridge Design Specifications, 2nd. ed., American Association of State Highway and Transportation Officials. Washington, D.C., 1998. With permission.

FIGURE 6.5 Solar radiation zones for the United States. (AASHTO LRFD Bridge Design Specifications 2nd. ed., American Association of State Highway and Transportation Officials. Washington, D.C., 1998. With permission.)

in this book. Temperature effects are dependent on the maximum potential temperature differential from the temperature at time of erection. Upper and lower bounds are shown in Table 6.6, where "moderate" and "cold" climates are defined as having fewer or more than 14 days with an average temperature below 0°C, respectively.

By using appropriate coefficients of thermal expansion, effects from temperature changes are calculated using basic structural analysis. More-refined analysis will consider the time lag between the surface and internal structure temperatures. The LRFD Specification identifies four zones in the United States and provides a linear relationship for the temperature gradient in steel and concrete. See Table 6.7 and Figures 6.5 and 6.6.

6.7 Exceptions to Code-Specified Design Loads

The designer is responsible not only for providing plans that accommodate design loads per the referenced Design Specifications, but also for any loads unique to the structure and bridge site. It is also the designer's responsibility to indicate all loading conditions designed for in the contract documents — preferably the construction plans. History seems to indicate that the next generation of bridge engineers will indeed be given the task of "improving" today's new structure. Therefore, the safety of future generations depends on today's designers doing a good job of documentation.

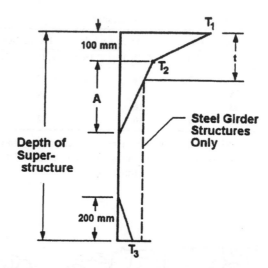

FIGURE 6.6 Positive vertical temperature gradient in concrete and steel superstructures. (AASHTO LRFD Bridge Design Specifications 2nd. ed., American Association of State Highway and Transportation Officials. Washington, D.C., 1998. With permission.)

References

1. AASHTO, LRFD Bridge Design Specifications 2nd. ed., American Association of State Highway and Transportation Officials. Washington, D.C., 1998.
2. FHWA Training Course, LRFD Design of Highway Bridges, Vol. 1, 1993.
3. AASHTO, American Association of State Highway and Transportation Officials, Washington, D.C. Officials, *Standard Specifications for Highway Bridges*, 16th ed., 1996, as amended by the Interim Specifications.
4. Caltrans, Bridge Design Specifications, California Department of Transportation, Sacramento, CA, 1999.

7
Structural Theory

Xila Liu
Tsinghua University, China

Leiming Zhang
Tsinghua University, China

7.1 Introduction

In this chapter, general forms of three sets of equations required in solving a solid mechanics problem and their extensions into structural theory are presented. In particular, a more generally used method, displacement method, is expressed in detail.

7.1.1 Basic Equations: Equilibrium, Compatibility, and Constitutive Law

In general, solving a solid mechanics problem must satisfy equations of equilibrium (static or dynamic), conditions of compatibility between strains and displacements, and stress–strain relations or material constitutive law (see Figure 7.1). The initial and boundary conditions on forces and displacements are naturally included.

From consideration of equilibrium equations, one can relate the stresses inside a body to external excitations, including body and surface forces. There are three equations of equilibrium relating the

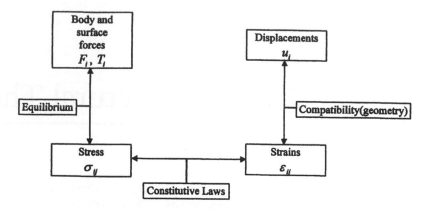

FIGURE 7.1   Relations of variables in solving a solid mechanics problem.

six components of stress tensor σ_{ij} for an infinitesimal material element which will be shown later in Section 7.2.1. In the case of dynamics, the equilibrium equations are replaced by equations of motion, which contain second-order derivatives of displacement with respect to time.

In the same way, taking into account geometric conditions, one can relate strains inside a body to its displacements, by six equations of kinematics expressing the six components of strain (ε_{ij}) in terms of the three components of displacement (u_i). These are known as the strain–displacement relations (see Section 7.3.1).

Both the equations of equilibrium and kinematics are valid regardless of the specific material of which the body is made. The influence of the material is expressed by constitutive laws in six equations. In the simplest case, not considering the effects of temperature, time, loading rates, and loading paths, these can be described by relations between stress and strain only.

Six stress components, six strain components, and three displacement components are connected by three equilibrium equations, six kinematics equations, and six constitutive equations. The 15 unknown quantities can be determined from the system of 15 equations.

It should be pointed out that the principle of superposition is valid only when small deformations and elastic materials are assumed.

7.1.2   Three Levels: Continuous Mechanics, Finite–Element Method, Beam–Column Theory

In solving a solid mechanics problem, the most direct method solves the three sets of equations described in the previous section. Generally, there are three ways to establish the basic unknowns, namely, the displacement components, the stress components, or a combination of both. The corresponding procedures are called the displacement method, the stress method, or the mixed method, respectively. But these direct methods are only practicable in some simple circumstances, such as those detailed in elastic theory of solid mechanics.

Many complex problems cannot be easily solved with conventional procedures. Complexities arise due to factors such as irregular geometry, nonhomogeneities, nonlinearity, and arbitrary loading conditions. An alternative now available is based on a concept of discretization. The finite-element method (FEM) divides a body into many "small" bodies called finite elements. Formulations by the FEM on the laws and principles governing the behavior of the body usually result in a set of simultaneous equations that can be solved by direct or iterative procedures. And loading effects such as deformations and stresses can be evaluated within certain accuracy. Up to now, FEM has been the most widely used structural analysis method.

In dealing with a continuous beam, the size of the three sets of equations is greatly reduced by assuming characteristics of beam members such as plane sections remain plane. For framed structures or structures constructed using beam–columns, structural mechanics gives them a more pithy and practical analysis.

7.1.3 Theoretical Structural Mechanics, Computational Structural Mechanics, and Qualitative Structural Mechanics

Structural mechanics deals with a system of members connected by joints which may be pinned or rigid. Classical methods of structural analysis are based on principles such as the principle of virtual displacement, the minimization of total potential energy, the minimization of total complementary energy, which result in the three sets of governing equations. Unfortunately, conventional methods are generally intended for hand calculations and developers of the FEM took great pains to minimize the amount of calculations required, even at the expense of making the methods somewhat unsystematic. This made the conventional methods unattractive for translation to computer codes.

The digital computer called for a more systematic method of structural analysis, leading to computational structural mechanics. By taking great care to formulate the tools of matrix notation in a mathematically consistent fashion, the analyst achieved a systematic approach convenient for automatic computation: matrix analysis of structures. One of the hallmarks of structural matrix analysis is its systematic nature, which renders digital computers even more important in structural engineering.

Of course, the analyst must maintain a critical, even skeptical, attitude toward computer results. In any event, computer results must satisfy our intuition of what is "reasonable." This qualitative judgment requires that the analyst possess a full understanding of structural behavior, both that being modeled by the program and that which can be expected in the actual structures. Engineers should decide what approximations are reasonable for the particular structure and verify that these approximations are indeed valid, and know how to design the structure so that its behavior is in reasonable agreement with the model adopted to analyze it. This is the main task of a structural analyst.

7.1.4 Matrix Analysis of Structures: Force Method and Displacement Method

Matrix analysis of structures was developed in the early 1950s. Although it was initially used on fuselage analysis, this method was proved to be pertinent to any complex structure. If internal forces are selected as basic unknowns, the analysis method is referred to as force method; in a similar way, the displacement method refers to the case where displacements are selected as primary unknowns. Both methods involve obtaining the joint equilibrium equations in terms of the basic internal forces or joint displacements as primary unknowns and solving the resulting set of equations for these unknowns. Having done this, one can obtain internal forces by backsubstitution, since even in the case of the displacement method the joint displacements determine the basic displacements of each member, which are directly related to internal forces and stresses in the member.

A major feature evident in structural matrix analysis is an emphasis on a systematic approach to the statement of the problem. This systematic characteristic together with matrix notation makes it especially convenient for computer coding. In fact, the displacement method, whose basic unknowns are uniquely defined, is generally more convenient than the force method. Most general-purpose structural analysis programs are displacement based. But there are still cases where it may be more desirable to use the force method.

7.2 Equilibrium Equations

7.2.1 Equilibrium Equation and Virtual Work Equation

For any volume V of a material body having A as surface area, as shown in Figure 7.2, it has the following conditions of equilibrium:

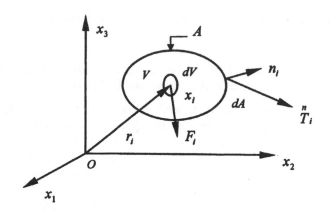

FIGURE 7.2   Derivation of equations of equilibrium.

At surface points

$$T_i = \sigma_{ji} n_j \qquad (7.1a)$$

At internal points

$$\sigma_{ji,j} + F_i = 0 \qquad (7.1b)$$

$$\sigma_{ji} = \sigma_{ij} \qquad (7.1c)$$

where n_i represents the components of unit normal vector **n** of the surface; T_i is the stress vector at the point associated with **n**; $\sigma_{ji,j}$ represents the first derivative of σ_{ij} with respect to x_j; and F_i is the body force intensity. Any set of stresses σ_{ij}, body forces F_i, and external surface forces T_i that satisfies Eqs. (7.1a-c) is a statically admissible set.

Equations (7.1b and c) may be written in (x,y,z) notation as

$$\frac{\partial \sigma_x}{\partial x} + \frac{\partial \tau_{xy}}{\partial y} + \frac{\partial \tau_{xz}}{\partial z} + F_x = 0$$

$$\frac{\partial \tau_{yx}}{\partial x} + \frac{\partial \sigma_y}{\partial y} + \frac{\partial \tau_{yz}}{\partial z} + F_y = 0 \qquad (7.1d)$$

$$\frac{\partial \tau_{zx}}{\partial x} + \frac{\partial \tau_{zy}}{\partial y} + \frac{\partial \sigma_z}{\partial z} + F_z = 0$$

and

$$\tau_{xy} = \tau_{yx}, \quad \text{etc.} \qquad (7.1e)$$

where σ_x, σ_y, and σ_z are the normal stress in (x,y,z) direction respectively; τ_{xy}, τ_{yx}, and so on, are the corresponding shear stresses in (x,y,z) notation; and F_x, F_y, and F_z are the body forces in (x,y,z) direction, respectively.

The principle of virtual work has proved a very powerful technique of solving problems and providing proofs for general theorems in solid mechanics. The equation of virtual work uses two

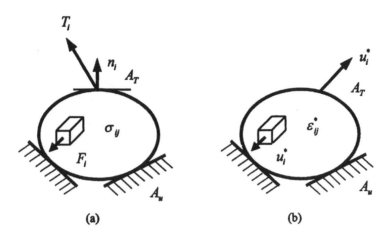

FIGURE 7.3 Two independent sets in the equation of virtual work.

independent sets of *equilibrium* and *compatible* (see Figure 7.3, where A_u and A_T represent displacement and stress boundary, respectively), as follows:

compatible set

$$\int_A T_i u_i^* dA + \int_V F_i u_i^* dV = \int_V \sigma_{ij} \varepsilon_{ij}^* dV \tag{7.2}$$

equilibrium set

or

$$\delta W_{\text{ext}} = \delta W_{\text{int}} \tag{7.3}$$

which states that the *external* virtual work (δW_{ext}) equals the *internal* virtual work (δW_{int}).

Here the integration is over the whole area A, or volume V, of the body. The stress field σ_{ij}, body forces F_i, and external surface forces T_i are a statically admissible set that satisfies Eqs. (7.1a–c). Similarly, the strain field ε_{ij}^* and the displacement u_i^* are a compatible kinematics set that satisfies displacement boundary conditions and Eq. (7.16) (see Section 7.3.1). This means the principle of virtual work applies only to small strain or small deformation.

The important point to keep in mind is that, neither the admissible equilibrium set σ_{ij}, F_i, and T_i (Figure 7.3a) nor the compatible set ε_{ij}^* and u_i^* (Figure 7.3b) need be the actual state, nor need the equilibrium and compatible sets be related to each other in any way. In the other words, these two sets are completely independent of each other.

7.2.2 Equilibrium Equation for Elements

For an infinitesimal material element, equilibrium equations have been summarized in Section 7.2.1, which will transfer into specific expressions in different methods. As in ordinary FEM or the displacement method, it will result in the following element equilibrium equations:

$$\{\overline{F}\}^e = [\overline{k}]^e \{\overline{d}\}^e \tag{7.4}$$

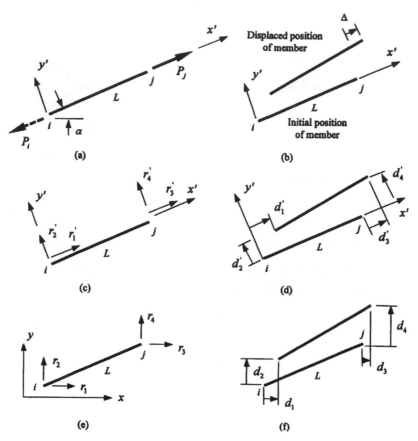

FIGURE 7.4 Plane truss member–end forces and displacements. (*Source*: Meyers, V.J., *Matrix Analysis of Structures*, New York: Harper & Row, 1983. With permission.)

where $\{\overline{F}\}^e$ and $\{\overline{d}\}^e$ are the element nodal force vector and displacement vector, respectively, while $[\overline{k}]^e$ is element stiffness matrix; the overbar here means in local coordinate system.

In the force method of structural analysis, which also adopts the idea of discretization, it is proved possible to identify a basic set of independent forces associated with each member, in that not only are these forces independent of one another, but also all other forces in that member are directly dependent on this set. Thus, this set of forces constitutes the minimum set that is capable of completely defining the stressed state of the member. The relationship between basic and local forces may be obtained by enforcing overall equilibrium on one member, which gives

$$\{\overline{F}\}^e = [L]\{P\}^e \tag{7.5}$$

where $[L]$ = the element force transformation matrix and $\{P\}^e$ = the element primary forces vector. It is important to emphasize that the physical basis of Eq. (7.5) is member overall equilibrium.

Take a conventional plane truss member for exemplification (see Figure 7.4), one has

$$\{\overline{k}\}^e = \begin{bmatrix} EA/l & 0 & -EA/l & 0 \\ 0 & 0 & 0 & 0 \\ -EA/l & 0 & EA/l & 0 \\ 0 & 0 & 0 & 0 \end{bmatrix} \tag{7.6}$$

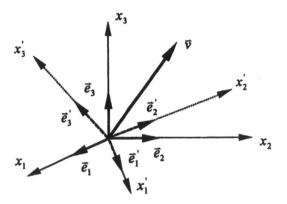

FIGURE 7.5 Coordinate transformation.

and

$$\{\overline{F}\}^e = \{r_1' \quad r_2' \quad r_3' \quad r_4'\}^T$$

$$\{\overline{d}\}^e = \{d_1' \quad d_2' \quad d_3' \quad d_4'\}^T \tag{7.7}$$

$$[L] = \{-1 \quad 0 \quad 1 \quad 0\}^T$$

$$\{P\}^e = \{P\}$$

where EA/l = axial stiffness of the truss member and P = axial force of the truss member.

7.2.3 Coordinate Transformation

The values of the components of vector **V**, designated by v_1, v_2, and v_3 or simply v_i, are associated with the chosen set coordinate axes. Often it is necessary to reorient the reference axes and evaluate new values for the components of **V** in the new coordinate system. Assuming that **V** has components v_i and v_i' in two sets of right-handed Cartesian coordinate systems x_i (old) and x_i' (new) having the same origin (see Figure 7.5), and $\vec{e}_i$, $\vec{e}_i'$ are the unit vectors of x_i and x_i', respectively. Then

$$v_i' = l_{ij} v_j \tag{7.8}$$

where $l_{ji} = \vec{e}_j' \cdot \vec{e}_i = \cos(x_j', x_i)$, that is, the cosines of the angles between x_i' and x_j axes for i and j ranging from 1 to 3; and $[\alpha] = (l_{ij})_{3\times3}$ is called coordinate transformation matrix from the old system to the new system.

It should be noted that the elements of l_{ij} or matrix $[\alpha]$ are not symmetrical, $l_{ij} \neq l_{ji}$. For example, l_{12} is the cosine of angle from x_1' to x_2 and l_{21} is that from x_2' to x_1 (see Figure 7.5). The angle is assumed to be measured from the primed system to the unprimed system.

For a plane truss member (see Figure 7.4), the transformation matrix from local coordinate system to global coordinate system may be expressed as

$$[\alpha] = \begin{bmatrix} \cos\alpha & -\sin\alpha & 0 & 0 \\ \sin\alpha & \cos\alpha & 0 & 0 \\ 0 & 0 & \cos\alpha & -\sin\alpha \\ 0 & 0 & \sin\alpha & \cos\alpha \end{bmatrix} \tag{7.9}$$

where α is the inclined angle of the truss member which is assumed to be measured from the global to the local coordinate system.

7.2.4 Equilibrium Equation for Structures

For discretized structure, the equilibrium of the whole structure is essentially the equilibrium of each joint. After assemblage,

For ordinary FEM or displacement method

$$\{F\} = [K]\{D\} \tag{7.10}$$

For force method

$$\{F\} = [A]\{P\} \tag{7.11}$$

where $\{F\}$ = nodal loading vector; $[K]$ = total stiffness matrix; $\{D\}$ = nodal displacement vector; $[A]$ = total forces transformation matrix; $\{P\}$ = total primary internal forces vector.

It should be noted that the coordinate transformation for each element from local coordinates to the global coordinate system must be done before assembly.

In the force method, Eq. (7.11) will be adopted to solve for internal forces of a statically determinate structure. The number of basic unknown forces is equal to the number of equilibrium equations available to solve for them and the equations are linearly independent. For statically unstable structures, analysis must consider their dynamic behavior. When the number of basic unknown forces exceeds the number of equilibrium equations, the structure is said to be statically indeterminate. In this case, some of the basic unknown forces are not required to maintain structural equilibrium. These are "extra" or "redundant" forces. To obtain a solution for the full set of basic unknown forces, it is necessary to augment the set of independent equilibrium equations with elastic behavior of the structure, namely, the force–displacement relations of the structure. Having solved for the full set of basic forces, we can determine the displacements by backsubstitution.

7.2.5 Influence Lines and Surfaces

In the design and analysis of bridge structures , it is necessary to study the effects intrigued by loads placed in various positions. This can be done conveniently by means of diagrams showing the effect of moving a unit load across the structures. Such diagrams are commonly called influence lines (for framed structures) or influence surfaces (for plates). Observe that whereas a moment or shear diagram shows the variation in moment or shear along the structure due to some particular position of load, an influence line or surface for moment or shear shows the variation of moment or shear at a *particular* section due to a unit load placed anywhere along the structure.

Exact influence lines for statically determinate structures can be obtained analytically by statics alone. From Eq. (7.11), the total primary internal forces vector $\{P\}$ can be expressed as

$$\{P\} = [A]^{-1}\{F\} \tag{7.12}$$

by which given a unit load at one node, the excited internal forces of all members will be obtained, and thus Eq. (7.12) gives the analytical expression of influence lines of all member internal forces for discretized structures subjected to moving nodal loads.

For statically indeterminate structures, influence values can be determined directly from a consideration of the geometry of the deflected load line resulting from imposing a unit deformation corresponding to the function under study, based on the principle of virtual work. This may better be demonstrated by a two-span continuous beam shown in Figure 7.6, where the influence line of internal bending moment M_B at section B is required.

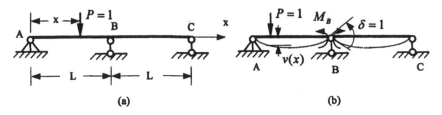

FIGURE 7.6   Influence line of a two-span continuous beam.

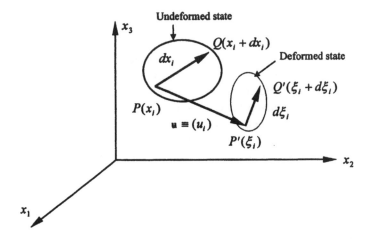

FIGURE 7.7   Deformation of a line element for Lagrangian and Eluerian variables.

Cutting section B to expose M_B and give it a unit relative rotation $\delta = 1$ (see Figure 7.6) and employing the principle of virtual work gives

$$M_B \cdot \delta = -P \cdot v(x) \tag{7.13}$$

Therefore,

$$M_B = -v(x) \tag{7.14}$$

which means the influence value of M_B equals to the deflection $v(x)$ of the beam subjected to a unit rotation at joint B (represented by dashed line in Figure 7.6b). Solving for $v(x)$ can be carried out easily referring to material mechanics.

7.3   Compatibility Equations

7.3.1   Large Deformation and Large Strain

Strain analysis is concerned with the study of deformation of a continuous body which is unrelated to properties of the body material. In general, there are two methods of describing the deformation of a continuous body, Lagrangian and Eulerian. The Lagrangian method employs the coordinates of each particle in the initial position as the independent variables. The Eulerian method defines the independent variables as the coordinates of each material particle at the time of interest.

Let the coordinates of material particle P in a body in the initial position be denoted by x_i (x_1, x_2, x_3) referred to the fixed axes x_i, as shown in Figure 7.7. And the coordinates of the particle after deformation are denoted by ξ_i (ξ_1, ξ_2, ξ_3) with respect to axes x_i. As for the independent variables, Lagrangian formulation uses the coordinates (x_i) while Eulerian formulation employs the coordinates (ξ_i). From motion analysis of line element PQ (see Figure 7.7), one has

For Lagrangian formulation, the Lagrangian strain tensor is

$$\varepsilon_{ij} = \frac{1}{2} \left(u_{i,j} + u_{j,i} + u_{r,i} u_{r,j} \right) \tag{7.15}$$

where $u_{i,j} = \partial u_i / \partial x_j$ and all quantities are expressed in terms of (x_i).

For Eulerian formulation, the Eulerian strain tensor is

$$E_{ij} = \frac{1}{2} \left(u_{i/j} + u_{j/i} + u_{r/i} u_{r/j} \right) \tag{7.16}$$

where $u_{i/j} = \partial u_i / \partial \xi_j$ and all quantities are described in terms of (ξ_i).

If the displacement derivatives $u_{i,j}$ and $u_{i/j}$ are not so small that their nonlinear terms cannot be neglected, it is called large deformation, and the solving of u_i will be rather difficult since the nonlinear terms appear in the governing equations.

If both the displacements and their derivatives are small, it is immaterial whether the derivatives in Eqs. (7.15) and (7.16) are calculated using the (x_i) or the (ξ_i) variables. In this case both Lagrangian and Eulerian descriptions yield the same strain–displacement relationship:

$$\varepsilon_{ij} = E_{ij} = \frac{1}{2} \left(u_{i,j} + u_{j,i} \right) \tag{7.17}$$

which means small deformation, the most common in structural engineering.

For given displacements (u_i) in strain analysis, the strain components (ε_{ij}) can be determined from Eq. (7.17). For prescribed strain components (ε_{ij}), some restrictions must be imposed on it in order to have single-valued continuous displacement functions u_i, since there are six equations for three unknown functions. Such restrictions are called compatibility conditions, which for a simply connected region may be written as

$$\varepsilon_{ij,kl} + \varepsilon_{kl,ij} - \varepsilon_{ik,jl} - \varepsilon_{ji,ik} = 0 \tag{7.18a}$$

or, expanding these expressions in the (x, y, z) notations, it gives

$$\frac{\partial^2 \varepsilon_x}{\partial y^2} + \frac{\partial^2 \varepsilon_y}{\partial x^2} = 2 \frac{\partial^2 \varepsilon_{xy}}{\partial x \partial y}$$

$$\frac{\partial^2 \varepsilon_y}{\partial z^2} + \frac{\partial^2 \varepsilon_z}{\partial y^2} = 2 \frac{\partial^2 \varepsilon_{yz}}{\partial y \partial z}$$

$$\frac{\partial^2 \varepsilon_z}{\partial x^2} + \frac{\partial^2 \varepsilon_x}{\partial z^2} = 2 \frac{\partial^2 \varepsilon_{zx}}{\partial z \partial x}$$

$$\frac{\partial}{\partial x} \left(-\frac{\partial \varepsilon_{yz}}{\partial x} + \frac{\partial \varepsilon_{zx}}{\partial y} + \frac{\partial \varepsilon_{xy}}{\partial z} \right) = \frac{\partial^2 \varepsilon_x}{\partial y \partial z} \tag{7.18b}$$

$$\frac{\partial}{\partial y} \left(-\frac{\partial \varepsilon_{zx}}{\partial y} + \frac{\partial \varepsilon_{xy}}{\partial z} + \frac{\partial \varepsilon_{yz}}{\partial x} \right) = \frac{\partial^2 \varepsilon_y}{\partial z \partial x}$$

$$\frac{\partial}{\partial z} \left(-\frac{\partial \varepsilon_{xy}}{\partial z} + \frac{\partial \varepsilon_{yz}}{\partial x} + \frac{\partial \varepsilon_{zx}}{\partial y} \right) = \frac{\partial^2 \varepsilon_z}{\partial x \partial y}$$

Any set of strains ε_{ij} and displacements u_i, that satisfies Eqs. (7.17) and (7.18a) or (7.18b), as well as displacement boundary conditions, is a kinematics admissible set, or a compatible set.

7.3.2 Compatibility Equation for Elements

For ordinary FEM, compatibility requirements are self-satisfied in the formulating procedure. As for equilibrium equations, a basic set of independent displacements can be identified for each member, and the kinematics relationships between member basic displacements and member–end displacements of one member can be given as follows:

$$\{\Delta\}^e = [L]^T \{\bar{d}\}^e \tag{7.19}$$

where $\{\Delta\}^e$ is element primary displacement vector, $[L]$ and $\{\bar{d}\}^e$ have been shown in Section 7.2.2. For plane truss member, $\{\Delta\}^e = \{\Delta\}$, where Δ is the relative displacement of the member (see Figure 7.5). It should also be noted that the physical basis of Eq. (7.19) is the overall compatibility of the element.

7.3.3 Compatibility Equation for Structures

For the whole structure, one has the following equation after assembly process:

$$\{\Delta\} = [A]^T \{D\} \tag{7.20}$$

where $\{\Delta\}$ = total primary displacement vector; $\{D\}$ = total nodal displacement vector; and $[A]^T$ = the transposition of $[A]$ described in Section 7.2.4.

A statically determinate structure is kinematically determinate. Given a set of basic member displacements, there are a sufficient number of compatibility relationships available to allow the structure nodal displacements to be determined. In addition to their application to settlement and fabrication error loading, thermal loads can also be considered for statically determinate structures. External forces on a structure cause member distortions and, hence, nodal displacements, but before such problems can be solved, the relationships between member forces and member distortions must be developed. These will be shown in Section 7.5.1.

7.3.4 Contragredient Law

During the development of the equilibrium and compatibility relationships, it has been noticed that various corresponding force and displacement transformations are the transposition of each other, as shown not only in Eqs. (7.5) and (7.19) of element equilibrium and compatibility relations, but also in Eqs. (7.11) and (7.20) of global equilibrium and compatibility relations, although each pair of these transformations was obtained independently of the other in the development. These special sets of relations are termed the contragredient law which was established on the basis of virtual work concepts. Therefore, after a particular force transformation matrix is obtained, the corresponding displacement transformation matrix would be immediately apparent, and it remains valid to the contrary.

7.4 Constitutive Equations

7.4.1 Elasticity and Plasticity

A material body will produce deformation when subjected to external excitations. If upon the release of applied actions the body recovers its original shape and size, it is called an *elastic* material, or

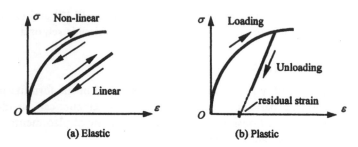

FIGURE 7.8　Sketches of behavior of elastic and plastic materials.

one can say the material has the characteristic of *elasticity*. Otherwise, it is a *plastic* material or a material with *plasticity*. For an elastic body, the current state of stress depends only on the current state of deformation; that is, the constitutive equations for elastic material are given by

$$\sigma_{ij} = F_{ij}(\varepsilon_{kl}) \tag{7.21}$$

where F_{ij} is called the elastic response function. Thus, the elastic material behavior described by Eq. (7.21) is reversible and path independent (see Figure 7.8a), in which case the material is usually termed *Cauchy elastic* material.

Reversibility and path independence are not exhibited by plastic materials (see Figure 7.8b). In general, a plastic material does not return to its original shape; *residual* deformation and stresses remain inside the body even when all external tractions are removed. As a result, it is necessary for plasticity to extend the elastic stress–strain relations into the plastic range where permanent plastic stain is possible. It makes the solution of a solid mechanics problem more complicated.

7.4.2　Linear Elastic and Nonlinear Elastic Behavior

Just as the term *linear* implies, linear elasticity means the elastic response function F_{ij} of Eq. (7.21) is a linear function, whose most general form for a Cauchy elastic material is given by

$$\sigma_{ij} = B_{ij} + C_{ijkl}\varepsilon_{kl} \tag{7.22}$$

where B_{ij} = components of initial stress tensor corresponding to the *initial strain-free* state (i.e., $\varepsilon_{ij} = 0$), and C_{ijkl} = tensor of material elastic constants.

If it is assumed that $B_{ij} = 0$, Eq. (7.22) will be reduced to

$$\sigma_{ij} = C_{ijkl}\varepsilon_{kl} \tag{7.23}$$

which is often referred to as the generalized Hook's law.

For an *isotropic* linear elastic material, the elastic constants in Eq. (7.23) must be the same for all directions and thus C_{ijkl} must be an isotropic fourth-order tensor, which means that there are only two independent material constants. In this case, Eq. (7.23) will reduce to

$$\sigma_{ij} = \lambda\varepsilon_{kk}\delta_{ij} + 2\mu\varepsilon_{ij} \tag{7.24}$$

where λ and μ are the two material constants, usually called *Lame's constants*; δ_{ij} = *Kronecker delta* and ε_{kk} = the summation of the diagonal terms of ε_{ij} according to the *summation convention*, which means that, whenever a subscript occurs twice in the same term, it is understood that the subscript is to be summed from 1 to 3.

If the elastic response function F_{ij} in Eq. (7.21) is not linear, it is called nonlinear elastic, and the material exhibits nonlinear mechanical behavior even when sustaining small deformation. That is, the material elastic "constants" do not remain constant any more, whereas the deformation can still be reversed completely.

7.4.3 Geometric Nonlinearity

Based on the sources from which it arises, nonlinearity can be categorized into material nonlinearity (including nonlinear elasticity and plasticity) and geometric nonlinearity. When the nonlinear terms in the strain–displacement relations cannot be neglected (see Section 7.3.1) or the deflections are large enough to cause significant changes in the structural geometry, it is termed geometric non-linearity. It is also called large deformation, and the principle of superposition derived from small deformations is no longer valid. It should be noted that for accumulated large displacements with small deformations, it could be linearized by a step-by-step procedure.

According to the different choice of reference frame, there are two types of Lagrangian formulation: the total Lagrangian formulation, which takes the original unstrained configuration as the reference frame, and the updated Lagrangian formulation based on the latest-obtained configuration, which are usually carried out step by step. Whatever formulation one chooses, a geometric stiffness matrix or initial stress matrix will be introduced into the equations of equilibrium to take account of the effects of the initial stresses on the stiffness of the structure. These depend on the magnitude or conditions of loading and deformations, and thus cause the geometric nonlinearity. In beam–column theory, this is well known as the second-order or the P–Δ effect. For detailed discussions, see Chapter 36.

7.5 Displacement Method

7.5.1 Stiffness matrix for elements

In displacement method, displacement components are taken as primary unknowns. From Eqs. (7.5) and (7.19) the equilibrium and compatibility requirements on elements have been acquired. For a statically determinate structure, no subsidiary conditions are needed to obtain internal forces under nodal loading or the displaced position of the structure given the basic distortion such as support settlement or fabrication errors. For a statically indeterminate structure, however, supplementary conditions, namely, the constitutive law of materials constructing the structure, should be incorporated for the solution of internal forces as well as nodal displacements.

From structural mechanics, the basic stiffness relationships for a member between basic internal forces and basic member–end displacements can be expressed as

$$\{P\}^e = [k]^e \{\Delta\}^e \tag{7.25}$$

where $[k]^e$ is the element basic stiffness matrix, which can be termed $[EA/l]$ for a conventional plane truss member (see Figure 7.4).

Substitution of Eqs. (7.19) and (7.25) into Eq. (7.5) yields

$$\{\overline{F}\}^e = [L][k]^e[L]^T\{\overline{d}\}^e$$
$$= [\overline{k}]^e\{\overline{d}\}^e \tag{7.26}$$

where

$$[\overline{k}]^e = [L][k]^e[L]^T \tag{7.27}$$

is called the element stiffness matrix, the same as in Eq. (7.4). It should be kept in mind that the element stiffness matrix $[\bar{k}]^e$ is symmetric and singular, since given the member–end forces, member–end displacements cannot be determined uniquely because the member may undergo rigid body movement.

7.5.2 Stiffness Matrix for Structures

Our final aim is to obtain equations that define approximately the behavior of the whole body or structure. Once the element stiffness relations of Eq. (7.26) is established for a generic element, the global equations can be constructed by an assembling process based on the law of compatibility and equilibrium, which are generally expressed in matrix notation as

$$\{F\} = [K]\{D\} \tag{7.28}$$

where $[K]$ is the stiffness matrix for the whole structure. It should be noted that the basic idea of assembly involves a minimization of *total* potential energy, and the assembled stiffness matrix $[K]$ is *symmetric* and *banded* or *sparsely populated*.

Eq. (7.28) tells us the capabilities of a structure to withstand applied loading rather than the true behavior of the structure if boundary conditions are not introduced. In other words, without boundary conditions, there can be an infinite number of possible solutions since stiffness matrix $[K]$ is singular; that is, its determinant vanishes. Hence, Eqs. (7.28) should be modified to reflect boundary conditions and the final modified equations are expressed by inserting overbars as

$$\{\bar{F}\} = [\bar{K}]\{\bar{D}\} \tag{7.29}$$

7.5.3 Matrix Inversion

It has been shown that sets of simultaneous algebraic equations are generated in the application of both the displacement method and the force method in structural analysis, which are usually linear. The coefficients of the equations are constant and do not depend on the magnitude or conditions of loading and deformations, since linear Hook's law is generally assumed valid and small strains and deformations are used in the formulation. Solving Eq. (7.29) is, namely, to invert the modified stiffness matrix $[\bar{K}]$. This requires tremendous computational efforts for large-scale problems. The equations can be solved by using direct, iterative, or other methods. Two steps of elimination and backsubstitution are involved in the direct procedures, among which are Gaussian elimination and a number of its modifications. These are some of the most widely used sets of direct methods because of their better accuracy and small number of arithmetic operations.

7.5.4 Special Consideration

In practice, a variety of special circumstances, ranging from loading to internal member conditions and supporting conditions, should be given due consideration in structural analysis.

Initially strains, which are not directly associated with stresses, result from two causes, thermal loading or fabrication error. If the member with initial strains is unconstrained, there will be a set of initial member–end displacements associated with these initial strains, but nevertheless no initial member–end forces. For a member constrained to act as part of a structure, the general member force–displacement relationships will be modified as follows:

$$\{\bar{F}\}^e = [\bar{k}]^e \left(\{\bar{d}\}^e - \{\bar{d_0}\}^e \right) \tag{7.30a}$$

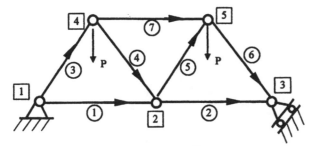

(a) Dimensions and Loading

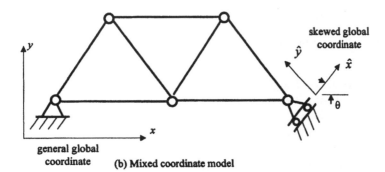

general global
coordinate   (b) Mixed coordinate model

FIGURE 7.9   Plane truss with skewed support.

or

$$\{\overline{F}\}^e = [\overline{k}]^e \{\overline{d}\}^e + \{R_{F0}\}^e \tag{7.30b}$$

where

$$\{R_{F0}\}^e = -[\overline{k}]^e \{\overline{d_0}\}^e \tag{7.31}$$

are fixed-end forces, and $\{\overline{d_0}\}^e$ a vector of initial member–end displacements for the member.

It is interesting to note that a support settlement may be regarded as an initial strain. Moreover, initial strains including thermal loading and fabrication errors, as well as support settlements, can all be treated as external excitations. Hence, the corresponding fixed-end forces as well as the equivalent nodal loading can be obtained which makes the conventional procedure described previously still practicable.

For a skewed support which provides a constraint to the structure in a nonglobal direction, the effect can be given due consideration by adapting a skewed global coordinate (see Figure 7.9) by introducing a skewed coordinate at the skewed support. This can perhaps be better demonstrated by considering a specific example of a plane truss shown in Figure 7.9. For members jointed at a skewed support, the coordinate transformation matrix will takes the form of

$$[\alpha] = \begin{bmatrix} \cos\alpha_i & -\sin\alpha_i & 0 & 0 \\ \sin\alpha_i & \cos\alpha_i & 0 & 0 \\ 0 & 0 & \cos\alpha_j & -\sin\alpha_j \\ 0 & 0 & \sin\alpha_j & \cos\alpha_j \end{bmatrix} \tag{7.32}$$

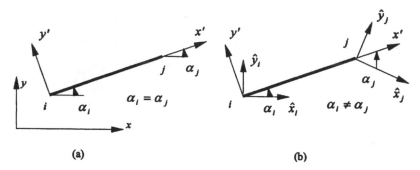

FIGURE 7.10 Plane truss member coordinate transformation. (a) Normal global coordinate; (b) skewed global coordinate.

where α_i and α_j are inclined angles of truss member in skewed global coordinate (see Figure 7.10), say, for member 2 in Figure 7.9, $\alpha_i = 0$ and $\alpha_j = -\theta$.

For other special members such as inextensional or variable cross section ones, it may be necessary or convenient to employ special member force–displacement relations in structural analysis. Although the development and programming of a stiffness method general enough to take into account all these special considerations is formidable, more important perhaps is that the application of the method remains little changed. For more details, readers are referred to Reference. [5].

7.6 Substructuring and Symmetry Consideration

For highly complex or large-scale structures, one is required to solve a very large set of simultaneous equations, which are sometimes restricted by the computation resources available. In that case, special data-handling schemes like static condensation are needed to reduce the number of unknowns by appropriately numbering nodal displacement components and disposition of element force–displacement relations. Static condensation is useful in dynamic analysis of framed structures since the rotatory moment of inertia is usually neglected.

Another scheme physically partitions the structure into a collection of smaller structures called "substructures," which can be processed by parallel computers. In static analysis, the first step of substructuring is to introduce imaginary fixed inner boundaries, and then release all inner boundaries simultaneously, which gives rise to a subsequent analysis of these substructure series in a smaller scale. It is essentially the patitioning of Eq. (7.28) as follows. For the rth substructure, one has

Case (α) : *Introducing inner fixed boundaries*

$$\begin{bmatrix} K_{bb} & K_{bi} \\ K_{ib} & K_{ii} \end{bmatrix}^{(r)} \begin{Bmatrix} 0 \\ D_i^\alpha \end{Bmatrix}^{(r)} = \begin{Bmatrix} F_b^\alpha \\ F_i \end{Bmatrix}^{(r)} \tag{7.33}$$

Case (β) : *Releasing all inner fixed boundaries*

$$\begin{bmatrix} K_{bb} & K_{bi} \\ K_{ib} & K_{ii} \end{bmatrix}^{(r)} \begin{Bmatrix} D_b \\ D_i^\beta \end{Bmatrix}^{(r)} = \begin{Bmatrix} F_b^\beta \\ 0 \end{Bmatrix}^{(r)} \tag{7.34}$$

where subscripts b and i denote inner fixed and free nodes, respectively.

Combining Eqs. (7.33) and (7.34) gives the force–displacement relations for enlarged elements — substructures which may be expressed as

$$\left[K_b\right]^{(r)}\left\{D_b\right\}^{(r)} = \left\{F_b\right\}^{(r)} \tag{7.35}$$

which is analogous to Eq. (7.26) and $\left\{F_b\right\}^{(r)} = \left\{F_b^{(r)}\right\} - \left[K_{bi}^{(r)}\right]\left[K_{ii}^{(r)}\right]^{-1}\left\{F_i^{(r)}\right\}$. And thereby the conventional procedure is still valid.

Similarly, in the cases of structural symmetry of geometry and material, proper consideration of loading symmetry and antisymmetry can give rise to a much smaller set of governing equations.

For more details, please refer to the literature on structural analysis.

References

1. Chen, W.F. and Saleeb, A.F., *Constitutive Eqs. for Engineering Materials,* Vols. 1 & 2, Elsevier Science Ltd., New York, 1994.
2. Chen, W.F., *Plasticity in Reinforced Concrete,* McGraw-Hill, New York, 1982.
3. Chen, W.F. and T. Atsuta, *Theory of Beam-Columns,* Vol. 1, *In-Plane Behavior and Design,* McGraw-Hill, New York, 1976.
4. Desai, C.S., *Elementary Finite Element Method,* Prentice-Hall, Englewood Cliffs, NJ, 1979.
5. Meyers, V.J., *Matrix Analysis of Structures,* Harper & Row, New York, 1983.
6. Michalos, J., *Theory of Structural Analysis and Design,* Ronald Press, New York, 1958.
7. Hjelmstad, K.D., *Fundamentals of Structural Mechanics,* Prentice-Hall College Div., Upper Saddle River, NJ, 1996.
8. Fleming, J.F., *Analysis of Structural Systems,* Prentice-Hall College Div., Upper Saddle River, NJ, 1996.
9. Dadeppo, D.A. *Introduction to Structural Mechanics and Analysis,* Prentice-Hall College Div., Upper Saddle River, NJ, 1998.

8

Structural Modeling

Alexander Krimotat
SC Solutions, Inc.

Li-Hong Sheng
California Department of Transportation

8.1   Introduction

Prior to construction of any structural system, an extensive engineering design and analysis process must be undertaken. During this process, many engineering assumptions are routinely used in the application of engineering principles and theories to practice. A subset of these assumptions is used in a multitude of analytical methods available to structural analysts. In the modern engineering office, with the proliferation and increased power of personal computers, increasing numbers of engineers depend on structural analysis computer software to solve their engineering problems. This modernization of the engineering design office, coupled with an increased demand placed on the accuracy and efficiency of structural designs, requires a more-detailed understanding of the basic principles and assumptions associated with the use of modern structural analysis computer programs. The most popular of these programs are GT STRUDL, STAADIII, SAP2000, as well as some more powerful and complex tools such as ADINA, ANSYS, NASTRAN, and ABAQUS.

The objective of the analysis effort is to investigate the most probable responses of a bridge structure due to a range of applied loads. The results of these investigations must then be converted to useful design data, thereby providing designers with the information necessary to evaluate the performance of the bridge structure and to determine the appropriate actions in order to achieve the most efficient design configuration. Additionally, calculation of the structural system capacities is an important aspect in determining the most reliable design alternative. Every effort must be made to ensure that all work performed during any analytical activity enables designers to produce a set of quality construction documents including plans, specifications, and estimates.

The purpose of this chapter is to present basic modeling principles and suggest some guidelines and considerations that should be taken into account during the structural modeling process. Additionally, some examples of numerical characterizations of selected bridge structures and their components are provided. The outline of this chapter follows the basic modeling process. First, the selection of modeling methodology is discussed, followed by a description of the structural geometry, definition of the material and section properties of the components making up the structure, and description of the boundary conditions and loads acting on the structure.

8.2   Theoretical Background

Typically, during the analytical phase of any bridge design, finite-element-based structural analysis programs are used to evaluate the structural integrity of the bridge system. Most structural analysis programs employ sound, well-established finite-element methodologies and algorithms to solve the analytical problem. Others employ such methods as moment distribution, column analogy, virtual work, finite difference, and finite strip, to name a few. It is of utmost importance for the users of these programs to understand the theories, assumptions, and limitations of numerical modeling using the finite-element method, as well as the limitations on the accuracy of the computer systems used to execute these programs. Many textbooks [1, 4, 6] are available to study the theories and application of finite-element methodologies to practical engineering problems. It is strongly recommended that examination of these textbooks be made prior to using finite-element-based computer programs for any project work. For instance, when choosing the types of elements to use from the finite-element library, the user must consider some important factors such as the basic set of assumptions used in the element formulation, the types of behavior that each element type captures, and the limitations on the physical behavior of the system.

Other important issues to consider include numerical solution techniques used in matrix operations, computer numerical precision limitations, and solution methods used in a given analysis. There are many solution algorithms that employ direct or iterative methods, and sparse solver technology for solving the same basic problems; however, selecting these solution methods efficiently requires the user to understand the best conditions in which to apply each method and the basis or assumptions involved with each method. Understanding the solution parameters such as tolerances for iterative methods and how they can affect the accuracy of a solution are also important, especially during the nonlinear analysis process.

Dynamic analysis is increasingly being required by many design codes today, especially in regions of high seismicity. Response spectrum analysis is frequently used and easily performed with today's analysis tools; however, a basic understanding of structural dynamics is crucial for obtaining the proper results efficiently and interpreting analysis responses. Basic linear structural dynamics theory can be found in many textbooks [2,3]. While many analysis tools on the market today can perform very sophisticated analyses in a timely manner, the user too must be more savvy and knowledgeable to control the overall analysis effort and optimize the performance of such tools.

8.3   Modeling

8.3.1   Selection of Modeling Methodology

The technical approach taken by the engineer must be based on a philosophy of providing practical analysis in support of the design effort. Significant importance must be placed on the analysis procedures by the entire design team. All of the analytical modeling, analysis, and interpretation of results must be based on sound engineering judgment and a solid understanding of fundamental engineering principles. Ultimately, the analysis must validate the design.

Many factors contribute to determination of the modeling parameters. These factors should reflect issues such as the complexity of the structure under investigation, types of loads being examined, and, most importantly, the information needed to be obtained from the analysis in the most efficient and "design-friendly" formats. This section presents the basic principles and considerations for structural modeling. It also provides examples of modeling options for the various bridge structure types.

A typical flowchart of the analysis process is presented in Figure 8.1. The technical approach to computer modeling is usually based on a logical progression. The first step in achieving a reliable computer model is to define a proper set of material and soil properties, based on published data and site investigations. Second, critical components are assembled and tested numerically where

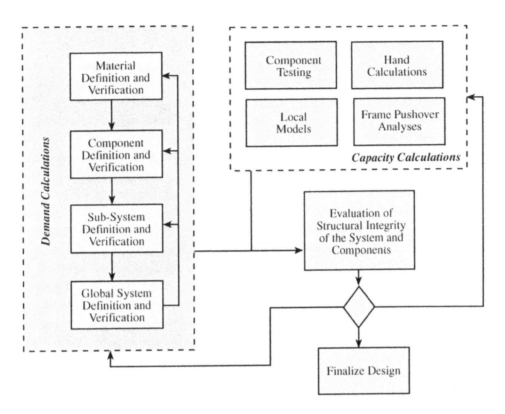

FIGURE 8.1 Typical analysis process.

validation of the performance of these components is considered important to the global model response. Closed-form solutions or available test data are used for these validations.

The next step is the creation and numerical testing of subsystems such as the bridge towers, superstructure elements, or individual frames. Again, as in the previous step, simple procedures are used in parallel to validate computer models. Last, a full bridge model consisting of the bridge subsystems is assembled and exercised. This final global model should include appropriate representation of construction sequence, soil and foundation boundary conditions, structural component behavior, and connection details.

Following the analysis and after careful examination of the analytical results, the data is postprocessed and provided to the designers for the purpose of checking the design and determining suitable design modifications, as necessary. Postprocessing might include computation of deck section resultant forces and moments, determination of extreme values of displacements for columns or towers and deck, and recovery of forces of constraint between structural components. The entire process may be repeated to validate any modifications made, depending on the nature and significance of such modifications.

An important part of the overall analytical procedure is determination of the capacities of the structural members. A combination of engineering calculations, computer analyses, and testing is utilized in order to develop a comprehensive set of component and system capacities. The evaluation of the structural integrity of the bridge structure, its components, and their connections are then conducted by comparing capacities with the demands calculated from the structural analysis.

Depending on the complexity of the structure under investigation and the nature of applied loads, two- or three-dimensional models can be utilized. In most cases, beam elements can be used to model structural elements of the bridge (Figure 8.2), so the component responses are presented in the form of force and moment resultants. These results are normally associated with individual element coordinate systems, thus simplifying the evaluations of these components. Normally, these

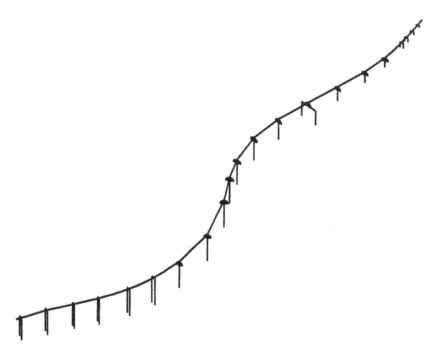

FIGURE 8.2 Typical beam model.

force resultants describe axial, shear, torsion, and bending actions at a given model location. Therefore, it is very important during the initial modeling stages to determine key locations of interest, so the model can be assembled such that important results can be obtained at these locations. While it is convenient to use element coordinate systems for the evaluation of the structural integrity of individual components, nodal results such as displacements and support reactions are usually output in the global coordinate systems. Proper refinement of the components must also be considered since different mesh size can sometimes cause significant variations in results. A balance between mesh refinement and reasonable element aspect ratios must be maintained so that the behavioral characteristics of the computer model is representative of the structure it simulates. Also, mesh refinement considerations must be made in conjunction with the cost to model efficiency. Higher orders of accuracy in modeling often come at a cost of analysis turnaround time and overall model efficiency. The analyst must use engineering judgment to determine if the benefits of mesh refinement justify the costs. For example, for the convenience in design of bridge details such as reinforcement bar cutoff, prestressing cable layouts, and section changes, the bridge superstructure is usually modeled with a high degree of refinement in the dead- and live-load analyses to achieve a well-defined force distribution. The same refinement may not be necessary in a dynamic analysis. Quite often, coarser models (at least four elements per span for the superstructure and three elements per column) are used in the dynamic analyses. These refinements are the minimum guidelines for discrete lumped mass models in dynamic analysis to maintain a reasonable mass distribution during the numerical solution process.

For more complex structures with complicated geometric configurations, such as curved plate girder bridges (Figure 8.3), or bridges with highly skewed supports (Figure 8.4), more-detailed finite-element models should be considered, especially if individual components within the superstructure need to be evaluated, which could not be facilitated with a beam superstructure representation. With the increasing speed of desktop computers, and advances in finite-element modeling tools, these models are becoming increasingly more popular. The main reason for their increased popularity is the improved accuracy, which in turn results in more efficient and cost-effective design.

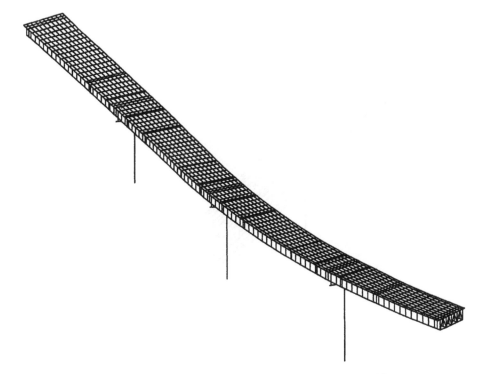

FIGURE 8.3   Steel plate-girder bridge—finite-element model.

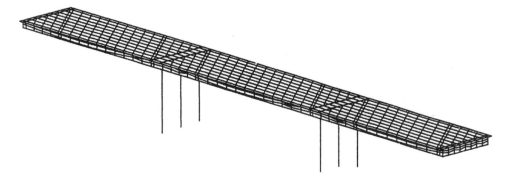

FIGURE 8.4   Concrete box girder with 45° skewed supports   finite-element model.

   More complex models, however, require a significantly higher degree of engineering experience and expertise in the theory and application of the finite-element method. In the case of a complex model, the engineer must determine the degree of refinement of the model. This determination is usually made based on the types of applied loads as well as the behavioral characteristics of the structure being represented by the finite-element model. It is important to note that the format of the results obtained from detailed models, such as shell and three-dimensional (3D) continuum models) is quite different from the results obtained from beam (or stick) models. Stresses and strains are obtained for each of the bridge components at a much more detailed level; therefore, calculation of a total force applied to the superstructure, for example, becomes a more difficult, tedious task. However, evaluation of local component behavior, such as cross frames, plate girder sections, or bridge deck sections, can be accomplished directly from the analysis results of a detailed finite-element model.

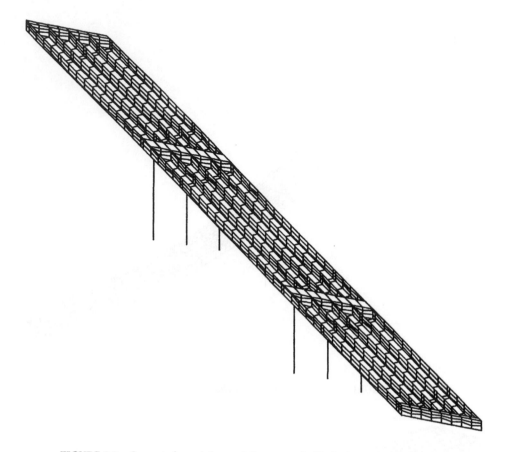

FIGURE 8.5 Concrete box-girder modeling example (deck elements not shown).

8.3.2 Geometry

After selecting an appropriate modeling methodology, serious considerations must be given to proper representation of the bridge geometric characteristics. These geometric issues are directly related to the behavioral characteristics of the structural components as well as the overall global structure. The considerations must include not only the global geometry of the bridge structure, i.e., horizontal alignment, vertical elevation, superelevation of the roadway, and severity of the support skews, but local geometric characterizations of connection details of individual bridge components as well. Such details include representations of connection regions such as column-to-cap beam, column-to-box girder, column-to-pile cap, cap beam to superstructure, cross frames to plate girder, gusset plates to adjacent structural elements, as well as various bearing systems commonly used in bridge engineering practice. Some examples of some modeling details are demonstrated in Figures 8.5 through 8.11.

Specifically, Figure 8.5 demonstrates how a detailed model of a box girder bridge structure can be assembled via use of shell elements (for girder webs and soffit), truss elements (for post-tensioning tendons), 3D solid elements (for internal diaphragms), and beam elements (for columns). Figure 8.6 illustrates some details of the web, deck, and abutment modeling for the same bridge structure. Additionally, spring elements are used to represent abutment support conditions for the vertical as well as back-wall directions. An example of a column and its connection to the superstructure in an explicit finite-element model is presented in the Figure 8.7. Three elements are used to represent the full length of the column. A set of rigid links connects the superstructure to each of the supporting columns (Figures 8.8 and 8.9). This is necessary to properly transmit bending

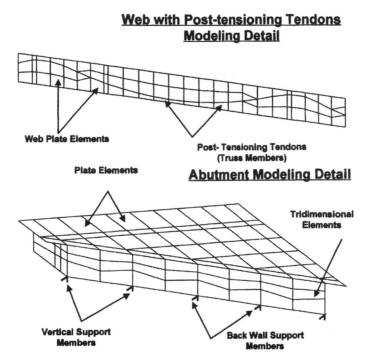

Web with Post-tensioning Tendons Modeling Detail

Web Plate Elements

Plate Elements

Post- Tensioning Tendons
(Truss Members)

Abutment Modeling Detail

Tridimensional
Elements

Vertical Support
Members

Back Wall Support
Members

FIGURE 8.6   Selected modeling details.

action of these components, since the beam elements (columns) are characterized by six degrees of freedom per node, while 3D solids (internal diaphragms) carry only three degrees of freedom per node (translations only). In this example post-tensioning tendons are modeled explicitly, via truss elements with the proper drape shape (Figure 8.9). This was done so that accurate post-tensioning load application was achieved and the effects of the skews were examined in detail. However, when beam models are used for the dynamic analysis (Figure 8.2), special attention must be given to the beam column joint modeling. For a box girder superstructure, since cap beams are monolithic to the superstructure, considerations must be given to capture proper dynamic behavior of this detail through modification of the connection properties. It is common to increase the section properties of the cap beam embedded in the superstructure to simulate high stiffness of this connection.

Figure 8.10 illustrates the plate girder modeling approach for a section of superstructure. Plate elements are used to model deck sections and girder webs, while beams are used to characterize flanges, haunches, cross frame members, as well as columns and cap beams (Figure 8.11). Proper offsets are used to locate the centerlines of these components in their proper locations.

8.3.3 Material and Section Properties

One of the most important aspects of capturing proper behavior of the structure is the determination of the material and section properties of its components. Reference [5] is widely used for calculating section properties for a variety of cross-sectional geometry. For 3D solid finite element, the material constitutive law is the only thing to specify whereas for other elements consideration of modification of material properties are needed to match the actual structural behavior. Most structural theories are based on homogeneous material such as steel. While this means structural behavior can be directly calculated using the actual material and section properties, it also indicates that nonhomogeneous material such as reinforced concrete may subject certain limitation. Because of the composite nonlinear performance nature of reinforced concrete, section properties need to be adjusted for the objective of analysis. For elastic analysis, if strength requirement is the objective, section

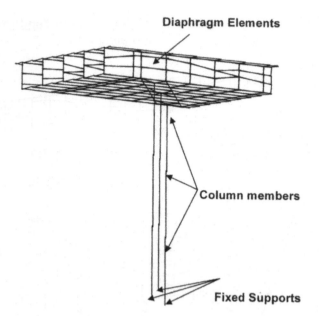

FIGURE 8.7 Bent region modeling detail.

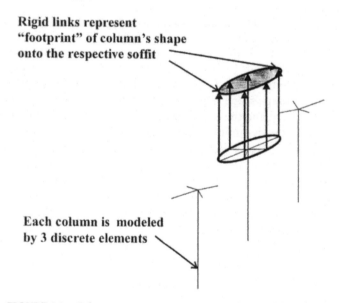

FIGURE 8.8 Column-to-superstructure connection modeling detail.

properties are less important as long as relative stiffness is correct. Section properties become most critical when structure displacement and deformation are objectives. Since concrete cracks beyond certain deformation, section properties need to be modified for this behavior. In general, if ultimate deformation is expected, then effective stiffness should be the consideration in section properties. It is common to use half value of the moment of inertia for reinforced concrete members and full value for prestressed concrete members. To replicate a rigid member behavior such as cap beams, section properties need to be amplified 100 times to eliminate local vibration problems in dynamic analysis.

Nonlinear behaviors are most difficult to handle in both complex and simple finite-element models. When solid elements are used, the constitutive relationships describing material behavior

Post-tensioning Tendons and Diaphragms

Columns with Rigid Link Connectors

FIGURE 8.9 Post-tensioning tendons, diaphragms, and column-to-diaphragm connection modeling examples.

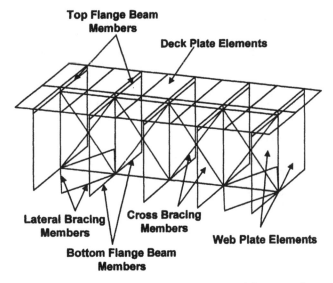

FIGURE 8.10 Plate girder superstructure modeling example.

should be utilized. These properties should be calibrated by the data obtained from the available test experiments. For beam–column-type elements, however, it is essential that the engineer properly estimates performance of the components either by experiments or theoretical detailed analysis. Once member performance is established, a simplified inelastic model can be used to simulate the expected member behavior. Depending on the complexity of the member, bilinear, or multilinear material representations may be used extensively. If member degradation needs to be incorporated in the analysis, then the Takeda model may be used. While a degrading model can correlate theoretical behavior with experimental results very well, elastic–plastic or bilinear models can give the engineer a good estimate of structural behavior without detailed material property parameters.

When a nonlinear analysis is performed, the engineer needs to understand the sensitivity issue raised by such analysis techniques. Without a good understanding of member behavior, it is very

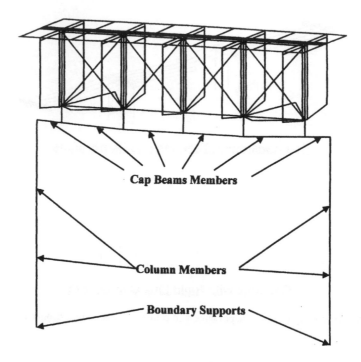

FIGURE 8.11   Plate girder bent region modeling example.

easy to fall into the "garbage in, garbage out" mode of operation. It is essential for the engineer to verify member behavior with known material properties before any production analyses are conducted. For initial design, all material properties should be based on the nominal values. However, it is important to verify the design with the expected material properties.

8.3.4   Boundary Conditions

Another key ingredient for the success of the structural analysis is the proper characterization of the boundary conditions of the structural system. Conditions of the columns or abutments at the support (or ground) points must be examined by engineers and properly implemented into the structural analysis model. This can be accomplished via several means based on different engineering assumptions. For example, during most of the static analysis, it is common to use a simple representation of supports (e.g., fixed, pinned, roller) without characterizations of the soil/foundation stiffness. However, for a dynamic analysis, proper representation of the soil/foundation system is essential (Figure 8.12). Most finite-element programs will accept a [6 × 6] stiffness matrix input for such system. Other programs require extended [12 × 12] stiffness matrix input describing the relationship between the ground point and the base of the columns. Prior to using these matrices, it is important that the user investigate the internal workings of the finite-element program, so the proper results are obtained by the analysis.

In some cases it is necessary to model the foundation/soil system with greater detail. Nonlinear modeling of the system can be accomplished via nonlinear spring/damper representation (Figure 8.13) or, in the extreme case, by explicit modeling of subsurface elements and plasticity-based springs representing surrounding soil mass (Figure 8.14). It is important that if this degree of detail is necessary, the structural engineer works very closely with the geotechnical engineers to determine proper properties of the soil springs. **As a general rule it is essential to set up small models to test behavior and check the results via hand calculations.**

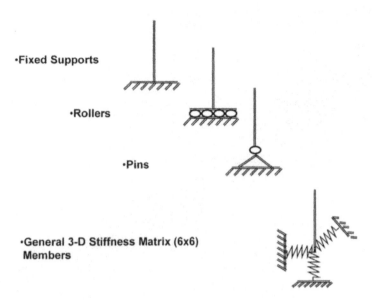

•Fixed Supports

•Rollers

•Pins

•General 3-D Stiffness Matrix (6x6) Members

FIGURE 8.12   Examples of foundation modeling.

FIGURE 8.13   Nonlinear spring/damper model.

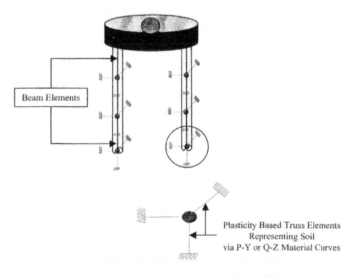

Beam Elements

Plasticity Based Truss Elements Representing Soil via P-Y or Q-Z Material Curves

FIGURE 8.14   Soil–structure interaction modeling.

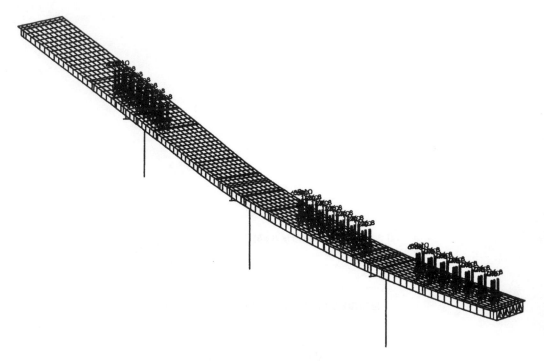

FIGURE 8.15 Truck load application example.

8.3.5 Loads

During engineering design activities, computer models are used to evaluate bridge structures for various service loads, such as traffic, wind, thermal, construction, and other service loads. These service loads can be represented by a series of static load cases applied to the structural model. Some examples of application of the truck loads are presented in Figures 8.15 and 8.16.

In many cases, especially in high seismic zones, dynamic loads control many bridge design parameters. In this case, it is very important to understand the nature of these loads, as well as the theory that governs the behavior of structural systems subjected to these dynamic loads. In high seismic zones, a multimode response spectrum analysis is required to evaluate the dynamic response of bridge structures. In this case, the response spectrum loading is usually described by the relationship of the structural period vs. ground acceleration, velocity, or displacement for a given structural damping. In some cases, usually for more complex bridge structures, a time history analysis is required. During these analytical investigations, a set of time history loads (normally, displacement or acceleration vs. time) is applied to the boundary nodes of the structure. Reference [3] is the most widely used theoretical reference related to the seismic analysis methodology for either response spectrum or time history analysis.

8.4 Summary

In summary, the analysis effort should support the overall design effort by verifying the design and addressing any issues with respect to the efficiency and the viability of the design. Before modeling commences, the engineer must define the scope of the problem and ask what key results and types of data he or she is interested in obtaining from the analytical model. With these basic parameters in mind, the engineer can then apply technical knowledge to formulate the simplest, most elegant model to represent the structure properly and provide the range of solutions that are accurate and fundamentally sound. The engineer must bound the demands on the structure by looking at limiting

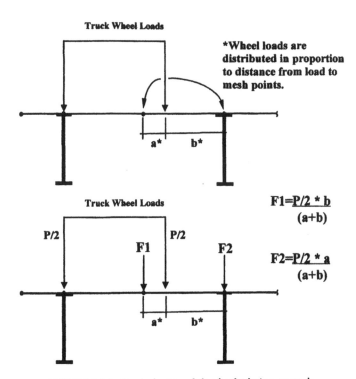

FIGURE 8.16 Equivalent truck load calculation example.

load cases and modifying the structure parameters, such as boundary conditions or material properties. Rigorous testing of components, hand calculations, local modeling, and sound engineering judgment must be used to validate the analytical model at all levels. Through a rigorous analytical methodology and proper use of today's analytical tools, structural engineers can gain a better understanding of the behavior of the structure, evaluate the integrity of the structure, and validate and optimize the structural design.

References

1. Bathe, K.J., *Finite Element Procedures*, 2nd ed., Prentice-Hall, Englewood Cliffs, NJ, 1996.
2. Chopra, A. K., *Dynamics of Structures, Theory and Applications to Earthquake Engineering*, Prentice-Hall, Englewood Cliffs, NJ, 1995.
3. Clough, R. W. and J. Penzien, *Dynamics of Structures*, 2nd ed., McGraw-Hill, New York, 1993.
4. Priestley, M.J.N., F. Seible, and G.M. Calvi, *Seismic Design and Retrofit of Bridges*, John Wiley & Sons, New York, 1996.
5. Young, W.C., *Roark's Formulas for Stress and Strain*, 6th ed., McGraw-Hill, New York, 1989.
6. Zienkiewicz, O.C. and R.L.Taylor, *The Finite Element Method*, Vol. 1, *Basic Formulation and Linear Problems*, 4th ed., McGraw-Hill, Berkshire, England, 1994.

Section II
Superstructure Design

9
Reinforced Concrete Bridges

Jyouru Lyang
*California Department
of Transportation*

Don Lee
*California Department
of Transportation*

John Kung
*California Department
of Transportation*

9.1 Introduction

The raw materials of concrete, consisting of water, fine aggregate, coarse aggregate, and cement, can be found in most areas of the world and can be mixed to form a variety of structural shapes. The great availability and flexibility of concrete material and reinforcing bars have made the reinforced concrete bridge a very competitive alternative. Reinforced concrete bridges may consist of precast concrete elements, which are fabricated at a production plant and then transported for erection at the job site, or cast-in-place concrete, which is formed and cast directly in its setting location. Cast-in-place concrete structures are often constructed monolithically and continuously. They usually provide a relatively low maintenance cost and better earthquake-resistance performance. Cast-in-place concrete structures, however, may not be a good choice when the project is on a fast-track construction schedule or when the available falsework opening clearance is limited. In this chapter, various structural types and design considerations for conventional cast-in-place, reinforced concrete highway bridge are discussed. Two design examples of a simply supported slab bridge and a two-span box girder bridge are also presented. All design specifications referenced in this chapter are based on 1994 AASHTO LRFD (Load and Resistance Factor Design) Bridge Design Specifications [1].

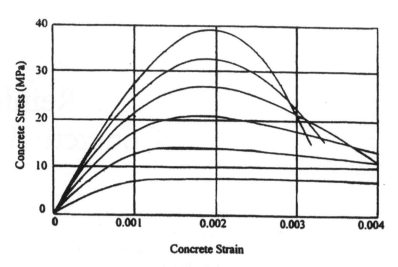

FIGURE 9.1 Typical stress–strain curves for concrete under uniaxial compression loading.

9.2 Materials

9.2.1 Concrete

1 Compressive Strength

The compressive strength of concrete (f_c') at 28 days after placement is usually obtained from a standard 150-mm-diameter by 300-mm-high cylinder loaded longitudinally to failure. Figure 9.1 shows typical stress–strain curves from unconfined concrete cylinders under uniaxial compression loading. The strain at the peak compression stress f_c' is approximately 0.002 and maximum usable strain is about 0.003. The concrete modulus of elasticity, E_c, may be calculated as

$$E_c = 0.043\gamma_c^{1.5}\sqrt{f_c'}\ \text{MPa} \qquad (9.1)$$

where γ_c is the density of concrete (kg/m³) and f_c' is the specified strength of concrete (MPa). For normal-weight concrete ($\gamma_c = 2300$ kg/m³), E_c may be calculated as $4800\sqrt{f_c'}$ MPa.

The concrete compressive strength or class of concrete should be specified in the contract documents for each bridge component. A typical specification for different classes of concrete and their corresponding specified compressive strengths is shown in Table 9.1. These classes are intended for use as follows:

- Class A concrete is generally used for all elements of structures and specially for concrete exposed to salt water.
- Class B concrete is used in footings, pedestals, massive pier shafts, and gravity walls.
- Class C concrete is used in thin sections under 100 mm in thickness, such as reinforced railings and for filler in steel grid floors.
- Class P concrete is used when strengths exceeding 28 MPa are required.
- Class S concrete is used for concrete deposited under water in cofferdams to seal out water.

Both concrete compressive strengths and water–cement ratios are specified in Table 9.1 for different concrete classes. This is because the water–cement ratio is a dominant factor contributing to both durability and strength, while simply obtaining the required concrete compressive strength to satisfy the design assumptions may not ensure adequate durability.

TABLE 9.1 Concrete Mix Characteristics by Class[1]

Class of Concrete	Minimum Cement Content (kg/m³)	Maximum Water–Cement Ratio (kg/kg)	Air Content Range, %	Coarse Aggregate per AASHTO M43 (square size of openings, mm)	28-day Compressive Strength, f_c' MPa
A	362	0.49	—	25 to 4.75	28
A(AE)	362	0.45	6.0 ± 1.5	25 to 4.75	28
B	307	0.58	—	50 to 4.75	17
B(AE)	307	0.55	5.0 ± 1.5	50 to 4.75	17
C	390	0.49	—	12.5 to 4.75	28
C(AE)	390	0.45	7.0 ± 1.5	12.5 to 4.75	28
P	334	0.49	As specified elsewhere	25 to 4.75 or 19 to 4.75	As specified elsewhere
S	390	0.58	—	25 to 4.75	—
Low-density	334	As specified in the contract documents			

Notes:
1. AASHTO Table C5.4.2.1-1 (From AASHTO LRFD Bridge Design Specifications, ©1994 by the American Association of State Highway and Transportation Officials, Washington, D.C. With permission.)
2. Concrete strengths above 70 MPa need to have laboratory testing verification. Concrete strengths below 16 MPa should not be used.
3. The sum of portland cement and other cementitious materials should not exceed 475 kg/m3.
4. Air-entrained concrete (AE) can improve durability when subjected to freeze–thaw action and to scaling caused by chemicals applied for snow and ice removal.

2. Tensile Strength

The tensile strength of concrete can be measured directly from tension loading. However, fixtures for holding the specimens are difficult to apply uniform axial tension loading and sometimes will even introduce unwanted secondary stresses. The direct tension test method is therefore usually used to determine the cracking strength of concrete caused by effects other than flexure. For most regular concrete, the direct tensile strength may be estimated as 10% of the compressive strength.

The tensile strength of concrete may be obtained indirectly by the split tensile strength method. The splitting tensile stress (f_s) at which a cylinder is placed horizontally in a testing machine and loaded along a diameter until split failure can be calculated as

$$f_s = 2P/(\pi LD) \qquad (9.2)$$

where P is the total applied load that splits the cylinder, L is the length of cylinder, and D is the diameter of the cylinder.

The tensile strength of concrete can also be evaluated by means of bending tests conducted on plain concrete beams. The flexural tensile stress, known as the modulus of rupture (f_r) is computed from the flexural formula M/S, where M is the applied failure bending moment and S is the elastic section modulus of the beam. Modulus of rupture (f_r) in MPa can be calculated as

$$f_r = \begin{cases} 0.63\sqrt{f_c'} & \text{for normal-weight concrete} \\ 0.52\sqrt{f_c'} & \text{for sand–low-density concrete} \\ 0.45\sqrt{f_c'} & \text{for all–low-density concrete} \end{cases} \qquad (9.3)$$

TABLE 9.2 Steel Deformed Bar Sizes and Weight
(ASTM A615M and A706M)

Bar Number	Nominal Dimensions		Unit Weight, kg/m
	Diameter, mm	Area, mm²	
10	9.5	71	0.560
13	12.7	129	0.994
16	15.9	199	1.552
19	19.1	284	2.235
22	22.2	387	3.042
25	25.4	510	3.973
29	28.7	645	5.060
32	32.3	819	6.404
36	35.8	1006	7.907
43	43.0	1452	11.38
57	57.3	2581	20.24

Both the splitting tensile stress (f_s) and flexural tensile stress (f_r) overestimate the tensile cracking stress determined by a direct tension test. However, concrete in tension is usually ignored in strength calculations of reinforced concrete members because the tensile strength of concrete is low. The modulus of elasticity for concrete in tension may be assumed to be the same as in compression.

3. Creep and Shrinkage

Both creep and shrinkage of concrete are time-dependent deformations and are discussed in Chapter 10.

9.2.2 Steel Reinforcement

Deformed steel bars are commonly employed as reinforcement in most reinforced concrete bridge construction. The surface of a steel bar is rolled with lugs or protrusions called deformations in order to restrict longitudinal movement between the bars and the surrounding concrete. Reinforcing bars, rolled according to ASTM A615/A615M specifications (billet steel) [2], are widely used in construction. ASTM A706/A706M low-alloy steel deformed bars (Grade 420 only) [2] are specified for special applications where extensive welding of reinforcement or controlled ductility for earthquake-resistant, reinforced concrete structures or both are of importance.

1. Bar Shape and Size

Deformed steel bars are approximately numbered based on the amount of millimeters of the nominal diameter of the bar. The nominal dimensions of a deformed bar are equivalent to those of a plain round bar which has the same mass per meter as the deformed bar. Table 9.2 lists a range of deformed bar sizes according to the ASTM specifications.

2. Stress–Strain Curve

The behavior of steel reinforcement is usually characterized by the stress–strain curve under uniaxial tension loading. Typical stress–strain curves for steel Grade 300 and 420 are shown in Figure 9.2. The curves exhibit an initial linear elastic portion with a slope calculated as the modulus of elasticity of steel reinforcement $E_s = 200,000$ MPa; a yield plateau in which the strain increases (from ε_y to ε_h) with little or no increase in yield stress (f_y); a strain-hardening range in which stress again increases with strain until the maximum stress (f_u) at a strain (ε_u) is reached; and finally a range in which the stress drops off until fracture occurs at a breaking strain of ε_b.

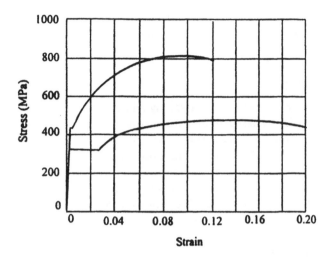

FIGURE 9.2   Typical stress–strain curves for steel reinforcement.

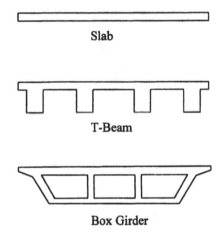

FIGURE 9.3   Typical reinforced concrete sections in bridge superstructures.

9.3   Bridge Types

Reinforced concrete sections, used in the bridge superstructures, usually consist of slabs, T-beams (deck girders), and box girders (Figure 9.3). Safety, cost-effectiveness, and aesthetics are generally the controlling factors in the selection of the proper type of bridges [3]. Occasionally, the selection is complicated by other considerations such as the deflection limit, life-cycle cost, traffic maintenance during construction stages, construction scheduling and worker safety, feasibility of falsework layout, passage of flood debris, seismicity at the site, suitability for future widening, and commitments made to officials and individuals of the community. In some cases, a prestressed concrete or steel bridge may be a better choice.

9.3.1   Slab Bridges

Longitudinally reinforced slab bridges have the simplest superstructure configuration and the neatest appearance. They generally require more reinforcing steel and structural concrete than do girder-type bridges of the same span. However, the design details and formworks are easier and less expensive. It has been found economical for simply supported spans up to 9 m and for continuous spans up to 12 m.

9.3.2 T-Beam Bridges

The T-beam construction consists of a transversely reinforced slab deck which spans across to the longitudinal support girders. These require a more-complicated formwork, particularly for skewed bridges, compared to the other superstructure forms. T-beam bridges are generally more economical for spans of 12 to 18 m. The girder stem thickness usually varies from 35 to 55 cm and is controlled by the required horizontal spacing of the positive moment reinforcement. Optimum lateral spacing of longitudinal girders is typically between 1.8 and 3.0 m for a minimum cost of formwork and structural materials. However, where vertical supports for the formwork are difficult and expensive, girder spacing can be increased accordingly.

9.3.3 Box-Girder Bridges

Box-girder bridges contain top deck, vertical web, and bottom slab and are often used for spans of 15 to 36 m with girders spaced at 1.5 times the structure depth. Beyond this range, it is probably more economical to consider a different type of bridge, such as post-tensioned box girder or steel girder superstructure. This is because of the massive increase in volume and materials. They can be viewed as T-beam structures for both positive and negative moments. The high torsional strength of the box girder makes it particularly suitable for sharp curve alignment, skewed piers and abutments, superelevation, and transitions such as interchange ramp structures.

9.4 Design Considerations

9.4.1 Basic Design Theory

The AASHTO LRFD Specifications (1994) [1] were developed in a reliability-based limit state design format. Limit state is defined as the limiting condition of acceptable performance for which the bridge or component was designed. In order to achieve the objective for a safe design, each bridge member and connection is required to examine some, or all, of the service, fatigue, strength, and extreme event limit states. All applicable limit states shall be considered of equal importance. The basic requirement for bridge design in the LRFD format for each limit state is as follows:

$$\eta \ \Sigma \gamma_i \ Q_i \le \phi \ R_n \tag{9.4}$$

where η = load modifier to account for bridge ductility, redundancy, and operational importance, γ_i = load factor for load component i, Q_i = nominal force effect for load component i, ϕ = resistance factor, and R_n = nominal resistance. The margin of safety for a bridge design is provided by ensuring the bridge has sufficient capacity to resist various loading combinations in different limit states.

The load factors, γ, which often have values larger than one, account for the loading uncertainties and their probabilities of occurrence during bridges design life. The resistance factors, ϕ, which are typically less than unity at the strength limit state and equal to unity for all other limit states, account for material variabilities and model uncertainties. Table 9.3 lists the resistance factors in the strength limit state for conventional concrete construction. The load modifiers, η, which are equal to unity for all non-strength-limit states, account for structure ductility, redundancy, and operational importance. They are related to the bridge physical strength and the effects of a bridge being out of service. Detailed load resistance factor design theory and philosophy are discussed in Chapter 5.

TABLE 9.3 Resistance Factors φ in the Strength Limit State for Conventional Construction

Strength Limit State	Resistance Factors φ
For flexural and tension of reinforced concrete	0.90
For shear and torsion	
Normal weight concrete	0.90
Lightweight concrete	0.70
For axial compression with spirals and ties	0.75
(except for Seismic Zones 3 and 4 at the extreme event limit state)	
For bearing on concrete	0.79
For compression in strut-and-tie models	0.70

Notes:

1. AASHTO 5.5.4.2.1 (From AASHTO LRFD Bridge Design Specifications, ©1994 by the American Association of State Highway and Transportation Officials, Washington, D.C. With permission.)

2. For compression members with flexural, the value of φ may be increased linearly to the value for flexural as the factored axial load resistance, ϕP_n, decreases from 0.10 $f'_c A_g$ to 0.

9.4.2 Design Limit States

1. Service Limit States

For concrete structures, service limit states correspond to the restrictions on cracking width and deformations under service conditions. They are intended to ensure that the bridge will behave and perform acceptably during its service life.

a. Control of Cracking

Cracking may occur in the tension zone for reinforced concrete members due to the low tensile strength of concrete. Such cracks may occur perpendicular to the axis of the members under axial tension or flexural bending loading without significant shear force, or inclined to the axis of the members with significant shear force. The cracks can be controlled by distributing steel reinforcements over the maximum tension zone in order to limit the maximum allowable crack widths at the surface of the concrete for given types of environment. The tensile stress in the steel reinforcement (f_s) at the service limit state should not exceed

$$f_{sa} = \frac{Z}{\left(d_c A\right)^{1/3}} \le 0.6 f_y \tag{9.5}$$

where d_c (mm) is the concrete cover measured from extreme tension fiber to the center of the closest bars and should not to be taken greater than 50 mm; A (mm²) is the concrete area having the same centroid as the principal tensile reinforcement divided by the number of bars; Z (N/mm) should not exceed 30,000 for members in moderate exposure conditions, 23,000 in severe exposure conditions, and 17,500 for buried structures. Several smaller tension bars at moderate spacing can provide more effective crack control by increasing f_{sa} rather than installing a few larger bars of equivalent area.

When flanges of reinforced concrete T-beams and box girders are in tension, the flexural tension reinforcement should be distributed over the lesser of the effective flange width or a width equal to ¹⁄₁₀ of the span in order to avoid the wide spacing of the bars. If the effective flange width exceeds ¹⁄₁₀ of the span length, additional longitudinal reinforcement, with an area not less than 0.4% of the excess slab area, should be provided in the outer portions of the flange.

For flexural members with web depth exceeding 900 mm, longitudinal skin reinforcements should be uniformly distributed along both side faces for a height of $d/2$ nearest the flexural tension reinforcement for controlling cracking in the web. Without such auxiliary steel, the width of the

TABLE 9.4   Traditional Minimum Depths for Constant Depth Superstructures

	Minimum Depth (Including Deck)	
Bridge Types	Simple Spans	Continuous Spans
Slabs	$\dfrac{1.2\,(S+3000)}{30}$	$\dfrac{(S+3000)}{30} \geq 165 \text{ mm}$
T-beams	$0.070L$	$0.065L$
Box beams	$0.060L$	$0.055L$
Pedestrian structure beams	$0.035L$	$0.033L$

Notes:
1. AASHTO Table 2.5.2.6.3-1 (From AASHTO LRFD Bridge Design Specifications, ©1994 by the American Association of State Highway and Transportation Officials, Washington, D.C. With permission.)
2. S (mm) is the slab span length and L (mm) is the span length.
3. When variable-depth members are used, values may be adjusted to account for change in relative stiffness of positive and negative moment sections.

cracks in the web may greatly exceed the crack widths at the level of the flexural tension reinforcement. The area of skin reinforcement (A_{sk}) in mm²/mm of height on each side face should satisfy

$$A_{sk} \geq 0.001\,(d_e - 760) \leq \frac{A_s}{1200} \tag{9.6}$$

where d_e (mm) is the flexural depth from extreme compression fiber to the centroid of the tensile reinforcement and A_s (mm²) is the area of tensile reinforcement and prestressing steel. The maximum spacing of the skin reinforcement shall not exceed $d/6$ or 300 mm.

b. Control of Deformations

Service-load deformations in bridge elements need to be limited to avoid the structural behavior which differs from the assumed design conditions and to ease the psychological effects on motorists. Service-load deformations may not be a potential source of collapse mechanisms but usually cause some undesirable effects, such as the deterioration of wearing surfaces and local cracking in concrete slab which could impair serviceability and durability. AASHTO LRFD [1] provides two alternative criteria for controlling the deflections:

Limiting Computed Deflections (AASHTO 2.5.2.6.2):

Vehicular load, general	Span length/800
Vehicular and/or pedestrian loads	Span length/1000
Vehicular load on cantilever arms	Span length/300
Vehicular and/or pedestrian loads on cantilever arms	Span length/1000

Limiting Span-to-Depth Ratios (AASHTO 2.5.2.6.3): For superstructures with constant depth, Table 9.4 shows the typical minimum depth recommendation for a given span length.

Deflections of bridges can be estimated in two steps: (1) instantaneous deflections which occur at the first loading and (2) long-time deflections which occur with time due to the creep and shrinkage of the concrete.

Instantaneous deflections may be computed by using the elastic theory equations. The modulus of elasticity for concrete can be calculated from Eq. (9.1). The moment of inertia of a section can be taken as either the uncracked gross moment of inertia (I_g) for uncracked elements or the effective moment of inertia (I_e) for cracked elements. The effective moment of inertia can be calculated as

$$I_e = \left(\frac{M_{cr}}{M_a}\right)^3 I_g + \left[1-\left(\frac{M_{cr}}{M_a}\right)^3\right] I_{cr} \leq I_g \tag{9.7}$$

and

$$M_{cr} = f_r \frac{I_g}{y_t} \tag{9.8}$$

where M_{cr} is the moment at first cracking, f_r is the modulus of rupture, y_t is the distance from the neutral axis to the extreme tension fiber, I_{cr} is the moment of inertia of the cracked section transformed to concrete (see Section 9.4.6), and M_a is the maximum moment in a component at the stage for which deformation is computed. For prismatic members, the effective moment of inertia may be calculated at midspan for simple or continuous bridges and at support for cantilevers. For continuous nonprismatic members, the moment of inertia may be calculated as the average of the critical positive and negative moment sections.

Long-time deflections may be calculated as the instantaneous deflection multiplied by the following:

If the instantaneous deflection is based on I_g: 4.0
If the instantaneous deflection is based on I_e: $3.0 - 1.2 \left(A_s' / A_s \right) \geq 1.6$

where A_s' is area of compression reinforcement and A_s is the area of tension reinforcement.

2. Fatigue Limit States

Fatigue limit states are used to limit stress in steel reinforcements to control concrete crack growth under repetitive truck loading in order to prevent early fracture failure before the design service life of a bridge. Fatigue loading consists of one design truck with a constant spacing of 9000 mm between the 145-kN axles. Fatigue is considered at regions where compressive stress due to permanent loads is less than two times the maximum tensile live-load stress resulting from the fatigue-load combination. Allowable fatigue stress range in straight reinforcement is limited to

$$f_f = 145 - 0.33 f_{min} + 55 \left(\frac{r}{h} \right) \tag{9.9}$$

where f_{min} (MPa) is the minimum stress in reinforcement from fatigue loading (positive for tension and negative for compression stress) and r/h is the ratio of the base radius to the height of rolled-on transverse deformations (0.3 may be used if the actual value in not known).

The cracked section properties should be used for fatigue. Gross section properties may be used when the sum of stresses, due to unfactored permanent loads, plus 1.5 times the fatigue load is not to exceed the tensile stress of $0.25\sqrt{f_c'}$.

3. Strength Limit States and Extreme Event Limit States

For reinforced concrete structures, strength and extreme event limit states are used to ensure that strength and stability are provided to resist specified statistically significant load combinations. A detailed discussion for these limit states is covered in Chapter 5.

9.4.3 Flexural Strength

Figure 9.4 shows a doubly reinforced concrete beam when flexural strength is reached and the depth of neutral axis falls outside the compression flange ($c > h_f$). Assume that both tension and compression steel are yielding and the concrete compression stress block is in a rectangular shape. ε_{cu} is the maximum strain at the extreme concrete compression fiber and is about 0.003 for unconfined concrete.

Concrete compression force in the web;

$$C_w = 0.85 \, f_c' \, ab_w = 0.85\beta_1 \, f_c' \, cb_w \tag{9.10}$$

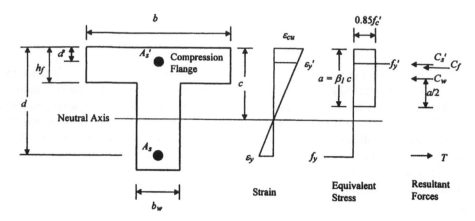

FIGURE 9.4   Reinforced concrete beam when flexural strength is reached.

where

$$a = c\,\beta_1 \tag{9.11}$$

Concrete compression force in the flange:

$$C_f = 0.85\beta_1\ f'_c\ (b-b_w)h_f \tag{9.12}$$

Compression force in the steel:

$$C'_s = A'_s f'_y \tag{9.13}$$

Tension force in the steel:

$$T = A_s f_y \tag{9.14}$$

From the equilibrium of the forces in the beam, we have

$$C_w + C_f + C'_s = T \tag{9.15}$$

The depth of the neutral axis can be solved as

$$c = \frac{A_s f_y - A'_s f'_y - 0.85\beta_1 f'_c (b - b_w)h_f}{0.85\beta_1 f'_c b_w} \geq h_f \tag{9.16}$$

The nominal flexural strength is

$$M_n = A_s f_y \left(d - \frac{a}{2}\right) + A'_s f'_y \left(\frac{a}{2} - d'\right) + 0.85\beta_1 f'_c (b - b_w)h_f \left(\frac{a}{2} - \frac{h_f}{2}\right) \tag{9.17}$$

where A_s is the area of tension steel, A'_s is the area of compression steel, b is the width of the effective flange, b_w is the width of the web, d is the distance between the centroid of tension steel and the most compressed concrete fiber, d' is the distance between the centroid of compression steel and the most compressed concrete fiber, and h_f is the thickness of the effective flange. The concrete stress factor, β_1 can be calculated as

$$\beta_1 = \begin{cases} 0.85 & \text{for } f_c' \le 28 \text{ MPa} \\ 0.85 - 0.05\left(\dfrac{f_c' - 28}{7}\right) & \text{for } 28 \text{ MPa} \le f_c' \le 56 \text{ MPa} \\ 0.65 & \text{for } f_c' \ge 56 \text{ MPa} \end{cases} \tag{9.18}$$

Limits for reinforcement are

- Maximum tensile reinforcement:

$$\frac{c}{d} \le 0.42 \tag{9.19}$$

When Eq. (9.19) is not satisfied, the reinforced concrete sections become overreinforced and will have sudden brittle compression failure if they are not well confined.

- Minimum tensile reinforcement:

$$\rho_{min} \ge 0.03 \frac{f_c'}{f_y}, \quad \text{where } \rho_{min} = \text{ratio of tension steel to gross area} \tag{9.20}$$

When Eq. (9.20) is not satisfied, the reinforced concrete sections become underreinforced and will have sudden tension steel fracture failure.

The strain diagram can be used to verify compression steel yielding assumption.

$$f_s' = f_y' \quad \text{if} \quad \varepsilon_s' = \varepsilon_{cu}\left(\frac{c - d'}{c}\right) \ge \frac{f_y'}{E_s} \tag{9.21}$$

If compression steel is not yielding as checked from Eqs. (9.21). The depth of neutral axis, c, and value of nominal flexural strength, M_n, calculated from Eqs. (9.16) and (9.17) are incorrect. The actual forces applied in compression steel reinforcement can be calculated as

$$C_s' = A_s' f_s' = A_s \varepsilon_s' E_s' = A_s \varepsilon_{cu}\left(\frac{d - c}{c}\right) E_s' \tag{9.22}$$

The depth of neutral axis, c, can be solved by substituting Eqs. (9.22) into forces equilibrium Eq. (9.15). The flexural strength, M_n, can then be obtained from Eq. (9.17) with the actual applied compression steel forces. In a typical beam design, the tension steel will always be yielding and the compression steel is close to reaching yielding strength as well.

If the depth of the neutral axis falls within the compression flange ($x \le h_f$) or for sections without compression flange, then the depth of the neutral axis, c, and the value of nominal flexural strength, M_n, can be calculated by setting b_w equal to b.

9.4.4 Shear Strength

1. Strut-and-Tie Model

The strut-and-tie model should be used for shear and torsion designs of bridge components at locations near discontinuities, such as regions adjacent to abrupt changes in the cross section, openings, and dapped ends. The model should also be used for designing deep footings and pile

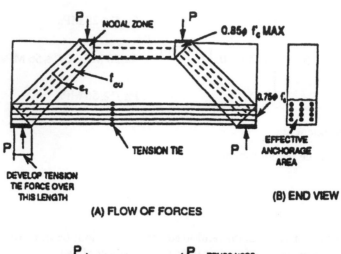

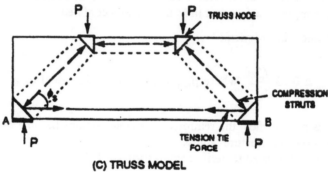

FIGURE 9.5 Strut-and-tie model for a deep beam. (*Source*: AASHTO LRFD Bridge Design Specifications, Figure 5.6.3.2-1, © 1994 by the American Association of State Highway and Transportation Officials, Washington, D.C. With permission.)

caps or in other situations where the distance between the centers of the applied load and the supporting reactions is less than about twice the member thickness. Figure 9.5 shows a strut-and-tie model for a deep beam that is composed of steel tension ties and concrete compressive struts. These are interconnected at nodes to form a truss capable of carrying all applied loads to the supports.

2. Sectional Design Model

The sectional design model can be used for the shear and torsion design for regions of bridge members where plane sections remain plane after loading. It was developed by Collins and Mitchell [4] and is based on the modified compression field theory. The general shear design procedure for reinforced concrete members, containing transverse web reinforcement, is as follows:

- Calculate the effective shear depth d_v:

   Effective shear depth is calculated between the resultants of the tensile and compressive forces due to flexure. This should not be less than the greater of $0.9d_e$ or $0.72h$, where d_e is the effective depth from extreme compression fiber to the centroid of the tensile reinforcement and h is the overall depth of a member.

- Calculate shear stress:

$$v = \frac{V_u}{\phi b_v d_v} \tag{9.23}$$

where b_v is the equivalent web width and V_u is the factored shear demand envelope from the strength limit state.

- Calculate v/f_c', if this ratio is greater than 0.25, then a larger web section needs to be used.
- Assume an angle of inclination of the diagonal compressive stresses, θ, and calculate the strain in the flexural tension reinforcement:

$$\varepsilon_x = \frac{\dfrac{M_u}{d_v} + 0.5 V_u \cot \theta}{E_s A_s} \qquad (9.24)$$

where M_u is the factored moment demand. It is conservative to take M_u enveloped from the strength limit state that will occur at that section, rather than a moment coincident with V_u.

- Use the calculated v/f_c' and ε_x to find θ from Figure 9.6 and compare it with the value assumed. Repeat the above procedure until the assumed θ is reasonably close to the value found from Figure 9.6. Then record the value of β, a factor which indicates the ability of diagonally cracked concrete to transmit tension.

- Calculate the required transverse web reinforcement strength, V_s:

$$V_s = \frac{V_u}{\phi} - V_c = \frac{V_u}{\phi} - 0.083\beta\sqrt{f_c'}\, b_v d_v \qquad (9.25)$$

where V_c is the nominal concrete shear resistance.

- Calculate the required spacing for the transverse web reinforcement:

$$s \leq \frac{A_v f_y d_v \cot \theta}{V_s} \qquad (9.26)$$

where A_v is the area of a transverse web reinforcement within distance s.

Check for the minimum transverse web reinforcement requirement:

$$A_v \geq 0.083\sqrt{f_c'}\,\frac{b_v S}{f_y} \quad \text{or} \quad s \leq \frac{A_v f_y}{0.083\sqrt{f_c'}\, b_v} \qquad (9.27)$$

Check for the maximum spacing requirement for transverse web reinforcements:

$$\text{if } V_u < 0.1\, f_c'\, b_v d_v, \quad \text{then } s \leq 0.8 d_v \leq 600 \text{ mm} \qquad (9.28)$$

$$\text{if } V_u \geq 0.1\, f_c'\, b_v d_v, \quad \text{then } s \leq 0.4 d_v \leq 300 \text{ mm} \qquad (9.29)$$

- Check the adequacy of the longitudinal reinforcements to avoid yielding due to the combined loading of moment, axial load, and shear.

$$A_s f_y \geq \frac{M_u}{d_v \phi} + \left(\frac{V_u}{\phi} - 0.5 V_s\right)\cot \theta \qquad (9.30)$$

If the above equation is not satisfied, then you need either to add more longitudinal reinforcement or to increase the amount of transverse web reinforcement.

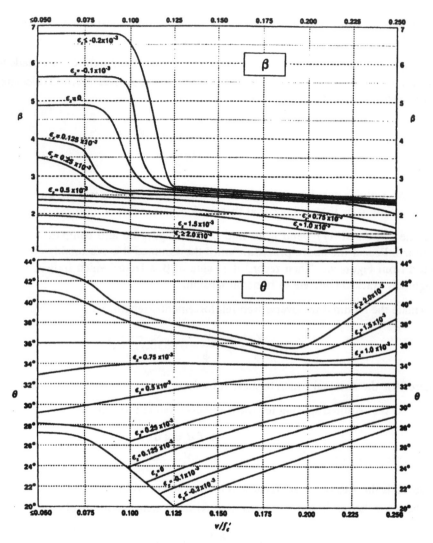

FIGURE 9.6 Values of θ and β for sections with transverse web reinforcement. (*Source:* AASHTO LRFD Bridge Design Specifications, Figure 5.8.3.4.2-1, ©1994 by the American Association of State Highway and Transportation Officials, Washington, D.C. With permission.)

9.4.5 Skewed Concrete Bridges

Shear, in the exterior beam at the obtuse corner of the bridge, needs to be adjusted when the line of support is skewed. The value of the correction factor obtained from AASHTO Table 4.6.2.2.3c-1, needs to be applied to live-load distribution factors for shear. In determining end shear in multibeam bridges, all beams should be treated like the beam at the obtuse corner, including interior beams.

   Moment load distribution factors in longitudinal beams on skew supports may be reduced according to AASHTO Table 4.6.2.2.2e-1, when the line supports are skewed and the difference between skew angles of two adjacent lines of supports does not exceed 10°.

9.4.6 Design Information

1. Stress Analysis at Service Limit States [5]

A reinforced concrete beam subject to flexural bending moment is shown in Figure 9.7 and x is the distance between the neutral axis and the extreme compressed concrete fiber. Assume the neutral axis

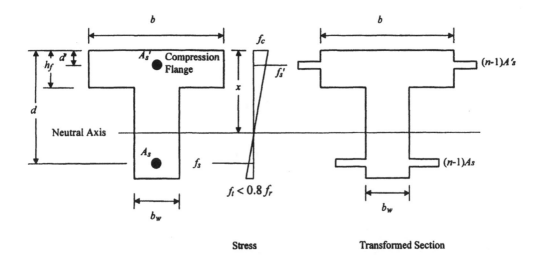

(a) Stress and Transformed Section Before Cracking

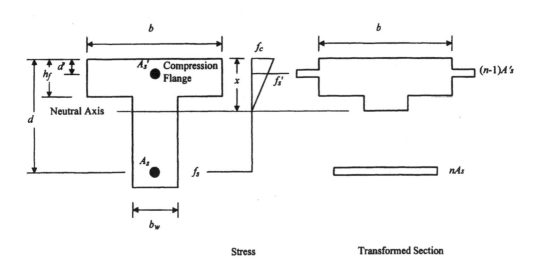

(b) Stress and Transformed Section After Cracking

FIGURE 9.7 Reinforced concrete beam for working stress analysis.

falls within the web ($x > h_f$) and the stress in extreme tension concrete fiber is greater than 80% of the concrete modulus of rupture ($f_t \geq 0.8\,f_r$). The depth of neutral axis, x, can be solved through the following quadratic equation by using the cracked transformed section method (see Figure 9.7).

$$b(x)\left(\frac{x}{2}\right) - (b - b_w)(x - h_f)\left(\frac{x - h_f}{2}\right) + (n-1)A_s'(x - d') = nA_s(d - x) \qquad (9.31)$$

$$x = \sqrt{B^2 + C} - B \qquad (9.32)$$

where

$$B = \frac{1}{b_w}\left[h_f(b - b_w) + nA_s + (n-1)A_s'\right] \tag{9.33}$$

$$C = \frac{2}{b_w}\left[\frac{h_f^2}{2}(b - b_w) + ndA_s + (n-1)d'A_s'\right] \tag{9.34}$$

and the moment of inertia of the cracked transformed section about the neutral axis:

$$I_{cr} = \frac{1}{3}bx^3 - \frac{1}{3}(b - b_w)(x - h_f)^3 + nA_s(d - x)^2 + (n-1)A_s'(x - d')^2 \tag{9.35}$$

if the calculated neutral axis falls within the compression flange ($x \le h_f$) or for sections without compression flange, the depth of neutral axis, x, and cracked moment of inertia, I_{cr}, can be calculated by setting b_w equal to b.

Stress in extreme compressed concrete fiber:

$$f_c = \frac{Mx}{I_{cr}} \tag{9.36}$$

Stress in compression steel:

$$f_s' = \frac{nM(x - d')}{I_{cr}} = nf_c\left(1 - \frac{d'}{x}\right) \tag{9.37}$$

Stress in tension steel:

$$f_s = \frac{nM(d - x)}{I_{cr}} = nf_c\left(\frac{d}{x} - 1\right) \tag{9.38}$$

where

$$n = \frac{E_s}{E_c} \tag{9.39}$$

and M is moment demand enveloped from the service limit state.

2. Effective Flange Width (AASHTO 4.6.2.6)

When reinforced concrete slab and girders are constructed monolithically, the effective flange width (b_{eff}) of a concrete slab, which will interact with girders in composite action, may be calculated as

For interior beams:

$$b_{eff}^I = \text{the smallest of} \begin{cases} \dfrac{l_{eff}}{4} \\[2mm] 12t_s + b_w \\[2mm] \text{the average spacing of adjacent beams} \end{cases} \tag{9.40}$$

TABLE 9.5 Cover for Unprotected Main Reinforcing Steel (mm)

Situation	Cover (mm)
Direct exposure to salt water	100
Cast against earth	75
Coastal	75
Exposure to deicing salt	60
Deck surface subject to tire stud or chain wear	60
Exterior other than above	50
Interior other than above	
• Up to No. 36 Bar	40
• No. 43 and No. 57 Bars	50
Bottom of CIP slab	
• Up to No. 36 Bar	25
• No. 43 and No. 57 Bars	50

Notes:
1. Minimum cover to main bars, including bars protected by epoxy coating, shall be 25 mm.
2. Cover to epoxy-coated steel may be used as interior exposure situation.
3. Cover to ties and stirrups may be 12 mm less than the value specified here, but shall not be less than 25 mm.
4. Modification factors for water:cement ratio, w/c, shall be the following:

   for $w/c \leq 0.40$ modification factor = 0.8
   for $w/c \geq 0.40$ modification factor = 1.2

Source: AASHTO Table C5.12.3-1. (From AASHTO LRFD Bridge Design Specifications, ©1994 by the American Association of State Highway and Transportation Officials, Washington, D.C. With permission.)

For exterior beams:

$$b_{eff}^{E} = \frac{1}{2} b_{eff}^{I} + \text{the smallest of} \begin{cases} \dfrac{l_{eff}}{8} \\ 6t_s + \dfrac{b_w}{2} \\ \text{the width of overhang} \end{cases} \tag{9.41}$$

where the effective span length (l_{eff}) may be calculated as the actual span for simply supported spans. Also, the distance between the points of permanent load inflection for continuous spans of either positive or negative moments (t_s) is the average thickness of the slab, and b_w is the greater of web thickness or one half the width of the top flange of the girder.

3. Concrete Cover (AASHTO 5.12.3)

Concrete cover for unprotected main reinforcing steel should not be less than that specified in Table 9.5 and modified for the water:cement ratio.

9.4.7 Details of Reinforcement

Table 9.6 shows basic tension, compression, and hook development length for Grade 300 and Grade 420 deformed steel reinforcement (AASHTO 5.11.2). Table 9.7 shows the minimum center-to-center spacing between parallel reinforcing bars (AASHTO 5.10.3).

TABLE 9.6 Basic Rebar Development Lengths for Grade 300 and 420 (AASHTO 5.11.2)

Bar Size	28 MPa			35 MPa			42 MPa		
	Tension	Compression	Hook	Tension	Compression	Hook	Tension	Compression	Hook
Grade 300, f_y = 300 MPa									
13	230	175	240	230	170	215	230	170	200
16	290	220	300	290	210	270	290	210	245
19	345	260	365	345	255	325	345	255	295
22	440	305	420	400	295	375	400	295	345
25	580	350	480	520	335	430	475	335	395
29	735	395	545	655	380	485	600	380	445
32	930	440	610	835	430	550	760	430	500
36	1145	490	680	1020	475	605	935	475	555
43	1420	585	815	1270	570	730	1160	570	665
57	1930	780	1085	1725	765	970	1575	760	885
Grade 420, f_y = 420 MPa									
13	320	245	255	320	235	225	320	235	210
16	405	305	320	405	295	285	405	295	260
19	485	365	380	485	355	340	485	355	310
22	615	425	445	560	410	395	560	410	360
25	810	485	505	725	470	455	665	470	415
29	1025	550	570	920	530	510	840	530	465
32	1300	615	645	1165	600	575	1065	600	525
36	1600	685	710	1430	665	635	1305	665	580
43	1985	820	855	1775	795	765	1620	795	700
57	2700	1095	1140	2415	1060	1020	2205	1060	930

Notes:
1. Numbers are rounded up to nearest 5 mm.
2. Basic hook development length has included reinforcement yield strength modification factor.
3. Minimum tension development length (AASHTO 5.11.2.1). Maximum of (1) basic tension development length times appropriate modification factors (AASHTO 5.11.2.1.2 and 5.11.2.1.3) and (2) 300 mm.
4. Minimum compression development length (AASHTO 5.11.2.2). Maximum of (1) basic compression development length times appropriate modification factors (AASHTO 5.11.2.2.2) and (2) 200 mm.
5. Minimum hook development length (AASHTO 5.11.2.4). Maximum of (1) basic hook development length times appropriate modification factors (AASHTO 5.11.2.4.2), (2) eight bar diameters, and (3) 150 mm.

Except at supports of simple spans and at the free ends of cantilevers, reinforcement (AASHTO 5.11.1.2) should be extended beyond the point at which it is no longer required to resist the flexural demand for a distance of

$$\text{the largest of} \begin{cases} \text{the effective depth of the member} \\ \text{15 times the nominal diameter of a bar} \\ \text{0.05 times the clear span length} \end{cases} \tag{9.42}$$

Continuing reinforcement shall extend not less than the development length beyond the point where bent or terminated tension reinforcement is no longer required for resisting the flexural demand.

TABLE 9.7 Minimum Rebar Spacing for CIP Concrete (mm) (AASHTO 5.10.3)

Bar Size	Minimum Spacing			
13	51	51	63	63
16	54	56	70	70
19	57	68	76	83
22	60	78	82	96
25	64	90	90	110
29	72	101	101	124
32	81	114	114	140
36	90	127	127	155
43	108	152	152	
57	143	203	203	

Notes:
1. Clear distance between bars should not be less than 1.5 times the maximum size of the course aggregate.
2. Note 1 does not need to be verified when maximum size of the course aggregate grading is less than 25 mm.
3. Bars spaced less than $3d_b$ on center require modification of development length (AASHTO 5.11.2.1.2).

For negative moment reinforcement, in addition to the above requirement for bar cutoff, it must be extended to a length beyond the inflection point for a distance of

$$\text{the largest of} \begin{cases} \text{the effective depth of the member} \\ \text{12 times the nominal diameter of a bar} \\ \text{0.0625 times the clear span length} \end{cases} \tag{9.43}$$

9.5 Design Examples

9.5.1 Solid Slab Bridge Design

Given
A simple span concrete slab bridge with clear span length (S) of 9150 mm is shown in Figure 9.8. The total width (W) is 10,700 mm, and the roadway is 9640 wide (W_R) with 75 mm (d_W) of future wearing surface.

The material properties are as follows: Density of wearing surface $\rho_w = 2250$ kg/m³; concrete density $\rho_c = 2400$ kg/m³; concrete strength $f_c' = 28$ MPa, $E_c = 26\ 750$ MPa; reinforcement $f_y = 420$ MPa, $E_s = 200,000$ MPa; $n = 8$.

Requirements
Design the slab reinforcement base on AASHTO-LRFD (1994) Strength I and Service I (cracks) Limit States.

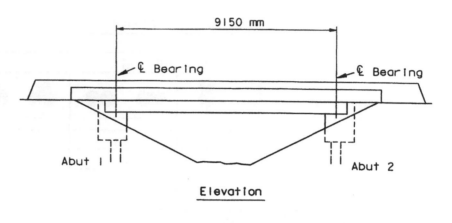

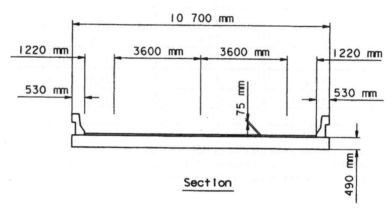

FIGURE 9.8   Solid slab bridge design example.

Solution

1. **Select Deck Thickness (Table 9.4)**

$$h_{min} = 1.2\left(\frac{S+3000}{30}\right) = 1.2\left(\frac{9150+3000}{30}\right) = 486\,mm$$

Use $h = 490$ mm

2. **Determine Live Load Equivalent Strip Width (AASHTO 4.6.2.3 and 4.6.2.1.4b)**
   a. *Interior strip width:*
      i. Single-lane loaded:

$$E_{interior} = 250 + 0.42\sqrt{L_1 W_1}$$

L_1 = lesser of actual span length and 18,000 mm
W_1 = lesser of actual width or 9000 mm for single lane loading or 18,000 mm for multilane loading

$$E_{interior} = 250 + 0.42\sqrt{(9150)(9000)} = 4061\text{ mm}$$

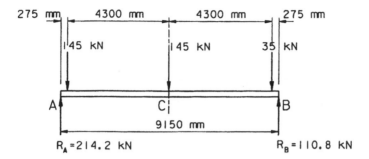

FIGURE 9.9 Position of design truck for maximum moment.

ii. Multilane loaded:

$$N_L = INT\left(\frac{W}{3600}\right) = INT\left(\frac{10,700}{3600}\right) = 2$$

$$\frac{W}{N_L} = \frac{10,700}{2} = 5350 \text{ mm}$$

$$E_{interior} = 2100 + 0.12\sqrt{L_1 W_1} = 2100 + \sqrt{(9150)(10,700)} = 3287 \text{ mm} < 5350 \text{ mm}$$

Use $E_{interior} = 3287$ mm

b. *Edge strip width*:

E_{edge} = the distance between the edge of the deck and the inside face of the barrier
      + 300 mm + ½ strip width < full strip or 1800 mm

$$E_{edge} = 530 + 300 + \frac{3287}{2} = 2324 \text{ mm} > 1800 \text{ mm}$$

Use $E_{edge} = 1800$ mm

3. **Dead Load**
   Slab: $W_{slab} = (0.49)(2400)(9.81)(10^{-3}) = 11.54 \text{ kN/m}^2$
   Future wearing: $W_{fw} = (0.075)(2250)(9.81)(10^{-3}) = 1.66 \text{ kN/m}^2$
   Assume 0.24 m³ concrete per linear meter of concrete barrier
   Concrete barrier: $W_{barrier} = (0.24)(2400)(9.81)(10^{-3}) = 5.65 \text{ kN/m}^2$

4. **Calculate Live-Load Moments**
   Moment at midspan will control the design.
   a. *Moment due to the design truck* (see Figure 9.9):

$$M_{LL\text{-}Truck} = (214.2)(4.575) - (145)(4.3) = 356.47 \text{ kN·m}$$

   b. *Moment due to the design tandem* (see Figure 9.10):

$$M_{LL\text{-}Tandem} = (95.58)(4.575) = 437.28 \text{ kN·m.}$$

Design Tandem Controls

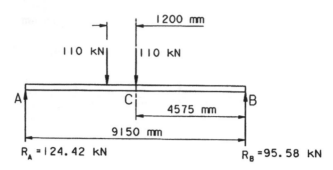

FIGURE 9.10 Position of tandem for maximum moment.

c. *Moment due to lane load:*

$$M_{\text{LL-Lane}} = \frac{(9.3)(9.15)^2}{8} = 97.32 \text{ kN·m}$$

5. **Determine Load Factors (AASHTO Table 3.4.1-1) and Load Combinations (AASHTO 1.3.3-5)**

   a. *Strength I Limit State load factors:*

   Weight of superstructure (DC): 1.25
   Weight of wearing surface (DW): 1.50
   Live Load (LL): 1.75
   $\eta_d = 0.95$, $\eta_R = 1.05$, $\eta_I = 0.95$
   $\eta = (0.95)(1.05)(0.95) = 0.948 \leq 0.95$

   Use $\eta = 0.95$

   b. *Interior strip moment* (1 m wide) (AASHTO 3.6.2.1 and 3.6.1.2.4):

   Dynamic load factor   IM = 0.33

   Lane load $M_{\text{LL-Lane}} = \left(\dfrac{97.32}{3.287}\right) = 29.61 \text{ kN·m}$

   Live load $M_{\text{LL+IM}} = (1+0.33)\left(\dfrac{437.28}{3.287}\right) + 29.61 = 206.54 \text{ kN·m}$

   Future wearing $M_{\text{DW}} = \dfrac{W_{fw}L^2}{8} = \dfrac{(1.66)(9.15)^2}{8} = 17.37 \text{ kN·m}$

   Dead load $M_{\text{DC}} = \dfrac{W_{slab}L^2}{8} = \dfrac{(11.54)(9.15)^2}{8} = 120.77 \text{ kN·m}$

   Factored moment $M_U = \eta[1.25(M_{\text{DC}}) + 1.50(M_{\text{DW}}) + 1.75(M_{\text{LL+IM}})]$
                 $= (0.95)[1.25(120.77) + (1.50)(17.37) + (1.75)(206.54)]$
                 $= 511.54 \text{ kN·m}$

   c. *Edge strip moment* (1 m wide) (AASHTO Table 3.6.1.1.2-1):

   End strip is limited to half lane width, use multiple presence factor 1.2 and half design lane load.

Lane load $\qquad M_{\text{LL-Lane}} = (1.2)\left(\dfrac{1}{2}\right)\left(\dfrac{97.3}{1.8}\right) = 32.44 \ \text{kN·m}$

Live load $\qquad M_{\text{LL+IM}} = (1 + 0.33)(1.2)\left(\dfrac{1}{2}\right)\left(\dfrac{437.28}{1.8}\right) + 32.44 = 226.3 \ \text{kN·m}$

Dead load $\qquad M_{\text{DC}} = \left(11.54 + \dfrac{5.65}{1.8}\right)\left(\dfrac{9.15^2}{8}\right) = 153.63 \ \text{kN·m}$

Future wearing $\qquad M_{\text{DW}} = (1.66)\left(\dfrac{1.8 - 0.53}{1.8}\right)\left(\dfrac{9.15^2}{8}\right) = 12.25 \ \text{kN·m}$

Factored moment $\qquad M_U = (0.95)[(1.25)\,(153.63) + (1.50)(12.25) + (1.75)(226.3)] =$
$\qquad\qquad\qquad\quad 579.12 \ \text{kN·m}$

6. Reinforcement Design
a. *Interior strip:*

Assume No. 25 bars, $d = 490 - 25 - \left(\dfrac{25}{2}\right) = 452.5 \ \text{mm}$.

The required reinforcements are calculated using Eqs. (9.11), (9.16), and (9.17).

Neglect the compression steel and set $b_w = b$ for sections without compression flange.

$$M_u = \phi A_s f_y \left(d - \dfrac{a}{2}\right) \quad \text{and} \quad a = c\beta_1 = \dfrac{A_s f_y}{0.85 f_c' b_w}$$

A_s can be solved by substituting a into M_u or

$$R_u = \dfrac{M_u}{\phi b d^2} = \dfrac{511.54 \times 10^6}{(0.9)(1000)(452.5)^2} = 2.766 \ \text{N/mm}$$

$$m = \dfrac{f_y}{(0.85) f_c'} = \dfrac{420}{(0.85)(28)} = 17.647$$

$$\rho = \dfrac{1}{m}\left[1 - \sqrt{1 - \dfrac{2mR_u}{f_y}}\right] = \dfrac{1}{17.647}\left[1 - \sqrt{1 - \dfrac{2(17.647)(2.776)}{420}}\right] = 0.00705$$

Required reinforced steel $A_s = \rho b d = (0.00705)(1000)(452.5) = 3189 \ \text{mm}^2/\text{m}$.

Maximum allowed spacing of No. 25 bar $= 510/3189 = 0.160 \ \text{m}$.

Try No. 25 bars at 150 mm.
i. Check limits for reinforcement:

$\beta_1 = 0.85$ for $f_c' = 28 \ \text{MPa}$; see Eq. (9.18)

$$c = \dfrac{A_s f_y}{0.85 \beta_1 f_c' b_w} = \dfrac{(510)(420)}{0.85(0.85)(28)(150)} = 70.6 \ \text{mm}$$

from Eqs. (9.19),

$$\dfrac{c}{d} = \dfrac{70.79}{452.5} = 0.156 \le 0.42 \qquad\qquad \text{OK}$$

from Eqs. (9.20),

$$\rho_{min} = \frac{510}{(150)(452.5)} = 0.007\,51 \geq (0.03)\left(\frac{28}{420}\right) = 0.002 \qquad \text{OK}$$

ii. Check crack control:

Service load moment $M_{sa} = 1.0[1.0(M_{DC}) + 1.0(M_{DW}) + 1.0(M_{LL+IM})]$
$\qquad\qquad\qquad\qquad = [120.77 + 17.37 + (176.93 + 29.61)]$
$\qquad\qquad\qquad\qquad = 344.68$ kN·m

$$0.8f_r = 0.8(0.63\sqrt{f_c'}) = 0.8(0.63)\sqrt{28} = 2.66 \text{ MPa}$$

$$f_c = \frac{M_{sa}}{S} = \frac{344,680}{\frac{1}{6}(490)^2} = 8.61\,\text{MPa} \geq 0.8f_r;\, , \text{Section is cracked}$$

Cracked moment of inertia can be calculated by using Eqs. (9.32) to (9.35).

$n = 8$, $b = 150.0$ mm, $A_s = 510$ mm, $d = 452.5$ mm.

$$B = \frac{1}{b}(nA_s) = \frac{1}{150}(8)(510) = 27.2$$

$$C = \frac{2}{b}(ndA_s) = \frac{2}{150}(8)(452.5)(510) = 24616$$

$$x = \sqrt{B^2 + C} - B = \sqrt{(27.2)^2 + (24616)} - (27.2) = 132 \text{ mm}$$

$$I_{cr} = \frac{1}{3}bx^3 + nA_s(d - x)^2 = \frac{1}{3}(150)(132)^3 + (8)(510)(452.5 - 132)^2 = 534.1 \times 10^6 \text{mm}^4$$

From Eq. (9.38) $f_s = n\dfrac{M_{sa}(d - x)}{I_{cr}} = (8)\dfrac{(344,680)(452.5 - 132)}{534.1 \times 10^6} = 248$ MPa

Allowable tensile stress in the reinforcement can be calculated from Eq. (9.5) with $Z = 23,000$ N/mm for moderate exposure and

$$d_c = 25 + \frac{25}{2} = 37.5 \text{ mm}$$

$$A = 2d_c \times \text{bar spacing} = (2)(37.5)(150) = 11,250 \text{ mm}$$

$$f_{sa} = \frac{Z}{(d_c A)^{1/3}} \leq 0.6f_y$$

$$f_{sa} = \frac{23,000}{[(37.5)(11,250)]^{1/3}} = \frac{23,000}{75} = 307 \text{ MPa} \geq 0.6f_y = 0.6\,(420) = 252 \text{ MPa}$$

$$f_s = 248\,\text{MPa} \leq f_{sa} = 252\,\text{MPa}, \qquad\qquad\qquad \text{OK}$$

Use No. 25 Bar @ 150 mm for interior strip

b. *Edge strip*:

By similar procedure, Edge Strip Use No. 25 bar at 125 mm

7. **Determine Distribution Reinforcement (AASHTO 5.14.4.1)**
The bottom transverse reinforcement may be calculated as a percentage of the main reinforcement for positive moment:

$$\frac{1750}{\sqrt{L}} \leq 50\%, \text{ that is, } \frac{1750}{\sqrt{9150}} = 18.3\% \leq 50\%$$

a. *Interior strip:*
Main reinforcement: No. 25 at150 mm,

$$A_s = \frac{510}{150} = 3.40 \text{ mm}^2/\text{mm}.$$

Required transverse reinforcement = $(0.183)(3.40) = 0.622$ mm²/mm

Use No. 16 @ 300 mm transverse bottom bars,

$$A_s = \frac{199}{300} = 0.663 \text{ mm}^2/\text{mm}$$

b. *End strip:*
Main reinforcement: No. 25 at 125 mm,

$$A_s = \frac{510}{125} = 4.08 \text{ mm}^2/\text{mm}$$

Required transverse reinforcement = $(0.183)(4.08) = 0.746$ mm²/mm

Use No. 16 at 250 mm, $A_s = 0.79$ mm²/mm.

For construction consideration, Use No. 16 @250 mm across entire width of the bridge.

8. **Determine Shrinkage and Temperature Reinforcement (AASHTO 5.10.8)**
Temperature

$$A_s \geq 0.75 \frac{A_g}{f_y} = 0.75 \frac{(1)(490)}{420} = 0.875 \text{ mm}^2/\text{mm in each direction}$$

Top layer = $0.875/2 = 0.438$ mm²/mm

Use No. 13 @ 300 mm transverse top bars, $A_s = 0.430$ mm²/mm

9. **Design Sketch**
See Figure 9.11 for design sketch in transverse section.

10. **Summary**
To complete the design, loading combinations for all limit states need to be checked. Design practice should also give consideration to long-term deflection, cracking in the support area for longer or continuous spans. For large skew bridges, alteration in main rebar placement is essential.

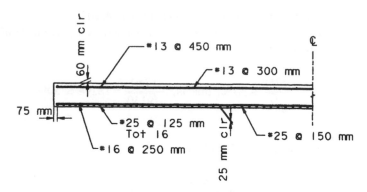

FIGURE 9.11   Slab reinforcement detail.

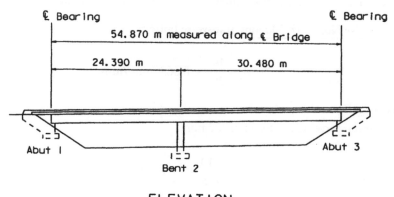

ELEVATION

FIGURE 9.12   Two-span reinforced box girder bridge.

9.5.2   Box-Girder Bridge Design

Given

A two-span continuous cast-in-place reinforced concrete box girder bridge, with span length of 24 390 mm (L_1) and 30 480 mm (L_2), is shown in Figure 9.12. The total superstructure width (W) is 10 800 mm, and the roadway width (W_R) is 9730 mm with 75 mm (d_W) thick of future wearing surface.

The material properties are assumed as follows: Density of wearing surface ρ_w = 2250 kg/m³; concrete density ρ_c = 2400 kg/m³; concrete strength f_c' = 28 MPa, E_c = 26 750 MPa; reinforcement f_y = 420 MPa, E_s = 200 000 MPa.

Requirements

Design flexural and shear reinforcements for an exterior girder based on AASHTO-LRFD (1994) Limit State Strength I, Service I (cracks and deflection), and Fatigue Limit States.

Solution

   1. Determine Typical Section (see Figure 9.13)

      a. *Section dimensions*:

         Try the following dimensions:

         Overall Structural Thickness, h = <u>1680 mm</u> (Table 9.4)

         Effective length, s = 2900 − 205 = 2695 mm

         Design depth (deck slab),

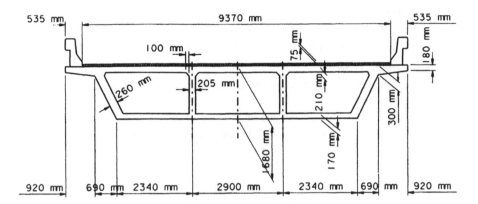

FIGURE 9.13 Typical section.

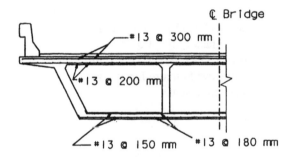

FIGURE 9.14 Slab reinforcement.

$t_{\text{top}} = \underline{210 \text{ mm}}$

$$> \frac{1}{20}(2900 - 205 - 100 \cdot 2) = 124.8 \text{ mm} \quad \text{(AASHTO 5.14.1.3)}$$

$$\frac{s}{t_{\text{top}}} = \frac{2695}{210} = 12.8 < 18 \quad \text{(AASHTO 9.7.2.4)}$$

Bottom flange depth,

$t_{\text{bot}} = \underline{170 \text{ mm}} > 140 \text{ mm (AASHTO 5.14.1.3)}$

$$> \frac{1}{16}(2900 - 205 - 100 \cdot 2) = 156 \text{ mm} \quad \text{(AASHTO 5.14.1.3)}$$

Web thickness, $b_w = \underline{205 \text{ mm}} > 200 \text{ mm for ease of construction (AASHTO 5.14.1.3)}$

b. *Deck slab reinforcement:*
   The detail slab design procedure is covered in Chapter 15 of this handbook. The slab design for this example, using the empirical method, is shown in Figure 9.14.

2. **Calculate Design Loads**
   The controlling load case is assumed to be Strength Limit State I.
   a. *Permanent load:*
      It is assumed that the self-weight of the box girder and the future wearing surface are equally distributed to each girder. The weight of the barrier rails is, however, distributed to the exterior girders only.
      Dead load of box girder = (0.000 023 57)(4 938 600) = 116.4 N/mm
      Dead load of the concrete barriers = 5.65(2) = 11.3 N/mm
      Dead load of the future wearing surface = (0.0000221)(729 750) = 16.12 N/mm
   b. *Live loads:*
      i. Vehicle live loads:
         A standard design truck (AASHTO 3.6.1.2.2), a standard design tandem (AASHTO 3.6.1.2.3), and the design lane load (AASHTO 3.6.1.2.4) are used to compute the extreme force effects.
      ii. Multiple presence factors (AASHTO 3.6.1.1.2 and AASHTO Table 3.6.1.1.2-1):
          No. of traffic lanes = INT (9730/3600) = 2 lanes
          The multiple presence factor, m = 1.0
      iii. Dynamic load allowance (AASHTO 3.6.2.1 and AASHTO Table 3.6.2.1-1):
           IM = 15% for Fatigue and Fracture Limit State
           IM = 33% for Other Limit States
   c. *Load modifiers:*
      For Strength Limit State:

$$\eta_D = 0.95; \quad \eta_R = 0.95; \quad \eta_I = 1.05; \quad \text{and} \quad \eta = \eta_D\eta_R\eta_I = 0.95 \text{ (AASHTO 1.3.2)}$$

   For Service Limit State:

$$\eta_D = 1.0; \quad \eta_R = 1.0; \quad \eta_I = 1.0; \quad \text{and} \quad \eta = \eta_D\eta_R\eta_I = 1.0 \text{ (AASHTO 1.3.2)}$$

   d. *Load factors:*

$$\gamma_{DC} = 0.9 \sim 1.25; \quad \gamma_{DW} = 0.65 \sim 1.50; \quad \gamma_{LL} = 1.75$$

   e. *Distribution factors for live-load moment and shear* (AASHTO 4.6.2.2.1):
      i. Moment distribution factor for exterior girders:
         For Span 1 and Span 2:

$$W_e = \frac{2900}{2} + 1211 = 2661 \text{ mm} \; < \; S = 2900 \text{ mm}$$

$$g_m^E = \frac{W_e}{4300} = \frac{2661}{4300} = 0.619$$

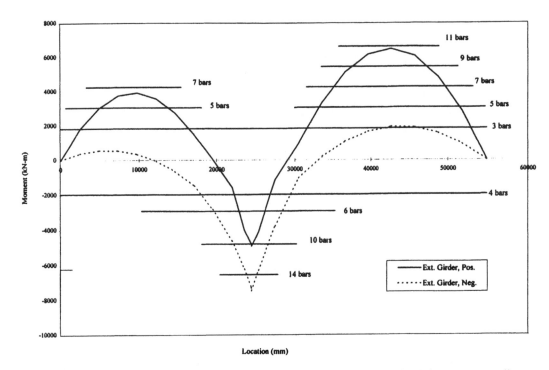

FIGURE 9.15 Design moment envelope and provided moment capacity with reinforcement cut-off.

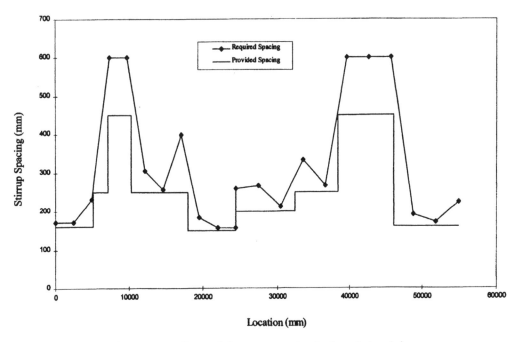

FIGURE 9.16 Shear reinforcement spacing for the exterior girder.

TABLE 9.8 Moment Envelope Summary for Exterior Girder at Every $1/10$ of Span Length of Span 1 and Span 2

Span	Distance (mm)	Unfactored Moment Envelope (kN-m)												Factored Moment Envelope (kN-m)	
		One Design Lane Load		One Truck		Train	Live Load Envelope		DC	DW	Exterior Girder		Exterior Girder	
		Positive	Negative	Positive	Negative	Negative	Positive	Negative			LL (Pos.)	LL (Neg.)	Positive	Negative
0.0 L_1	0	0	0	0	0	0	0	0	0	0	0	0	0	0
0.1 L_1	2439	216	-55	568	-93	-93	971	-179	607	70	601	-111	1819	378
0.2 L_1	4878	377	-110	955	-187	-187	1647	-359	1005	116	1020	-222	3053	561
0.3 L_1	7317	482	-165	1203	-281	-281	2082	-539	1196	137	1289	-333	3758	553
0.4 L_1	9756	533	-220	1303	-375	-375	2266	-719	1178	135	1403	-445	3923	350
0.5 L_1	12195	528	-275	1278	-468	-469	2228	-897	951	109	1379	-556	3577	-43
0.6 L_1	14634	467	-331	1157	-562	-563	2006	-1078	517	59	1242	-668	2762	-632
0.7 L_1	17073	352	-386	912	-656	-657	1565	-1258	-126	-15	969	-779	1494	-1466
0.8 L_1	19512	181	-441	570	-750	-751	939	-1439	-976	-113	581	-890	62	-2800
0.9 L_1	21951	38	-580	220	-844	-1055	331	-1785	-2035	-234	205	-1105	-1545	-4587
0.96 L_1	23414	-216	-748	29	-903	-1396	-177	-2344	-2810	-324	-110	-1451	-3981	-6211
1.0 L_1	24390	-326	-878	0	-938	-1616	-326	-2725	-3303	-380	-202	-1686	-4799	-7267
0.0 L_2	24390	-281	-904	0	-1059	-1643	-281	-2780	-3400	-392	-174	-1721	-4885	-7457
0.03 L_2	25304	-273	-754	0	-868	-1433	-273	-2394	-2835	-327	-169	-1482	-4112	-6295
0.1 L_2	27438	41	-466	224	-527	-960	339	-1569	-1597	-184	210	-971	-1130	-3772
0.2 L_2	30486	202	-234	640	-468	-468	1053	-856	-119	-14	652	-530	974	-1042
0.3 L_2	33534	470	-196	1072	-410	-410	1896	-741	1035	119	1173	-459	3349	195
0.4 L_2	36582	662	-168	1403	-351	-351	2528	-635	1862	214	1565	-393	5118	1071
0.5 L_2	39630	768	-140	1591	-293	-293	2884	-530	2365	272	1785	-328	6163	1645
0.6 L_2	42678	787	-112	1640	-234	-234	2968	-423	2543	292	1837	-262	6490	1919
0.7 L_2	45726	720	-84	1532	-176	-176	2758	-318	2395	275	1707	-197	6073	1890
0.8 L_2	48774	566	-56	1210	-117	-117	2175	-212	1922	221	1347	-131	4835	1562
0.9 L_2	51822	326	-28	708	-59	-59	1268	-106	1123	129	785	-66	2822	930
1.0 L_2	54870	0	0	0	0	0	0	0	0	0	0	0	0	0

TABLE 9.9 Shear Envelope Summary at Every 1/10 of Span 1 and Span 2

Span	Distance (mm)	One Design Lane Load Positive	One Design Lane Load Negative	One Truck Positive	One Truck Negative	Live Load Envelope Positive	Live Load Envelope Negative	DC	DW	Exterior Girder LL (Pos.)	Exterior Girder LL (Neg.)	Factored Shear Envelope Exterior Girder Positive	Factored Shear Envelope Exterior Girder Negative
0.0 L_1	0	100	-23	272	-38	462	-74	292	34	353	-56	981	176
0.1 L_1	2439	77	-23	233	-38	387	-74	206	24	296	-56	770	97
0.2 L_1	4878	55	-23	195	-38	314	-74	120	14	240	-56	562	18
0.3 L_1	7317	32	-23	159	-70	243	-116	36	4	186	-89	358	-115
0.4 L_1	9756	9	-23	124	-105	174	-163	-50	-6	133	-124	175	-274
0.5 L_1	12195	-13	-36	92	-143	109	-226	-136	-16	84	-173	14	-471
0.6 L_1	14634	-23	-59	64	-178	62	-296	-221	-26	48	-226	-126	-675
0.7 L_1	17073	-23	-81	41	-212	32	-363	-306	-35	24	-278	-243	-875
0.8 L_1	19512	-23	-104	23	-243	8	-427	-391	-45	6	-327	-353	-1072
0.9 L_1	21951	-23	-127	8	-271	-12	-487	-477	-55	-9	-373	-660	-1264
1.0 L_1	24390	-23	-149	0	-295	-23	-541	-563	-65	-16	-414	-787	-1449
0.0 L_2	24390	171	9	299	0	569	9	645	74	442	6	1606	882
0.1 L_2	27438	143	9	277	-3	511	5	539	62	383	4	1365	734
0.2 L_2	30486	115	9	250	-15	448	-11	432	50	335	-8	1141	386
0.3 L_2	33534	86	9	220	-38	379	-42	325	38	284	-31	910	249
0.4 L_2	36582	58	9	186	-59	305	-69	218	25	229	-52	675	115
0.5 L_2	39630	30	9	149	-87	228	-107	112	13	171	-80	435	-30
0.6 L_2	42678	9	-8	111	-119	157	-166	5	1	117	-125	202	-202
0.7 L_2	45726	9	-36	75	-154	109	-241	-102	-12	81	-180	41	-437
0.8 L_2	48774	9	-65	48	-192	73	-320	-209	-24	55	-240	-103	-681
0.9 L_2	51822	9	-93	20	-231	36	-400	-316	-36	27	-300	-248	-925
1.0 L_2	54870	9	-121	19	-272	34	-483	-422	-49	26	-362	-348	-1171

ii. Shear distribution factor for exterior girders:

Design Lane	Span 1	Span 2
One design lane loaded	$g_v^E = \dfrac{0.5(1015+2815)}{\left(\dfrac{\sqrt{5}}{2}\right)(2884)} = 0.594$	$g_v^E = \dfrac{0.5(1015+2815)}{\left(\dfrac{\sqrt{5}}{2}\right)(2884)} = 0.594$
Two or more design lanes loaded	$d_e = 1066 - 535 = 531 < 1500$	$d_e = 1066 - 535 = 531 < 1500$
	$e = 0.64 + \dfrac{531}{3800} = 0.78$	$e = 0.64 + \dfrac{531}{3800} = 0.78$
	$g_v^E = 0.78\left(\dfrac{2900}{2200}\right)^{0.9}\left(\dfrac{1680}{24,385}\right)^{0.1}$	$g_v^E = 0.78\left(\dfrac{2900}{2200}\right)^{0.9}\left(\dfrac{1680}{30,480}\right)^{0.1}$
	$= 0.765$	$= 0.749$
Govern	0.765	0.749

f. *Factored moment envelope and shear envelope:*
   The moment and shear envelopes for the exterior girder, unfactored and factored based on Strength Limit State I, are listed in Tables 9.8 and 9.9. Figures 9.15 and 9.16 show the envelope diagram for moments and shears based on Strength Limit State I, respectively.

3. **Flexural Design**
   a. *Determine the effective flange width* (Section 9.4.6):
      i. Effective compression flange for positive moments:
          Span 1:
             For interior girder,

$$b_{top}^I = \text{the smallest of} \begin{cases} \dfrac{1}{4}L_{1,eff} = \dfrac{1}{4}(0.65)(24,390) = 3963 \text{ mm} \\[2mm] 12t_{top} + b_w = 12(210) + (205) = \underline{2725 \text{ mm}} \text{ } governs \\[2mm] \text{the average spacing of adjacent beams} = 2900 \text{ mm} \end{cases}$$

             For exterior girder,

$$b_{top}^E = \dfrac{1}{2}b_{top}^I + \text{the smallest of}$$

$$\begin{cases} \dfrac{1}{8}L_{1,eff} = \dfrac{1}{8}(0.65)(24,390) = 1982 \text{ mm} \\[3mm] 6t_{top} + \dfrac{1}{2}b_w = (6)(210) + \dfrac{1}{2}(291) = 1405 \text{ mm} \\[3mm] \text{the width of the overhang} = 920 + \dfrac{291}{2} = \underline{1065 \text{ mm}} \text{ } governs \end{cases}$$

$$= \dfrac{1}{2}(2724) + 1065$$

$$= 2427 \text{ mm}$$

          Span 2: The effective flange widths for Span 2 turns out to be the same as those in Span 1.

ii. Effective compression flange for negative moments:
Span 1:
For interior girder,

$$b'_{bot} = \text{the smallest of} \begin{cases} \frac{1}{4}L_{eff} = \frac{1}{4}[(0.5)(24\ 390) + (0.25)(30\ 480) = 4954 \text{ mm} \\[2mm] 12t_{bot} + b_w = 12(170) + (205) = \underline{2245 \text{ mm}} \quad governs \\[2mm] \text{the average spacing of adjacent beams} = 2900 \text{ mm} \end{cases}$$

For exterior girder,

$$b^E_{bot} = \frac{1}{2}b'_{bot} + \text{the smallest of}$$

$$\begin{cases} \frac{1}{8}L_{eff} = \frac{1}{8}[(0.5)(24,390) + (0.25)(30,480) = 2477 \text{ mm} \\[3mm] 6t_{bot} + \frac{1}{2}b_w = (6)(170) + \frac{1}{2}(291) = 1166 \text{ mm} \\[3mm] \text{the width of the overhang} = 0 + \frac{291}{2} = \underline{146 \text{ mm}} \quad governs \end{cases}$$

$$= \frac{1}{2}(2245) + 146$$

$$= 1268 \text{ mm}$$

Span 2: The effective flange widths are the same as those in Span 1.
b. *Required flexural reinforcement:*
   The required reinforcements are calculated using Eqs. (9.16) and (9.17), neglecting the compression steel
   The minimum reinforcement required, based on Eq. (9.20), is

$$\rho_{min} \geq 0.03\frac{f'_c}{f_y} = 0.03\left(\frac{28}{420}\right) = 0.002$$

$$A_g(\text{Exterior girder}) = 1\ 103\ 530 \text{ mm}^2$$

$$A_{s_{min}}(\text{Exterior girder}) = (0.002)(1\ 103\ 530) = 2207 \text{ mm}^2$$

Use $A_{s_{min}} = \underline{2500 \text{ mm}^2}$
The required and provided reinforcements for sections located at ⅒ of each span interval and the face of the bent cap are listed in Table 9.10.

TABLE 9.10 Section Reinforcement Design for Exterior Girder

Section	Distance from Abut. 1 (mm)	Positive Moment					Negative Moment				
		M_u (kN-m)	A_s Required (mm²)	No. of Reinf. Bars Use #36	A_s (provided) (mm²)	ϕM_n (provided) (kN-m)	M_u (kN-m)	A_s Required (mm²)	No. of Reinf. Bars Use #32	A_s (Provided) (mm²)	ϕM_n (Provided) (kN-m)
0.0 L_1	0	0	0	3	3018	1841	0	0	4	3276	1954
0.1 L_1	2439	1819	2979	3	3018	1841	0	0	4	3276	1954
0.2 L_1	4878	3053	5020	5	5030	3055	0	0	4	3276	1954
0.3 L_1	7317	3758	6194	7	7042	4257	0	0	4	3276	1954
0.4 L_1	9756	3923	6469	7	7042	4257	0	0	4	3276	1954
0.5 L_1	12195	3577	5892	7	7042	4257	43	72	6	4914	2910
0.6 L_1	14634	2762	4537	5	5030	3055	632	1048	6	4914	2910
0.7 L_1	17073	1494	2444	3	3018	1841	1466	2448	6	4914	2910
0.8 L_1	19512	62	101	3	3018	1841	2800	4724	6	4914	2910
0.9 L_1	21951	0	0	3	3018	1841	4587	7848	10	8190	4780
0.96 L_1	23414	0	0	3	3018	1841	6211	10767	14	11466	6544
0.03 L_2	25304	0	0	3	3018	1841	6295	10920	14	11466	6544
0.1 L_2	27438	0	0	3	3018	1841	3772	6412	10	8190	4780
0.2 L_2	30486	974	1590	3	3018	1841	1042	1735	6	4914	2910
0.3 L_2	33534	3349	5512	7	7042	4257	0	0	6	4914	2910
0.4 L_2	36582	5118	8475	9	9054	5449	0	0	4	3276	1954
0.5 L_2	39630	6163	10243	11	11066	6629	0	0	4	3276	1954
0.6 L_2	42678	6490	10799	11	11066	6629	0	0	4	3276	1954
0.7 L_2	45726	6073	10091	11	11066	6629	0	0	4	3276	1954
0.8 L_2	48774	4835	7999	9	9054	5449	0	0	4	3276	1954
0.9 L_2	51822	2822	4637	5	5030	3055	0	0	4	3276	1954
1.0 L_2	54870	0	0	3	3018	1841	0	0	4	3276	1954

c. *Reinforcement layout:*
  i. Reinforcement cutoff (Section 9.4.7):
   • The extended length at cutoff for positive moment reinforcement, No. 36, is

   $$\text{the largest of} \begin{cases} \text{Effective depth of the section} = \underline{1625 \text{ mm}} \quad \textit{governs} \\ 15 \, d_b = 537 \text{ mm} \\ 0.05 \text{ of span length} = 0.05 \, (24 \, 390) = 1220 \text{ mm} \end{cases}$$

   From Table 9.6, the stagger lengths for No. 36 and No. 32 bars are

   $$l_d \text{ of No. 36 bars} = 1600 \text{ mm}$$

   $$l_d \text{ of No. 32 bars} = 1300 \text{ mm}$$

   • The extended length at cutoff for negative moment reinforcement, No. 32, is

   $$\text{the largest of} \begin{cases} \text{Effective depth of the section} = \underline{1601 \text{ mm}} \quad \textit{governs} \\ 15 \, d_b = 485 \text{ mm} \\ 0.05 \text{ of span length} = 0.05 \, (30 \, 480) = 1524 \text{ mm} \end{cases}$$

   • Negative moment reinforcements, in addition to the above requirement for bar cutoff, have to satisfy Eq. (9.43). The extended length beyond the inflection point has to be the largest of the following:

   $$\begin{cases} d = \underline{1601 \text{ mm}} \quad \textit{governs for Span 1} \\ 12d_b = 387.6 \text{ mm} \\ 0.0625 \times (\text{clear span length}) = (0.0625)(24 \, 390) = 1524 \text{ mm} \\ \quad \text{or} \qquad\qquad\qquad\qquad = (0.0625)(30 \, 480) = \underline{1905 \text{ mm}} \textit{ governs for Span 2} \end{cases}$$

  ii. Reinforcement distribution (Section 9.4.2):

   $$\frac{1}{10} (\text{average adjacent span length}) = \frac{1}{10} (30 \, 480 + 24 \, 385) = 2743 \text{ mm}$$

   $$b_{\text{top}}^E = 2427 \text{ mm} < 2743 \text{ mm}$$

   All tensile reinforcements should be distributed within the effective tension flange width.

  iii. Side reinforcements in the web, Eq. (9.6)

   $$A_{sk} \geq 0.001(d_e - 760) = 0.001(1625 - 760) = 0.865 \text{ mm}^2/\text{mm of height}$$

   $$A_{sk} \leq \frac{A_s}{1200} = \frac{13,462}{1200} = 11.21 \text{ mm}^2/\text{mm of height}$$

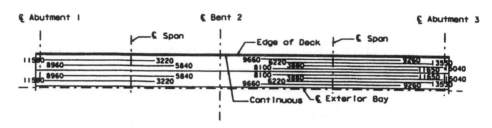

FIGURE 9.17   Bottom slab reinforcement of exterior girder.

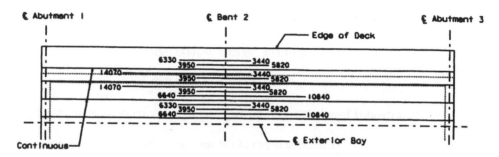

FIGURE 9.18   Top deck reinforcement of exterior girder.

$$A_{sk} = 0.865(250) = 216 \text{ mm}^2 \quad \underline{\text{Use No. 19 at 250 mm on each side face of the web}}$$

The reinforcement layout for bottom slab and top deck of exterior girder are shown in Figure 9.17 and 9.18, respectively. The numbers next to the reinforcing bars indicate the bar length extending beyond either the centerline of support or span.

4. **Shear Design**

From Table 9.9, it is apparent that the maximum shear demand is located at the critical section near Bent 2 in Span 2.

a. *Determine the critical section near Bent 2 in Span 2:*

$$A_s = 11,466 \text{ mm}^2, \ b = 1268 \text{ mm}$$

$$a = \frac{A_s f_y}{0.85 f_c' b} = \frac{(11,466)(420)}{0.85(28)(1268)} = 160 \text{ mm}$$

$$d_v = \text{ the largest of } \begin{cases} d_e - \dfrac{a}{2} = 1601 - \dfrac{160}{2} = \underline{1521 \text{ mm}} \ \textit{governs} \\[2mm] 0.9 d_e = 0.9(1601) = 1441 \text{ mm} \\[2mm] 0.72h = 0.72(1680) = 1210 \text{ mm} \end{cases}$$

The critical section is at a distance of d_v from the face of the support, i.e., distance between centerline of Bent 2 and the critical section = 600 + 1521 = 2121 mm = $0.07L_2$.

b. *At the above section, find M_u and V_u, using interpolation from Tables 9.8 and 9.9:*

$$M_u = 3772 + (6295 - 3772)(0.03/0.07) = 4853 \text{ kN·m}$$

$$V_u = 1365 + (1606 - 1365)(0.03/0.1) = 1437 \text{ kN}$$

$$v = \frac{V_u}{\phi_v b_v d_v} = \frac{1437 \cdot (1000)}{(0.9)(291)(1521)} = 3.61 \text{ MPa}$$

$$\frac{v}{f_c'} = \frac{3.61}{28} = 0.129 < 0.25 \quad \text{O.K.}$$

c. *Determine θ and β, and required shear reinforcement spacing:*
   Try $\theta = 37.5°$, $\cot \theta = 1.303$, $A_s = 11\ 466 \text{ mm}^2$, $E_c = 200 \text{ GPa}$, from Eq. (9.24).

$$\varepsilon_x = \frac{\dfrac{4\ 853\ 000}{1521} + 0.5(1437)(1.303)}{200(11\ 466)} = 1.80 \times 10^{-3}$$

   From Figure 9.6, we obtain $\theta = 37.5°$, which agrees with the assumption.
   Use $\theta = 37.5°$, $\beta = 1.4$, from Eq. (9.25)

$$V_s = \frac{1437}{0.9} - 0.083(1.4)\sqrt{28}(291)(1521) \times 10^{-3} = 1325 \text{ kN}$$

   Use No. 16 rebars, $A_v = 199(2) = 398 \text{ mm}^2$, from Eq. (9.26)

$$\text{Required spacing, } s \le \frac{(398)(420)(1521)}{1325 \times 10^3}(1.303) = 250 \text{ mm}$$

d. *Determine the maximum spacing required:*
   Note that $V_u = 1437 \text{ kN} > 0.1\ f_c' b_v' d_v = 0.1(28)(291)(1521) \times 10^{-3} = 1239 \text{ kN}$
   From Eqs. (9.27) and (9.29):

$$s_{max} = \text{the smallest of} \begin{cases} \dfrac{(398)(420)}{0.083\sqrt{28}(291)} = 1307 \text{ mm} \\[2mm] 0.4(1521) = 608 \text{ mm} \\[2mm] \underline{300 \text{ mm}} \quad \text{governs} \end{cases}$$

   Use $s = 250 \text{ mm} < 300 \text{ mm}$ OK.
e. *Check the adequacy of the longitudinal reinforcements, using Eq. (9.30):*

$$A_s f_y = (11\ 466)(420) = 4\ 815\ 720 \text{ N}$$

$$\frac{M_u}{d_v \phi_f} + \left(\frac{V_u}{\phi_v} - 0.5 V_s \right) \cot \theta = \frac{4853 \times 10^6}{(1521)(0.9)} + \left(\frac{1437 \times 10^3}{0.9} - 0.5(1325 \times 10^3) \right)(1.303)$$

$$= 4\ 762\ 401 \text{ N} < A_s f_y \quad \text{O.K.}$$

Using the above procedure, the shear reinforcements, i.e., stirrups in the web, for each section can be obtained. Figure 9.16 shows the shear reinforcements required and provided in the exterior girder for both spans.

6. **Crack Control Check (Section 9.4.2)**

For illustration purpose, we select the section located at midspan of Span 1 in this example, i.e. at $0.5 \, L_1$

a. *Check if the section is cracked:*

Service load moment, $M_{pos} = (1.0)(M_{DC} + M_{DW} + M_{LL+IM})$
$$= (1.0)(951 + 109 + 1379)$$
$$= 2439 \text{ kN-m}$$

Modulus of rupture $f_r = 0.63\sqrt{f_c'} = 0.63\sqrt{28} = 3.33 \text{ MPa}, \ 0.8f_r = 2.66 \text{ MPa}$

$b_{top} = 2427 \text{ mm}, \ b_{bot} = 1268 \text{ mm}$, obtain

$I_g = 4.162 \times 10^{11} \text{ mm}^4$ and $\bar{y} = 655 \text{ mm}$,

where $\bar{y}$ is the distance from the most compressed concrete fiber to the neutral axis

$$S = \frac{I_g}{(d - \bar{y})} = \frac{4.162 \times 10^{11}}{(1680 - 655)} = 4.06 \times 10^8 \text{ mm}^3$$

$$f_c = \frac{M_{pos}}{S} = \frac{2439 \times 10^6}{4.06 \times 10^8} = 6.01 \text{ MPa} > 0.8f_r = 2.66 \text{ MPa}$$

The section is cracked.

b. *Calculate tensile stress of the reinforcement:*

Assuming the neutral axis is located in the web, thus applying Eqs. (9.31) through (9.34) with $A_s = 7042 \text{ mm}^2$, $A_s' = 0$, and $\beta_1 = 0.85$, solve for x

$$x = 239 \text{ mm} > h_f = b_{top} = 210 \text{ mm} \quad \text{O.K.}$$

From Eq. (9.35), obtain

$$I_{cr} = \frac{1}{3}(2427)(239)^3 - \frac{1}{3}(2427 - 291)(239 - 210)^3 + 7(7042)(1625 - 239)^2$$

$$= 1.057 \times 10^{11} \text{ mm}^4$$

and from Eq. (9.38), the tensile stress in the longitudinal reinforcement is

$$f_s = \frac{7(2439 \times 10^6)(1625 - 239)}{1.057 \times 10^{11}} = 224 \text{ MPa}$$

c. *The allowable stress can be obtained using Eq. (9.5), with Z = 30 000 for moderate exposure and d_c = 50 mm*

$$f_{sa} = \frac{Z}{(d_c A)^{\frac{1}{3}}} = \frac{30,000}{\left((50)\dfrac{(50 \cdot 2 \cdot 1268)}{7} \right)^{\frac{1}{3}}} = 310 \text{ MPa} > 0.6f_y = 252 \text{ MPa}$$

$\underline{\text{Use } f_{sa} = 252 \text{ MPa} > f_s = 223 \text{ MPa}} \qquad \text{O.K.}$

The other sections can be checked following the same procedure described above.

7. **Check Deflection Limit**

Based on the Service Limit State, we can compute the I_e for sections at $\frac{1}{10}$ of the span length interval. For illustration, let the section be at $0.4L_2$

Deflection distribution factor = (no. of design lanes)/(no. of supporting beams) = 2/4 = 0.5
Note that b_{top} = 2424 mm, t_{top} = 210 mm, b_w = 291 mm, h = 1680 mm, d = 1625 mm, b_{bot} = 1268 mm, t_{bot} = 170 mm, and neglecting compression steel

$$A_g = (2427)(210) + (1680 - 210 - 170)(291) + (1268)(170) = 1\ 103\ 530\ \text{mm}^2$$

$$y_t = \left[\frac{(509\ 670)\left(1680-\frac{210}{2}\right)+(378\ 300)\left(170+\frac{1300}{2}\right)+(215\ 560)\left(\frac{170}{2}\right)}{1\ 103\ 530}\right] = 1025\ \text{mm}$$

$$I_g = \frac{1}{12}(2427)(210)^3 + (509\ 670)(550)^2 + \frac{1}{12}(291)(1300)^3 + (378\ 300)(205)^2$$

$$+ \frac{1}{12}(1268)(170)^3 + (215\ 560)(940)^2$$

$$= 4.16 \times 10^{11}\ \text{mm}^4$$

$$M_{cr} = f_r \frac{I_g}{y_t} = (3.33)\frac{4.16 \times 10^{11}}{1025} = 1.35 \times 10^9\ \text{N-mm}$$

Use Eqs. (9.31) throuth (9.35) to solve for x and I_{cr} with A_s = 9054 mm² and $A_s' = 0$, we obtain

$$x = 272\ \text{mm}, \quad I_{cr} = 1.32 \times 10^{11}\ \text{mm}^4$$

From Table 9.8:

$$M_a = 1862 + 214 + (0.5)(1565) = 2859\ \text{kN-m}$$

$$\frac{M_{cr}}{M_a} = \frac{1.35 \times 10^9}{2.86 \times 10^9} = 0.47$$

$$I_e = \left(\frac{M_{cr}}{M_a}\right)^3 I_g + \left[1 - \left(\frac{M_{cr}}{M_a}\right)^3\right] I_{cr} = (0.47)^3(4.16\times 10^{11}) + \left[1-(0.47)^3\right](1.32\times 10^{11}) = 1.61\times 10^{11}\ \text{mm}^4$$

The above computation can be repeated to obtain I_e for other sections. It is assumed that the maximum deflection occurs where the maximum flexural moment is. To be conservative, the minimum I_e is used to calculate the deflection.

$$\Delta_{max} = \begin{cases} 19\ \text{mm} & \text{truck load} \\ 13\ \text{mm} & \text{lane + 25\% of truck load} \end{cases} < \frac{L_2}{800} = \frac{30\ 480}{800} = 38\ \text{mm} \quad \text{O.K.}$$

8. Check Fatigue Limit State
For illustration purpose, check the bottom reinforcements for the section at $0.7L_1$. For positive moment at this section, A_s = 4024 mm², A_s' = 4095 mm², d = 1625 mm, and d' = 79 mm. Note that the maximum positive moment due to the assigned truck is 757 kN-m, while the largest negative moment 598 kN-m.

$$M_{max}\ \text{due to fatigue load} = 0.75(0.619)(757)(1 + 0.15) = 404\ \text{kN-m}$$

Use Eqs. (9.31) through (9.35) and (9.38) to obtain the maximum tensile stress in the main bottom reinforcements as

$$f_{max} = 64 \text{ MPa}$$

The negative moment at this section is

M_{min} due to fatigue load = $0.75(0.619)(-598)(1 + 0.15) = -319$ kN-m

Using Eqs. (9.31) through (9.35) and (9.38), with $A_s = 4095$ mm², $A'_s = 4024$ mm², $d = 1601$ mm, and $d' = 55$ mm, we obtain the maximum compressive stress in the main bottom reinforcements as

$$f_{min} = -7.0 \text{ MPa}$$

Thus, the stress range for fatigue

$$f_{max} - f_{min} = 64 - (-7.0) = 71 \text{ MPa}$$

From Eq. (9.9), allowable stress range

$$f_r = 145 - 0.33(-7.0) + 55(0.3) = 164 \text{ MPa} > 71 \text{ MPa} \qquad\qquad \text{OK}$$

Other sections can be checked in the same fashion described above.

9. Summary

The purpose of the above example is mainly to illustrate the design procedure for flexural and shear reinforcement for the girder. It should be noted that, in reality, the controlling load case may not be the Strength Limit State; therefore, all the load cases specified in the AASHTO should be investigated for a complete design. It should also be noted that the interior girder design can be achieved by following the similar procedures described herein.

References

1. AASHTO, *AASHTO LRFD Bridge Design Specifications*, American Association of State Highway and Transportation Officials, Washington, D.C., 1994.
2. ASTM, *Annual Book of ASTM Standards*, American Society for Testing and Materials, Philadelphia, 1996.
3. Caltrans, *Bridge Design Aids Manual*, California Department of Transportation, Sacramento, 1994.
4. Collins, M. P. and Mitchell, D., *Prestressed Concrete Structures*, Prentice-Hall, Englewood Cliffs, NJ, 1991.
5. Caltrans, *Bridge Design Practices*, California Department of Transportation, Sacramento, 1995.
6. ACI Committee 318, Building Code Requirements for Reinforced Concrete (ACI 318-95), American Concrete Institute, 1995.
7. Barker, R. M. and Puckett, J. A., *Design of Highway Bridges*, John Wiley & Sons, New York, 1997.
8. Park, R. and Paulay, T., *Reinforced Concrete Structures*, John Wiley & Sons, New York, 1975.
9. Xanthakos, P. P., *Theory and Design of Bridges*, John Wiley & Sons, New York, 1994.

10

Prestressed Concrete Bridges

Lian Duan
California Department of Transportation

Kang Chen
MG Engineering, Inc.

Andrew Tan
Everest International Consultants, Inc.

10.1 Introduction

Prestressed concrete structures, using high-strength materials to improve serviceability and durability, are an attractive alternative for long-span bridges, and have been used worldwide since the 1950s. This chapter focuses only on conventional prestressed concrete bridges. Segmental concrete bridges will be discussed in Chapter 11. For more detailed discussion on prestressed concrete, references are made to textbooks by Lin and Burns [1], Nawy [2], Collins and Mitchell [3].

10.1.1 Materials

10.1.1.1 Concrete

A 28-day cylinder compressive strength (f_c') of concrete 28 to 56 MPa is used most commonly in the United States. A higher early strength is often needed, however, either for the fast precast method used in the production plant or for the fast removal of formwork in the cast-in-place method. The modulus of elasticity of concrete with density between 1440 and 2500 kg/m³ may be taken as

$$E_c = 0.043 w_c \sqrt{f_c'} \tag{10.1}$$

where w_c is the density of concrete (kg/m³). Poisson's ratio ranges from 0.11 to 0.27, but 0.2 is often assumed.

The modulus of rupture of concrete may be taken as [4]

$$
f_r = \begin{cases}
0.63\sqrt{f_c'} & \text{for normal weight concrete — flexural} \\[2mm]
0.52\sqrt{f_c'} & \text{for sand - lightweight concrete — flexural} \\[2mm]
0.44\sqrt{f_c'} & \text{for all - lightweight concrete — flexural} \\[2mm]
0.1 f_c' & \text{for direct tension}
\end{cases} \qquad (10.2)
$$

Concrete shrinkage is a time-dependent material behavior and mainly depends on the mixture of concrete, moisture conditions, and the curing method. Total shrinkage strains range from 0.0004 to 0.0008 over the life of concrete and about 80% of this occurs in the first year.

For moist-cured concrete devoid of shrinkage-prone aggregates, the strain due to shrinkage ε_{sh} may be estimated by [4]

$$
\varepsilon_{sh} = -k_s k_h \left(\frac{t}{35+t} \right) 0.51 \times 10^{-3} \qquad (10.3)
$$

$$
K_s = \left[\frac{\dfrac{t}{26 e^{0.0142(V/S)} + t}}{\dfrac{t}{45+t}} \right] \left[\frac{1064 - 3.7(V/S)}{923} \right] \qquad (10.4)
$$

where t is drying time (days); k_s is size factor and k_h is humidity factors may be approximated by $K_h = (140\text{-}H)/70$ for $H < 80\%$; $K_h = 3(100\text{-}H)/70$ for $H \geq 80\%$; and V/S is volume to surface area ratio. If the moist-cured concrete is exposed to drying before 5 days of curing, the shrinkage determined by Eq. (10.3) should be increased by 20%.

For stem-cured concrete devoid of shrinkage-prone aggregates:

$$
\varepsilon_{sh} = -k_s k_h \left(\frac{t}{55+t} \right) 0.56 \times 10^{-3} \qquad (10.5)
$$

Creep of concrete is a time-dependent inelastic deformation under sustained load and depends primarily on the maturity of the concrete at the time of loading. Total creep strain generally ranges from about 1.5 to 4 times that of the "instantaneous" deformation. The creep coefficient may be estimated as [4]

$$
\psi(t, t_1) = 3.5 K_c K_f \left(1.58 - \frac{H}{120} \right) t_i^{-0.118} \frac{(t - t_i)^{0.6}}{10 + (t - t_i)^{0.6}} \qquad (10.6)
$$

$$
K_f = \frac{62}{42 + f_c'} \qquad (10.7)
$$

$$
K_s = \left[\frac{\dfrac{t}{26 e^{0.0142(V/S)} + t}}{\dfrac{t}{45+t}} \right] \left[\frac{1.8 + 1.77 e^{-0.0213(V/S)}}{2.587} \right] \qquad (10.8)
$$

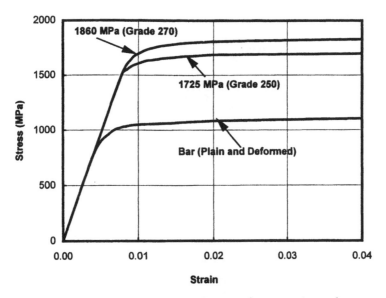

FIGURE 10.1 Typical stress–strain curves for prestressing steel.

where H is relative humidity (%); t is maturity of concrete (days); t_i is age of concrete when load is initially applied (days); K_c is the effect factor of the volume-to-surface ratio; and K_f is the effect factor of concrete strength.

Creep, shrinkage, and modulus of elasticity may also be estimated in accordance with CEB-FIP Mode Code [15].

10.1.1.2 Steel for Prestressing

Uncoated, seven-wire stress-relieved strands (AASHTO M203 or ASTM A416), or low-relaxation seven-wire strands and uncoated high-strength bars (AASHTO M275 or ASTM A722) are commonly used in prestresssed concrete bridges. Prestressing reinforcement, whether wires, strands, or bars, are also called *tendons*. The properties for prestressing steel are shown in Table 10.1.

TABLE 10.1 Properties of Prestressing Strand and Bars

Material	Grade and Type	Diameter (mm)	Tensile Strength f_{pu} (MPa)	Yield Strength f_{py} (MPa)	Modulus of Elasticity E_p (MPa)
Strand	1725 MPa (Grade 250)	6.35–15.24	1725	80% of f_{pu} except 90% of f_{pu}	197,000
	1860 MPa (Grade 270)	10.53–15.24	1860	for low relaxation strand	
Bar	Type 1, Plain	19 to 25	1035	85% of f_{pu}	
	Type 2, Deformed	15 to 36	1035	80% of f_{pu}	207,000

Typical stress–strain curves for prestressing steel are shown in Figure 10.1. These curves can be approximated by the following equations:

For Grade 250 [5]:

$$f_{ps} = \begin{cases} 197,000\,\varepsilon_{ps} & \text{for } \varepsilon_{ps} \leq 0.008 \\ 1710 - \dfrac{0.4}{\varepsilon_{ps} - 0.006} < 0.98 f_{pu} & \text{for } \varepsilon_{ps} > 0.008 \end{cases} \qquad (10.9)$$

For Grade 270 [5]:

$$f_{ps} = \begin{cases} 197,000\,\varepsilon_{ps} & \text{for } \varepsilon_{ps} \le 0.008 \\ 1848 - \dfrac{0.517}{\varepsilon_{ps} - 0.0065} < 0.98 f_{pu} & \text{for } \varepsilon_{ps} > 0.008 \end{cases} \qquad (10.10)$$

For Bars:

$$f_{ps} = \begin{cases} 207,000\,\varepsilon_{ps} & \text{for } \varepsilon_{ps} \le 0.004 \\ 1020 - \dfrac{0.192}{\varepsilon_{ps} - 0.003} < 0.98 f_{pu} & \text{for } \varepsilon_{ps} > 0.004 \end{cases} \qquad (10.11)$$

10.1.1.3 Advanced Composites for Prestressing

Advanced composites–fiber-reinforced plastics (FPR) with their high tensile strength and good corrosion resistance work well in prestressed concrete structures. Application of advanced composites to prestressing have been investigated since the 1950s [6–8]. Extensive research has also been conducted in Germany and Japan [9]. The Ulenbergstrasse bridge, a two-span (21.3 and 25.6 m) solid slab using 59 fiberglass tendons, was built in 1986 in Germany. It was the first prestressed concrete bridge to use advanced composite tendons in the world [10].

FPR cables and rods made of araramid, glass, and carbon fibers embedded in a synthetic resin have an ultimate tensile strength of 1500 to 2000 MPa, with the modulus of elasticity ranging from 62,055 MPa to 165,480 MPa [9]. The main advantages of FPR are (1) a high specific strength (ratio of strength to mass density) of about 10 to 15 times greater than steel; (2) a low modulus of elasticity making the prestress loss small; (3) good performance in fatigue; tests show [11] that for CFRP, at least three times the higher stress amplitudes and higher mean stresses than steel are achieved without damage to the cable over 2 million cycles.

Although much effort has been given to exploring the use of advanced composites in civil engineering structures (see Chapter 51) and the cost of advanced composites has come down significantly, the design and construction specifications have not yet been developed. Time is still needed for engineers and bridge owners to realize the cost-effectiveness and extended life expectancy gained by using advanced composites in civil engineering structures.

10.1.1.4 Grout

For post-tensioning construction, when the tendons are to be bound, grout is needed to transfer loads and to protect the tendons from corrosion. Grout is made of water, sand, and cements or epoxy resins. AASHTO-LRFD [4] requires that details of the protection method be indicated in the contract documents. Readers are referred to the *Post-Tensioning Manual* [12].

10.1.2 Prestressing Systems

There are two types of prestressing systems: pretensioning and post-tensioning systems. Pretensioning systems are methods in which the strands are tensioned before the concrete is placed. This method is generally used for mass production of pretensioned members. Post-tensioning systems are methods in which the tendons are tensioned after concrete has reached a specified strength. This technique is often used in projects with very large elements (Figure 10.2). The main advantage of post-tensioning is its ability to post-tension both precast and cast-in-place members. Mechanical prestressing–jacking is the most common method used in bridge structures.

FIGURE 10.2 A post–tensioned box–girder bridge under construction.

10.2 Section Types

10.2.1 Void Slabs

Figure 10.3a shows FHWA [13] standard precast prestressed voided slabs. Sectional properties are listed in Table 10.2. Although the cast-in-place prestressed slab is more expensive than a reinforced concrete slab, the precast prestressed slab is economical when many spans are involved. Common spans range from 6 to 15 m. Ratios of structural depth to span are 0.03 for both simple and continuous spans.

10.2.2 I-Girders

Figures 10.3b and c show AASHTO standard I-beams [13]. The section properties are given in Table 10.3. This bridge type competes well with steel girder bridges. The formwork is complicated, particularly for skewed structures. These sections are applicable to spans 9 to 36 m. Structural depth-to-span ratios are 0.055 for simple spans and 0.05 for continuous spans.

10.2.3 Box Girders

Figure 10.3d shows FHWA [13] standard precast box sections. Section properties are given in Table 10.4. These sections are used frequently for simple spans of over 30 m and are particularly suitable for widening bridges to control deflections.

The box-girder shape shown in Figure 10.3e is often used in cast-in-place prestressed concrete bridges. The spacing of the girders can be taken as twice the depth. . This type is used mostly for spans of 30 to 180 m. Structural depth-to-span ratios are 0.045 for simple spans, and 0.04 for continuous spans. The high torsional resistance of the box girder makes it particularly suitable for curved alignment (Figure 10.4) such as those needed on freeway ramps.

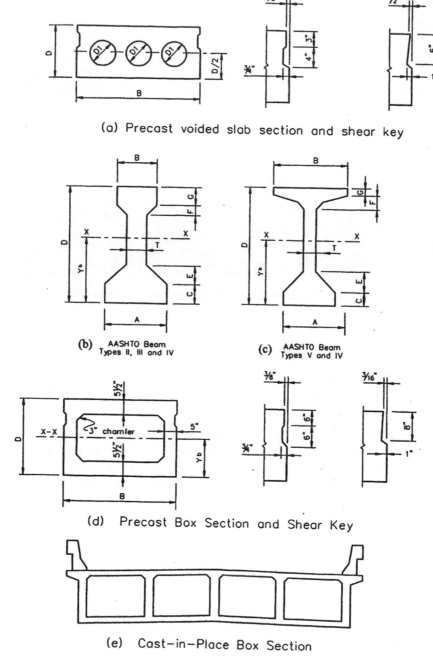

(a) Precast voided slab section and shear key

(b) AASHTO Beam
Types II, III and IV

(c) AASHTO Beam
Types V and IV

(d) Precast Box Section and Shear Key

(e) Cast-in-Place Box Section

FIGURE 10.3   Typical cross sections of prestressed concrete bridge superstructures.

10.3 Losses of Prestress

Loss of prestress refers to the reduced tensile stress in the tendons. Although this loss does affect the service performance (such as camber, deflections, and cracking), it has no effect on the ultimate strength of a flexural member unless the tendons are unbounded or the final stress is less than $0.5f_{pu}$ [5]. It should be noted, however, that an accurate estimate of prestress loss is more pertinent in some prestressed concrete members than in others. Prestress losses can be divided into two categories:

TABLE 10.2 Precast Prestressed Voided Slabs Section Properties (Fig. 10.3a)

Span Range, ft (m)	Section Dimensions				Section Properties		
	Width B in. (mm)	Depth D in. (mm)	D1 in. (mm)	D2 in. (mm)	A in.² (mm² 10⁶)	I_x in.⁴ (mm⁴ 10⁹)	S_x in.³ (mm³ 10⁶)
25	48	12	0	0	576	6,912	1,152
(7.6)	(1,219)	(305)	(0)	(0)	(0.372)	(2.877)	(18.878)
30~35	48	15	8	8	569	12,897	1,720
(10.1~10.70)	(1,219)	(381)	(203)	(203)	(0.362)	(5.368)	(28.185)
40~45	48	18	10	10	628	21,855	2,428
(12.2~13.7)	(1,219)	(457)	(254)	(254)	(0.405)	(10.097)	(310.788)
50	48	21	12	10	703	34,517	3,287
(15.2)	(1,219)	(533)	(305)	(254)	(0.454)	(1.437)	(53.864)

TABLE 10.3 Precast Prestressed I-Beam Section Properties (Figs. 10.3b and c)

AASHTO Beam Type	Section Dimensions, in. (mm)							
	Depth D	Bottom Width A	Web Width T	Top Width B	C	E	F	G
II	36 (914)	18 (457)	6 (152)	12 (305)	6 (152)	6 (152)	3 (76)	6 (152)
III	45 (1143)	22 (559)	7 (178)	16 (406)	7 (178)	7.5 (191)	4.5 (114)	7 (178)
IV	54 (1372)	26 (660)	8 (203)	20 (508)	8 (203)	9 (229)	6 (152)	8 (203)
V	65 (1651)	28 (711)	8 (203)	42 (1067)	8 (203)	10 (254)	3 (76)	5 (127)
VI	72 (1829)	28 (711)	8 (203)	42 (1067)	8 (203)	10 (254)	3 (76)	5 (127)

	Section Properties					
	A in.² (mm² 10ᵇ)	Y_b in. (mm)	I_x in.⁴ (mm⁴ 10⁹)	S_b in.³ (mm⁴ 10⁶)	S_t in.³ (mm⁴ 10⁶)	Span Ranges, ft (m)
II	369	15.83	50,980	3220	2528	40 ~ 45
	(0.2381)	(402.1)	(21.22)	(52.77)	(41.43)	(12.2 ~ 13.7)
III	560	20.27	125,390	6186	5070	50 ~ 65
	(0.3613)	(514.9)	(52.19)	(101.38)	(83.08)	(15.2 ~ 110.8)
IV	789	24.73	260,730	10543	8908	70 ~ 80
	(0.5090)	(628.1)	(108.52)	(172.77)	(145.98)	(21.4 ~ 24.4)
V	1013	31.96	521,180	16307	16791	90 ~ 100
	(0.6535)	(811.8)	(216.93)	(267.22)	(275.16)	(27.4 ~ 30.5)
VI	1085	36.38	733,340	20158	20588	110 ~ 120
	(0.7000)	(924.1)	(305.24)	(330.33)	(337.38)	(33.5 ~ 36.6)

- Instantaneous losses including losses due to anchorage set (Δf_{pA}), friction between tendons and surrounding materials (Δf_{pF}), and elastic shortening of concrete (Δf_{pES}) during the construction stage;
- Time-dependent losses including losses due to shrinkage (Δf_{pSR}), creep (Δf_{pCR}), and relaxation of the steel (Δf_{pR}) during the service life.

The total prestress loss (Δf_{pT}) is dependent on the prestressing methods.

For pretensioned members:

$$\Delta f_{pT} = \Delta f_{pES} + \Delta f_{pSR} + \Delta f_{pCR} + \Delta f_{pR} \tag{10.12}$$

For post-tensioned members:

$$\Delta f_{pT} = \Delta f_{pA} + \Delta f_{pF} + \Delta f_{pES} + \Delta f_{pSR} + \Delta f_{pCR} + \Delta f_{pR} \tag{10.13}$$

FIGURE 10.4   Prestressed box–girder bridge (I-280/110 Interchange, CA).

TABLE 10.4   Precast Prestressed Box Section Properties (Fig. 10.3d)

	Section Dimensions		Section Properties				
Span ft (m)	Width B in. (mm)	Depth D in. (mm)	A in.^{2} (mm^2 10^6)	Y_b in. (mm)	I_x in^4 (mm^4 10^9)	S_b in.^{3} (mm^3 10^6)	S_t in.^{3} (mm^3 10^6)
50	48	27	693	13.37	65,941	4,932	4,838
(15.2)	(1,219)	(686)	(0.4471)	(3310.6)	(27.447)	(80.821)	(710.281)
60	48	33	753	16.33	110,499	6,767	6,629
(18.3)	(1,219)	(838)	(0.4858)	(414.8)	(45.993)	(110.891)	(108.630)
70	48	39	813	110.29	168,367	8,728	8,524
(21.4)	(1,219)	(991)	(0.5245)	(490.0)	(70.080)	(143.026)	(1310.683)
80	48	42	843	20.78	203,088	9,773	9,571
(24.4)	(1,219)	(1,067)	(0.5439)	(527.8)	(84.532)	(160.151)	(156.841)

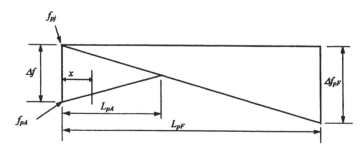

FIGURE 10.5   Anchorage set loss model.

TABLE 10.5 Friction Coefficients for Post-Tensioning Tendons

Type of Tendons and Sheathing	Wobble Coefficient K $(1/mm) \times (10^{-6})$	Curvature Coefficient μ (1/rad)
Tendons in rigid and semirigid galvanized ducts, seven-wire strands	0.66	0.05 ~ 0.15
Pregreased tendons, wires and seven-wire strands	0.98 ~ 6.6	0.05 ~ 0.15
Mastic-coated tendons, wires and seven-wire strands	3.3 ~ 6.6	0.05 ~ 0.15
Rigid steel pipe deviations	66	0.25, lubrication required

Source: AASHTO LRFD Bridge Design Specifications, 1st Ed., American Association of State Highway and Transportation Officials. Washington, D.C. 1994. With permission.

10.3.1 Instantaneous Losses

10.3.1.1 Anchorage Set Loss

As shown in Figure 10.5, assuming that the anchorage set loss changes linearly within the length (L_{pA}), the effect of anchorage set on the cable stress can be estimated by the following formula:

$$\Delta f_{pA} = \Delta f \left(1 - \frac{x}{L_{pA}}\right) \tag{10.14}$$

$$L_{pA} = \sqrt{\frac{E\,(\Delta L)\,L_{pF}}{\Delta f_{pF}}} \tag{10.15}$$

$$\Delta f = \frac{2\,\Delta f_{pF}\,L_{pA}}{L_{pF}} \tag{10.16}$$

where ΔL is the thickness of anchorage set; E is the modulus of elasticity of anchorage set; Δf is the change in stress due to anchor set; L_{pA} is the length influenced by anchor set; L_{pF} is the length to a point where loss is known; and x is the horizontal distance from the jacking end to the point considered.

10.3.1.2 Friction Loss

For a post-tensioned member, friction losses are caused by the tendon profile *curvature effect* and the local deviation in tendon profile *wobble effects.* AASHTO-LRFD [4] specifies the following formula:

$$\Delta f_{pF} = f_{pj}\left(1 - e^{-(Kx + \mu\alpha)}\right) \tag{10.17}$$

where K is the wobble friction coefficient and μ is the curvature friction coefficient (see Table 10.5); x is the length of a prestressing tendon from the jacking end to the point considered; and α is the sum of the absolute values of angle change in the prestressing steel path from the jacking end.

10.3.1.3 Elastic Shortening Loss Δf_{pES}

The loss due to elastic shortening can be calculated using the following formula [4]:

$$\Delta f_{pES} = \begin{cases} \dfrac{E_p}{E_{ci}}\,f_{cgp} & \text{for pretensioned members} \\[2em] \dfrac{N-1}{2N}\dfrac{E_p}{E_{ci}}\,f_{cgp} & \text{for post-tensioned members} \end{cases} \tag{10.18}$$

TABLE 10.6 Lump Sum Estimation of Time-Dependent Prestress Losses

Type of Beam Section	Level	For Wires and Strands with f_{pu} = 1620, 1725, or 1860 MPa	For Bars with f_{pu} = 1000 or 1100 MPa
Rectangular beams and solid slab	Upper bound	200 + 28 PPR	130 + 41 PPR
	Average	180 + 28 PPR	
Box girder	Upper bound	145 + 28 PPR	100
	Average	130 + 28 PPR	
I-girder	Average	$230\left[1.0-0.15\dfrac{f'_c-41}{41}\right]+41\,\text{PPR}$	130 + 41 PPR
Single–T, double–T hollow core and voided slab	Upper bound	$230\left[1.0-0.15\dfrac{f'_c-41}{41}\right]+41\,\text{PPR}$	$230\left[1.0-0.15\dfrac{f'_c-41}{41}\right]+41\,\text{PPR}$
	Average	$230\left[1.0-0.15\dfrac{f'_c-41}{41}\right]+41\,\text{PPR}$	

Note:
1. PPR is partial prestress ratio = $(A_{ps}f_{py})/(A_{ps}f_{py} + A_s f_y)$.
2. For low-relaxation strands, the above values may be reduced by
   - 28 MPa for box girders
   - 41 MPa for rectangular beams, solid slab and I-girders, and
   - 55 MPa for single–T, double–T, hollow–core and voided slabs.

Source: AASHTO LRFD Bridge Design Specifications, 1st Ed., American Association of State Highway and Transportation Officials. Washington, D.C. 1994. With permission.

where E_{ci} is modulus of elasticity of concrete at transfer (for pretensioned members) or after jacking (for post-tensioned members); N is the number of identical prestressing tendons; and f_{cgp} is sum of the concrete stress at the center of gravity of the prestressing tendons due to the prestressing force at transfer (for pretensioned members) or after jacking (for post-tensioned members) and the self-weight of members at the section with the maximum moment. For post-tensioned structures with bonded tendons, f_{cgp} may be calculated at the center section of the span for simply supported structures, at the section with the maximum moment for continuous structures.

10.3.2 Time-Dependent Losses

10.3.2.1 Lump Sum Estimation

AASHTO-LRFD [4] provides the approximate lump sum estimation (Table 10.6) of time-dependent loses Δf_{pTM} resulting from shrinkage and creep of concrete, and relaxation of the prestressing steel. While the use of lump sum losses is acceptable for "average exposure conditions," for unusual conditions, more-refined estimates are required.

10.3.2.2 Refined Estimation

a. *Shrinkage Loss:* Shrinkage loss can be determined by formulas [4]:

$$\Delta f_{pSR} = \begin{cases} 93 - 0.85H & \text{for pretensioned members} \\ 11 - 1.03H & \text{for post-tensioned members} \end{cases} \tag{10.19}$$

where H is average annual ambient relative humidity (%).

b. *Creep Loss:* Creep loss can be predicted by [4]:

$$\Delta f_{pCR} = 12 f_{cgp} - 7\Delta f_{cdp} \geq 0 \tag{10.20}$$

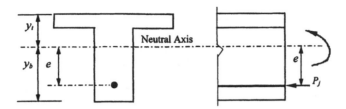

FIGURE 10.6 Prestressed concrete member section at Service Limit State.

where f_{cgp} is concrete stress at center of gravity of prestressing steel at transfer, and Δf_{cdp} is concrete stress change at center of gravity of prestressing steel due to permanent loads, except the load acting at the time the prestressing force is applied.

c. *Relaxation Loss:* The total relaxation loss (Δf_{pR}) includes two parts: relaxation at time of transfer Δf_{pR1} and after transfer Δf_{pR2}. For a pretensioned member initially stressed beyond $0.5 f_{pu}$, AASHTO-LRFD [4] specifies

$$
\Delta f_{pR1} = \begin{cases} \dfrac{\log 24t}{10}\left[\dfrac{f_{pi}}{f_{py}} - 0.55\right]f_{pi} & \text{for stress-relieved strand} \\[4mm] \dfrac{\log 24t}{40}\left[\dfrac{f_{pi}}{f_{py}} - 0.55\right]f_{pi} & \text{for low-relaxation strand} \end{cases} \tag{10.21}
$$

For stress-relieved strands

$$
\Delta f_{pR2} = \begin{cases} 138 - 0.4\Delta f_{pES} - 0.2(\Delta f_{pSR} + \Delta f_{pCR}) & \text{for pretensioning} \\[2mm] 138 - 0.3\Delta f_{pF} - 0.4\Delta f_{pES} - 0.2(\Delta f_{pSR} + \Delta f_{pCR}) & \text{for post-tensioning} \end{cases} \tag{10.22}
$$

where t is time estimated in days from testing to transfer. For low-relaxation strands, Δf_{pR2} is 30% of those values obtained from Eq. (10.22).

10.4 Design Considerations

10.4.1 Basic Theory

Compared with reinforced concrete, the main distinguishing characteristics of prestressed concrete are that

- The stresses for concrete and prestressing steel and deformation of structures at each stage, i.e., during prestressing, handling, transportation, erection, and the service life, as well as stress concentrations, need to be investigated on the basis of elastic theory.
- The prestressing force is determined by concrete stress limits under service load.
- Flexure and shear capacities are determined based on the ultimate strength theory.

For the prestressed concrete member section shown in Figure 10.6, the stress at various load stages can be expressed by the following formula:

$$
f = \frac{P_j}{A} \pm \frac{P_j e y}{I} \pm \frac{My}{I} \tag{10.23}
$$

TABLE 10.7 Stress Limits for Prestressing Tendons

Stress Type	Prestressing Method	Prestressing Tendon Type		
		Stress Relieved Strand and Plain High-Strength Bars	Low Relaxation Strand	Deformed High-Strength Bars
At jacking, f_{pj}	Pretensioning	$0.72f_{pu}$	$0.78f_{pu}$	—
	Post-tensioning	$0.76f_{pu}$	$0.80f_{pu}$	$0.75f_{pu}$
After transfer, f_{pt}	Pretensioning	$0.70f_{pu}$	$0.74f_{pu}$	—
	Post-tensioning — at anchorages and couplers immediately after anchor set	$0.70f_{pu}$	$0.70f_{pu}$	$0.66f_{pu}$
	Post-tensioning — general	$0.70f_{pu}$	$0.74f_{pu}$	$0.66f_{pu}$
At Service Limit State, f_{pe}	After all losses	$0.80f_{py}$	$0.80f_{py}$	$0.80f_{py}$

Source: AASHTO LRFD Bridge Design Specifications, 1st Ed., American Association of State Highway and Transportation Officials. Washington, D.C. 1994. With permission.

TABLE 10.8 Temporary Concrete Stress Limits at Jacking State before Losses due to Creep and Shrinkage — Fully Prestressed Components

Stress Type	Area and Condition	Stress (MPa)
Compressive	Pretensioned	$0.60 f'_{ci}$
	Post-tensioned	$0.55 f'_{ci}$
Tensile	Precompressed tensile zone without bonded reinforcement	N/A
	Area other than the precompressed tensile zones and without bonded auxiliary reinforcement	$0.25\sqrt{f'_{ci}} \leq 1.38$
	Area with bonded reinforcement which is sufficient to resist 120% of the tension force in the cracked concrete computed on the basis of uncracked section	$0.58\sqrt{f'_{ci}}$
	Handling stresses in prestressed piles	$0.415\sqrt{f'_{ci}}$

Note: Tensile stress limits are for nonsegmental bridges only.
Source: AASHTO LRFD Bridge Design Specifications, 1st Ed., American Association of State Highway and Transportation Officials. Washington, D.C. 1994. With permission.

where P_j is the prestress force; A is the cross-sectional area; I is the moment of inertia; e is the distance from the center of gravity to the centroid of the prestressing cable; y is the distance from the centroidal axis; and M is the externally applied moment.

Section properties are dependent on the prestressing method and the load stage. In the analysis, the following guidelines may be useful:

- Before bounding of the tendons, for a post-tensioned member, the net section should be used theoretically, but the gross section properties can be used with a negligible tolerance.
- After bounding of tendons, the transformed section should be used, but gross section properties may be used approximately.
- At the service load stage, transformed section properties should be used.

10.4.2 Stress Limits

The stress limits are the basic requirements for designing a prestressed concrete member. The purpose for stress limits on the prestressing tendons is to mitigate tendon fracture, to avoid inelastic tendon deformation, and to allow for prestress losses. Tables 10.7 lists the AASHTO-LRFD [4] stress limits for prestressing tendons.

TABLET 10.9 Concrete Stress Limits at Service Limit State after All Losses — Fully Prestressed Components

Stress Type	Area and Condition		Stress (MPa)
Compressive	Nonsegmental bridge at service stage		$0.45\,f_c'$
	Nonsegmental bridge during shipping and handling		$0.60\,f_c'$
	Segmental bridge during shipping and handling		$0.45\,f_c'$
Tensile	Precompressed tensile zone assuming uncracked section	With bonded prestressing tendons other than piles	$0.50\sqrt{f_c'}$
		Subjected to severe corrosive conditions	$0.25\sqrt{f_c'}$
		With unbonded prestressing tendon	No tension

Note: Tensile stress limits are for nonsegmental bridges only.
Source: AASHTO LRFD Bridge Design Specifications, 1st Ed., American Association of State Highway and Transportation Officials. Washington, D.C. 1994. With permission.

The purpose for stress limits on the concrete is to ensure no overstressing at jacking and after transfer stages and to avoid cracking (fully prestressed) or to control cracking (partially prestressed) at the service load stage. Tables 10.8 and 10.9 list the AASHTO-LRFD [4] stress limits for concrete.

A prestressed member that does not allow cracking at service loads is called a fully prestressed member, whereas one that does is called a partially prestressed member. Compared with full prestress, partial prestress can minimize camber, especially when the dead load is relatively small, as well as provide savings in prestressing steel, in the work required to tension, and in the size of end anchorages and utilizing cheaper mild steel. On the other hand, engineers must be aware that partial prestress may cause earlier cracks and greater deflection under overloads and higher principal tensile stresses under service loads. Nonprestressed reinforcement is often needed to provide higher flexural strength and to control cracking in a partially prestressed member.

10.4.3 Cable Layout

A cable is a group of prestressing tendons and the center of gravity of all prestressing reinforcement. It is a general design principle that the maximum eccentricity of prestressing tendons should occur at locations of maximum moments. Although straight tendons (Figure 10.7a) and harped multi-straight tendons (Figure 10.7b and c) are common in the precast members, curved tendons are more popular for cast-in-place post-tensioned members. Typical cable layouts for bridge superstructures are shown in Figure 10.7.

To ensure that the tensile stress in extreme concrete fibers under service does not exceed code stress limits [4, 14], cable layout envelopes are delimited. Figure 10.8 shows limiting envelopes for simply supported members. From Eq. (10.23), the stress at extreme fiber can be obtained

$$f = \frac{P_j}{A} \pm \frac{P_j eC}{I} \pm \frac{MC}{I} \tag{10.24}$$

where C is the distance of the top or bottom extreme fibers from the center gravity of the section (y_b or y_t as shown in Figure 10.6).

When no tensile stress is allowed, the limiting eccentricity envelope can be solved from Eq. (10.24) with

$$e_{limit} = \frac{I}{AC} \pm \frac{M}{IP_j} \tag{10.25}$$

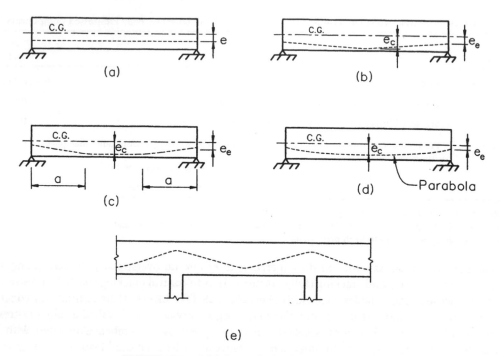

FIGURE 10.7   Cable layout for bridge superstructures.

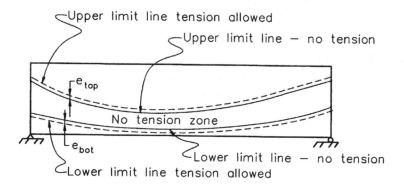

FIGURE 10.8   Cable layout envelopes.

For limited tension stress f_t, additional eccentricities can be obtained:

$$e' = \frac{f_t I}{P_j C} \tag{10.26}$$

10.4.4   Secondary Moments

The primary moment ($M_1 = P_j e$) is defined as the moment in the concrete section caused by the eccentricity of the prestress for a statically determinate member. The secondary moment M_s (Figure 10.9d) is defined as moment induced by prestress and structural continuity in an indeterminate member. Secondary moments can be obtained by various methods. The resulting moment is simply the sum of the primary and secondary moments.

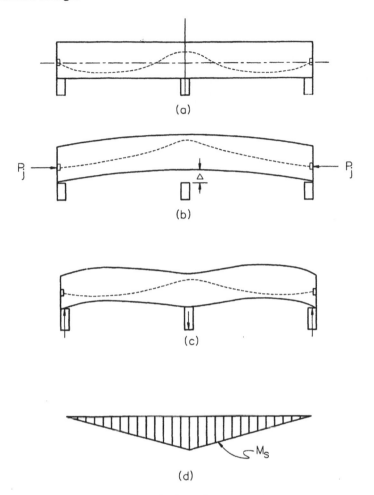

(a)

(b)

(c)

(d)

FIGURE 10.9   Secondary moments.

10.4.5   Flexural Strength

Flexural strength is based on the following assumptions [4]:

- For members with bonded tendons, strain is linearly distributed across a section; for members with unbonded tendons, the total change in tendon length is equal to the total change in member length over the distance between two anchorage points.
- The maximum usable strain at extreme compressive fiber is 0.003.
- The tensile strength of concrete is neglected.
- A concrete stress of 0.85 f_c' is uniformly distributed over an equivalent compression zone.
- Nonprestressed reinforcement reaches the yield strength, and the corresponding stresses in the prestressing tendons are compatible based on plane section assumptions.

For a member with a flanged section (Figure 10.10) subjected to uniaxial bending, the equations of equilibrium are used to give a nominal moment resistance of

$$M_n = A_{ps}f_{ps}\left(d_p - \frac{a}{2}\right) + A_s f_y\left(d_s - \frac{a}{2}\right)$$

$$-A_s' f_y'\left(d_s' - \frac{a}{2}\right) + 0.85\, f_c'(b - b_w)\beta_1 h_f\left(\frac{a}{2} - \frac{h_f}{2}\right)$$

(10.27)

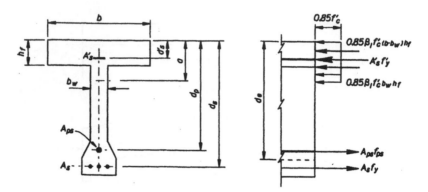

FIGURE 10.10   A flanged section at nominal moment capacity state.

$$a = \beta_1 c \tag{10.28}$$

For bonded tendons:

$$c = \frac{A_{ps}f_{pu} + A_s f_y - A'_s f'_y - 0.85\beta_1 f'_c (b - b_w)h_f}{0.85\beta_1 f'_c b_w + kA_{ps}\dfrac{f_{pu}}{d_p}} \geq h_f \tag{10.29}$$

$$f_{ps} = f_{pu}\left(1 - k\frac{c}{d_p}\right) \tag{10.30}$$

$$k = 2\left(1.04 - \frac{f_{py}}{f_{pu}}\right) \tag{10.31}$$

$$0.85 \geq \beta_1 = 0.85 - \frac{(f'_c - 28)(0.05)}{7} \geq 0.65 \tag{10.32}$$

where A represents area; f is stress; b is the width of the compression face of member; b_w is the web width of a section; h_f is the compression flange depth of the cross section; d_p and d_s are distances from extreme compression fiber to the centroid of prestressing tendons and to centroid of tension reinforcement, respectively; subscripts c and y indicate specified strength for concrete and steel, respectively; subscripts p and s mean prestressing steel and reinforcement steel, respectively; subscripts ps, py, and pu correspond to states of nominal moment capacity, yield, and specified tensile strength of prestressing steel, respectively; superscript ′ represents compression. The above equations also can be used for rectangular section in which $b_w = b$ is taken.

For unbound tendons:

$$c = \frac{A_{ps}f_{pu} + A_s f_y - A'_s f'_y - 0.85\beta_1 f'_c (b - b_w)h_f}{0.85\beta_1 f'_c b_w} \geq h_f \tag{10.33}$$

$$f_{ps} = f_{pe} + \Omega_u E_p \varepsilon_{cu}\left(\frac{d_p}{c} - 1.0\right)\frac{L_1}{L_2} \leq 0.94 f_{py} \tag{10.34}$$

where L_1 is length of loaded span or spans affected by the same tendons; L_2 is total length of tendon between anchorage; Ω_u is the bond reduction coefficient given by

$$\Omega_u = \begin{cases} \dfrac{3}{L/d_p} & \text{for uniform and near third point loading} \\[3mm] \dfrac{1.5}{L/d_p} & \text{for near midspan loading} \end{cases} \qquad (10.35)$$

in which L is span length.

Maximum reinforcement limit:

$$\frac{c}{d_e} \le 0.42 \qquad (10.36)$$

$$d_e = \frac{A_{ps}f_{ps}d_p + A_s f_y d_s}{A_{ps}f_{ps} + A_s f_y} \qquad (10.37)$$

Minimum reinforcement limit:

$$\phi M_n \ge 1.2 \ M_{cr} \qquad (10.38)$$

in which ϕ is flexural resistance factor 1.0 for prestressed concrete and 0.9 for reinforced concrete; M_{cr} is the cracking moment strength given by the elastic stress distribution and the modulus of rupture of concrete.

$$M_{cr} = \frac{I}{y_t}\left(f_r + f_{pe} - f_d\right) \qquad (10.39)$$

where f_{pe} is compressive stress in concrete due to effective prestresses; and f_d is stress due to unfactored self-weight; both f_{pe} and f_d are stresses at extreme fiber where tensile stresses are produced by externally applied loads.

10.4.6 Shear Strength

The shear resistance is contributed by the concrete, the transverse reinforcement and vertical component of prestressing force. The modified compression field theory-based shear design strength [3] was adopted by the AASHTO-LRFD [4] and has the formula:

$$V_n = \text{the lesser of} \begin{cases} V_c + V_s + V_p \\[2mm] 0.25 f_c' b_v d_v + V_p \end{cases} \qquad (10.40)$$

where

$$V_c = 0.083\beta\sqrt{f_c'}\, b_v d_v \qquad (10.41)$$

$$V_s = \frac{A_v f_y d_v (\cos\theta + \cot\alpha)\sin\alpha}{s} \qquad (10.42)$$

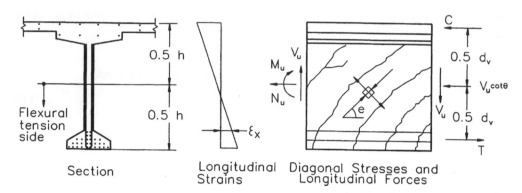

FIGURE 10.11 Illustration of A_c for shear strength calculation. (*Source*: AASHTO LRFD Bridge Design Specifications, 1st Ed., American Association of State Highway and Transportation Officials. Washington, D.C. 1994. With permission.)

TABLE 10.10 Values of θ and β for Sections with Transverse Reinforcement

$\dfrac{v}{f_c'}$	Angle (degree)	$\varepsilon_x \times 1000$										
		−.02	−0.15	−0.1	0	0.125	0.25	0.50	0.75	1.00	1.50	2.00
≤ 0.05	θ	27.0	27.0	27.0	27.0	27.0	28.5	29.0	33.0	36.0	41.0	43.0
	β	6.78	6.17	5.63	4.88	3.99	3.49	2.51	2.37	2.23	1.95	1.72
0.075	θ	27.0	27.0	27.0	27.0	27.0	27.5	30.0	33.5	36.0	40.0	42.0
	β	6.78	6.17	5.63	4.88	3.65	3.01	2.47	2.33	2.16	1.90	1.65
0.100	θ	23.5	23.5	23.5	23.5	24.0	26.5	30.5	34.0	36.0	38.0	39.0
	β	6.50	5.87	5.31	3.26	2.61	2.54	2.41	2.28	2.09	1.72	1.45
0.127	θ	20.0	21.0	22.0	23.5	26.0	28.0	31.5	34.0	36.0	37.0	38.0
	β	2.71	2.71	2.71	2.60	2.57	2.50	2.37	2.18	2.01	1.60	1.35
0.150	θ	22.0	22.5	23.5	25.0	27.0	29.0	32.0	34.0	36.0	36.5	37.0
	β	2.66	2.61	2.61	2.55	2.50	2.45	2.28	2.06	1.93	1.50	1.24
0.175	θ	23.5	24.0	25.0	26.5	28.0	30.0	32.5	34.0	35.0	35.5	36.0
	β	2.59	2.58	2.54	2.50	2.41	2.39	2.20	1.95	1.74	1.35	1.11
0.200	θ	25.0	25.5	26.5	27.5	29.0	31.0	33.0	34.0	34.5	35.0	36.0
	β	2.55	2.49	2.48	2.45	2.37	2.33	2.10	1.82	1.58	1.21	1.00
0.225	θ	26.5	27.0	27.5	29.0	30.5	32.0	33.0	34.0	34.5	36.5	39.0
	β	2.45	2.38	2.43	2.37	2.33	2.27	1.92	1.67	1.43	1.18	1.14
0.250	θ	28.0	28.5	29.0	30.0	31.0	32.0	33.0	34.0	35.5	38.5	41.5
	β	2.36	2.32	2.36	2.30	2.28	2.01	1.64	1.52	1.40	1.30	1.25

(*Source*: AASHTO LRFD Bridge Design Specifications, 1st Ed., American Association of State Highway and Transportation Officials. Washington, D.C. 1994. With permission.)

where b_v is the effective web width determined by subtracting the diameters of ungrouted ducts or one half the diameters of grouted ducts; d_v is the effective depth between the resultants of the tensile and compressive forces due to flexure, but not to be taken less than the greater of $0.9d_e$ or $0.72h$; A_v is the area of transverse reinforcement within distance s; s is the spacing of stirrups; α is the angle of inclination of transverse reinforcement to longitudinal axis; β is a factor indicating ability of diagonally cracked concrete to transmit tension; θ is the angle of inclination of diagonal compressive stresses (Figure 10.11). The values of β and θ for sections with transverse reinforcement are given in Table 10.10. In using this table, the shear stress v and strain ε_x in the reinforcement on the flexural tension side of the member are determined by

$$v = \frac{V_u - \phi V_p}{\phi b_v d_v}$$

(10.43)

$$\varepsilon_x = \frac{\dfrac{M_u}{d_v} + 0.5N_u + 0.5V_u \cot\theta - A_{ps}f_{po}}{E_s A_s + E_p A_{ps}} \le 0.002 \qquad (10.44)$$

where M_u and N_u are factored moment and axial force (taken as positive if compressive) associated with V_u and f_{po} is stress in prestressing steel when the stress in the surrounding concrete is zero and can be conservatively taken as the effective stress after losses f_{pe}. When the value of ε_x calculated from the above equation is negative, its absolute value shall be reduced by multiplying by the factor F_ε, taken as

$$F_\varepsilon = \frac{E_s A_s + E_p A_{ps}}{E_c A_c + E_s A_s + E_p A_{ps}} \qquad (10.45)$$

where E_s, E_p, and E_c are modulus of elasticity for reinforcement, prestressing steel, and concrete, respectively; A_c is area of concrete on the flexural tension side of the member as shown in Figure 10.11.

Minimum transverse reinforcement:

$$A_{v\,min} = 0.083\sqrt{f_c'}\,\frac{b_v s}{f_y} \qquad (10.46)$$

Maximum spacing of transverse reinforcement:

$$\text{For } V_u < 0.1 f_c' b_v d_v \qquad s_{max} = \text{the smaller of} \begin{cases} 0.8 d_v \\ 600 \text{ mm} \end{cases} \qquad (10.47)$$

$$\text{For } V_u \ge 0.1 f_c' b_v d_v \qquad s_{max} = \text{the smaller of} \begin{cases} 0.4 d_v \\ 300 \text{ mm} \end{cases} \qquad (10.48)$$

10.4.7 Camber and Deflections

As opposed to load deflection, camber is usually referred to as reversed deflection and is caused by prestressing. A careful evaluation of camber and deflection for a prestressed concrete member is necessary to meet serviceability requirements. The following formulas developed by the moment–area method can be used to estimate midspan immediate camber for simply supported members as shown in Figure 10.7.

For straight tendon (Figure 10.7a):

$$\Delta = \frac{L^2}{8E_c I} M_e \qquad (10.49)$$

For one-point harping tendon (Figure 10.7b):

$$\Delta = \frac{L^2}{8E_c I}\left(M_c + \frac{2}{3}M_e\right) \qquad (10.50)$$

For two-point harping tendon (Figure 10.7c):

$$\Delta = \frac{L^2}{8E_cI}\left(M_c + M_e - \frac{M_e}{3}\left(\frac{2a}{L}\right)^2\right) \tag{10.51}$$

For parabola tendon (Figure 10.7d):

$$\Delta = \frac{L^2}{8E_cI}\left(M_e + \frac{5}{6}M_c\right) \tag{10.52}$$

where M_e is the primary moment at end, P_je_{end}, and M_c is the primary moment at midspan P_je_c. Uncracked gross section properties are often used in calculating camber. For deflection at service loads, cracked section properties, i.e., moment of inertia I_{cr}, should be used at the post-cracking service load stage. It should be noted that long term effect of creep and shrinkage shall be considered in the final camber calculations. In general, final camber may be assumed 3 times as great as immediate camber.

10.4.8 Anchorage Zones

In a pretensioned member, prestressing tendons transfer the compression load to the surrounding concrete over a length L_t gradually. In a post-tensioned member, prestressing tendons transfer the compression directly to the end of the member through bearing plates and anchors. The anchorage zone, based on the principle of St. Venant, is geometrically defined as the volume of concrete through which the prestressing force at the anchorage device spreads transversely to a more linear stress distribution across the entire cross section at some distance from the anchorage device [4].

For design purposes, the anchorage zone can be divided into general and local zones [4]. The region of tensile stresses is the general zone. The region of high compressive stresses (immediately ahead of the anchorage device) is the local zone. For the design of the general zone, a "strut-and-tie model," a refined elastic stress analysis or approximate methods may be used to determine the stresses, while the resistance to bursting forces is provided by reinforcing spirals, closed hoops, or anchoraged transverse ties. For the design of the local zone, bearing pressure is a major concern. For detailed requirements, see AASHTO-LRFD [4].

10.5 Design Example

Two-Span Continuous Cast-in-Place Box-Girder Bridge

Given
A two-span continuous cast-in-place prestressed concrete box-girder bridge has two equal spans of length 48 m with a single-column bent. The superstructure is 10.4 m wide. The elevation view of the bridge is shown in Figure 10.12a.

Material:
Initial concrete: $f'_{ci} = 24$ MPa, $E_{ci} = 24,768$ MPa
Final concrete: $f'_c = 28$ MPa, $E_c = 26,752$ MPa
Prestressing steel: $f_{pu} = 1860$ MPa low relaxation strand, $E_p = 197,000$ MPa
Mild steel: $f_y = 400$ MPa, $E_s = 200,000$ MPa

Prestressing:
Anchorage set thickness = 10 mm
Prestressing stress at jacking $f_{pj} = 0.8 f_{pu} = 1488$ MPa
The secondary moments due to prestressing at the bent are $M_{DA} = 1.118 P_j$, $M_{DG} = 1.107 P_j$

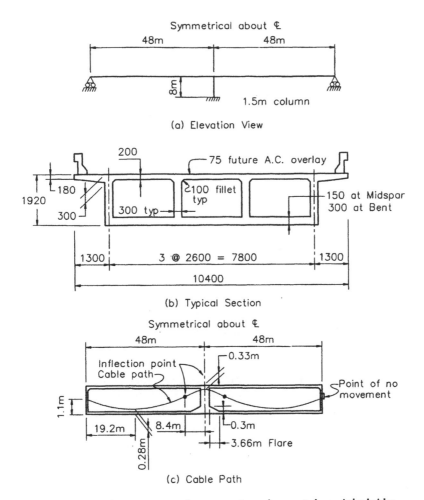

FIGURE 10.12   A two–span continuous prestressed concrete box–girder bridge.

Loads:
   Dead Load = self-weight + barrier rail + future wearing 75 mm AC overlay
   Live Load = AASHTO HL-93 Live Load + dynamic load allowance

Specification:
   AASHTO-LRFD [4] (referred as AASHTO in this example)

Requirements

1. Determine cross section geometry
2. Determine longitudinal section and cable path
3. Calculate loads
4. Calculate live load distribution factors for interior girder
5. Calculate unfactored moment and shear demands for interior girder
6. Determine load factors for Strength Limit State I and Service Limit State I
7. Calculate section properties for interior girder
8. Calculate prestress losses
9. Determine prestressing force P_j for interior girder

10. Check concrete strength for interior girder — Service Limit State I
11. Flexural strength design for interior girder — Strength Limit State I
12. Shear strength design for interior girder — Strength Limit State I

Solution

1. **Determine Cross Section Geometry**

   a. *Structural depth — d:*
   For prestressed continuous spans, the structural depth d can be determined using a depth-to-span ratio (d/L) of 0.04 (AASHTO LRFD Table 2.5.2.6.3-1).

   $$d = 0.04L = 0.04(48) = 1.92 \text{ m}$$

   b. *Girder spacing — S:*
   The spacing of girders is generally taken no more than twice their depth.

   $$S_{max} < 2\ d = 2\ (1.92) = 3.84 \text{ m}$$

   By using an overhang of 1.2 m, the center-to-center distance between two exterior girders is 10.4 m − (2)(1.2 m) = 8 m.

Try three girders and two bays, $S = 8$ m/2 = 4 m > 3.84 m	NG
Try four girders and three bays, $S = 8$ m/3 = 2.67 m < 3.84 m	OK
Use a girder spacing $S = 2.6$ m	

   c. *Typical section:*
   From past experience and design practice, we select that a thickness of 180 mm at the edge and 300 mm at the face of exterior girder for the overhang. The web thickness is chosen to be 300 mm at normal section and 450 mm at the anchorage end. The length of the flare is usually taken as $\frac{1}{10}$ of the span length, say 4.8 m. The deck and soffit thickness depends on the clear distance between adjacent girders; 200 and 150 mm are chosen for the deck and soffit thickness, respectively. The selected box-girder section configurations for this example are shown in Figure 10.12b. The section properties of the box girder are as follows:

Properties	Midspan	Bent (face of support)
A (m^2)	5.301	6.336
I (m^4)	2.844	3.513
y_b (m)	1.102	0.959

2. **Determine Longitudinal Section and Cable Path**

   To lower the center of gravity of the superstructure at the face of the bent cap in the CIP post-tensioned box girder, the thickness of soffit is flared to 300 mm as shown in Figure 10.12c. A cable path is generally controlled by the maximum dead-load moment and the position of the jack at the end section. Maximum eccentricities should occur at points of maximum dead load moments and almost no eccentricity should be present at the jacked end section. For this example, the maximum dead-load moments occur at three locations: at the bent cap, at the locations close to 0.4L for Span 1 and 0.6L for Span 2. A parabolic cable path is chosen as shown in Figure 10.12c.

3. **Calculate Loads**

   a. *Component dead load — DC:*
   The component dead load DC includes all structural dead loads with the exception of the future wearing surface and specified utility loads. For design purposes, two parts of the DC are defined as:

   DC1 — girder self-weight (density 2400 kg/m³) acting at the prestressing stage
   DC2 — barrier rail weight (11.5 kN/m) acting at service stage after all losses.
 b. *Wearing surface load — DW*:
  The future wearing surface of 75 mm with a density 2250 kg/m³

$$DW = \text{(deck width} - \text{barrier width) (thickness of wearing surface) (density)}$$
$$= [10.4 \text{ m} - 2(0.54 \text{ m})](0.075 \text{ m})(2250 \text{ kg/m}^3)(9.8066 \text{ m/s}^2) = 15{,}423 \text{ N/m}$$
$$= 15.423 \text{ kN/m}$$

 c. *Live-Load LL and Dynamic Load Allowance — IM*:
  The design live load *LL* is the AASHTO HL-93 vehicular live loading. To consider the wheel-load impact from moving vehicles, the dynamic load allowance *IM* = 33% [AASHTO LRFD Table 3.6.2.1-1] is applied to the design truck.

4. Calculate Live Load Distribution Factors
 AASHTO [1994] recommends that approximate methods be used to distribute live load to individual girders (AASHTO-LRFD 4.6.2.2.2). The dimensions relevant to this prestressed box girder are: depth $d = 1920$ mm, number of cells $N_c = 3$, spacing of girders $S = 2600$ mm, span length $L = 48{,}000$ mm, half of the girder spacing plus the total overhang $W_e = 2600$ mm, and the distance between the center of an exterior girder and the interior edge of a barrier $d_e = 1300 - 535 = 765$ m. This box girder is within the range of applicability of the AASHTO approximate formulas. The live-load distribution factors are calculated as follows:

 a. *Live-load distribution factor for bending moments*:
  i. Interior girder (AASHTO Table 4.6.2.2.2b-1):
   • One design lane loaded:

$$g_M = \left(1.75 + \frac{S}{1100}\right)\left(\frac{300}{L}\right)^{0.35}\left(\frac{1}{N_c}\right)^{0.45}$$

$$= \left(1.75 + \frac{2600}{1100}\right)\left(\frac{300}{48{,}000}\right)^{0.35}\left(\frac{1}{3}\right)^{0.45} = 0.425 \text{ lanes}$$

   • Two or more design lanes loaded:

$$g_M = \left(\frac{13}{N_c}\right)^{0.3}\left(\frac{S}{430}\right)\left(\frac{1}{L}\right)^{0.25}$$

$$= \left(\frac{13}{3}\right)^{0.3}\left(\frac{2600}{430}\right)\left(\frac{1}{48{,}000}\right)^{0.25} = 0.634 \text{ lanes} \quad \text{(controls)}$$

  ii. Exterior girder (AASHTO Table 4.6.2.2.2d-1):

$$g_M = \frac{W_e}{4300} = \frac{2600}{4300} = 0.605 \text{ lanes}$$

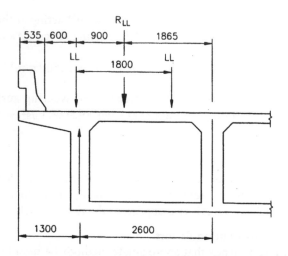

FIGURE 10.13 Live–load distribution for exterior girder — lever rule.

b. *Live-load distribution factor for shear:*

   i. Interior girder (AASHTO Table 4.62.2.3a-1):
      • One design lane loaded:

$$g_V = \left(\frac{S}{2900} \right)^{0.6} \left(\frac{d}{L} \right)^{0.1}$$

$$= \left(\frac{2600}{2900} \right)^{0.6} \left(\frac{1920}{48,000} \right)^{0.1} = 0.679 \ \text{lanes}$$

      • Two or more design lanes loaded:

$$g_V = \left(\frac{S}{2200} \right)^{0.9} \left(\frac{d}{L} \right)^{0.1}$$

$$= \left(\frac{2600}{2200} \right)^{0.9} \left(\frac{1920}{48,000} \right)^{0.1} = 0.842 \ \text{lanes} \quad \text{(controls)}$$

   ii. Exterior girder (AASHTO Table 4.62.2.3b-1):
      • One design lane loaded — Lever rule:
        The lever rule assumes that the deck in its transverse direction is simply supported by the girders and uses statics to determine the live-load distribution to the girders. AASHTO-LRFD [4] also requires that when the lever rule is used, the multiple presence factor *m* should apply. For a one design lane loaded, *m* = 1.2. The lever rule model for the exterior girder is shown in Figure 10.13. From static equilibrium:

$$R = \frac{965 + 900}{2600} = 0.717$$

$$g_v = mR = 1.2(0.717) = 0.861 \quad \text{(controls)}$$

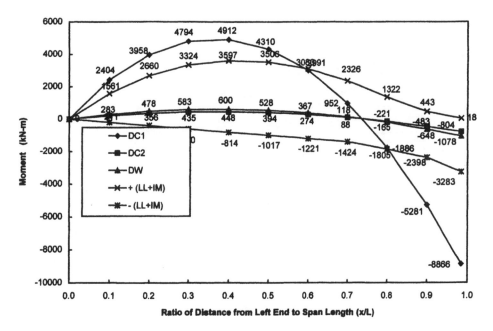

FIGURE 10.14 Moment envelopes for Span 1.

- Two or more design lanes loaded — Modify interior girder factor by e:

$$g_V = eg_{V(\text{interior girder})} = \left(0.64 + \frac{d_e}{3800}\right)g_{V(\text{interior girder})}$$

$$= \left(0.64 + \frac{765}{3800}\right)(0.842) = 0.708 \text{ lanes}$$

- The live load distribution factors at the strength limit state:

Strength Limit State I	Interior Girder	Exterior Girder
Bending moment	0.634 lanes	0.605 lanes
Shear	0.842 lanes	0.861 lanes

5. **Calculate Unfactored Moments and Shear Demands for Interior Girder**

   It is practically assumed that all dead loads are carried by the box girder and equally distributed to each girder. The live loads take forces to the girders according to live load distribution factors (AASHTO Article 4.6.2.2.2). Unfactored moment and shear demands for an interior girder are shown in Figures 10.14 and 10.15. Details are listed in Tables 10.11 and 10.12. Only the results for Span 1 are shown in these tables and figures since the bridge is symmetrical about the bent.

6. **Determine Load Factors for Strength Limit State I and Service Limit State I**

   a. *General design equation* (ASHTO Article 1.3.2):

$$\eta \sum \gamma_i Q_i \leq \phi R_n \tag{10.53}$$

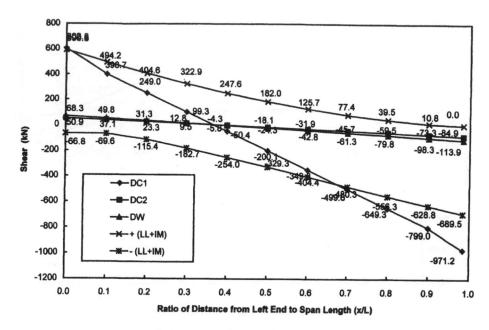

FIGURE 10.15 Shear envelopes for Span 1

TABLE 10.11 Moment and Shear due to Unfactored Dead Load for the Interior Girder

| | | Unfactored Dead Load | | | | | |
| | | DC1 | | DC2 | | DW | |
Span	Location (x/L)	M_{DC1} (kN-m)	V_{DC1} (kN)	M_{DC2} (kN-m)	V_{DC2} (kN)	M_{DW} (kN-m)	V_{DW} (kN)
	0.0	0	603	0	51	0	68
	0.1	2404	399	211	37	283	50
	0.2	3958	249	356	23	478	31
	0.3	4794	99	435	10	583	13
1	0.4	4912	−50	448	−4	600	−6
	0.5	4310	−200	394	−18	528	−24
	0.6	2991	−350	274	−32	367	−43
	0.7	952	−500	88	−46	118	−61
	0.8	−1805	−649	−165	−59	−221	−80
	0.9	−5281	−799	−483	−73	−648	−98
	Face of column	−8866	−971	−804	−85	−1078	−114

Note:
1. *DC1* — interior girder self-weight.
2. *DC2* — barrier self-weight.
3. *DW* — wearing surface load.
4. Moments in Span 2 are symmetrical about the bent.
5. Shears in span are anti-symmetrical about the bent.

where γ_i are load factors and ϕ a resistance factor; Q_i represents force effects; R_n is the nominal resistance; η is a factor related to the ductility, redundancy, and operational importance of that being designed and is defined as:

$$\eta = \eta_D \eta_R \eta_I \geq 0.95 \tag{10.54}$$

TABLE 10.12 Moment and Shear Envelopes and Associated Forces for the Interior Girder due to AASHTO HL-93 Live Load

Span	Location (x/L)	Positive Moment and Associated Shear		Negative Moment and Associated Shear		Shear and Associated Moment	
		M_{LL+IM} (kN-m)	V_{LL+IM} (kN)	V_{LL+IM} (kN)	M_{LL+IM} (kN-m)	V_{LL+IM} (kN)	M_{LL+IM} (kN-m)
1	0.0	0	259	0	−255	497	0
	0.1	1561	312	−203	−42	416	1997
	0.2	2660	249	−407	−42	341	3270
	0.3	3324	47	−610	−42	272	3915
	0.4	3597	108	−814	−42	−214	3228
	0.5	3506	−25	−1017	−42	−277	3272
	0.6	3080	−81	−1221	−42	−341	2771
	0.7	2326	−258	−1424	−42	−404	1956
	0.8	1322	−166	−1886	−68	−468	689
	0.9	443	−112	−2398	−141	−529	−945
	Face of column	18	−97	−3283	−375	−581	−1850

Note:
1. *LL* + *IM* — AASHTO HL-93 live load plus dynamic load allowance.
2. Moments in Span 2 are symmetrical about the bent.
3. Shears in Span 2 are antisymmetrical about the bent.
4. Live load distribution factors are considered.

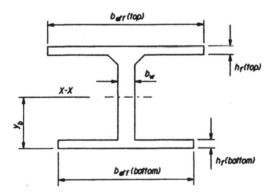

FIGURE 10.16 Effective flange width of interior girder.

For this bridge, the following values are assumed:

Limit States	Ductility η_D	Redundancy η_R	Importance η_I	η
Strength limit state	0.95	0.95	1.05	0.95
Service limit state	1.0	1.0	1.0	1.0

  b. *Load factors and load combinations:*
    The load factors and combinations are specified as (AASHTO Table 3.4.1-1):
      Strength Limit State I: $1.25(DC1 + DC2) + 1.5(DW) + 1.75(LL + IM)$
      Service Limit State I: $DC1 + DC2 + DW + (LL + IM)$

7. Calculate Section Properties for Interior Girder
  For an interior girder as shown in Figure 10.16, the effective flange width b_{eff} is determined (AASHTO Article 4.6.2.6) by:

TABLE 10.13 Effective Flange Width and Section Properties
for Interior Girder

Location	Dimension	Mid span	Bent (face of support)
Top flange	h_f (mm)	200	200
	$L_{eff}/4$ (mm)	9,000	11,813
	$12h_f + b_w$ (mm)	2,700	2,700
	S (mm)	2,600	2,600
	b_{eff} (mm)	**2,600**	**2,600**
Bottom flange	h_f (mm)	150	300
	$L_{eff}/4$ (mm)	9,000	11,813
	$12h_f + b_w$ (mm)	2,100	3,900
	S (mm)	2,600	2,600
	b_{eff} (mm)	**2,100**	**2,600**
Area	A (m²)	1.316	1.736
Moment of inertia	I (m⁴)	0.716	0.968
Center of gravity	y_b (m)	1.085	0.870

Note: L_{eff} = 36.0 m for midspan; L_{eff} = 47.25 m for the bent; b_w = 300 mm.

$$b_{eff} = \text{the lesser of} \begin{cases} \dfrac{L_{eff}}{4} \\ 12h_f + b_w \\ S \end{cases} \qquad (10.55)$$

where L_{eff} is the effective span length and may be taken as the actual span length for simply supported spans and the distance between points of permanent load inflection for continuous spans; h_f is the compression flange thickness and b_w is the web width; and S is the average spacing of adjacent girders. The calculated effective flange width and the section properties are shown in Table 10.13 for the interior girder.

8. **Calculate Prestress Losses**
   For a CIP post-tensioned box girder, two types of losses, instantaneous losses (friction, anchorage set, and elastic shortening) and time-dependent losses (creep and shrinkage of concrete, and relaxation of prestressing steel), are significant. Since the prestress losses are not symmetrical about the bent for this bridge, the calculation is performed for both spans.
   a. *Frictional loss Δf_{pF}:*

$$\Delta f_{pF} = f_{pj}\left(1 - e^{-(Kx + \mu\alpha)}\right) \qquad (10.56)$$

where K is the wobble friction coefficient = 6.6×10^{-7}/mm and μ is the coefficient of friction = 0.25 (AASHTO Article 5.9.5.2.2b); x is the length of a prestressing tendon from the jacking end to the point considered; α is the sum of the absolute values of angle change in the prestressing steel path from the jacking end.

For a parabolic cable path (Figure 10.17), the angle change is $\alpha = 2e_p/L_p$, where e_p is the vertical distance between two control points and L_p is the horizontal distance between two control points. The details are given in Table 10.14.

   b. *Anchorage set loss Δf_{pA}:*
   For an anchor set thickness of ΔL = 10 mm and E = 200,000 MPa, consider the point D where L_{pF} = 48 m and Δf_{pF} = 96.06 MPa:

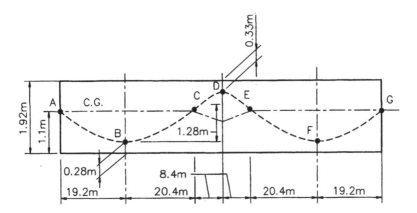

FIGURE 10.17 Parabolic cable path.

TABLE 10.14 Prestress Frictional Loss

Segment	e_p (mm)	L_p (m)	α (rad)	$\Sigma\alpha$ (rad)	ΣL_b (m)	Point	Δf_{pF} (Mpa)
A	0.00	0	0	0	0	A	0.00
AB	820	19.2	0.0854	0.0854	19.2	B	31.44
BC	926	20.4	0.0908	0.1762	39.6	C	64.11
CD	381	8.4	0.0908	0.2669	48.0	D	96.06
DE	381	8.4	0.0908	0.3577	56.4	E	127.28
EF	926	20.4	0.0908	0.4484	76.8	F	157.81
FG	820	19.2	0.0854	0.5339	96.0	G	185.91

$$L_{pA} = \sqrt{\frac{E\,(\Delta L)\,L_{pF}}{\Delta f_{pF}}} = \sqrt{\frac{200,000(10)(48,000)}{96.06}} = 31\,613 \text{ mm} = 31.6 \text{ m} \quad < 48 \text{ m} \qquad \text{OK}$$

$$\Delta f = \frac{2\,\Delta f_{pF}\,L_{pA}}{L_{pF}} = \frac{2(96.06)(31.6)}{48} = 126.5 \text{ MPa}$$

$$\Delta f_{pA} = \Delta f\left(1 - \frac{x}{L_{pA}}\right) = 126.5\left(1 - \frac{x}{31.6}\right)$$

c. *Elastic shortening loss* Δf_{pES}:
The loss due to elastic shortening in post-tensioned members is calculated using the following formula (AASHTO Article 5.9.5.2.3b):

$$\Delta f_{pES} = \frac{N-1}{2N}\frac{E_p}{E_{ci}}f_{cgp} \qquad (10.57)$$

To calculate the elastic shortening loss, we assume that the prestressing jack force for an interior girder $P_j = 8800$ kN and the total number of prestressing tendons $N = 4$. f_{cgp} is calculated for face of support section:

$$f_{cgp} = \frac{P_j}{A} + \frac{P_j e^2}{I_x} + \frac{M_{DC1} e}{I_x}$$

$$= \frac{8800}{1.736} + \frac{8800\,(0.714)^2}{0.968} + \frac{(-8866)(0.714)}{0.968}$$

$$= 5069 + 4635 - 6540 = 3164 \ \text{kN}/\text{m}^2 = 3.164\,\text{MPa}$$

$$\Delta f_{pES} = \frac{N-1}{2N}\frac{E_p}{E_{ci}}f_{cgp} = \frac{4-1}{2(4)}\frac{197{,}000}{24768}(3.164) = 9.44\ \text{MPa}$$

d. *Time-dependent losses Δf_{pTM}:*

AASHTO provides a table to estimate the accumulated effect of time-dependent losses resulting from the creep and shrinkage of concrete and the relaxation of the steel tendons. From AASHTO Table 5.9.5.3-1:

$$\Delta f_{pTM} = 145 \ \text{MPa} \quad \text{(upper bound)}$$

e. *Total losses Δf_{pT}:*

$$\Delta f_{pT} = \Delta f_{pF} + \Delta f_{pA} + \Delta f_{pES} + \Delta f_{pTM}$$

Details are given in Table 10.15.

9. **Determine Prestressing Force P_j for Interior Girder**

Since the live load is not in general equally distributed to girders, the prestressing force P_j required for each girder may be different. To calculate prestress jacking force P_j, the initial prestress force coefficient F_{pCI} and final prestress force coefficient F_{pCF} are defined as:

$$F_{pCI} = 1 - \frac{\Delta f_{pF} + \Delta f_{pA} + \Delta f_{pES}}{f_{pj}} \tag{10.58}$$

$$F_{pCF} = 1 - \frac{\Delta f_{pT}}{f_{pj}} \tag{10.59}$$

The secondary moment coefficients are defined as:

$$M_{sC} = \begin{cases} \dfrac{x}{L}\dfrac{M_{DA}}{P_j} & \text{for Span 1} \\[2ex] \left(1-\dfrac{x}{L}\right)\dfrac{M_{DG}}{P_j} & \text{for Span 2} \end{cases} \tag{10.60}$$

where x is the distance from the left end for each span. The combined prestressing moment coefficients are defined as:

$$M_{psCI} = F_{pCI}(e) + M_{sC} \tag{10.61}$$

TABLE 10.15   Cable Path and Prestress Losses

Span	Location (x/L)	Prestress Losses (MPa)					Force Coefficient	
		Δf_{pF}	Δf_{pA}	Δf_{pES}	Δf_{pTM}	Δf_{pT}	F_{pCI}	F_{pCF}
	0.0	0.00	126.50			280.94	0.909	0.811
	0.1	7.92	107.28			269.65	0.916	0.819
	0.2	15.80	88.07			258.31	0.924	0.826
	0.3	23.64	68.85			246.94	0.931	0.834
	0.4	31.44	49.64			235.52	0.939	0.842
1	0.5	39.19	30.42	9.44	145.00	224.06	0.947	0.849
	0.6	46.91	11.21			212.56	0.955	0.857
	0.7	54.58	0.00			209.02	0.957	0.860
	0.8	62.21	0.00			216.65	0.952	0.854
	0.9	77.89	0.00			232.33	0.941	0.844
	1.0	96.06	0.00			250.50	0.929	0.832
	0.0	96.06				250.50	0.929	0.832
	0.1	113.99				268.43	0.917	0.820
	0.2	129.10				283.54	0.907	0.809
	0.3	136.33				290.77	0.902	0.805
2	0.4	143.53	0.00	9.44	145.00	297.97	0.897	0.800
	0.5	150.69				305.13	0.892	0.795
	0.6	157.81				312.25	0.888	0.790
	0.7	164.89				319.33	0.883	0.785
	0.8	171.94				326.38	0.878	0.781
	0.9	178.94				333.38	0.873	0.776
	1.0	185.91				340.35	0.869	0.771

Note: $F_{pCI} = 1 - \dfrac{\Delta f_{pF} + \Delta f_{pA} + \Delta f_{pES}}{f_{pj}}$

$F_{pCF} = 1 - \dfrac{\Delta f_{pT}}{f_{pj}}$

$$M_{psCF} = F_{pCF}(e) + M_{sC} \tag{10.62}$$

where e is the distance between the cable and the center of gravity of a cross section; positive values of e indicate that the cable is above the center of gravity, and negative ones indicate the cable is below the center of gravity of the section.

The prestress force coefficients and the combined moment coefficients are calculated and tabled in Table 10.16. According to AASHTO, the prestressing force P_j can be determined using the concrete tensile stress limit in the precompression tensile zone (see Table 10.5):

$$f_{DC1} + f_{DC2} + f_{DW} + f_{LL+IM} + f_{psF} \geq -0.5\sqrt{f_c'} \tag{10.63}$$

in which

$$f_{DC1} = \frac{M_{DC1}C}{I_x} \tag{10.64}$$

$$f_{DC2} = \frac{M_{DC2}C}{I_x} \tag{10.65}$$

TABLE 10.16 Prestress Force and Moment Coefficients

Span	Location (x/L)	Cable Path e (m)	Force Coefficients		Moment Coefficients (m)				
			F_{pCI}	F_{pCF}	$F_{pCI}e$	$F_{pCF}e$	M_{sC}	M_{psCI}	M_{psCF}
	0.0	0.015	0.909	0.811	0.014	0.012	0.000	0.014	0.012
	0.1	−0.344	0.916	0.819	−0.315	−0.281	0.034	−0.281	−0.247
	0.2	−0.600	0.924	0.826	−0.554	−0.496	0.068	−0.486	−0.428
	0.3	−0.754	0.931	0.834	−0.702	−0.629	0.102	−0.600	−0.526
	0.4	−0.805	0.939	0.842	−0.756	−0.678	0.136	−0.620	−0.541
1	0.5	−0.754	0.947	0.849	−0.714	−0.640	0.171	−0.543	−0.470
	0.6	−0.600	0.955	0.857	−0.573	−0.514	0.205	−0.368	−0.310
	0.7	−0.344	0.957	0.860	−0.329	−0.295	0.239	−0.090	−0.057
	0.8	0.015	0.952	0.884	0.014	0.013	0.273	0.287	0.286
	0.9	0.377	0.941	0.844	0.355	0.318	0.307	0.662	0.625
	1.0	0.717	0.929	0.832	0.666	0.596	0.341	1.007	0.937
	0.0	0.717	0.929	0.832	0.666	0.596	0.347	1.013	0.943
	0.1	0.377	0.917	0.820	0.346	0.309	0.312	0.658	0.622
	0.2	0.015	0.907	0.809	0.014	0.012	0.278	0.291	0.290
	0.3	−0.344	0.902	0.805	−0.310	−0.277	0.243	−0.067	−0.034
	0.4	−0.600	0.897	0.800	−0.538	−0.480	0.208	−0.330	−0.272
2	0.5	−0.754	0.892	0.795	−0.673	−0.599	0.174	−0.499	−0.426
	0.6	−0.805	0.888	0.790	−0.715	−0.636	0.139	−0.576	−0.497
	0.7	−0.754	0.883	0.785	−0.665	−0.592	0.104	−0.561	−0.488
	0.8	−0.600	0.878	0.781	−0.527	−0.468	0.069	−0.457	−0.399
	0.9	−0.344	0.873	0.776	−0.300	−0.267	0.035	−0.266	−0.232
	1.0	0.015	0.869	0.771	0.013	0.012	0.000	0.013	0.012

Note: e is distance between cable path and central gravity of the interior girder cross section, positive means cable is above the central gravity, and negative indicates cable is below the central gravity.

$$f_{DW} = \frac{M_{DW}C}{I_x} \tag{10.66}$$

$$f_{LL+IM} = \frac{M_{LL+IM}C}{I_x} \tag{10.67}$$

$$f_{psF} = \frac{P_{pe}}{A} + \frac{(P_{pe}e)C}{I_x} + \frac{M_sC}{I_x} = \frac{F_{pCF}P_j}{A} + \frac{M_{psCF}P_jC}{I_x} \tag{10.68}$$

where $C (= y_b$ or $y_t)$ is the distance from the extreme fiber to the center of gravity of the cross section. f'_c is in MPa and P_{pe} is the effective prestressing force after all losses have been incurred. From Eqs. (10.63) and (10.68), we have

$$P_j = \frac{-f_{DC1} - f_{DC2} - f_{DW} - f_{LL+IM} - 0.5\sqrt{f'_c}}{\dfrac{F_{pCF}}{A} + \dfrac{M_{psCF}C}{I_x}} \tag{10.69}$$

Detailed calculations are given in Table 10.17. Most critical points coincide with locations of maximum eccentricity: 0.4L in Span 1, 0.6L in Span 2, and at the bent. For this bridge, the controlling section is through the right face of the bent. Herein, $P_j = 8741$ kN. Rounding P_j up to 8750 kN gives a required area of prestressing steel of $A_{ps} = P_j/f_{pj} = 8750/1488 \, (1000) = 5880$ mm².

TABLE 10.17 Determination of Prestressing Jacking Force for an Interior Girder

Span	Location (x/L)	Top Fiber Stress (MPa) f_{DC1}	f_{DC2}	f_{DW}	f_{LL+IM}	Jacking Force, P_j (kN)	Bottom Fiber Stress (MPa) f_{DC1}	f_{DC2}	f_{DW}	f_{LL+IM}	Jacking Force P_j (kN)
	0.0	0.000	0.000	0.000	0.000	—	0.000	0.000	0.000	0.000	0
	0.1	2.803	0.246	0.330	1.820	—	−3.642	−0.320	−0.429	−2.365	4405
	0.2	4.616	0.415	0.557	3.103	—	−5.998	−0.540	−0.724	−4.032	6778
	0.3	5.591	0.507	0.680	3.876	—	−7.265	−0.659	−0.884	−5.037	7824
	0.4	5.728	0.522	0.700	4.195	—	−7.442	−0.678	−0.910	−5.450	8101
1	0.5	5.027	0.459	0.616	4.089	—	−6.532	−0.597	−0.800	−5.313	7807
	0.6	3.488	0.319	0.428	3.591	—	−4.532	−0.415	−0.557	−4.667	6714
	0.7	1.110	0.102	0.137	2.712	—	−1.443	−0.133	−0.178	−3.524	3561
	0.8	−2.105	−0.192	−0.258	1.542	2601	2.736	0.250	0.335	−2.004	—
	0.9	−6.159	−0.565	−0.756	0.516	5567	8.003	0.733	0.982	−0.671	—
	1.0	−9.617	−0.872	−1.169	0.020	8406	7.968	0.722	0.969	−0.016	—
	0.0	−9.617	−0.872	−1.169	0.020	8370	7.968	0.722	0.969	−0.016	—
	0.1	−.6.159	−0.564	−0.756	0.516	5661	8.003	0.733	0.982	−0.671	—
	0.2	−2.105	−0.192	−0.258	1.542	2681	2.736	0.250	0.335	−2.004	—
	0.3	1.110	0.102	0.137	2.712	—	−1.443	−0.133	−0.178	−3.524	3974
	0.4	3.488	0.319	0.428	3.591	—	−4.532	−0.415	−0.557	−4.667	7381
2	0.5	5.027	0.459	0.616	4.089	—	−6.532	−0.597	−0.800	−5.313	8483
	0.6	5.728	0.522	0.700	4.195	—	−7.443	−0.678	−0.910	−5.450	**8741**
	0.7	5.591	0.507	0.680	3.876	—	−7.265	−0.659	−0.884	−5.037	8382
	0.8	4.616	0.415	0.557	3.103	—	−5.998	−0.540	−0.724	−4.032	7220
	0.9	2.803	0.246	0.330	1.820	—	−3.642	−0.320	−0.429	−2.365	4666
	1.0	0.000	0.000	0.000	0.000	—	0.000	0.000	0.000	0.000	0

Notes:
1. Positive stress indicates compression and negative stress indicates tension.
2. P_j are obtained by Eq. (10.69).

10. Check Concrete Strength for Interior Girder — Service Limit State I

Two criteria are imposed on the level of concrete stresses when calculating required concrete strength (AASHTO Article 5.9.4.2):

$$\begin{cases} f_{DC1} + f_{psI} \leq 0.55 f_{ci}' & \text{at prestressing state} \\ f_{DC1} + f_{DC2} + f_{DW} + f_{LL+IM} + f_{psF} \leq 0.45 f_c' & \text{at service state} \end{cases} \qquad (10.70)$$

$$f_{psI} = \frac{P_{jI}}{A} + \frac{(P_{jI}e)C}{I_x} + \frac{M_{sI}C}{I_x} = \frac{F_{pCI}P_j}{A} + \frac{M_{psCI}P_jC}{I_x} \qquad (10.71)$$

The concrete stresses in the extreme fibers (after instantaneous losses and final losses) are given in Tables 10.18. and 10.19. For the initial concrete strength in the prestressing state, the controlling location is the top fiber at 0.8L section in Span 1. From Eq. (10.70), we have

$$f_{ci,reg}' \geq \frac{f_{DC1} + f_{psI}}{0.55} = \frac{7.15}{0.55} = 13 \,\text{MPa}$$

$$\therefore \quad \underline{\text{use } f_{ci}' = 24 \text{ MPa}} \qquad\qquad \text{OK}$$

TABLE 10.18 Concrete Stresses after Instantaneous Losses for the Interior Girder

Span	Location (x/L)	Top Fiber Stress (MPa)					Bottom Fiber Stress (MPa)				
		f_{DC1}	F_{pC1}, P_j/A	M_{psC1}, $P_j^*Y_t/I$	f_{psI}	Total Initial Stress	f_{DC1}	F_{pC1}, P_j/A	M_{psC1}, $P_j^*Y_t/I$	f_{psI}	Total Initial Stress
1	0.0	0.00	6.04	0.14	6.18	6.18	0.00	6.04	−0.18	5.86	5.86
	0.1	2.80	6.09	−2.87	3.23	6.03	−3.64	6.09	3.72	9.82	6.17
	0.2	4.62	6.14	−4.96	1.18	5.80	−6.00	6.14	6.45	12.59	6.59
	0.3	5.59	6.19	−6.12	0.07	5.66	−7.27	6.19	7.95	14.15	6.88
	0.4	5.73	6.24	−6.32	−0.08	5.65	−7.44	6.24	8.22	14.46	7.02
	0.5	5.03	6.30	−5.54	0.75	5.78	−6.53	6.30	7.20	13.50	6.97
	0.6	3.49	6.35	−3.76	2.59	6.08	−4.53	6.35	4.88	11.23	6.70
	0.7	1.11	6.36	−0.92	5.44	6.55	−1.44	6.36	1.20	7.56	6.12
	0.8	−2.11	6.33	2.93	9.26	7.15	2.74	6.33	−3.81	2.52	5.26
	0.9	−6.16	6.26	6.76	13.02	6.86	8.00	6.26	−8.78	−2.52	5.48
	1.0	−9.62	4.68	9.56	14.24	4.62	7.97	4.68	−7.92	−3.24	4.73
2	0.0	−9.62	4.68	9.62	14.30	4.68	7.97	4.68	−7.97	−3.28	4.68
	.1	−6.16	6.10	6.72	12.82	6.66	8.00	6.10	−8.73	−2.63	5.37
	0.2	−2.11	6.03	2.97	9.00	6.90	2.74	6.03	−3.86	2.17	4.90
	0.3	1.11	6.00	−0.69	5.31	6.42	−1.44	6.00	0.89	6.89	5.45
	0.4	3.49	5.97	−3.37	2.60	6.08	−4.53	5.97	4.38	10.34	5.81
	0.5	5.03	5.93	−5.09	0.84	5.87	−6.53	5.93	6.62	12.55	6.02
	0.6	5.73	5.90	−5.87	0.03	5.75	−7.44	5.90	7.63	13.54	6.09
	0.7	5.59	5.87	−5.73	0.14	5.73	−7.27	5.87	7.44	13.31	6.05
	0.8	4.62	5.84	−4.67	1.71	5.79	−6.00	5.84	6.07	11.90	5.91
	0.9	2.80	5.81	−2.71	3.10	5.90	−3.64	5.81	3.52	9.33	5.69
	1.0	0.00	5.78	0.13	5.91	5.91	0.00	5.78	−0.17	5.60	5.60

Note: Positive stress indicates compression and negative stress indicates tension

For the final concrete strength at the service limit state, the controlling location is in the top fiber at 0.6L section in Span 2. From Eq. (10.70), we have

$$f'_{c,req} \geq \frac{f_{DC1} + f_{DC2} + f_{DW} + f_{LL+IM} + f_{psF}}{0.45} = \frac{11.32}{0.45} = 21.16 \text{ MPa} < 28 \text{ MPa}$$

$$\therefore \quad \underline{\text{choose } f'_c = 28 \text{ MPa}} \qquad\qquad\qquad \text{OK}$$

11. **Flexural Strength Design for Interior Girder — Strength Limit State I**
    AASHTO [4] requires that for the Strength Limit State I

$$M_u \leq \phi M_n$$

$$M_u = \eta \sum \gamma_i M_i = 0.95[1.25(M_{DC1} + M_{DC2}) + 1.5 M_{DW} + 1.75 M_{LLH}] + M_{ps}$$

where ϕ is the flexural resistance factor 1.0 and M_{ps} is the secondary moment due to prestress. Factored moment demands M_u for the interior girder in Span 1 are calculated in Table 10.20. Although the moment demands are not symmetrical about the bent (due to different secondary prestress moments), the results for Span 2 are similar and the differences will not be considered in this example. The detailed calculations for the flexural resistance ϕM_n are shown in Table 10.21. It is seen that no additional mild steel is required.

TABLE 10.19 Concrete Stresses after Total Losses for the Interior Girder

		Top Fiber Stress (MPa)					Bottom Fiber Stress (MPa)				
Span	Location (x/L)	f_{LOAD}	$\dfrac{F_{pCF^*}}{P_{j/A}}$	$\dfrac{M_{psCF^*}}{P_{j^*y_t/I}}$	f_{psF}	Total Final Stress	f_{LOAD}	$\dfrac{F_{pCF^*}}{P_{j/A}}$	$\dfrac{M_{psCF^*}}{P_{j^*y_b/I}}$	f_{psF}	Total Final Stress
	0.0	0.00	5.39	0.12	5.52	5.52	0.00	5.39	−0.16	5.23	5.23
	0.1	5.20	5.44	−2.52	2.92	8.12	−6.76	5.44	3.28	8.72	1.97
	0.2	8.69	5.49	−4.36	1.13	9.82	−11.29	5.49	5.67	11.16	−0.13
	0.3	10.66	5.55	−5.37	0.17	10.83	−13.85	5.55	6.98	12.52	−1.32
	0.4	11.14	5.60	−5.52	0.07	11.22	−14.48	5.60	7.18	12.77	−1.71
1	0.5	10.19	5.65	−4.79	0.85	11.05	−13.24	5.65	6.23	11.88	−1.37
	0.6	7.83	5.70	−3.16	2.54	10.37	−10.17	5.70	4.11	9.81	−0.36
	0.7	4.06	5.71	−0.58	5.14	9.20	−5.28	5.71	0.75	6.47	1.19
	0.8	−4.75	5.68	2.91	8.60	3.84	6.18	5.68	−3.79	1.89	8.07
	0.9	−10.28	5.61	6.38	11.99	1.72	13.35	5.61	−8.29	−2.68	10.67
	1.0	−15.22	4.19	8.90	13.09	−2.13	12.61	4.19	−7.37	−3.18	9.43
	0.0	−15.22	4.19	8.95	13.14	−2.07	12.61	4.19	−7.42	−3.23	9.38
	0.1	−10.28	5.45	6.34	11.79	1.52	13.35	5.45	−8.24	−2.79	10.56
	0.2	−4.75	5.38	2.96	8.34	3.58	6.18	5.38	−3.84	1.54	7.72
	0.3	4.06	5.35	0.34	5.01	9.07	−5.28	5.35	0.45	5.80	0.52
	0.4	7.83	5.32	−2.77	2.55	10.37	−10.17	5.32	3.60	8.92	−1.25
2	0.5	10.19	5.29	−4.34	0.94	11.13	−13.24	5.29	5.64	10.93	−2.31
	0.6	11.14	5.25	−5.07	0.18	**11.32**	−14.48	5.25	6.59	11.85	**−2.63**
	0.7	10.66	5.22	−4.98	0.24	10.90	−13.85	5.22	6.47	11.69	−2.15
	0.8	8.69	5.19	−4.07	1.12	9.81	−11.29	5.19	5.29	10.48	−0.81
	0.9	5.20	5.16	−2.37	2.79	7.99	−6.76	5.16	3.08	8.24	1.48
	1.0	0.00	5.13	0.12	5.25	5.25	0.00	5.13	−0.15	4.97	4.97

Notes:
1. $f_{LOAD} = f_{DC1} + f_{DC2} + f_{DW} + f_{LL+IM}$.
2. Positive stress indicates compression and negative stress indicates tension.

TABLE 10.20 Factored Moments for an Interior Girder

		M_{DC1} (kN-m)	M_{DC2} (kN-m)	M_{DW} (kN-m)	M_{LL+IM} (kN-m)		M_{ps} (kN-m)	M_u (kN-m)	
Span	Location (x/L)	Dead Load 1	Dead Load 2	Wearing Surface	Positive	Negative	P/S	Positive	Negative
	0.0	0	0	0	0	0	0	**0**	0
	0.1	2404	211	283	1561	−203	298	**6,402**	3,469
	0.2	3958	356	478	2660	−407	597	**10,824**	5,725
	0.3	4794	435	583	3324	−610	895	**13,462**	6,922
	0.4	4912	448	600	3597	−814	1194	**14,393**	7,060
1	0.5	4310	395	528	3506	−1017	1492	**13,660**	6,140
	0.6	2991	274	367	3080	−1221	1790	**11,310**	4,161
	0.7	952	88	118	2326	−1424	2089	**7,358**	1,124
	0.8	−1805	−165	−221	1322	−1886	2387	1,931	**3,403**
	0.9	−5281	−483	−648	443	−2398	2685	−4.348	9,071
	1.0	−8866	−804	−1078	18	−3283	2984	−10,005	−15,492

Note: $M_u = 0.95[1.25(M_{DC1} + M_{DC2}) + 1.5M_{DW} + 1.75M_{LL+IM}] + M_{pr}$.

12. Shear Strength Design for Interior Girder — Strength Limit State I

AASHTO [4] requires that for the strength limit state I

$$V_u \le \phi V_n$$

$$V_u = \eta \sum \gamma_i V_i = 0.95[1.25(V_{DC1} + V_{DC2}) + 1.5V_{DW} + 1.75V_{LL+IM}] + V_{ps}$$

TABLE 10.21   Flexural Strength Design for Interior Girder — Strength Limit State I

Span	Location (x/L)	A_{ps} mm²	d_p mm	A_s mm²	d_s mm	b mm	c mm	f_{ps} Mpa	d_e mm	a mm	ϕM_n Mpa	M_u kN-m
	0.0		32.16	0	72.06	104	7.14	253.2	32.16	6.07	5,206	0
	0.1		46.09	0	72.06	104	7.27	258.1	46.09	6.18	7,833	4,009
	0.2		56.04	0	72.06	104	7.33	260.1	56.04	6.23	9,717	6,820
	0.3		61.54	0	72.06	104	7.35	261.0	61.54	6.25	10,759	8,469
	0.4		64.00	0	72.06	104	7.36	261.3	64.00	6.26	11,226	9,012
1	0.5	8.47	62.29	0	72.06	104	7.36	261.1	62.29	6.25	10,903	8,494
	0.6		57.20	0	72.06	104	7.34	260.3	57.20	6.24	9,937	6,942
	0.7		48.71	0	72.06	104	7.29	258.7	48.71	6.20	8,328	4,392
	0.8		38.20	0	71.06	82.5	21.19	228.1	38.20	18.01	−4,965	−1,397
	0.9		53.48	0	71.06	82.5	23.36	237.0	53.48	19.86	−7,822	−5,906
	1.0		62.00	0	71.06	104	8.13	261.0	62.00	6.25	−10,848	−10,716

TABLE 10.22   Factored Shear for an Interior Girder

Span	Location (x/L)	V_{DC1} (kN) Dead Load 1	V_{DC2} (kN) Dead Load 2	V_{DW} (kN) Wearing Surface	V_{LL+IM} (kN) Envelopes	M_{LL+IM} (kN-m) Associated	V_{ps} (kN) P/S	V_u (kN)	M_u (kN-m) Associated
	0.0	602.8	50.9	68.3	497.0	0.0	62.2	1762.0	0
	0.1	398.7	37.1	49.8	416.1	1997.4	62.2	1342.5	7,128
	0.2	249.0	23.3	31.3	340.7	3270.3	62.2	996.5	11,838
	0.3	99.3	9.5	12.8	271.9	3915.3	62.2	661.6	14,446
	0.4	−50.4	−4.3	−5.8	−213.9	3228.4	62.2	−366.6	13,780
1	0.5	−200.1	−18.1	−24.3	−277.3	3271.7	62.2	−692.6	13,270
	0.6	−349.8	−31.9	−42.8	−340.5	2771.1	62.2	−1018.2	10,797
	0.7	−499.6	−45.7	−61.3	−404.4	1955.7	62.2	−1345.0	6,742
	0.8	−649.3	−59.5	−79.8	−468.4	689.4	62.2	−1671.9	879
	0.9	−799.0	−73.3	−98.3	−529.5	−945.3	62.2	−1994.0	−6,655
	1.0	−971.2	−84.9	−113.9	−580.6	−1849.7	62.2	−2319.5	−13,110

Note: $V_y = 0.95[1.25(V_{DC1} + V_{DC2}) + 1.5V_{DW} + 1.75V_{LL+IM}] + V_{ps}$.

TABLE 10.23   Shear Strength Design for Interior Girder   Strength Limit State I

Span	Location (x/L)	d_v (mm)	y' (rad)	V_p (kN)	v/f_c'	ε_x (1000)	θ (°)	β	V_c (kN)	S (mm)	ϕV_n (kN)	$|V_u|$ (kN)
	0.0	1382	0.085	606	0.133	−0.256	21.0	2.68	428	100	1860	1762
	0.1	1382	0.064	459	0.101	−0.382	27.0	5.60	894	300	1513	1342
	0.2	1382	0.043	309	0.078	−6.241	33.0	2.37	378	200	1036	996
	0.3	1503	0.021	156	0.052	−6.299	38.0	2.10	365	300	753	662
	0.4	1555	0.000	0	0.036	−6.357	36.0	2.23	400	600	511	367
1	0.5	1503	0.021	159	0.055	−6.415	36.0	2.23	387	400	710	693
	0.6	1382	0.043	320	0.080	−6.473	30.0	2.48	396	200	1076	1018
	0.7	1382	0.064	482	0.099	−0.401	27.0	5.63	899	300	1538	1345
	0.8	1382	0.085	639	0.120	−0.398	23.5	6.50	1038	300	1813	1672
	0.9	1382	0.091	670	0.152	−6.372	23.5	3.49	557	100	2017	1994
	1.0	1502	0.000	0	0.233	−6.280	36.0	1.00	173	40	2343	2319

where ϕ is shear resistance factor 0.9 and V_{ps} is the secondary shear due to prestress. Factored shear demands V_u for the interior girder are calculated in Table 10.22. To determine the effective web width, assume that the VSL post-tensioning system of 5 to 12 tendon units [VLS, 1994] will be used with a grouted duct diameter of 74 mm. In this example, $b_v = 300 − 74/2 = 263$ mm. Detailed calculations of the shear resistance ϕV_n (using two-leg #15M stirrups $A_v = 400$ mm²) for Span 1 are shown in Table 10.23. The results for Span 2 are similar to Span 1 and the calculations are not repeated for this example.

References

1. Lin, T. Y. and Burns, N. H., *Design of Prestressed Concrete Structure*, 3rd ed., John Wiley & Sons, New York, 1981.
2. Nawy, E. G., *Prestressed Concrete: A Fundamental Approach*, 2nd ed., Prentice-Hall, Englewood Cliffs, NJ, 1996.
3. Collins, M. P. and Mitchell, D., *Prestressed Concrete Structures*, Prentice-Hall, Englewood Cliffs, NJ, 1991.
4. AASHTO, *AASHTO LRFD Bridge Design Specifications*, 1st ed., American Association of State Highway and Transportation Officials, Washington, D.C., 1994.
5. PCI, *PCI Design Handbook – Precast and Prestressed Concrete*, 3rd ed., Prestressed Concrete Institute, Chicago, IL, 1985.
6. Eubunsky, I. A. and Rubinsky, A., A preliminary investigation of the use of fiberglass for prestressed concrete, *Mag. Concrete Res.*, Sept., 71, 1954.
7. Wines, J. C. and Hoff, G. C., Laboratory Investigation of Plastic — Glass Fiber Reinforcement for Reinforced and Prestressed Concrete, *Report 1*, U.S. Army Corps of Engineers, Waterway Experimental Station, Vicksburg, MI, 1966.
8. Wines, J. C., Dietz, R. J., and Hawly, J. L., Laboratory Investigation of Plastic — Glass Fiber Reinforcement for Reinforced and Prestressed Concrete, *Report 2*, U.S. Army Corps of Engineers, Waterway Experimental Station, Vicksburg, MI, 1966.
9. Iyer, S.I. and Anigol, M., Testing and evaluating fiberglass, graphite, and steel prestressing cables for pretensioned beams, in *Advanced Composite Materials in Civil Engineering Structures*, Iyer, S. I. and Sen, R., Eds., ASCE, New York, 1991, 44.
10. Miesseler, H. J. and Wolff, R., Experience with fiber composite materials and monitoring with optical fiber sensors, in *Advanced Composite Materials in Civil Engineering Structures*, Iyer, S. I. and Sen, R., Eds., ASCE, New York, 1991, 167–182.
11. Kim, P. and Meier, U., CFRP cables for large structures, in *Advanced Composite Materials in Civil Engineering Structures*, Iyer, S. I. and Sen, R., Eds., ASCE, New York, 1991, 233–244.
12. PTI, *Post-Tensioning Manual*, 3rd ed., Post-Tensioning Institute, Phoenix, AZ, 1981.
13. FHWA, *Standard Plans for Highway Bridges, Vol. I, Concrete Superstructures*, U.S. Department of Transportation, FHWA, Washington, D.C., 1990.
14. ACI, *Building Code Requirements for Structural Concrete (ACI318-95) and Commentary* (ACI318R-95), American Concrete Institute, Farmington Hills, MI, 1995.
15. CEB-FIP, *Model Code for concrete structures*. (MC-90). Comité Euro-international du Béton (CEB)-Fédération Internationale de la précontrainte (FIP) (1990). Thomas Telford, London, U.K. 1993.

11

Segmental Concrete Bridges

Gerard Sauvageot
J. Muller International

0-8493-7434-0/00/$0.00+$.50
© 2000 by CRC Press LLC

11.1 Introduction

Before the advent of segmental construction, concrete bridges would often be made of several precast girders placed side by side, with joints between girders being parallel to the longitudinal axis of the bridge. With the modern segmental concept, the segments are slices of a structural element between joints which are perpendicular to the longitudinal axis of the structure.

When segmental construction first appeared in the early 1950s, it was either cast in place as used in Germany by Finsterwalder et al., or precast as used in France by Eugène Freyssinet and Jean Muller. The development of modern segmental construction is intertwined with the development of balanced cantilever construction.

By the use of the term *balanced cantilever construction,* we are describing a phased construction of a bridge superstructure. The construction starts from the piers cantilevering out to both sides in such a way that each phase is tied to the previous ones by post-tensioning tendons, incorporated into the permanent structure, so that each phase serves as a construction base for the following one.

The first attempts to use balanced cantilever construction, in its pure form, were made by Baumgart, who in 1929 built the Río Peixe Bridge in Brazil in reinforced concrete, casting the 68-m-long main span in free cantilevering. The method did not really prosper, however, until the post-tensioning technique had been sufficiently developed and generally recognized to allow crack-free concrete cantilever construction.

From 1950, several large bridges were built in Germany with the use of balanced cantilever construction with a hinge at midspan, using cast-in-place segments, such as

- Moselbrücke Koblenz, 1954: Road bridge, 20 m wide, with three spans of 101, 114, 123 m plus short ballasted end spans hidden in large abutments; the cross section is made up of twin boxes of variable depth, connected by the top slab.
- Rheinbrücke Bendorf, 1964: Twin motorway bridges, 1,031 m long, with three main river spans of 71, 208, 71 m, built-in free cantilever construction with variable depth box sections.

In France, the cantilever construction took a different direction, emphasizing the use of precast segments.

Precast segments were used by Eugène Freyssinet for construction of the well-known six bridges over the Marne River in France (1946 to 1950). The longitudinal frames were assembled from precast segments, which were prestressed vertically and connected by dry-packed joints and longitudinal post-tensioning tendons. Precast segments were also used by Jean Muller for the execution of a girder bridge in upstate New York, where longitudinal girders were precast in three segments each, which were assembled by dry-packed joints and longitudinal post-tensioning tendons.

From 1960, Jean Muller systematically applied precast segments to cantilever construction of bridges. It is characteristic for precast segmental construction, in its purest form, that segments are match cast, which means that each segment is cast against the previous one so that the end face of one segment will be an imprint of the neighbor segment, ensuring a perfect fit at the erection. The early milestones were as follows:

- Bridge over the Seine at Choisy-le-Roi in France, 1962: Length 37+55+37 = 130 m; the bridge is continuous at midspan, with glued joints between segments (first precast segmental bridge).
- Viaduc d'Oleron in France, 1964 to 1966: Total length 2862 m, span lengths generally 79 m, with hinges in the quarterpoint of every fourth span; the segments were cast on a long bench (long-line method); erection was by self-launching overhead gantry (first large-scale, industrialized precast bridge construction).

In the same period, precast segmental construction was adopted by other designers for bridge construction with cast-in-place joints. Some outstanding structures deserve mention:

- Ager Brücke in Austria, 1959 to 1962: Precast segments placed on scaffold, cast-in-place joints.
- Río Caroni in Venezuela, 1962 to 1964: Bridge with multiple spans of 96-m each. Precast segments 9.2 m long, were connected by 0.40-m-wide cast-in-place joints to constitute the 480-m-long bridge deck weighing 8400 tons, which was placed by incremental launching with temporary intermediate supports.
- Oosterschelde Bridge in The Netherlands, 1962 to 1965: Precast segmental bridge with a total length of 5 km and span lengths of 95 m; the precast segments are connected by cast-in-place, 0.4-m-wide joints and longitudinal post-tensioning.

Since the 1960s, the construction method has undergone refinements, and it has been developed further to cover many special cases, such as progressive construction of cantilever bridges, span-by-span construction of simply supported or continuous spans, and precast-segmental construction of frames, arches, and cable-stayed bridge decks.

In 1980, precast segmental construction was applied to the Long Key and Seven Mile Bridges in the Florida Keys in the United States. The Long Key Bridge has 100 spans of 36 m each, with continuity in groups of eight spans. The Seven Mile Bridge has 270 spans of 42 m each with continuity in groups of seven spans. The spans were assembled from 5.6-m-long precast segments placed on erection girders and made self-supporting by the stressing of longitudinal post-tensioning tendons. The construction method became what is now known as span-by-span construction.

Comparing cast-in-place segmental construction with precast segmental construction, the following features come to mind:

- Cast-in-place segmental construction is a relatively slow construction method. The work is performed *in situ*, i.e., exposed to weather conditions. The time-dependent deformations of the concrete become very important as a result of early loading of the young concrete. This method requires a relatively low degree of investment (travelers).
- Precast segmental construction is a fast construction method determined by the time required for the erection. The major part of the work is performed in the precasting yard, where it can be protected against inclement weather. Precasting can start simultaneously with the foundation work. The time-dependent deformations of the concrete become less important, as the concrete may have reached a higher age by the time the segments are placed in the structure. This method requires relatively important investments in precasting yard, molds, lifting gear, transportation, and erection equipment. Therefore, this method requires a certain volume of work to become economically viable. Typically, the industrialized execution of the structure leads to higher quality of the finished product.

Since the 1960s, the precast segmental construction method has won widespread recognition and is used extensively throughout the world. Currently, very comprehensive bridge schemes, with more than 20,000 segments in one scheme, are being built as large urban and suburban viaducts for road or rail. It is reasonable to expect that the precast segmental construction method, as introduced by Jean Muller, will contribute extensively to meet the infrastructure needs of humankind well into the next millennium.

11.2 Balanced Cantilever Girder Bridges

11.2.1 Overview

Balanced cantilever segmental construction for concrete box-girder bridges has long been recognized as one of the most efficient methods of building bridges without the need for falsework. This method has great advantages over other forms of construction in urban areas where temporary shoring

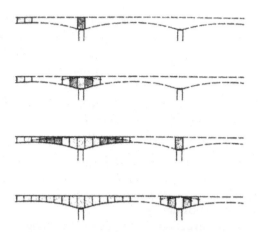

FIGURE 11.1 Balanced cantilever construction.

would disrupt traffic and services below, in deep gorges, and over waterways where falsework would not only be expensive but also a hazard. Construction commences from the permanent piers and proceeds in a "balanced" manner to midspan (see Figure 11.1). A final closure joint connects cantilevers from adjacent piers. The structure is hence self-supporting at all stages. Nominal out-of-balance forces due to loads on the cantilever can be resisted by several methods where any temporary equipment is reusable from pier to pier.

The most common methods are as follows:

- Monolithic connection to the pier if one is present for the final structure;
- Permanent, if present, or temporary double bearings and vertical temporary post-tensioning;
- A simple prop/tie down to the permanent pile cap;
- A prop against an overhead gantry if one is mobilized for placing segments or supporting formwork.

The cantilevers are usually constructed in 3- to 6-m-long segments. These segments may be cast in place or precast in a nearby purpose-built yard, transported to the specific piers by land, water, or on the completed viaduct, and erected into place. Both methods have merit depending on the specific application.

It is usually difficult to justify the capital outlay for the molds, casting yard, and erection equipment required for precast segmental construction in a project with a deck area of less than 5000 m². The precasting technique may be viable for smaller projects provided existing casting yard and molds can be mobilized and the segments could be erected by a crane.

11.2.2 Span Arrangement and Typical Cross Sections

Typical internal span-to-depth ratios for constant-depth girders are between 18 and 22. However, box girders shallower than 2 m in depth introduce practical difficulties for stressing operations inside the box and girders shallower than 1.5 m become very difficult to form. This sets a minimum economical span for this type of construction of 25 to 30 m. Constant-depth girders deeper than 2.5 to 3.0 m are unusual and therefore for spans greater than 50 m consideration should be given to varying-depth girders through providing a curved soffit or haunches. For haunch lengths of 20 to 25% of the span from the pier, internal span-to-depth ratios of 18 at the pier and as little as 30 at midspan are normally used.

Single-cell box girders provide the most efficient section for casting – these days multicell boxes are rarely used in this method of construction. Inclined webs improve aesthetics but introduce added difficulties in formwork when used in combination with varying-depth girders. The area of

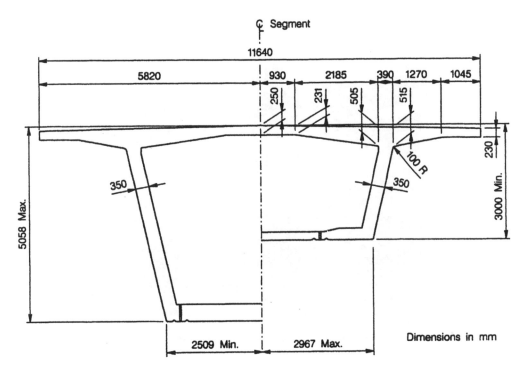

FIGURE 11.2 Typical cross section of a varying-depth girder for a 93-m span.

the bottom slab at the pier is determined by the modulus required to keep bottom fiber compressions below the allowable maximum at this location. In the case of internal tendons local haunches are used at the intersection of the bottom slab and the webs to provide sufficient space for accommodating the required number of tendon ducts at midspan. The distance between the webs at their intersection with the top slab is determined by achieving a reasonable balance between the moments at this node. Web thicknesses are determined largely by shear considerations with a minimum of 250 mm when no tendon ducts internal to the concrete are present and 300 mm in other cases. Figure 11.2 shows the typical dimensions of a varying-depth box girder.

11.2.3 Cast-in-Place Balanced Cantilever Bridges

The cast-in-place technique is preferred for long and irregular span lengths with few repetitions. Bridge structures with one long span and two to four smaller spans usually have a varying-depth girder to carry the longer span, hence making the investment in a mold which accommodates varying-depth segments even more uneconomical. A prime example of application of balanced cantilevering in an urban environment to avoid disruption to existing road services below is the structure of the Bangkok Light Rail Transit System, where it crosses the Rama IV Flyover (see Figure 11.3). The majority of the 26-km viaduct structure is precast, but at this intersection a 60-m span was required to negotiate the existing road at a third level with the flyover in service below. A three-span, 30-, 60-, 30-m structure was utilized with a box-girder depth of 3.5 m at the pier and 2.0 m at midspan and a parabolic curved soffit. The flyover was only disrupted a few nights during concrete placement of the segments directly above as a precaution.

In the above example, the side spans were constructed by balanced cantilevering; however, ideal arrangement of spans normally provides end spans which are greater than half the internal spans. These, therefore, cannot be completed by balanced cantilevering, and various techniques are used to reach the abutments. The most economical and common method is the use of falsework; however,

FIGURE 11.3   Construction of the Bangkok Transit System over Rama IV Flyover, Thailand.

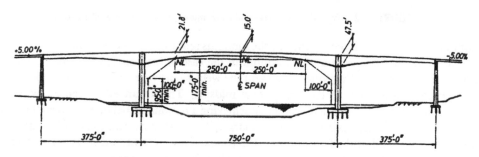

FIGURE 11.4   Houston Ship Channel Bridge, United States.

should the scale of the project justify use of an auxiliary truss to support the formwork during balanced cantilevering, then this could also be used for completing the end spans.

Another example of a cast-in-place balanced cantilever bridge is the Houston Ship Channel Bridge where a three-span, 114-, 229-, 114-m structure was used over the navigation channel (see Figure 11.4). A three-web box girder carrying four lanes of traffic is fixed to the main piers to make the structure a three-span rigid frame. Unusual span-to-depth ratios were dictated by the maximum allowable grade of the approach viaducts and the clearance required for the ship channel. The soffit was given a third-degree parabolic profile to increase the structural depth near the piers in order to compensate for the very limited height of the center portion of the main span. Maximum depth at the pier is 14.6 m, a span-to-depth ratio of 15.3 to enable a minimum depth at midspan of 4.6 m, and a span-to-depth ratio of 49. The box girder is post-tensioned in three dimensions: four 12.7-mm strands at 600-mm centers transversely in the top slab as well as longitudinal and vertical post-tensioning in the webs.

11.2.4   Precast Balanced Cantilever Bridges

Extending segmental construction to balanced cantilevering, and hence eliminating the need for falsework as well as substantial increases in the rate of construction, requires a huge leap in the technology of precasting: match casting. The very first bridge that benefited from match-casting technology was the Choisy-le-Roi Bridge near Paris, designed by Jean Muller and completed in 1964. This method has since grown in popularity and sophistication and is used throughout the world today. The essential feature of match casting is that successive segments are cast against the adjoining segment in the correct relative orientation with each other starting from the first segment away from the pier. The segments are subsequently erected on the pier in the same order, and hence no adjustments are necessary between segments during assembly. The joints are either left dry or made of a very thin layer of epoxy resin, which does not alter the match-cast geometry. Post-tensioning may proceed as early as practicable since there is no need for joints to cure.

The features of this method that provide significant advantages over the cast-in-place method, provided the initial investment in the required equipment is justified by the scale of the project, are immediately obvious and may be listed as follows:

- Casting the superstructure segments may be started at the beginning of the project and at the same time as the construction of the substructure. In fact, this is usually required since the speed of erection is much faster than production output of the casting yard and a stockpile of segments is necessary before erection begins.
- Rate of erection is usually 10 to 15 times the production achieved by the cast-in-place method. The time required for placing reinforcement and tendons and, most importantly, the waiting time for curing of the concrete is eliminated from the critical path.
- Segments are produced in an assembly-line factory environment, providing consistent rates of production and allowing superior quality control. The concrete of the segments is matured, and hence the effects of shrinkage and creep are minimized.

The success of this method relies heavily on accurate geometry control during match casting as the methods available for adjustments during erection offer small and uncertain results. The required levels of accuracy in surveying the segments match-cast against each other are higher than in other areas of civil engineering in order to assure acceptable tolerances at the tip of the cantilevers.

The size and weight of precast segments are limited by the capacity of transportation and placing equipment. For most applications segment weights of 40 to 80 tons are the norm, and segments above 250 tons are seldom economical. An exception to the above is the recent example of the main spans of the Confederation Bridge where complete 192.5-m-long balanced cantilevers weighing 7500 tons were lifted into place using specialized equipment (see Figure 11.5). The 250-m main spans of this fixed link in Atlantic Canada, connecting Cape Tormentine, New Brunswick, and Borden, Prince Edward Island, were constructed by a novel precasting method. The scale of the project was sufficiently large to justify precast segmental construction; however, adverse weather and site conditions provided grounds for constructing the balanced cantilevers, 14 m deep at the piers, in a similar method to cast-in-place construction but in a nearby casting yard. The completed balanced cantilevers were then positioned atop completed pier shafts in a single operation. A light template match-cast against the base of the pier segment allowed fast and accurate alignment control on the spans.

11.2.5　Loads on Substructure

The methods for supporting the nominal out-of-balance forces during balanced cantilevering were described earlier. The following forces should be considered in calculating the possible out-of-balance forces:

- In precast construction, one segment out of balance and the loss of a segment on the balancing cantilever as an ultimate condition;
- In precast construction, presence of a stressing platform (5 to 10 tons) on one cantilever only or the loss of the form traveler in the case of cast-in-place construction;

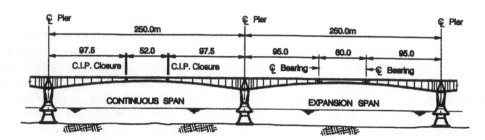

FIGURE 11.5   Main spans of the Confederation Bridge, Canada.

- Live loading on one side of 1.5 kN/m^2;
- Wind loading during construction;
- The possibility of one cantilever having a 2.5% higher dead weight than the other.

The loads on the substructure do not usually govern the design of these elements provided balanced cantilever construction is considered at the onset of the design stage. The out-of-balance forces may provide higher temporary longitudinal moments than for the completed structure; however, in the case of a piled foundation, this usually governs the arrangement and not the number of the piles.

11.2.6   Typical Post-Tensioning Layout

Post-tensioning tendons may be internal or external to the concrete section, but inside the box girder, housed in steel pipes, or both. External post-tensioning greatly simplifies the casting process and the reduced eccentricities available compared with internal tendons are normally compensated by lower frictional losses along the tendons and hence higher forces.

The choice of the size of the tendons must be made in relation to the dimensions of the box-girder elements. A minimum number of tendons would be required for the balanced cantilevering process, and these may be anchored on the face of the segments, on internal blisters, or a combination of both. After continuity of opposing cantilevers is achieved, the required number of midspan tendons may be installed across the closure joint and anchored on internal bottom blisters. Depending on the arrangement and length of the spans, economies may be made by arranging some of the tendons to cross two or more piers, deviating from the top at the piers to the bottom at midspan, thereby reducing the number of anchorages and stressing operations. External post-tensioning is best used for these continuity tendons which would allow longer tendon runs due to the reduced frictional losses. Where the tendons are external to the concrete elements, deviators at piers, quarterspan, and midspan are used to achieve the required profile. An example of a typical internal post-tensioning layout is shown in Figure 11.6.

11.2.7   Articulation and Hinges

The movements of the structure under the effects of cyclic temperature changes, creep, and shrinkage are traditionally accommodated by provision of halving joint-type hinges at the center of various spans. This practice is now discontinued due to the unacceptable creep deformations that occur at these locations. If such hinges are used, these are placed at contraflexure points to minimize the effects of long-term deflections. A development on simple halving joints is a moment-resisting joint, which allows longitudinal movements only. All types of permanent hinges that are more easily exposed to the elements of water and salt from the roadway provide maintenance difficulties and should be eliminated or reduced wherever possible.

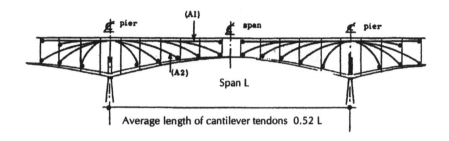

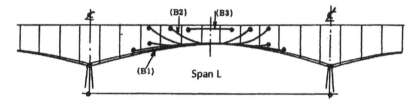

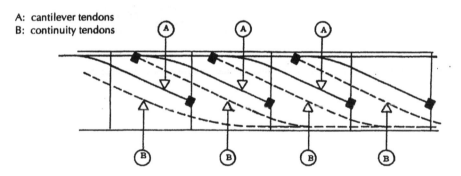

FIGURE 11.6 Typical post-tensioning layout for internal tendons.

If the piers are sufficiently flexible, then a fully continuous bridge may be realized with joints at abutments only. When seismic considerations are not a dominant design feature and a monolithic connection with the pier is not essential, bearings atop of the piers are preferred as they reduce maintenance and replacement cost. In addition, it will allow free longitudinal movements of the deck. A monolithic connection or a hinged bearing at one or more piers would provide a path for transmitting loads to suitable foundation locations.

11.3 Progressive and Span-by-Span Constructed Bridges

11.3.1 Overview

In progressive or span-by-span construction methods, construction starts at one end and proceeds continuously to the other end. Generally, progressive construction is used where access to the ground level is restricted either by physical constraints or by environmental concerns. Deck variable cross sections and span lengths up to 60 m are easily accommodated. In contrast, span-by-span precast segmental construction is used typically where speed of construction is of major concern. Span lengths up to 50 m are most economical as it minimizes the size of the erection equipment.

FIGURE 11.7 Fréburge Viaduct, France—erection with movable stay tower.

11.3.2 Progressive Construction

The progressive method step-by-step erection process is derived from cantilever construction, where segments are placed in a successive cantilever fashion. The method is valid for both precast and cast-in-place segments. Due to the excessively high bending moments the cantilever deck has to resist over the permanent pier during construction, either a temporary bent or a temporary movable tower–stay assembly would have to be used. As shown in Figure 11.7, for precast construction using a temporary tower and stay system, segments are transported over the erected portion of the bridge to the end of the completed portion. Using some type of lifting equipment, e.g., a swivel crane, the segment is placed in position and supported temporarily either by post-tensioning to the previous segment or by stays from a tower.

The advantages of this methods are

- Operations are conducted at deck level.
- Reactions on piers are vertical.
- The method can easily accommodate variable horizontal curves.

The disadvantages are

- The first span is erected on falsework.
- Forces in the superstructure during erection are different from those in the completed structure.
- The piers are temporarily subjected to higher reactions from dead load than in the final structure because of the length of the cantilever erected. However, considering the other loads in the final structure, this case is not generally controlling the pier design.

FIGURE 11.8 Completed Linn Cove Viaduct, United States.

FIGURE 11.9 Linn Cove Viaduct — pier being constructed from the deck level.

The Linn Cove Viaduct (1983) on the Blue Ridge Parkway in North Carolina shown in Figure 11.8, demonstrated the potential progressive placement when one is forced to overcome extreme environmental and physical constraints. Because access at the ground level was limited, the piers were constructed from the deck level, at the tip of an extended cantilever span. Temporary cable stays could not be used due to the extreme horizontal curvature in the bridge. Instead, temporary bent supports were erected between permanent piers. Figure 11.9 shows one temporary support in the background while a permanent precast pier is being erected from the deck level.

FIGURE 11.10   Lifting completed span of the Seven Mile Bridge, Florida, using an *overhead truss.*

11.3.3   Span-by-Span Construction

As with balanced cantilever and progressive placement, span-by-span construction activity is performed primarily at the deck level and typically implemented for long viaducts having numerous, but relatively short spans, e.g., <50 m. It was initially developed as a cast-in-place method of construction, on formwork, with construction joints at joint of contraflexure. The form traveler is supported either on the bridge piers, on the edge of the previously erected span and the next pier or, at times, even at the ground level. With the precast segmental method, segments are placed and adjusted on a steel erection girder spanning from pier to pier, then post-tensioned together in one operation. Although both the cast-in-place and the precast span-by-span construction methods continue to be used, precast segmental has become the method of choice for most applications.

Long Key and Seven Mile Bridges, United States: Two early applications of the precast span-by-span method are the Long Key Bridge (1977) and the Seven Mile Bridge (1978), both located in the Florida Keys. The shorter, 3000-m, 100-span Long Key Bridge is the first application of precast span-by-span construction with dry segment joints and external post-tensioning in the United States.

Essentially the same bridge design concept as Long Key — only much longer — the 10,931-m, 270-span Seven Mile Bridge utilized rectangular precast piers and an overhead truss, as shown in Figure 11.10. The overhead truss allowed easier repositioning from one span to the next one and thus improved overall erection speed.

Bang Na–Bang Pli–Bang Pakong Expressway, Thailand: A number of span-by-span highway and rail mega projects have been either completed recently or currently are under construction in Southeast Asia. Probably, the most innovative of these recent applications is the 54,000-m, 1300-span, Bang Na–Bang Pli–Bang Pakong Expressway. The girder supports segment assembly and span installation activities. This erection process can be regarded as "assembly-line" in that there is no requirement for disassembly and reassembly of the erection girder as it travels from pier to pier. The piers, although designed structurally for the construction process, can also be seen to provide an aesthetically pleasing, somewhat "floating," appearance to the six-lane, 27-m-wide box girder. Figure 11.11 shows one of the erection girders as it lifts a segment. With five erection girders erecting a span every 2 days or 780 m of superstructure per week, construction of the viaduct is expected to last approximately 2 years and be completed in 1999, without interruption of traffic below.

Roize, France: Another innovative example of span-by-span construction is the 112-m, three-span, prestressed composite truss Roize Bridge (1991) in the French Alps, shown in Figure 11.12. The deck is made of prestressed concrete and steel. Each factory-built tetrahedron module and

FIGURE 11.11   Bang Na Expressway, Thailand — launching of girder erection.

FIGURE 11.12   View of the Roize Bridge, France — space truss spans using tetrahedron modules.

precast pretensioned slab is placed on erection beams and adjusted into position. After welding the bottom member joints and casting the closure strips, the modules are post-tensioned together as a completed span. Due to the modular basis, this two-lane bridge represents a new class of super-lightweight, factory-built segments.

Channel Bridge, United States. The first precast, prestressed channel bridge in the United States was built in 1974 in San Diego, California, as a pedestrian crossing at San Diego State University. This concept was reused 18 years later as an experimental study for new bridge standards, initially

FIGURE 11.13   Channel Bridge, France, under construction.

by the French Highway Administration. Figure 11.13 shows the 54-m, two-span Champfeuillet Bridge (1992), under construction along the Rhône Alpine Motorway near Grenoble, France. The most innovative aspect of the concept is the use of the concrete parapets as part of the structure. With the primary longitudinal post-tensioning passing through the barriers, an extremely lightweight, shallow section is possible.

Research and implementation of the Channel Bridge, although continuing in Europe, also has begun recently in the United States. Initiated by the Federal Highway Administration (FHWA) and the Highway Innovative Technology Evaluation Center (HITEC), a branch of the Civil Engineering Research Foundation (CERF), at least two applications of the Channel Bridge concept have been completed recently in the United States for the New York State Department of Transportation (NYSDOT).

The primary benefits of the concept are as follows:

- Lightweight, easily placed segments.
- Fast erection times with small investment in erection equipment.
- Increased vertical clearance beneath the superstructure, because the load-carrying members are above the roadway slab, not below.
- A reduction in the number of bridge overpass piers required, which increases safety levels for traffic lanes below.

Span by span, as used today, utilizes post-tensioning tendons outside the concrete, but inside the box girder for ease of precasting and speed of installation together with dry joints, no epoxy, between segments. The post-tensioning tendons are continuous from pier segment to pier segment.

11.4 Incrementally Launched Bridges

11.4.1 Overview

The incremental launching technique has been used on bridges numbering in the hundreds since its introduction by Professor Fritz Leonhardt in 1961 for the Río Caroni Bridge in Venezuela. It is an effective alternative for the bridge designer to consider when the site meets its particular alignment requirements. The method entails casting the superstructure, or a portion thereof, at a stationary location behind one of the abutments. The completed or partially completed structure is then jacked into place horizontally, i.e., pushed along the bridge alignment. Subsequent segments can then be cast onto the already completed portion and in turn pushed onto the piers. Because all of the casting operations are concentrated at a location easily accessible from the ground, concrete quality of the same level expected from a precasting yard can be achieved. The procedure has the advantage that, like the balanced cantilever technique, it obviates the need for falsework to cast the girder. Moreover, heavy erection equipment, cranes, gantries, and the like, are not necessary, nor is the use of epoxy at segment joints. Usually, the only special equipment required is light steel truss work for a launching nose to reduce the cantilever moments during launching.

11.4.2 Special Requirements

There are two peculiarities associated with the technique, which must be appreciated by the designer. The first is that the alignment must be straight or, if it involves curves, the curvature must be constant. The second is that during launching, every section of the girder will be subjected to both the maximum and minimum moments of the span; and the leading cantilever portion will be subjected to slightly higher moments. This second constraint usually leads to slightly deeper sections, on the order of $\frac{1}{15}$ the span, than would otherwise be considered. The girders must also be of constant depth as each section will at sometime be supported on the temporary bearings. Other considerations include the necessity for a large area behind the abutment for the casting operations, the requirement to lift the bridge off of the temporary bearings, and place it on the permanent ones when launching is complete and the need for very careful control of geometry during casting.

Incremental launching is generally considered for long viaducts with many spans of the same length. Spans up to 100 m can be considered; the requirement for constant-depth girders makes longer spans uneconomical. A single long span in the center of a project can be achieved by launching from both abutments and finishing at the long span with two converging cantilevers. The practical length limit for launching in about 1000 m. Bridges of twice this length can be considered by launching from both abutments.

11.4.3 Typical Post-Tensioning Layout

During superstructure launching each section of the girder is subjected to constantly reversing bending moments as it proceeds from temporary support to midspan. Because of the sign change in the applied moments, the efficient use of draped tendons for launching load effects is impossible. The general procedure has therefore been to apply axial prestressing for the launching operation. These tendons are usually straight, being contained in the top and bottom slabs of the girder. The tendons for successive segments must be spliced to these with couplers or stressed in buttresses in an overlapping fashion. This prestressing is subsequently augmented with either draped tendons or short top- and bottom-slab tendons for respective negative and positive moment regions in the completed structure to meet service state requirements. In some instances, permanent draped prestressing has been placed in the configuration required for the final condition, and temporary tendons with an opposing drape are provided to counteract their bending effects during launching. These temporary tendons are then removed when launching is complete.

11.4.4 Techniques for Reducing Launching Moments

As suggested above, the launching moments in the leading spans, especially the first cantilever span, will be greater than those in the following interior spans. If the girder is simply launched to the first pier with no special provision to reduce these moments, they will in fact be on the order of six times the typical negative moment over a pier. The method used most frequently to overcome this problem has been a light structural-steel launching nose attached to the leading cantilever (see Figures 11.14 and 11.15). This nose supports the girder without the weight penalty of the heavier concrete section. In order to be effective, the nose must be both as light and as stiff as possible.

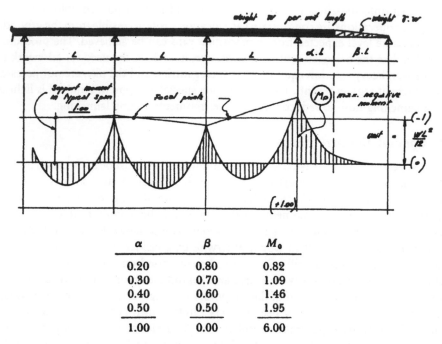

FIGURE 11.14   Critical negative moment during launching with nose. $M_1 = [WL^2/12] (6\alpha^3 + 6y) (1 - \alpha^3)$. Multiplier $= WL^2/12$. For $y = 0.11$.

α	β	M_0
0.20	0.80	0.82
0.30	0.70	1.09
0.40	0.60	1.46
0.50	0.50	1.95
1.00	0.00	6.00

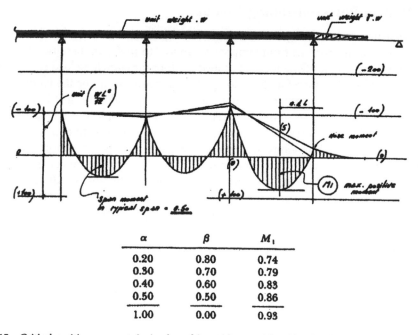

FIGURE 11.15   Critical positive moment during launching with nose. $M_1 = [(WL^2/12) (0.933 - 2.96y\beta^2)]$. Multiplier $= WL^2/12$. For $y = 0.11$.

α	β	M_1
0.20	0.80	0.74
0.30	0.70	0.79
0.40	0.60	0.83
0.50	0.50	0.86
1.00	0.00	0.93

For longer spans, the steel nose is not as effective, and other methods have been employed to reduce launching moments. Temporary piers are a viable solution when ground conditions are such that the foundation costs are relatively modest and the pier height is not too great. If either of these conditions is not found, the cost can escalate rapidly as a temporary pier will be required in every span.

One last method that has been employed successfully is a temporary pylon attached to the deck at the trailing end of the first span which supports stays connected to the leading end. This device is very efficient in reducing the cantilever moment in the leading span; however, it produces an undesirable positive moment when the pylon is at midspan. For this reason, the stays must be equipped with a jack to adjust the stay force as needed during the various stages of the launching operations.

11.4.5 Casting Bed and Launching Methods

Segment lengths for incrementally launched bridges are generally greater than for other types of segmental bridges. Typical segment lengths range from 15 to 40 m. Usually, a casting area twice the length of the segment is required for actual casting and the ancillary operations that must be conducted there. The casting bed is generally a significant structure itself, as the strict geometry-control requirements of the technique make settlement of the formwork unacceptable.

Launching has been accomplished in the past either by tendons attached to the girder and horizontal jacks bearing on the abutment or by a horizontal jack bearing on the abutment face connected to a vertical jack which slides on a bearing. The upper surface of the vertical jack is fitted with a friction device to bear on the soffit of the box girder. The vertical jack is inflated to provide the normal force required for transferring the launching force by friction.

11.5 Arches, Rigid Frames, and Truss Bridges

11.5.1 Arch Bridges

The first step toward the segmental construction of arches was taken shortly after World War I by Eugéne Freyssinet. He employed hydraulic jacks to lift the completed Villeneuve arch from its falsework by applying an internal thrust at its crown. This departure from the classical method of striking the centering to develop the thrust in the arch opened the door to modern arch construction techniques that do not rely on falsework. It also presented the opportunity to reduce the bending moments in the arch by eliminating the dead load bending associated with axial shortening of the ribs.

11.5.1.1 Arches Erected without Falsework

The development of stay-cable and form-traveler technology has made possible the erection of arches in cantilever fashion without a centering supported from below. One early example of this technique was the suite of viaducts built in Caracas, Venezuela, in 1952 (see Figure 11.16). The first quarter of the arch span was supported by light forms which were in turn supported by stay cables attached to a pilaster at the springing of the arch. The crown portion of the arch was then completed with a light centering supported on the already-completed portion of the arch so that no falsework was required in the valley below.

Several variations on this theme were subsequently developed. The methods employed varied, depending on site conditions, from the use of very high pylons with a single group of stays allowing construction of the arch all the way to the crown to those which used the permanent spandrel columns in conjunction with temporary stay diagonals to form a truss. These methods are summarized in Figure 11.17.

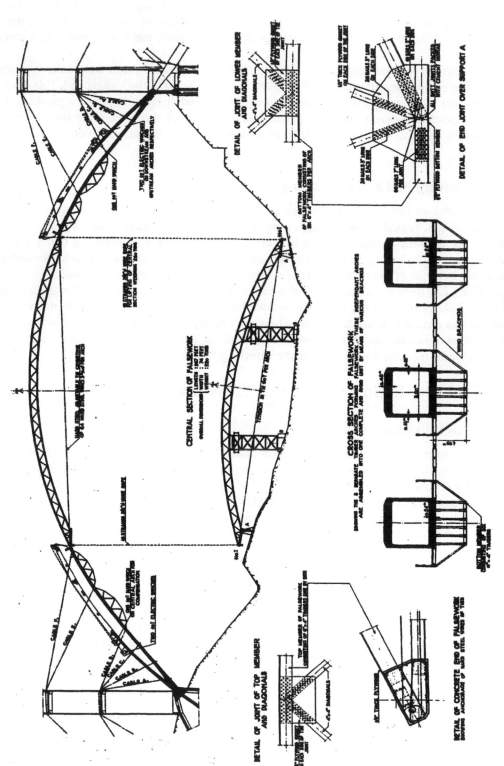

FIGURE 11.16   Caracas viaducts-erection of center portion of arch falsework.

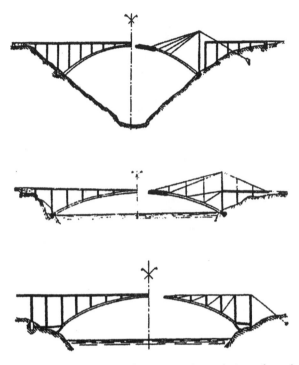

FIGURE 11.17   Various cantilevers—erection techniques for arches.

11.5.1.2   Precast Arches

The first precast segmental arch bridge was built in France in 1948. This bridge at Luzancy over the Marne River is composed of three box-section arches built up from 2.44-m-long precast segments. The finished span length is 55 m. Because of the severe clearance requirements, the arch has a very unusual span-to-rise ratio of 23 to 1. The segments were connected via 20-mm dry-packed joints and prestressing on the approach behind an abutment, with the resulting rib being moved to its final position by an aerial cableway.

Construction of concrete arches without falsework was employed almost exclusively in conjunction with the cast-in-place cantilever technique until the construction of the Natches Trace Bridge in Tennessee in 1993. This precast arch, which originally was designed for erection on a moveable falsework, was the first precast arch to be erected on stays. The unusual design, which omits spandrel columns, results in a slender appearance. There are two arches: one with a span of 177 m and a rise of 44 m, the other with a span of 141 m and a rise of 31 m. The arch segments are 4.9 m wide and vary in depth from 4 m at the springing to 3 m at the crown (see Figure 11.18).

11.5.2   Rigid Frames

Frame bridges can be considered a hybrid of arch and girder forms. They are an appropriate alternative to either of those types for intermediate span lengths. Rigid frame bridges are well suited to segmental construction techniques.

Rigid frame bridges often have some of the same site requirements as arch bridges. They are well suited to valleys and generally will require foundations capable of resisting large horizontal actions. Generally, some form of temporary support will be required until the frame is complete, meaning that construction techniques that eliminate falsework may need slight modification for these structures. One of the most aesthetically convincing applications of the rigid frame is the Bonhomme Bridge in Brittany, France (see Figure 11.19). This slant-leg frame was built using the cast-in-place balanced cantilever technique. Temporary piers were installed below the slant legs to support them

FIGURE 11.18   Natches truss arch—cantilever erection of ribs.

FIGURE 11.19   Bonhomme Bridge in Brittany, France.

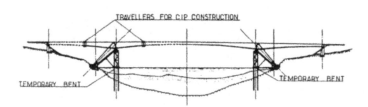

FIGURE 11.20   Temporary support for the Bonhomme Bridge.

before the thrust was developed in the frame (see Figure 11.20). Jacks under the temporary-support piers and at the midspan closure were used to adjust the geometry before closing the span.

11.5.3   Segmental Trusses

Although relatively few examples have been built, segmental trusses are interesting, especially for long spans, in that they offer very efficient use of materials. This economy translates directly into lighter elements and smaller loads to be dealt with during construction, as well as reduced material cost.

FIGURE 11.21 Cantilever erection of the Viaduct des Glacièrs, France.

One of the earliest segmental trusses was the Mangfallbrücke in Austria, which was constructed in 1959. It had a total length of 288 m with a maximum span of 108 m and was constructed by the cast-in-place segmental technique in conjunction with temporary piers.

Later examples were developed in precast segmental, of which the Viaduc de Sylans and the Viaduc des Glacieres are the most notable. These sister structures were constructed in balanced cantilever with a self-launching overhead truss (see Figure 11.21). The segments were prestressed in three directions with a combination of external and internal tendons. "X" members for the open webs were precast and subsequently placed in the molds prior to segment casting.

The most recent development in segmental trusses is the composite truss. This concept employs concrete for the top and bottom chords and steel sections for the open webs. In some cases, however, steel is used for the tension chord as well. An excellent example of this type of construction is the bridge over the Roize in France, built with the span-by-span method. This structure was conceived as a truss work of factory-produced steel truss work and precast slabs. These two elements were joined at the site by cast-in-place joints and external tendons (see Figure 11.22). The precast slabs served as the top (compression) chord while a hexagonal steel tube served as the bottom chord. The resulting structure is equally viable as the deck for short-span viaducts and stiffening girder for long-span cable-supported bridges.

11.6 Segmental Cable-Stayed Bridges

11.6.1 Overview

Theories on cable-stayed bridges are presented in another chapter. We shall address here cable-stayed bridges only as they relate to segmental construction. In the majority of segmental cable-stayed bridges, the methods of construction fall in the three following categories, by order of importance:

- Cantilever construction
- In-stage construction
- Push-out construction

FIGURE 11.22   The Roize Bridge, France — erection of steel bottom chord and webs.

The choice of material depends upon many factors and load conditions; it should be remembered that concrete is an excellent material for cable-stayed structures, because of its properties in resisting compression and its mass and damping characteristics in resisting aerodynamic vibrations. For the proposed Ceremonial Bridge in Malaysia, with a main span of 1000 m and a single plane of stays, concrete deck in the pylon area is associated with a composite cross section toward the center of the span. Comparative studies show that the replacement of the composite section with its concrete slab by an orthotropic slab would adversely affect the project because of its lack of mass.

11.6.2   Cantilever Construction

11.6.2.1   Design

It is important to keep the project simple and pay attention to details to achieve economy and efficiency during construction.

The length of segments must be equal and, depending upon the spacing of stays, the segment joints must be such that a stay always falls in the same location within a segment. If the segments are long, the stay should be located toward the free end of the segment. Cross sections must be kept constant as much as possible, the variations being limited to the web and bottom slab thickness. The post-tensioning layout must be repetitive from segment to segment (see Figure 11.23). Erection phases are critical in terms of stability and stresses. Wind effects on the partially built structure must be investigated for static and dynamic effects. A shorter return period is usually used during construction (10 years). Seismic effects must also be investigated in areas prone to earthquakes. To increase stability, temporary cables can be installed at a certain stage of completion.

Stresses in the main elements of the structure often reach a maximum during construction, and the final state of stresses in the finished structure depends greatly on the accuracy of construction. Figure 11.24 shows a typical erection cycle. It is important that all erection phases be reviewed to ensure that the stresses are within allowable limits at each stage.

Stay forces are large and applied on very localized areas of the deck, and their local effects must be analyzed in detail. For instance, the stays apply high, concentrated forces on the section, at the middle, in the case of a single plane of stays, or at the edges with two planes of stays. These forces are not immediately available in the whole cross section, but are spread out at approximately 45°. This shear lag effect is more critical during construction than in service. Construction phases should be checked, assuming a 45° distribution of the horizontal component of the stay force while the vertical component is effectively applied at the stay anchorage (a finite-element computer program will generate the exact cross section stress distribution).

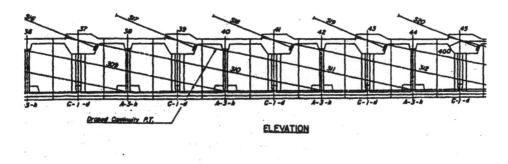

ELEVATION

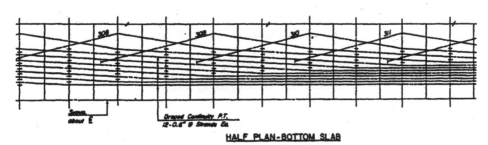

HALF PLAN-BOTTOM SLAB

FIGURE 11.23 Sunshine Skyway, Florida — stay cables and post-tensioning layout.

TYPICAL ERECTION PHASES

① LIFT SEGMENT

② RESTRESS STAY

③ LIFT SEGMENT

④ INSTALL AND STRESS STAY

FIGURE 11.24 Typical erection phases.

This analysis usually shows the necessity of adding a temporary post-tensioning system toward the end of the cantilever, in the area outside the stay centerline (see Figure 11.25).

When a stay is anchored in an already constructed deck, such as a backstay anchored in the side span, the horizontal component of the stay force is distributed half in compression in front of the anchor and half in tension in the back of the anchor. This is called the entrainment effect; care must be taken to have enough tension capacity behind the anchor, either rebars or available compression, to prevent cracking or opening of the joint in case of precast construction.

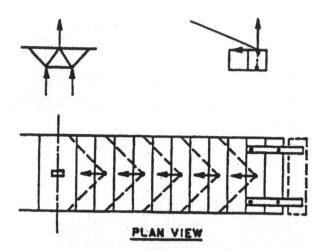

PLAN VIEW

FIGURE 11.25   Shear lag during construction.

11.6.2.2   Cantilever Cast-in-Place Construction

Cast-in-place stayed bridges are built according to the same general principle as a typical box-girder bridge. After the pylon has been built up to the first pylon stay anchorage points and the starting deck segment at the pylon cast, travelers can be installed and cantilever construction started. Temporary stays are sometimes necessary to carry the weight of the traveler plus the newly cast segment before the permanent stay pertaining to that segment can be installed and stressed especially for thin, small inertia decks. A more elegant way is to use the permanent stay, which can be anchored in a precast anchorage block secured to the traveler. The horizontal component of the stay force is carried either by the traveler (see Figure 11.26) or by a precast member, which becomes part of the future segment. The permanent stay can also be anchored in the final deck if the stay anchor structure is staggered ahead of the whole section. This was the case at the Isère Bridge shown in Figures 11.27 and 11.28, with the center spine where the stays are anchored was cast in a first phase and the remainder of the section in a second phase. The phases are as follows:

- Launching of traveler;
- Concreting of the center spine (8 m) and stressing stay to 35% of its final force;
- Launching side forms, connecting bottom slab, and stressing stay to 70% of its final force;
- Concreting top slab and stressing stay to 80% of its final force.

11.6.2.3   Cantilever Precast Construction

Precast segmental bridges become economically feasible for relatively large bridges where the cost associated with setting up a casting yard can be offset by the speed of casting segments and the speed of erection. It is very interesting if the approaches to the main span are also precast segmentally, because then the cost of equipment is written off on an even larger volume.

A great example is the Sunshine Skyway Bridge in Florida, with a main span of 366 m for a total length of 1220 m, where the same cross section is used throughout the high-level bridge (see Figures 11.29 and 11.30). The 120-ton segments were precast in a yard close to the site and delivered by barge. They were lifted into place by beam-and-winch assemblies mounted on the previously completed portion of the deck. The same lifting equipment was used for the high approaches to the main spans. The low-level approaches were made of two parallel box girders.

For the James River Bridge in Virginia, the same twin parallel precast box girders were used from one end of the bridge to the other. For the main span, a single plan of stays was used and the two boxes were connected by a transverse frame at each stay anchor location (see Figures 11.31 and

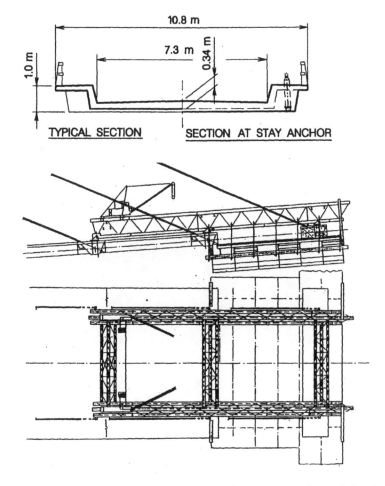

FIGURE 11.26 Santa Rosa Bridge, Bolivia – general view, cross section, and elevation.

11.32). With this scheme, construction can be carried out at deck level, with the segments erected by the span-by-span method. The main span can be built with crane-type lifting equipment mounted on the completed portion of the deck or, if desired, with cranes at ground level or on barges in the river.

11.6.2.4 Structural Steel Segmental Cantilever Construction

Cantilever construction can be applied to steel structures as well, the most recent example being the Normandie Bridge with an 856-m main span and 43.5-m approach spans. Concrete box girders are used for the approaches and part of the main span. The approaches were constructed by incremental launching and the first 116 m out from the pylon by segmental cast-in-place balanced cantilever techniques. Steel segmental construction is used for the remaining 640 m of the main span because of its light weight. The 19.65-m steel segments are barged to the site, lifted in place, secured against the previous segment, and then welded (see Figure 11.33).

11.6.3 In-Stage Construction

With this method, the deck is cast on a fixed soffit, with the side forms moving as the segments are cast. Stays can be installed during the casting, then stressed afterward. The advantage is that the bridge does not go through high-stress-level stages during erection and is practically built in its final stage. This method is only a variation of the cast-in-place scheme.

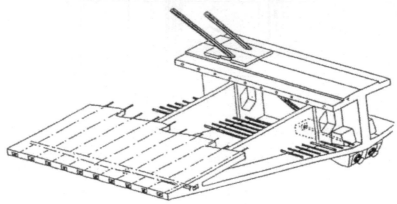

FIGURE 11.27   Isère Bridge, France — general and isometric views.

11.6.4   Push-Out Construction

This method is rarely used and not well adapted to cable-stayed bridges. Its use is restricted to sites where temporary supports can be installed. During pushing, the deck is subjected to large moment variations so steel decks are more suitable.

11.7   Design Considerations

11.7.1   Overview

The intent of this section is to present conditions that the designer should be aware of to produce a satisfactory design. The segmental technique is closely related to the method of construction and the structural system employed. It is usually identified with cantilever construction, but special attention must also be exercised with other methods, such as span-by-span, incremental launching, or progressive placement.

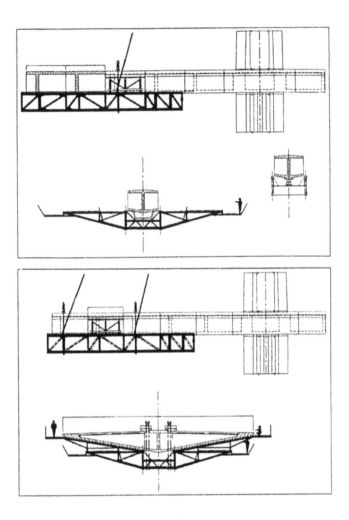

FIGURE 11.28 Isére Bridge — casting sequence.

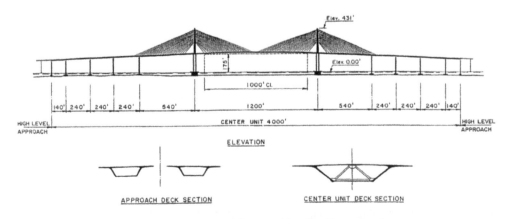

FIGURE 11.29 Sunshine Skyway Bridge, Florida — elevation.

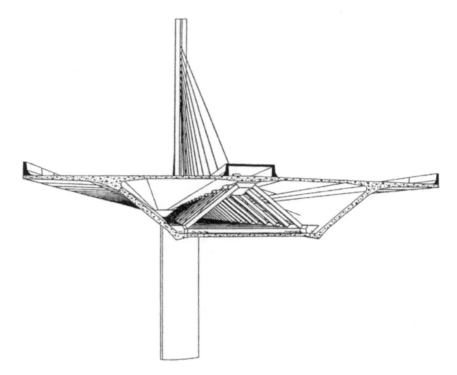

FIGURE 11.30 Sunshine Skyway Bridge — isometric view.

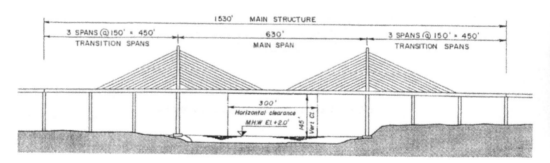

FIGURE 11.31 James River Bridge, Massachusetts – elevation.

11.7.2 Span Arrangement

11.7.1.1 Balanced Cantilever Construction

The span arrangement should avoid spans of significantly different lengths, if possible. This takes best advantage of the construction method by using cantilevers which are balanced about the column. The abutment spans of bridges built with this method are typically 60 to 65% of the central span length. These shorter end spans minimize the length of the bridge adjacent to the abutment, which must be built by using a different method, typically one employing falsework. Spans shorter than this may require a detail to resist uplift at the abutment resulting in live loading on the adjacent span (see Figure 11.34).

11.7.1.2 Span-by-Span Construction

For span-by-span construction the averaging of adjacent span lengths is not required, although it is advantageous to maintain similar span lengths adjacent to one another. The length of the abutment

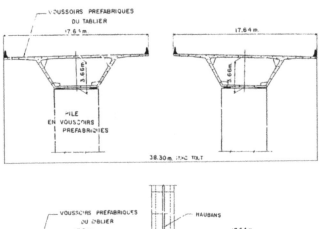

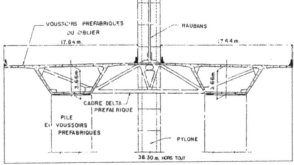

FIGURE 11.32   James River Bridge — cross sections.

FIGURE 11.33   Normandie Bridge, France — lifting of a a steel segment.

or end span is typically kept the same as the interior spans. This is reasonable for this type of construction, since the secondary moments due to post-tensioning in the end spans are less than for the interior spans, and the post-tensioning requirement is therefore similar.

11.7.1.3 Location of Expansion Joints

Concrete bridge decks have been built with a length up to 1220 m between expansion joints and have had acceptable performance. The placement of expansion joints within a longer viaduct may

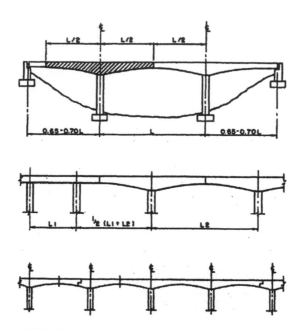

FIGURE 11.34   Balanced cantilever span arrangement.

be necessary to accommodate the change in length of structure due to creep, shrinkage, and thermal changes. The location of the expansion joint within a span will vary, depending on the method of construction.

For balanced cantilever construction, the expansion joints were initially located at the tip of the cantilevers, which is the middle of the span on the completed structure, for ease of construction. Creep effect under dead load plus post-tensioning drives the tip of the cantilever down, resulting in unacceptable angle break at midspan. This disposition is no longer used. An alternative solution is to place the joint at the point of contraflexure of the equivalent continuous span, thus very effectively reducing the angle of break under creep and live load. However, this technique requires expansion segments at midlength of the cantilever, making construction more difficult. The latest technique goes back to the joint at midspan, but with the addition of a stiffening steel beam across the joint, turning the hinged span into a continuous span with expansion capability. Further refinements are introduced such as the capability of controlling the deflection of the span by vertical jacking on the steel beam during the life of the bridge. This technique has been successfully used as it does not interfere with the cantilever erection process (see Figure 11.35).

For spans built with the use of the span-by-span method the expansion joints are typically located at the centerline of a column. The adjacent box-girder spans are both supported by the column with movement allowed between the spans. With this method, the angle break at the expansion joint is minimized, and there is no requirement for temporary moment restraint between the adjacent sections.

11.7.2   Cross-Section Dimensions

11.7.2.1   Overall Box-Girder Dimensions

The overall width of a concrete segmental box-girder bridge is quite adaptable to any requirement. Box-girder spans have been built with widths as low as 3.6 m and as great as 27.50 m, with the configuration of the box girder varying significantly.

The depth of precast segmental box girders is generally somewhat greater than that of similar spans with cast-in-place construction. This increased depth is necessary to offset more stringent requirements for extreme fiber axial stresses and restrictions on the locations of post-tensioning

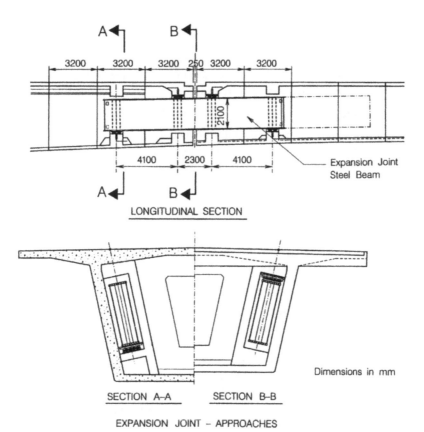

FIGURE 11.35 Expansion joint — approaches.

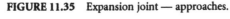

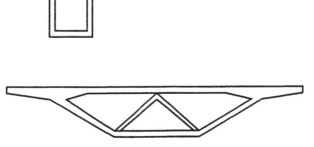

FIGURE 11.36 Various cross sections.

tendons. Multiple-cell, box-girder bridges will also have more webs to place tendons than a comparable width, single-cell segmental box girder. For span-by-span construction, the span-to-depth ratio should not exceed 25 to 1, and is more comfortable at 20 to 1. For balanced cantilever construction, the span-to-depth ratio at the support should not exceed 18 to 1. However, variable-depth box girders built in balanced cantilever fashion are quite common with straight haunched sections and parabolic extrados. The span-to-depth ratio at midspan of a variable-depth balanced cantilever bridge should not exceed 40 to 1 (see Figure 11.36).

11.7.2.2 Web Thickness

The thickness of the web is generally determined such that the required post-tensioning tendons may be placed without interfering with concrete placement or risking cracking during stressing of tendons. Principal stress values at service limit state for no cracking in concrete should be checked in the webs at the neutral axis and at the intersection with top and bottom flanges. This will give a good indication whether the thickness of the web is sufficient. Most design codes also place a limit on the ultimate shear capacity for a box girder to ensure that the web does not fail in diagonal compression prior to the yielding of stirrup reinforcing.

11.7.2.3 Slab Thickness

Slab thicknesses are generally determined to limit deflection under live loading and to provide the necessary flexural capacity. These limits are similar to those of slab thickness for bridge structures built with the use of more traditional construction methods. Span-to-thickness ratios should be in the range of 30 to 1. Since most segmental box girders have transversely post-tensioned top slabs, the minimum thickness of a top slab should be 200 mm, with possibly thicker values at the tendon anchorages. Bottom slab thickness may be less, down to 180 mm, if there is no longitudinal or transverse post-tensioning embedded in the slab.

11.7.3 Temperature Gradients

11.7.3.1 Linear Temperature Gradients

Temperature gradients are caused by the top or bottom surface of the structure being warmer than the other. The shape of the temperature distribution along the depth of the section is beyond the scope of this text. However, this distribution may be assumed to be linear or nonlinear with magnitudes given in relevant texts [3]. Due to its high thermal mass, concrete structures are more adversely affected by the thermal gradient than steel structures.

Effects of a linear temperature gradient can be easily evaluated using hand-calculation methods. Once the magnitude of the temperature gradient has been determined, the unrestrained curvature at any point along the span can be determined by

$$R = \frac{\Delta T \cdot \alpha \cdot E_c}{h} \tag{11.1}$$

where
R = radius of curvature
ΔT = linear temperature differential between top and bottom fibers of cross section
E_c = Modulus of elasticity of concrete
α = thermal expansion coefficient
h = depth of cross section

Once the unrestrained curvature along the structure is known, the final force distribution can be determined by evaluating the redundant support reactions. It is noted that for a statically determinate structure the linear temperature gradient results in zero effect on the structure.

11.7.3.2 Nonlinear Temperature Gradients

Nonlinear temperature gradients are more difficult to evaluate and are best handled by a well-suited computer program. The general theory is presented here; for a more detailed elaboration see Reference [1]. The nonlinear temperature distribution is determined by field measurements and thermodynamic principles. The general shape may be as shown in Figure 11.37. Assuming that the material has linear stress–strain properties, that plane sections will remain plane (Navier–Bernoulli hypothesis), and that temperature varies only with depth (two-dimensional problem), one can make the following theoretical derivation of the problem: the free thermally induced strain is proportional to the temperature distribution; however, this strain distribution violates the second assumptions above, namely, that plane sections remain plane. In order for the section to remain plane under the

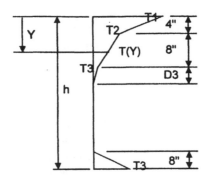

FIGURE 11.37 Nonlinear gradient.

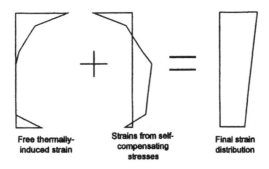

Free thermally-induced strain

Strains from self-compensating stresses

Final strain distribution

FIGURE 11.38 Self-compensating stresses.

effects of the applied temperature gradient, there must be some induced stress on said section. This is termed self-compensating stress. The final strain distribution on the section is, therefore, linear and is the sum of the free thermally induced strain and the strain induced by the self-compensating stresses (see Figure 11.38).

The self-compensating stresses can be derived as

$$\sigma(Y) = E_c \cdot \sigma \cdot T(Y) - \frac{P}{A} - \frac{M \cdot Y}{I} \tag{11.2}$$

where
Y = variable along depth of cross section
$T(Y)$ = temperature at abscissa Y
P = $\int_Y E \cdot a \cdot T(Y) \cdot b(Y) dY$
M = $\int_Y E \cdot a \cdot T(Y) \cdot b(Y) \cdot Y \cdot dY$
$b(Y)$ = width of section at abscissa Y

Similar to the linear gradient, there is now a free unrestrained curvature of the structure along its length. If the structure is continuous, this will result in reactions due to the restraint of the system. The unrestrained curvature at any point along the structure is

$$R = \frac{M}{E_c \cdot I} \tag{11.3}$$

Once the unrestrained curvature along the structure is known, the continuity force distribution can be determined by evaluating the redundant support reactions. The total stress on a section is, therefore, the summation of the self-compensating stresses and the continuity stresses. For a statically determinate structure, the stress on a section is not zero as for a linear temperature gradient; the continuity stresses are zero, but the self-compensating stresses may be significant.

11.7.4   Deflection

11.7.4.1   Dead Load and Creep Deflection

Global vertical deflections of segmental box-girder bridges due to the effects of dead load and post-tensioning as well as the long-term effect of creep are normally predicted during the design process by the use of a computer analysis program. The deflections are dependent, to a large extent, on the method of construction of the structure, the age of the segments when post-tensioned, and the age of the structure when other loads are applied. It can be expected, therefore, that the actual deflections of the structure would be different from that predicted during design due to changed assumptions. The deflections are usually recalculated by the contractor's engineer, based on the actual construction sequence.

11.7.4.2   Camber Requirements

The permanent deflection of the structure after all creep deflections have occurred, normally 10 to 15 years after construction, may be objectionable from the perspective of riding comfort for the users or for the confidence of the general public. Even if there is no structural problem with a span with noticeable sag, it will not inspire public confidence. For these reasons, a camber will normally be cast into the structure so that the permanent deflection of the bridge is nearly zero. It may be preferable to ignore the camber, if it is otherwise necessary to cast a sag in the structure during construction.

11.7.4.3   Global Deflection Due to Live Load

Most design codes have a limit on the allowable global deflection of a bridge span due to the effects of live load. The purpose of this limit is to avoid the noticeable vibration for the user and minimize the effects of moving load impact. When structures are used by pedestrians as well as motorists, the limits are further tightened.

11.7.4.4   Local Deflection Due to Live Load

Similar to the limits of global deflection of bridge spans, there are also limitations on the deflection of the local elements of the box-girder cross section. For example, the AASHTO Specifications limit the deflection of cantilever arms due to service live load plus impact to $\frac{1}{300}$ of the cantilever length, except where there is pedestrian use [1].

11.7.5   Post-Tensioning Layout

11.7.5.1   External Post-Tensioning

While most concrete bridges cast on falsework or precast beam bridges have utilized post-tensioning in ducts which are fully encased in the concrete section, other innovations have been made in precast segmental construction. Especially prevalent in structures constructed using the span-by-span method, post-tensioning has been placed inside the hollow cell of the box girder but not encased in concrete along its length. This is know as external post-tensioning. External post-tensioning is easily inspected at any time during the life of the structure, eliminates the problems associated with internal tendons, and eliminates the need for using expensive epoxy adhesive between precast segments. The problems associated with internal tendons are (1) misalignment of the tendons at segment joints, which causes spalling; (2) lack of sheathing at segment joints; and (3) tendon pull-through on spans with tight curvature (see Figure 11.39). External prestressing has been used on many projects in Europe, the United States, and Asia and has performed well.

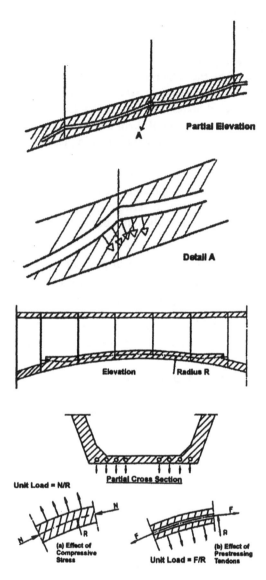

FIGURE 11.39   Problems with internal tendons.

11.7.5.2   Future Post-Tensioning

The provision for the addition of post-tensioning in the future in order to correct unacceptable creep deflections or to strengthen the structure for additional dead load, i.e., future wearing surface, is now required by many codes. Of the positive and negative moment post-tensioning, 10% is reasonable. Provisions should be made for access, anchorage attachment, and deviation of these additional tendons. External, unbonded tendons are used so that ungrouted ducts in the concrete are not left open.

11.8   Seismic Considerations

11.8.1   Design Aspects and Design Codes

Due to typical vibration characteristics of bridges, it is generally accepted that under seismic loads, some portion of the structure will be allowed to yield, to dissipate energy, and to increase the period of vibration of the system. This yielding is usually achieved by either allowing the columns to yield plastically (monolithic deck/superstructure connection), or by providing a yielding or a soft bearing system [6].

The same principles also apply to segmental structures, i.e., the segmental superstructure needs to resist the demands imposed by the substructure. Very few implementations of segmental structures are found in seismically active California, where most of the research on earthquake-resistant bridges is conducted in the United States. The Pine Valley Creek Bridge, Parrots Ferry Bridge, and Norwalk/El Segundo Line Overcrossing, all of them being in California, are examples of segmental structures; however, these bridges are all segmentally cast in place, with mild reinforcement crossing the segment joints.

Some guidance for the seismic design of segmental structures is provided in the latest edition of the AASHTO Guide Specifications for Design and Construction of Segmental Concrete Bridges [2], which now contains a chapter dedicated to seismic design. The guide allows precast-segmental construction without reinforcement across the joint, but specifies the following additional requirements for these structures:

- For Seismic Zones C and D [1], either cast-in-place or epoxied joints are required.
- At least 50% of the prestress force should be provided by internal tendons.
- The internal tendons alone should be able to carry 130% of the dead load.

For other seismic design and detailing issues, the reader is referred to the design literature provided by the California Department of Transportation, Caltrans, for cast-in-place structures [5-8].

11.8.2 Deck/Superstructure Connection

Regardless of the design approach adopted (ductility through plastic hinging of the column or through bearings), the deck/superstructure connection is a critical element in the seismic resistant system. A brief description of the different possibilities follows.

11.8.2.1 Monolithic Deck/Superstructure Connection

For the longitudinal direction, plastic hinging will form at the top and bottom of the columns. Since most of the testing has been conducted on cast-in-place joints, this continues to be the preferred option for these cases. For short columns and for solid columns, the detailing in this area can be readily adapted from standard Caltrans practice for cast-in-place structures, as shown on Figure 11.40. The joint area is then essentially detailed so it is no different from that of a fully cast-in-place bridge. In particular, a Caltrans requirement for positive moment reinforcement over the pier can be detailed with prestressing strand, as shown below. For large spans and tall columns, hollow column sections would be more appropriate. In these cases, care should be taken to confine the main column bars with closely spaced ties, and joint shear reinforcement should be provided according to Reference [3 or 7].

The use of fully precast pier segments in segmental superstructures would probably require special approval of the regulating government agency, since such a solution has not yet been tested for bridges and is not codified. Nevertheless, based upon first principles, and with the help of strut–tie models, it is possible to design systems that would work in practice [6]. The segmental superstructure should be designed to resist at least 130% of the column nominal moment using the strength reduction factors prescribed in Ref. [2].

Of further interest may be a combination of precast and cast-in-place joint as shown in Figure 11.41, which was adapted from Ref. [8]. Here, the precast segment serves as a form for the cast-in-place portion that fills up the remainder of the solid pier cap. Other ideas can also be derived from the building industry where some model testing has been performed. Of particular interest for bridges could be a system that works by leaving dowels in the columns and supplying the precast segment with matching formed holes, which are grouted after the segment is slipped over the reinforcement [9].

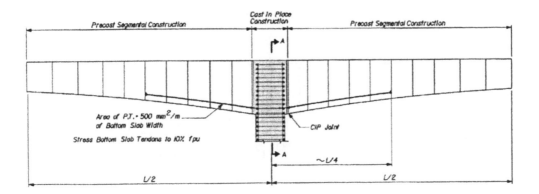

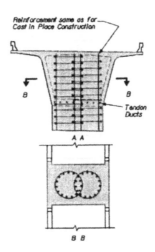

FIGURE 11.40   Deck/pier connection with cast-in-place joint.

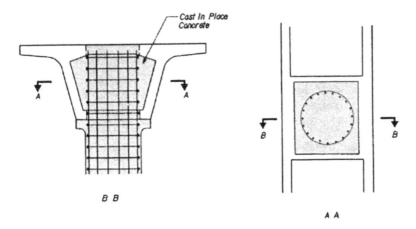

FIGURE 11.41   Combination of precast and cast-in-place joint.

11.8.2.2 Deck/Superstructure Connection via Bearings

Typically, for spans up to 45 m erected with the span-by-span method, the superstructure will be supported on bearings. For action in the longitudinal direction, elastomeric or isolation bearings are preferred to a fixed-end/expansion-end arrangement, since these better distribute the load

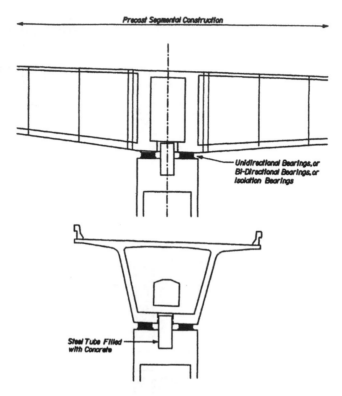

FIGURE 11.42 Deck/pier connection with bearings.

between the bearings. Furthermore, these bearings will increase the period of the structure, which results in an overall lower induced force level (beneficial for higher-frequency structures), and isolation bearings will provide some structural damping as well.

In the transverse direction, the bearings may be able to transfer load between super- and substructure by shear deformation; however, for the cases where this is not possible, shear keys can be provided as is shown in Figure 11.42. It should be noted that in regions of high seismicity, for structures with tall piers or soft substructures, the bearing demands may become excessive and a monolithic deck–superstructure connection may become necessary.

For the structure-on-bearings approach, the force level for the superstructure can be readily determined, since once the bearing demands are obtained from the analysis, they can be applied to the superstructure and substructure. The superstructure should resist the resulting forces at ultimate (using the applicable code force-reduction factors), whereas the substructure can be allowed to yield plastically if necessary.

11.8.2.3 Expansion Hinges

From the seismic point of view, it is desirable to reduce the number of expansion hinges (EH) to a minimum. If EHs are needed, the most beneficial location from the seismic point of view is at midspan. This can be explained by observing Figure 11.43, where the superstructure bending moments, resulting from column plastic hinging (M_p), have been plotted for the case of an EH at midspan and for an EH at quarterspan. For the latter, it can be seen that the moment at the face of the column varies within the range of $\pm\frac{3}{4} M_p$, whereas with the hinge at midspan, the values are only between $\pm\frac{1}{2} M_p$.

The location of expansion hinges within a span, and its characteristics, depends also on the stiffness of the substructure and the type of connection of the superstructure to the piers. Table 11.1 presents general guidelines intended to assist in the selection of location of expansion hinges.

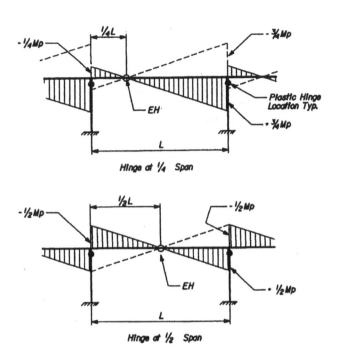

FIGURE 11.43 Longitudinal superstructure seismic moments with hinges at quarterspan and at midspan.

TABLE 11.1 Location of Expansion Hinges in Segmental Bridges

Span Support System	Location of EH		
	Over Pier	Intermediate Point	Midspan
On bearings	• Standard solution for simple spans • For continuous spans generates moderate superstructure moments at adjacent piers	• Complicated erection for cantilever construction • Generates moderate superstructure moments at adjacent piers • Moderate EH openings requiring restrainers and moderate seat widths, or lock-up devices	• Simplest location for cantilever construction • Will require continuity beam inside cross section • Minimizes superstructure seismic moments • Moderate EH openings requiring adequate gap between the end segments
Monolithic with pier	Not applicable	• Complicated erection for cantilever construction • Generates very large superstructure moments at adjacent piers • If substructure is stiff, expect relatively small EH movements; otherwise expect very large movements, requiring restrainers and large seat widths, or lock-up devices	• Simplest location for cantilever construction • Will require continuity beam inside cross section • Minimizes superstructure seismic moments • If substructure is stiff, expect relatively small EH movements; otherwise expect very large movements, requiring lock-up devices

11.8.2.4 Precast Segmental Piers

Precast segmental piers are usually hollow cross section to save weight. From research in other areas it can be extrapolated that the precast segments of the pier would be joined by means of unbonded prestressing tendons anchored in the footing. The advantage of unbonded over bonded tendons is that for the former, the prestress force would not increase significantly under high column displacement demands, and would therefore not cause inelastic yielding of the strand, which would otherwise lead to a loss of prestress.

The detail of the connection to the superstructure and foundation would require some insight into the dynamic characteristics of such a connection, which entails joint opening and closing — providing that dry joints are used between segments. This effect is similar to footing rocking, which is well known to be beneficial to the response of a structure in an earthquake. This is due to the period shift and the damping of the soil. The latter effect is clearly not available to the precast columns, but the period shift is. Details need to be developed for the bearing areas at the end of the columns, as well as the provision for clearance of the tendons to move relative to the pier during the event.

If the upper column segment is designed to be connected monolithically to the superstructure, yielding of the reinforcement should be expected. In this case, the expected plastic hinge length should be detailed ductile, using closely spaced ties [3,5].

11.9 Casting and Erection

11.9.1 Casting

There are obvious major differences in casting and erection when working with cast-in-place cantilever in travelers or in handling precast segments. There are also common features, which must be kept in mind in the design stages to keep the projects simple and thereby economic and efficient, such as

- Keeping the length of segments equal and segments straight, even in curved bridges;
- Maintaining constant cross section dimensions as much as possible;
- Minimizing the number of diaphragms and stiffeners, and avoiding dowels through formwork.

11.9.1.1 Cast-in-Place Cantilevers

Conventional Travelers

The conventional form traveler supports the weight of the fresh concrete of the new segment by means of longitudinal beams or frames extending out in cantilever from the last segment. These beams are tied down to the previous segment. A counterweight is used when launching the traveler forward. The main beams are subjected to some deflections, which may produce cracks in the joint between the old and new segments. Jacking of the form during casting is sometimes needed to avoid these cracks. The weight of a traveler is about 60% of the weight of the segment. The rate of construction is typically one segment per traveler per week. Precast concrete anchor blocks are used to speed up post-tensioning operations. In cold climates, curing can be accelerated by various heating processes.

Construction Camber Control

The most critical practical problem of cast-in-place construction is deflection control. There are five categories of deflections during and after construction:

- Deflection of traveler frame under the weight of the concrete segment;
- Deflection of the concrete cantilever arm during construction under the weight of segment plus post-tensioning;
- Deflection of cantilever arms after construction and before continuity;
- Short- and long-term deflections of the continuous structure;
- Short- and long-term pier shortenings and foundation settlements.

The sum of the various deflection values for the successive sections of the deck allows the construction of a camber diagram to be added to the theoretical profile of the bridge. A construction camber for setting the elevation of the traveler at each joint must also be developed.

11.9.1.2 Precast Segments

Opposite to the precast girder concept where the bridge is cut longitudinally in the precast segmental methods, the bridge is cut transversally, each slice being a segment. Segments are cast in a casting yard one at a time. Furthermore, the new segment is cast against the previously cast segment so that the faces in contact match perfectly. This is the match-cast principle. When the segments are reassembled at the bridge site, they will take the same relative position with regard to the adjacent segments that they had when they were cast. Accuracy of segment geometry is an absolute priority, and adequate surveying methods must be used to ensure follow-up of the geometry.

Match casting of the segments is a prerequisite for the application of glued joints, achieved by covering the end face of one or both of the meeting segments with epoxy at the erection. The epoxy serves as a lubricant during the assembly of the segments, and it ensures a watertight joint in the finished structure. Full watertightness is needed for corrosion protection of internal tendons (tendons inside the concrete). The tensile strength of the epoxy material is higher than that of the concrete, but, even so, the strength of the epoxy is not considered in the structural behavior of the joint. The required shear capacity is generally provided by shear keys, single or multiple, in combination with longitudinal post-tensioning.

With the introduction of external post-tensioning, where the tendons are installed in PE ducts, outside the concrete but inside the box girder, the joints are relieved of the traditional requirement of watertightness and are left dry. The introduction of external tendons in connection with dry joints greatly enhanced the efficiency of precasting.

11.9.1.3 Casting Methods

There are two methods for casting segments. The first one is the long-line method, where all the segments are cast in their correct position on a casting bed that reproduces the span. The second method, used most of the time, is the short-line method, where all segments are cast in the same place in a stationary form, and against the previously cast segment. After casting and initial curing, the previously cast segment is removed for storage, and the freshly cast segment is moved into place (see Figure 11.44).

11.9.1.4 Geometry Control

A pure translation of each segment between cast and match-cast position results in a straight bridge (Figure 11.45). To obtain a bridge with a vertical curve, the match-cast segment must first be translated and given a rotation α in the vertical plane (Figure 11.46). Practically, the bulkhead is left fixed and the mold bottom under the conjugate unit adjusted. To obtain a horizontal curvature, the conjugate unit is given a rotation β in the horizontal plane (see Figure 11.47). To obtain a variable superelevation, the conjugate unit is rotated around a horizontal axis located in the middle of the top slab (Figure 11.48).

All these adjustments of the conjugate unit can be combined to obtain the desired geometry of the bridge.

11.9.2 Erection

The type of erection equipment depends upon the erection scheme contemplated during the design process; the local conditions, either over water or land; the speed of erection and overall construction schedule. It falls into three categories, independent lifting equipment such as cranes, deck-mounted lifting equipment such as beam and winch or swivel crane, and launching girder equipment.

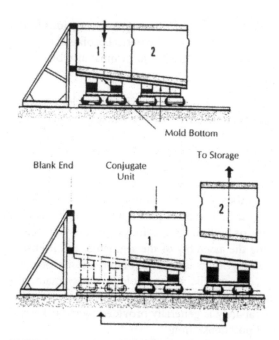

Mold Bottom

Blank End   Conjugate Unit   To Storage

FIGURE 11.44   Typical short-line precasting operation.

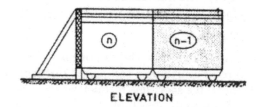

ELEVATION

FIGURE 11.45.   Straight bridge.

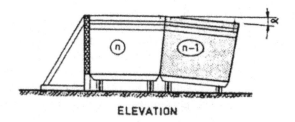

ELEVATION

FIGURE 11.46   Bridge with vertical curve.

11.9.2.1 Balanced Cantilever Method

The principle of the method is to erect or cast the pier segment first, then to place typical segments one by one from each side of the pier, or in pairs simultaneously from both sides. Each newly placed precast segment is fixed to the previous one with temporary PT bars, until the cantilever tendons are installed and stressed. The closure joint between cantilever tips is poured in place and continuity tendons installed and stressed.

In order to carry out this erection scheme, segments must be lifted and installed at the proper location. The simplest way is to use a crane, either on land or barge mounted. Many bridges have

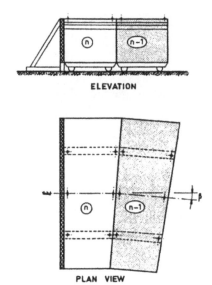

FIGURE 11.47   Bridge with horizontal curve.

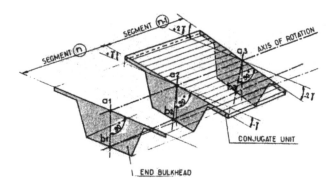

FIGURE 11.48   Bridge with superelevation.

been erected with cranes as they do not require an investment in special lifting equipment. This method is slow. Typically, two to four segments per day are placed. It is used on relatively short bridges. An alternative is to have a winch on the last segment erected. The winch is mounted on a beam fixed to the segment. It picks up segments from below, directly from truck or barge. After placing the segment, the beam and winch system is moved forward to pick up the next segment and so on. Usually, a beam-and-winch system is placed on each cantilever tip. This method is also slow; however, it does not require a heavy crane on the site, which is always very expensive, especially if the segments are heavy.

When bridges are long and the erection schedule short, the best method is the use of launching girders, which then take full advantage of the precast segmental concept for speed of erection.

There are two essential types of self-launching gantries developed for this erection method. The first type is a gantry with a length slightly longer than the typical span (see Figure 11.49). During erection of the cantilever, the center leg rests on the pier while the rear leg rests on the cantilever tip of the previously erected span, which must resist the corresponding reaction. Prior to launching, the back spans must be made continuous. Then, the center leg is moved to the forward cantilever

FIGURE 11.49   French Creek Viaduct, U.S.— single erection truss with portal legs.

tip, which must resist the weight of the gantry plus the weight of the pier segment. This stage controls the design of the gantry, which must be made as light as possible, and of the cantilever.

The second type of gantry has a length that is twice that of the typical span (see Figure 11.50). The reaction from the legs during the erection and launching of the next span is always applied on the piers, so there is no concentrated erection load on the cantilever tip. Each erection cycle consists of the erection of all typical segments of the cantilever and then the placement of the pier segment for the next cantilever, without changing the position of the truss.

The gantries can be categorized by their cross section: single truss, with portal-type legs, and two launching trusses with a gantry across. The twin box girders of the bridge in Hawaii were built with two parallel, but independent trusses (see Figure 11.51), with a typical span of 100.0 m, segment weights of 70 tons; the two bridge structures are 27.5 m apart with different elevations and longitudinal slopes. This system is a refinement of the first type of gantry applied to twin decks with variable geometry.

Normally, the balanced cantilever method is used for spans from 60 to 110 m, with a launching girder. One full, typical cycle of erection is placing segments, installing and stressing post-tensioning tendons, and launching the truss to its next position. It takes about 7 to 10 days, but may vary greatly according to the specifics of a project and the sophistication of the launching girder. With proper equipment and planning, erection of 16 segments per day has been achieved.

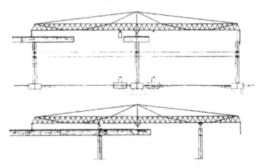

FIGURE 11.50 Río Niteroi, Brazil — two segments being erected simultaneously from one erection truss.

FIGURE 11.51 H-3 Windward Viaduct, Hawaii.

A modification of this method was used to build the 13-km-long Confederation Bridge (typical span 250 m long), linking Prince Edward Island with New Brunswick in Atlantic Canada. The main girder was constructed in a precasting yard as a precast, balanced cantilever. Then the 197-m-long main girder, with a self-weight of 7500 tons, was placed on the pier with a floating crane (see Figure 11.52).

FIGURE 11.52   Confederation Bridge, Canada — floating crane.

FIGURE 11.53   Bang Na–Bang Pli–Bang Pakong Expressway, Thailand — D6 segment erection.

11.9.2.2   Span-by-Span Construction

In the first stage, all precast segments for each span are assembled on an erection girder. The second stage is installing and stressing the tendons, and as a result the span becomes self-supported. In comparison with the balanced cantilever method, those girders have to be designed to carry the load of the entire span. Normally, the duration of one erection cycle is 2 to 3 days per span.

The 56-km-long Bang Na–Bang Pli–Bang Pakong Expressway in Bangkok, carrying six lanes of traffic, totaling 27 m in width, is assembled on erection girders. The girder is placed in the middle of a Y-shaped column. The segments, with self-weight up to 100 tons, are placed on the chassis with a swivel crane and then transported to their final position. Two schemes of erection were developed: a swivel crane mounted to the front of the girder picking up segments from trucks on the highway below and a swivel crane placed on the previously erected span with segments delivered over the deck already built (see Figure 11.53).

Another principle used in erection equipment for the span-by-span method, is the so-called overhang girder. In this case, the girder is above the superstructure, and the precast segments are hung from it.

11.9.2.3 Safety

Due to the inherent character of temporary structures, erection equipment is usually designed to take full advantage of the materials and care must be taken to analyze in-depth all construction stages, anticipating mistakes or shortcuts made on site that always occur, and stay within reasonable safety limits. Overall stability when resting on temporary supports or during launching, reversal of forces, bucking, etc., are the most common problems encountered in these structures. Lifting bars or tie-downs must always be designed with failure or mishandling of one of those elements in mind, and appropriate ultimate resisting paths incorporated in the concept.

11.10 Future of Segmental Bridges

Since their appearance in 1962, precast segmental technologies have been used worldwide in the design and construction of practically all types of bridges. Nevertheless, in the last 5 years, an important further development of these technologies has taken place in Southeast Asia which will have a decisive impact on the way bridges will be built in the next century.

11.10.1 The Challenge

The explosive development of the Southeast Asian economies has been forcing local governments to find new solutions for building infrastructures (build, operate, transfer), which can, when properly managed, become success stories for both the government agencies who organize the projects and for the private groups who develop them.

Privatization of such projects is now accepted by numerous countries as a viable solution for the challenges they face. Large infrastructure projects, worth billions of dollars, are at present being designed, built, financed, and operated by private companies in the region. This trend is expected to extend progressively to the global construction markets. The two key factors of such projects are the amount of toll to be paid by the users and the duration of the concession until the project is transferred to the government agency. For the roadway projects, the tolls vary anywhere from between $1 to $10. The duration of the concessions varies in general from 25 to 35 years.

While basically simple, the scheme presents very complex problems for its implementation. Multidisciplinary skills and a new vision for the design, construction, and operation of the roads and bridges are necessary in order to avoid technical or financial failures. The challenges that private organizations must be able to face can be summarized in just a few words: they need to design and build very competitive projects in the shortest possible time, ensuring the longest possible service life. In light of the experiences gained in the new markets, it is becoming more evident each day that precast segmental technologies often bring the right solutions to these challenges.

11.10.2 Concepts

When Jean Muller invented the precast concrete segmental technology in 1962, his vision was to create an industrialized construction system to build any type of bridge with standard modules, assembled with post-tensioning, without any cast-in-place concrete.

To achieve this objective, he developed the concept of match-cast joints, which allows the transverse slicing of concrete box girders and the assembly of such slices — the segments — in the same order as they were produced, without any need for additional *in situ* concrete to complete the bridge deck [10, 11].

In addition to epoxy-glued joints, the use of dry joints became widespread. In 1978, through the design of the Long Key Bridge in Florida, internal post-tensioning was replaced by external post-tensioning [12]. A number of other concepts invented by Jean Muller allowed further development of the modular construction concept: span-by-span assembly method (Long Key Bridge), progressive

FIGURE 11.54   Bang Na Expressway, Thailand — D6, six-traffic-lane segment at precast yard.

TABLE 11.2   Standard Segments

Segment	Lanes	Widths (m) Min-Max
D2	2	08–12
D3	3	10–15
D4	4	14–20
D6	6	18–30

placing (Linn Cove Bridge), precast segmental construction of the piers, D6 cable-stayed segments (Sunshine Skyway Bridge), delta frames (James River Bridge, C&D Canal Bridge), etc.

These concepts and others developed more recently for the large projects, which are being built in Canada and Thailand, allow the prefabrication of bridge structures with precast modules ranging from 20 tons, in channel-shaped overpasses, to 7500 tons for the main girders of the Confederation Bridge in Canada. In this project, one of the largest bridges ever built, no cast-in-place structural concrete was used in the construction of the main spans [13]. The construction modules are all manufactured in sophisticated and industrialized precasting plants that ensure an unequaled construction rate and quality (see Figure 11.54).

By further standardizing the segments with the number of traffic lanes that they carry, we have developed the modules in Table 11.2. These modules allow the construction of viaducts of any width, ranging from 7 to 8 m up to 30 m. Concurrently, with the effort to standardize the cross sections for precast segmental bridges, there has been significant development of design, shop drawing software, and geometry control systems. This gives us the capability to produce drawings by the thousands for viaducts, interchanges, and merging sections that give, for each segment, the detailed geometry and dimensions of concrete and rebar and the layout of the post-tensioning. Such shop drawings are an essential part of the system and must be integrated into the structural design; no standardization and, hence, no industrialization is possible without them (Figure 11.55).

11.10.3   New Developments

The dynamic business environment of Southeast Asian markets is quite favorable for the introduction of innovative concepts. In recent years, technologies that took over 20 years to develop in Europe and some 10 years to spread throughout the United States were absorbed by countries such

FIGURE 11.55 The Confederation Bridge, Canada — Prince Edward Island precasting yard.

as Thailand, which had limited prior experience in the field of bridge engineering, and already concepts never used before are being developed for new projects, thus giving these countries a leading position in construction innovation. By using the most innovative technologies, the developers involved in the private roadway projects are dramatically changing the very nature of the construction business that, until very recently, was considered one of the most conservative sectors of the industry. The innovative concepts that the industrialization of bridges is introducing cover different areas:

- Reduction of construction time and construction cost;
- Durability of the structures (25 to 50 to 100 years);
- Replaceability of components such as bearings, post-tensioning, stays;
- Earthquake resistance of the structures;
- Staged construction;
- Integrated inspection and surveillance systems;
- Users' comfort and safety.

It is evident that the multiplication of such private projects, where cost, time, and durability are the decisive factors, will open the way to innovation in the bridge business as never before.

11.10.4 Environmental Impact

To prevent private projects from turning into environmental nightmares, private developers need to comply with strict obligations with respect to aesthetics, rights of way, and maintenance of the structures during the duration of the concessions. Government agencies have been developing design, construction, and operation criteria that will progressively become the rules of the BOT projects. As an example, such rules may force structural engineers to conceive structures that can be built in or over crowded areas of cities, with a minimal impact on existing conditions. Or they may impose specific constraints on aesthetics, shapes, or dimensions of structural elements. Further, they may require maintenance costs to be budgeted.

The involvement of the communities in such projects, even if sometimes it may be difficult to manage and may require a profound knowledge of the interests and aspirations of those concerned by the project, is essential for the smooth development of the work. In general, projects that are not well integrated into the context of the local environment or not consistent with the users' expectations run the risk of finishing in disarray or remaining incomplete.

11.10.5 Industrial Production of Structures

The experience acquired in large- or medium-size projects demonstrates that the industrialization of the production of structural elements always brings clear advantages in terms of quality and construction time. What frequently has been less noticed is the advantage that such industrialization can offer as far as the cost of a specific project.

The major change in the contractual conditions of the BOT projects is that the cost of "design + construction time" can now be estimated very precisely. If the completion of a project is delayed by 1 month, for instance, in a project worth U.S.$800,000,000, the cost to the developer is approximately equal to the interest that must be paid on that amount. If the interest is 5%, this represents U.S.$40,000,000/year; thus, every month gained in the duration of the design + construction period represents U.S.$3,300,000, or roughly, U.S.$100,000/day.

In these large projects the industrialization of production and the use of sophisticated systems to transport, erect, and assemble the prefabricated modules is reducing the duration of cycles, which usually may take 6 years when managed by the government agencies, to some 3 years, when managed by private organizations in a fully integrated way.

The introduction of the "assembly-line" approach to bridge building was taken to its limits during the construction of the Northumberland Strait Crossing (Confederation Bridge), Canada, a major bridge project which extends over 13 km of icy strait, with extreme weather conditions. The actual assembly of the components that constitute the 43 spans, 250-m each, took place in only 12 months, whereas to build just a single cast-in-place span of 250 m by traditional means is a difficult venture that takes at least 2 years (see Figure 11.55).

11.10.6 The Assembly of Structures

The production of the structural modules for precast segmental projects represents half of the process. The other half relates to the transport of these modules to the site, to their erection, and to the assembly methods to constitute the structural integrity of the bridge.

Generally, for the transport of current segments weighing from 30 to 100 tons, equipment already available in the market has been used. The transport is commonly by road, using convoys of "low boys," or by water, using barges. In some projects currently being built, 30 to 40 segments weighing between 50 and 60 tons are transported every night from the casting yards situated some 100 km from the large metropolis, to the site in the center of the city. The segments are picked up directly from the trucks by the assembly gantries, between midnight and five o'clock in the morning, to avoid interfering with heavy city traffic during the day.

The erection and assembly of such segments are also performed in a highly industrialized environment. With the span-by-span construction method, spans of 40 m can be assembled in 2 days, with crews working after hours. The cycle is almost independent of the type of segment, from D2 to D6, and therefore, the method is ideal for spans from 30 to 45 m. For cantilever construction, special gantries have been developed to assemble two parallel viaducts mimicking the procedure used in Hawaii, achieving speeds of construction of 3 weeks to complete two double cantilevers of 100 m [14]. This method very competitively covers spans of 80 to 120 m. Progressive placing of segments, using a swivel crane, has also been improved for this type of construction, which allows construction of spans from 45 to 65 m. Finally, for large cable-stayed spans, the use of precast segmental technologies successfully tested in milestone projects, like the Sunshine Skyway and the James River Bridges, is now being developed to cover different cross sections for the segments and to combine space trusses and composite sections [15].

The use of gigantic floating cranes, such as the *Svanen*, to place units as large as 190 m and with weights of 7500 tons, opens new prospects for the construction of bridges over rivers and straits (see Figure 11.52). Bridges that previously were almost impossible to build competitively and within the common constraints of construction schedules can now be conceived, designed, and built in short periods of time, by intensive use of precast technologies.

FIGURE 11.56 View of the completed Second Expressway System Project, Thailand.

Clearly, this evolution is going to accelerate and will become global. Equipment designed to be used anywhere in the world will allow for the reduction of costs charged on a specific project. Furthermore, we can expect improvements in the performance and reliability of equipment specifically conceived to perform heavy lifting and assembly of bridge modules. Bridges will be designed which take into consideration the availability and the characteristics of these machines. Construction methods will then become, more than ever, a decisive factor in the design of structures.

11.10.7 Prospective

Design-and-build projects that were common in the 1960s provided some of the most innovative contributions to bridge engineering. Engineers and contractors working together produced competitive structures that paved the way for the development that has taken place all over the world during the last quarter century (Figure 11.56).

A new wave of innovative bridge concepts is already being generated by the privatization of roadway and bridge projects. This wave, which began in the vibrant business environment of Southeast Asia, will eventually reach the United States and the European markets. This time, engineers and contractors will be seconded by developers, finance specialists, and industrialists to shape the structures that will be built during the next century. The construction industry will also join other key industries in adopting high-technology and innovation as essential ingredients of its renewal [16].

References

1. AASHTO, *Standard Specifications for Highway Bridges*, 16th ed., American Association of State Highway and Transportation Officials, Washington, D.C., 1996.
2. AASHTO, *Guide Specifications for Design and Construction of Segmental Concrete Bridges*, Draft 2nd ed., American Association of State Highway and Transportation Officials, Washington, D.C., August 1997.

3. Imbsen, R. A. et al., *Thermal Effects in Concrete Bridge Superstructures*, National Cooperative Highway Research Program Report 276, Transportation Research Board, Washington, D.C., 1985.

4. Priestley, M. J. N. et al., *Seismic Design and Retrofit of Bridges*, John Wiley & Sons, New York, 1996.

5. California Department of Transportation, *Bridge Design Specifications*, Sacramento.

6. California Department of Transportation, *Bridge Memos to Designers*, Sacramento.

7. California Department of Transportation, *Seismic Design Memo*, Sacramento.

8. Riobóo Martin, J. M., A new dimension in precast prestressed concrete bridges for congested urban areas in high seismic zones, *PCI J.*, 37, (2), 1992.

9. Restrepo, J. I. et al., Design of connections of earthquake resisting precast reinforced concrete perimeter frames, *PCI J.*, 40, (5), 1995.

10. Muller, J., Ten years of experience in precast segmental construction, *J. Precast/Prestressed Concrete Insti.*, 20, (1), 28–61, 1975.

11. Podolny, W., et al., *Construction and Design of Prestressed Concrete Segmental Bridges*, John Wiley & Sons, New York, 1982.

12. Muller, J., *Evolution dans la Construction de Grands Ponts: Montage et Entretien*, IABSE, 11th Congress, Vienna, 1980.

13. Sauvageot, G., Northumberland Strait Crossing, Canada, *4th International Bridge Engineering Conference; Proceedings*, 7, Vol. 1, August, 1995, 238–248.

14. Dodson, B., *Bangkok Second Stage Expressway System Segmental Structures*, in *4th International Bridge Engineering Conference, Proceedings* 7, Vol. 2, August, 1995, 199–204.

15. Sauvageot, G., *Hawaii H-3 Precast Segmental Windward Viaduct*, FIP Congress, Washington, D.C., 1994.

16. Muller, J., *Reflections on cable-stayed bridges*, Rev. Gen. Routes Aérodromes, Paris, October, 1994.

12

Steel-Concrete Composite I-Girder Bridges

Lian Duan
*California Department
of Transportation*

Yusuf Saleh
*California Department
of Transportation*

Steve Altman
*California Department
of Transportation*

12.1 Introduction

An I-section is the simplest and most effective solid section of resisting bending and shear. In this chapter straight, steel–concrete composite I-girder bridges are discussed (Figure 12.1). Materials and components of I-section girders are described. Design considerations for flexural, shear, fatigue, stiffeners, shear connectors, diaphragms and cross frames, and lateral bracing with examples are presented. For a more detailed discussion, reference may be made to recent texts by Xanthakos [1], Baker and Puckett [2], and Taly [3].

12.2 Structural Materails

Four types of structural steels (structural carbon steel, high-strength low-alloy steel, heat-treated low-alloy steel, and high-strength heat-treated alloy steel) are commonly used for bridge structures. Designs are based on minimum properties such as those shown in Table 12.1. ASTM material property standards differ from AASHTO in notch toughness and weldability requirements. Steel meeting the AASHTO-M requirements is prequalified for use in welded bridges.

Concrete with 28-day compressive strength f'_c = 16 to 41 MPa is commonly used in concrete slab construction. The transformed area of concrete is used to calculate the composite section properties. The short-term modular ratio n is used for transient loads and long-term modular ratio

FIGURE 12.1 Steel–concrete composite girder bridge (I-880 Replacement, Oakland, California)

TABLE 12.1 Minimum Mechanic Properties of Structural Steel

Material	Structural Steel	High-Strength Low-Alloy Steel		Quenched and Tempered Low-Alloy Steel	High Yield Strength Quenched and TemperedLow-Alloy Steel	
AASHTO designation	M270 Grade 250	M270 Grade 345	M270 Grade 345W	M270 Grade 485W	M270 Grades 690/690W	
ASTM designation	A709M Grade 250	A709M Grade 345	A709M Grade 345W	A709M Grade 485W	M709M Grades 690/690W	
Thickness of plate (mm)	Up to 100 included				Up to 65 included	Over 65–100 included
Shapes	All Groups			Not Applicable		
F_u (MPa)	400	450	485	620	760	690
F_y (MPa)	250	345	485	485	690	620

F_y = minimum specified yield strength or minimum specified yield stress; F_u = minimum tensile strength; E = modulus of elasticity of steel (200,000 MPa).

Source: American Association of State Highway and Transportation Officials, AASHTO LRFD Bridge Design Specifications, Washington, D.C., 1994. With permission.

$3n$ for permanent loads. For normal-weight concrete the short-term ratio of modulus of elasticity of steel to that of concrete are recommended by AASHTO-LRFD [4]:

$$n = \begin{cases} 10 & \text{for } 16 \leq f_c' < 20 \text{ MPa} \\ 9 & \text{for } 20 \leq f_c' < 25 \text{ MPa} \\ 8 & \text{for } 25 \leq f_c' < 32 \text{ MPa} \\ 7 & \text{for } 32 \leq f_c' < 41 \text{ MPa} \\ 6 & \text{for } f_c' \leq 41 \text{ MPa} \end{cases} \qquad (12.1)$$

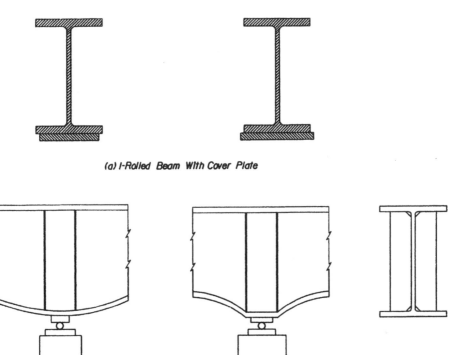

(a) I-Rolled Beam With Cover Plate

Fish Belly Haunch

Parabolic Haunch

Cross Section

(b) Built-Up Plate Girder with Haunches

FIGURE 12.2 Typical sections.

12.3 Structural Components

12.3.1 Classification of Sections

I-sectional shapes can be classified in three categories based on different fabrication processes or their structural behavior as discussed below:

1. A steel I-section may be a *rolled* section (*beam*, Figure 12.2a) with or without cover plates, or a *built-up* section (*plate girder*, Figure 12.2b) with or without haunches consisting of top and bottom flange plates welded to a web plate. Rolled steel I-beams are applicable to shorter spans (less than 30 m) and plate girders to longer span bridges (about 30 to 90 m). A plate girder can be considered as a deep beam. The most distinguishing feature of a plate girder is the use of the transverse stiffeners that provide tension-field action increasing the postbuckling shear strength. The plate girder may also require longitudinal stiffeners to develop inelastic flexural buckling strength.

2. I-sections can be classified as *composite* or *noncomposite*. A steel section that acts with the concrete deck to resist flexure is called a composite section (Figure 12.3a). A steel section disconnected from the concrete deck is noncomposite (Figure 12.3b). Since composite sections most effectively use the properties of steel and concrete, they are often the best choice. Steel–concrete composite girder bridges are recommended by AASHTO-LRFD [4] whereas noncomposite members are not and are less frequently used in the United States.

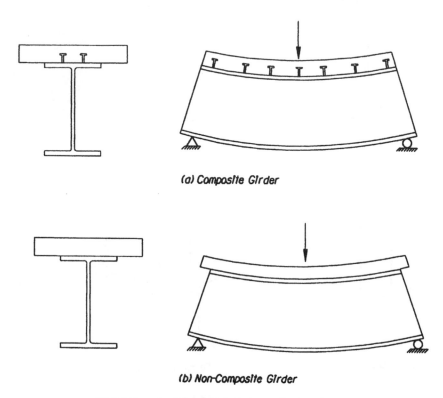

(a) Composite Girder

(b) Non-Composite Girder

FIGURE 12.3   Composite and noncomposite section.

3. Steel sections can also be classified as *compact*, *noncompact*, and *slender* element sections [4-6]. A qualified compact section can develop a full plastic stress distribution and possess a inelastic rotation capacity of approximately three times the elastic rotation before the onset of local buckling. Noncompact sections develop the yield stress in extreme compression fiber before buckling locally, but will not resist inelastic local buckling at the strain level required for a fully plastic stress distribution. Slender element sections buckle elastically before the yield stress is achieved.

12.3.2   Selection of Structural Sections

Figure 12.4 shows a typical portion of a composite I-girder bridge consisting of a concrete deck and built-up plate girder I-section with stiffeners and cross frames. The first step in the structural design of an I-girder bridge is to select an I-rolled shape or to size initially the web and flanges of a plate girder. This section presents the basic principles of selecting I-rolled shapes and sizing the dimensions of a plate girder.

The ratio of overall depth (steel section plus concrete slab) to the effective span length is usually about 1:25 and the ratio of depth of steel girder only to the effective span length is about 1:30. I-rolled shapes are standardized and can be selected from a manual such as the AISC-LRFD [7]. It should be noted that the web of a rolled section always meets compactness requirements while the flanges may not. To increase the flexural strength of a rolled section, it is common to add cover plates to the flanges. The I-rolled beams are usually used for simple-span length up to 30 m for highway bridges and 25 m for railway bridges. Plate girder sections provide engineers freedom and flexibility to proportion the flanges and web plates efficiently. Plate girders must have sufficient

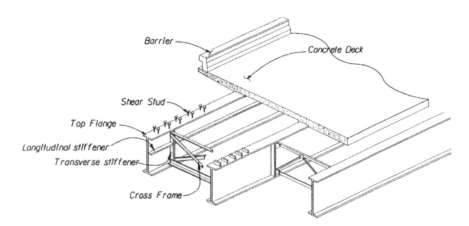

FIGURE 12.4 Typical components of composite I-girder bridge.

flexural and shear strength and stiffness. A practical choice of flange and web plates should not result in any unusual fabrication difficulties. An efficient girder is one that meets these requirements with the minimum weight. An economical one minimizes construction costs and may or may not correspond to the lowest weight alternative [8].

- *Webs:* The web mainly provides shear strength for the girder. The *web height* is commonly taken as $\frac{1}{18}$ to $\frac{1}{20}$ of the girder span length for highway bridges and slightly less for railway bridges. Since the web contributes little to the bending resistance, its thickness (t) should be as small as local buckling tolerance allows. Transverse stiffeners increase shear resistance by providing tension field action and are usually placed near the supports and large concentrated loads. Longitudinal stiffeners increase flexure resistance of the web by controlling lateral web deflection and preventing the web bending buckling. They are, therefore, attached to the compression side. It is usually recommended that sufficient web thickness be used to eliminate the need for longitudinal stiffeners as they can create difficulty in fabrication. Bearing stiffeners are also required at the bearing supports and concentrated load locations and are designed as compression members.

- *Flanges:* The flanges provide bending strength. The width and thickness are usually determined by choosing the area of the flanges within the limits of the width-to-thickness ratio, b/t, and the requirement as specified in the design specifications to prevent local buckling. Lateral bracing of the compression flanges is usually needed to prevent lateral torsional buckling during various load stages.

- *Hybrid Sections:* The hybrid section consisting of flanges with a higher yield strength than that of the web may be used to save materials; this is becoming more promoted because of the new high-strength steels.

- *Variable Sections:* Variable cross sections may be used to save material where the bending moment is smaller and/or larger near the end of a span (see Figure 12.2b). However, the manpower required for welding and fabrication may be increased. The cost of manpower and material must be balanced to achieve the design objectives. The designer should consult local fabricators to determine common practices in the construction of a plate girder.

Highway bridges in the United States are designed to meet the requirements under various limit states specified by AASHTO-LRFD [4,5] such as strength, fatigue and fracture, service, and extreme events (see Chapter 5). Constructibility must be considered. The following sections summarize basic concepts and AASHTO-LRFD [4,5] requirements for composite I-girder bridges.

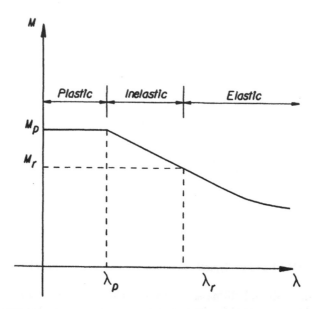

FIGURE 12.5   Three-range design format for steel flexural members.

12.4   Flexural Design

12.4.1   Basic Concept

The flexural resistance of a steel beam/girder is controlled by four failure modes or limit states: yielding, flange local buckling, web local buckling, and lateral-torsional buckling [9]. The moment capacity depends on the yield strength of steel (F_y), the slenderness ratio λ in terms of width-to-thickness ratio (b/t or h/t_w) for local buckling and unbraced length to the radius of gyration about strong axis ratio (L_b/r_y) for lateral-torsional buckling. As a general design concept for steel structural components, a three-range design format (Figure 12.5): plastic yielding, inelastic buckling, and elastic buckling are generally followed. In other words, when slenderness ratio λ is less than λ_p, a section is referred to as compact, plastic moment capacity can be developed; when $\lambda_p < \lambda < \lambda_r$, a section is referred to as noncompact, moment capacity less than M_p but larger than yield moment M_y can be developed; and when $\lambda > \lambda_r$, a section or member is referred to as slender and elastic buckling failure mode will govern. Figure 12.6 shows the dimensions of a typical I-girder. Tables 12.2 and 12.3 list the AASHTO-LRFD [4,5] design formulas for determination of flexural resistance in positive and negative regions.

12.4.2   Yield Moment

The yield moment M_y for a composite section is defined as the moment that causes the first yielding in one of the steel flanges. M_y is the sum of the moments applied separately to the steel section only, the short-term composite section, and the long-term composite section. It is based on elastic section properties and can be expressed as

$$M_y = M_{D1} + M_{D2} + M_{AD} \tag{12.6}$$

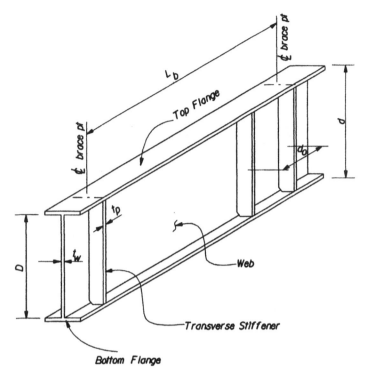

FIGURE 12.6 Typical girder dimensions.

where M_{D1} is moment due to factored permanent loads on steel section; M_{D2} is moment due to factored permanent loads such as wearing surface and barriers on long-term composite section; M_{AD} is additional live-load moment to cause yielding in either steel flange and can be obtained from the following equation:

$$M_{AD} = S_n \left[F_y - \frac{M_{D1}}{S_s} - \frac{M_{D2}}{S_{3n}} \right] \tag{12.7}$$

where S_s, S_n, and S_{3n} are elastic section modulus for steel, short-term composite, and long-term composite sections, respectively.

12.4.3 Plastic Moment

The plastic moment M_p for a composite section is defined as the moment that causes the yielding in the steel section and reinforcement and a uniform stress distribution of $0.85\ f'_c$ in compression concrete slab (Figure 12.7). In positive flexure regions, the contribution of reinforcement in concrete slab is small and can be neglected.

The first step of determining M_p is to find the plastic neutral axis (PNA) by equating total tension yielding forces in steel to compression yield in steel and/or concrete slab. The plastic moment is then obtained by summing the first moment of plastic forces in various components about the PNA. For design convenience, Table 12.4 lists the formulas for $\overline{Y}$ and M_p.

TABLE 12.2   AASHTO-LRFD Design Formulas of Positive Flexure Ranges for Composite Girders (Strength Limit State)

Items	Compact Section Limit, λ_p	Noncompact Section Limit, λ_r	Slender Sections
Web slenderness $2D_{cp}/t_w$	$3.76\sqrt{E/F_{yc}}$	$\alpha_{st}\sqrt{E/f_c}$	N/A
Compression flange slenderness b/t	No requirement at strength limit state		
Compression flange bracing L_b/r_t	No requirement at strength limit state, but should satisfy $1.76\sqrt{E/F_{yc}}$ for loads applied before concrete deck hardens		$> 1.76\sqrt{E/F_{yc}}$
Nominal flexural resistance	For simple spans and continuous spans with compact interior support section: For $D_p \le D'$, $M_n = M_p$ If $D' < D_p \le 5D'$ $M_n = \dfrac{5M_p - 0.85M_y}{4} + \dfrac{0.85M_y - M_p}{4}\left(\dfrac{D_p}{D'}\right)$ For continuous spans with noncompact interior support section: $M_n = 1.3R_h M_y$ but not taken greater than the applicable values from the above two equations. **Required section ductility** $D_p/D' \le 5$ $D' = \beta\left(\dfrac{d + t_s + t_h}{7.5}\right)$ $\beta = \begin{cases} 0.9 \text{ for } F_y = 250 \text{ MPa} \\ 0.7 \text{ for } F_y = 345 \text{ MPa} \end{cases}$	For compression flange: $F_n = R_b R_h F_{yc}$ For tension flange: $F_n = R_b R_h F_{yt}\sqrt{1 - 3\left(\dfrac{f_v}{F_{yt}}\right)}$ R_b = load shedding factor, for tension flange = 1.0; for compression flange = 1.0 if either a longitudinal stiffener is provided or $2D_c/t_w \le \lambda_b\sqrt{E/f_c}$ is satisfied; otherwise see Eq. (12.2)	Compression Flange: Eq. (12.4) Tension Flange: $F_n = R_b R_h F_{yt}$

A_{fc} = compression flange area
d = depth of steel section
D_{cp} = depth of the web in compression at the plastic moment
D_p = distance from the top of the slab to the plastic neutral axis
f_c = stress in compression flange due to factored load
f_v = maximum St. Venant torsional shear stress in the flange due to the factored load
F_n = nominal stress at the flange
F_{yc} = specified minimum yield strength of the compression flange
F_{yt} = specified minimum yield strength of the tension flange
M_p = plastic flexural moment

$$R_b = 1 - \left(\frac{a_r}{1200 + 300a_r}\right)\left(\frac{2D_c}{t_w} - \lambda_b\sqrt{\frac{E}{f_c}}\right) \qquad (12.2)$$

$$a_r = \frac{2D_c t_w}{A_{fc}} \qquad (12.3)$$

M_y = yield flexural moment
R_h = hybrid factor, 1.0 for homogeneous section, see AASHTO-LRFD 6.10.5.4
t_h = thickness of concrete haunch above the steel top flange
t_s = thickness of concrete slab; t_w = web thickness
α_{st} = 6.77 for web without longitudinal stiffeners and 11.63 with longitudinal stiffeners
λ_b = 5.76 for compression flange area $\ge$ tension flange area, 4.64 for compression area < tension area

TABLE 12.3 AASHTO-LRFD Design Formulas of Negative Flexure Ranges for Composite I Sections (Strength Limit State)

Items	Compact Section Limit, λ_p	Noncompact Section Limit λ_r	Slender Sections
Web slenderness, $2D_{cp}/t_w$	$3.76\sqrt{E/F_{yc}}$	$\alpha_{st}\sqrt{E/F_{yc}}$	N/A
Compression flange slenderness, $b_f/2t_f$	$0.382\sqrt{E/F_{yc}}$	$1.38\sqrt{\dfrac{E}{f_c\sqrt{\dfrac{2D_c}{t_w}}}}$	$>1.38\sqrt{\dfrac{E}{f_c\sqrt{\dfrac{2D_c}{t_w}}}}$
Compression flange unsupported length, L_b	$\left[0.124-0.0759\left(\dfrac{M_l}{M_p}\right)\right]\left[\dfrac{r_y E}{F_{yc}}\right]$	$1.76r_t\sqrt{\dfrac{E}{F_{yc}}}$	$>1.76r_t\sqrt{\dfrac{E}{F_{yc}}}$
Nominal flexural resistance	$M_n = M_p$	$F_n = R_b R_h F_{yf}$	Compression flange Eq. (12.4) Tension flange $F_n = R_b R_h F_{yt}$

b_f = width of compression flange
t_f = thickness of compression flange
M_l = lower moment due to factored loading at end of the unbraced length
r_y = radius of gyration of steel section with respect to the vertical axis (mm)
r_t = radius of gyration of compression flange of steel section plus one third of the web in compression with respect to the vertical axis (mm)

For lateral torsional buckling AASHTO-LRFD 6.10.5.5:

$$
F_n = \begin{cases}
C_b R_b R_h F_{yc}\left[1.33-0.18\left(\dfrac{L_b}{r_t}\right)\sqrt{\dfrac{F_{yc}}{E}}\right] \leq R_b R_h F_{yc} & \text{for } L_p < L_b < L_r \\[4mm]
C_b R_b R_h\left[\dfrac{9.86E}{\left(L_b/r_t\right)^2}\right] \leq R_b R_h F_{yc} & \text{for } L_b \geq L_r
\end{cases}
\tag{12.4}
$$

$$
C_b = 1.75-1.05\left(\dfrac{P_1}{P_2}\right)+0.3\left(\dfrac{P_1}{P_2}\right)^2 \leq 2.3
\tag{12.5}
$$

P_1 = smaller force in the compression flange at the braced point due to factored loading
P_2 = larger force in the compression flange at the braced point due to factored loading

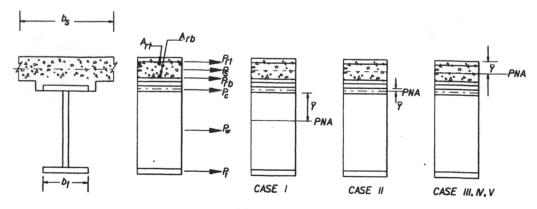

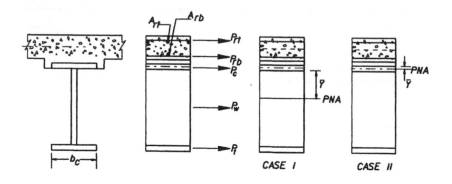

FIGURE 12.7 Plastic moments for composite sections.

Example 12.1: Three-Span Continuous Composite Plate-Girder Bridge

Given

A three-span continuous composite plate-girder bridge has two equal end spans of length 49.0 m and one midspan of 64 m. The superstructure is 13.4 m wide. The elevation, plan, and typical cross section are shown in Figure 12.8.

Structural steel:	A709 Grade 345; $F_{yw} = F_{yt} = F_{yc} = F_y = 345$ MPa
Concrete:	$f_c' = 280$ MPa ; $E_c = 25,000$ MPa; modular ratio $n = 8$
Loads:	Dead load = steel plate girder + concrete deck + barrier rail + future wearing 75 mm AC overlay
	Live load = AASHTO HL-93 + dynamic load allowance
Deck:	Concrete deck with thickness of 275 mm has been designed

Steel section in positive flexure region:

Top flange:	$b_{fc} = 460$ mm	$t_{fc} = 25$ mm
Web:	$D = 2440$ mm	$t_w = 16$ mm
Bottom flange:	$b_{ft} = 460$ mm	$t_{ft} = 45$ mm

Construction: Unshored; unbraced length for compression flange $L_b = 6.1$ m.

TABLE 12.4 Plastic Moment Calculation

Regions	Case	Condition and $\bar{Y}$	$\bar{Y}$ and M_p
	I — PNA in web	$P_t + P_w \ge P_c + P_s + P_{rb} + P_{rt}$ $\bar{Y} = \left(\dfrac{D}{2}\right)\left[\dfrac{P_t - P_c - P_s - P_{rt} - P_{rb}}{P_w} + 1\right]$	$M_p = \dfrac{P_w}{2D}\left[\bar{Y}^2 + (D - \bar{Y})^2\right]$ $+\left[P_s d_s + P_{rt} d_{rt} + P_{rb} d_{rb} + P_c d_c + P_t d_t\right]$
	II — PNA in top flange	$P_t + P_w + P_c \ge P_s + P_{rb} + P_{rt}$ $\bar{Y} = \left(\dfrac{t_c}{2}\right)\left[\dfrac{P_w + P_c - P_s - P_{rt} - P_{rb}}{P_c} + 1\right]$	$M_p = \dfrac{P_c}{2t_c}\left[\bar{Y}^2 + (t_c - \bar{Y})^2\right]$ $+\left[P_s d_s + P_{rt} d_{rt} + P_{rb} d_{rb} + P_w d_w + P_t d_t\right]$
Positive Figure 12.7a	III — PNA in slab, below P_{rb}	$P_t + P_w + P_c \ge \left(\dfrac{C_{rb}}{t_s}\right)P_s + P_{rb} + P_{rt}$ $\bar{Y} = (t_s)\left[\dfrac{P_w + P_c + P_s - P_{rt} - P_{rb}}{P_s}\right]$	$M_p = \left(\dfrac{\bar{Y}^2 P_s}{2t_s}\right)^2$ $+\left[P_{rt} d_{rt} + P_{rb} d_{rb} + P_c d_c + P_w d_w + P_t d_t\right]$
	IV — PNA in slab at P_{rb}	$P_t + P_w + P_c + P_{rb} \ge \left(\dfrac{C_{rb}}{t_s}\right)P_s + P_{rt}$ $\bar{Y} = C_{rb}$	$M_p = \left(\dfrac{\bar{Y}^2 P_s}{2t_s}\right)^2$ $+\left[P_{rt} d_{rt} + P_c d_c + P_w d_w + P_t d_t\right]$
	V — PNA in slab, above P_{rb}	$P_t + P_w + P_c + P_{rb} \ge \left(\dfrac{C_{tb}}{t_s}\right)P_s + P_{rt}$ $\bar{Y} = (t_s)\left[\dfrac{P_{rb} + P_c + P_w + P_t - P_{rb}}{P_s}\right]$	$M_p = \left(\dfrac{\bar{Y}^2 P_s}{2t_s}\right)^2$ $+\left[P_{rt} d_{rt} + P_{rb} d_{rb} + P_c d_c + P_w d_w + P_t d_t\right]$
Negative Figure 12.7b	I — PNA in web	$P_{cr} + P_w \ge P_c + P_{rb} + P_{rt}$ $\bar{Y} = \left(\dfrac{D}{2}\right)\left[\dfrac{P_c - P_{ct} - P_{rt} - P_{rb}}{P_s} + 1\right]$	$M_p = \dfrac{P_w}{2D}\left[\bar{Y}^2 + (D - \bar{Y})^2\right]$ $+\left[P_{rt} d_{rt} + P_{rb} d_{rb} + P_t d_t + P_c d_c\right]$
	II — PNA in top flange	$P_t + P_w + P_t \ge P_{rb} + P_{rt}$ $\bar{Y} = \left(\dfrac{t_t}{2}\right)\left[\dfrac{P_{rb} + P_c - P_w - P_{rb}}{P_t} + 1\right]$	$M_p = \dfrac{P_t}{2t_t}\left[\bar{Y}^2 + (t_t - \bar{Y})^2\right]$ $+\left[P_{rt} d_{rt} + P_{rb} d_{rb} + P_t d_t + P_c d_c\right]$

P_{rt} $= F_{yrt} A_{rt}$; $P_s = 0.85 f_c' b_s t_s$; $P_{rb} = F_{yrb} A_{rb}$

P_c $= F_{yc} b_c t_c$; $P_w = F_{yw} D t_w$; $P_t = F_{yt} b_t t_t$

A_{rb}, A_{rt} = reinforcement area of bottom and top layer in concrete deck slab

F_{yrb}, F_{yrt} = yield strength of reinforcement of bottom and top layers

b_c, b_t, b_s = width of compression, tension steel flange, and concrete deck slab

t_c, t_t, t_w, t_s = thickness of compression, tension steel flange, web, and concrete deck slab

F_{yt}, F_{yc}, F_{yw} = yield strength of tension flange, compression flange, and web

Source: American Association of State Highway and Transportation Officials, AASHTO LRFD Bridge Design Specifications, Washington, D.C., 1994. With permission.

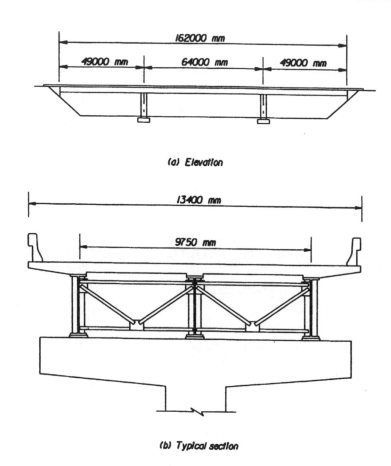

(a) Elevation

(b) Typical section

FIGURE 12.8   Three-spans continuous plate-girder bridge.

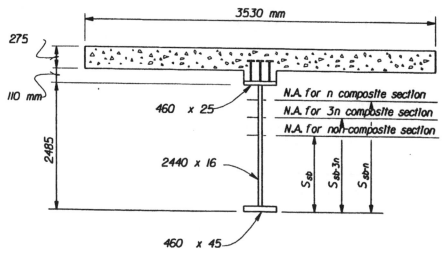

FIGURE 12.9   Cross section for positive flexure region.

Maximum positive moments in Span 1 due to factored loads applied to the steel section, and to the long-term composite section are $M_{D1} = 6859$ kN-m and $M_{D2} = 2224$ kN-m, respectively.

Requirement

Determine yield moment M_y, plastic moment M_p, and nominal moment M_n of an interior girder for positive flexure region.

Solutions

1. **Determine Effective Flange Width (AASHTO Article 4.6.2.6)**

   For an interior girder, the effective flange width is

$$b_{\text{eff}} = \text{the lesser of} \begin{cases} \dfrac{L_{\text{eff}}}{4} = \dfrac{35{,}050}{4} = 8763 \text{ mm} \\[3mm] 12t_s + \dfrac{b_f}{2} = (12)(275) + \dfrac{460}{2} = 3530 \text{ mm} \quad \text{(controls)} \\[3mm] S = 4875 \text{ mm} \end{cases}$$

   where L_{eff} is the effective span length and may be taken as the actual span length for simply supported spans and the distance between points of permanent load inflection for continuous spans (35.05 m); b_f is top flange width of steel girder.

2. **Calculate Elastic Composite Section Properties**

   For the section in the positive flexure region as shown in Figure 12.9, its elastic section properties for the noncomposite, the short-term composite ($n = 8$), and the long-term composite ($3n = 24$) are calculated in Tables 12.5 to 12.7.

   TABLE 12.5 Noncomposite Section Properties for Positive Flexure Region

Component	A (mm²)	y_i (mm)	$A_i y_i$ (mm³)	$y_i - y_{sb}$ (mm)	$A_i(y_i - y_{sb})^2$ (mm⁴)	I_o (in⁴)
Top flange 460 × 25	11,500	2498	28.7 (10)⁶	1395	22.4 (10)⁹	1.2 (10)⁶
Web 2440 × 16	39,040	1265	49.4 (10)⁶	162	3.0 (10)⁸	19.4 (10)⁹
Bottom flange 460 × 45	20,700	22.5	4.7 (10)⁵	−1081	24.2 (10)⁹	3.5 (10)⁶
Σ	71,240	—	78.6 (10)⁶		46.8 (10)⁹	19.4 (10)⁹

$$y_{sb} = \frac{\sum A_i y_i}{\sum A_i} = \frac{78.6(10)^6}{71240} = 1103 \text{ mm} \qquad y_{st} = (45 + 2440 + 25) - 1103 = 1407 \text{ mm}$$

$$I_{\text{girder}} = \sum I_o + \sum A_i(y_i - y_{sb})^2$$

$$= 19.4(10)^9 + 46.8(10)^9 = 66.2(10)^9 \text{ mm}^4$$

$$S_{sb} = \frac{I_{\text{girder}}}{y_{sb}} = \frac{66.2(10)^9}{1103} = 60.0(10)^6 \text{ mm}^3 \qquad S_{st} = \frac{I_{\text{girder}}}{y_{st}} = \frac{66.2(10)^9}{1407} = 47.1(10)^6 \text{ mm}^3$$

3. **Calculate Yield Moment M_y**

   The yield moment M_y corresponds to the first yielding of either steel flange. It is obtained by the following formula:

$$M_y = M_{D1} + M_{D2} + M_{AD}$$

TABLE 12.6   Short-Term Composite Section Properties ($n = 8$)

Component	A (mm^2)	y_i (mm)	A_iy_i (mm^3)	$y_i - y_{sb\text{-}n}$ (mm)	$A_i(y_i - y_{sb\text{-}n})^2$ (mm^4)	I_o (mm^4)
Steel section	71,240	1103	78.6 (10)^6	−1027	75.1 (10)^9	19.4 (10)^9
Concrete slab 3530/8 × 275	121,344	2733	3.3 (10)^8	603	44.1 (10)^9	2.3 (10)^8
Σ	192,584	—	4.1 (10)^8	—	119.2 (10)^9	19.6 (10)^9

$$y_{sb\text{-}n} = \frac{\sum A_i y_i}{\sum A_i} = \frac{4.1(10)^8}{192,584} = 2130 \text{ mm} \qquad y_{st\text{-}n} = (45 + 2440 + 25) - 2130 = 380 \text{ mm}$$

$$I_{com\text{-}n} = \sum I_o + \sum A_i(y_i - y_{sb\text{-}n})^2$$

$$= 19.6(10)^9 + 119.2(10)^9 = 138.8(10)^9 \text{ mm}^4$$

$$S_{sb\text{-}n} = \frac{I_{com\text{-}n}}{y_{sb\text{-}n}} = \frac{138.8(10)^9}{2130} = 65.2(10)^6 \text{ mm}^3 \qquad S_{st\text{-}n} = \frac{I_{com\text{-}n}}{y_{st\text{-}n}} = \frac{138.8(10)^9}{380} = 365.0(10)^6 \text{ mm}^3$$

TABLE 12.7   Long-Term Composite Section Properties ($3n = 24$)

Component	A (mm^2)	y_i (mm)	A_iy_i (mm^3)	$y_i - y_{sb\text{-}3n}$ (mm)	$A_i(y_i - y_{sb\text{-}3n})^2$ (mm^4)	I_o (mm^4)
Steel section	71,240	1103	78.6 (10^6)	−590	24.8 (10^9)	19.4 (10^9)
Concrete slab 3530/24 × 275	40,448	2733	1.1 (10^8)	1040	43.7 (10^9)	2.3 (10^8)
Σ	111,688	—	10846.4	—	68.5 (10^9)	19.6 (10^9)

$$y_{sb\text{-}3n} = \frac{\sum A_i y_i}{\sum A_i} = \frac{88.1(10)^9}{111,688} = 1693 \text{ mm} \qquad y_{st\text{-}3n} = (45 + 2440 + 25) - 1693 = 817 \text{ mm}$$

$$I_{com\text{-}3n} = \sum I_o + \sum A_i(y_i - y_{sb\text{-}3n})^2$$

$$= 19.6(10)^9 + 68.5(10)^9 = 88.1(10)^9 \text{ mm}^4$$

$$S_{sb\text{-}3n} = \frac{I_{con\text{-}3n}}{y_{sb\text{-}3n}} = \frac{88.1(10)^9}{1693} = 52.0(10)^6 \text{ mm}^3 \qquad S_{st\text{-}3n} = \frac{I_{com\text{-}3n}}{y_{st\text{-}3n}} = \frac{88.4(10)^9}{817} = 107.9(10)^6 \text{ mm}^3$$

$$M_{AD} = S_n \left(F_y - \frac{M_{D1}}{S_s} - \frac{M_{D2}}{S_{3n}} \right)$$

$$M_{D1} = 6859 \text{ kN-m}$$

$$M_{D2} = 2224 \text{ kN-m}$$

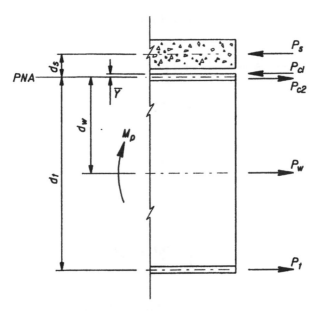

FIGURE 12.10 Plastic moment state.

For the top flange:

$$M_{AD} = (368.4)10^{-3}\left(345(10)^3 - \frac{6859}{47.1(10)^{-3}} - \frac{2224}{108.6(10)^{-3}}\right)$$

$$= 65,905 \text{ kN-m}$$

For the bottom flange:

$$M_{AD} = (65.2)10^{-3}\left(345(10)^3 - \frac{6859}{60.0(10)^{-3}} - \frac{2224}{52.1(10)^{-3}}\right)$$

$$= 12,257 \text{ kN-m} \quad \text{(controls)}$$

$$\therefore \quad M_y = 6859 + 2224 + 12,257 = 21,340 \text{ kN-m}$$

4. **Calculate Plastic Moment Capacity M_p**
   For clarification, the reinforcement in slab is neglected. We first determine the location of
   the PNA (see Figure 12.10 and Table 12.4).

$$P_s = 0.85f_c'b_{eff}t_s = 0.85(28)(3530)(275) = 23,104 \text{ kN}$$

$$P_{c1} = \overline{Y}b_{fc}F_{yc}$$

$$P_{c2} = A_cF_{yc} - P_{c1} = (t_c - \overline{Y})b_{fc}F_{yc}$$

$$P_c = P_{c1} + P_{c2} = A_{fc}F_{yc} = (460)(25)(345) = 3967 \text{ kN}$$

$$P_w = A_wF_{yw} = (2440)(16)(345) = 13,469 \text{ kN}$$

$$P_t = A_{ft}F_{yt} = (460)(45)(345) = 7141 \text{ kN}$$

Since $P_t + P_w + P_c > P_s$, the PNA is located within top of flange (Case II, Table 12.4).

$$\bar{Y} = \frac{t_c}{2}\left(\frac{P_w + P_t - P_s}{P_c} + 1\right)$$

$$= \frac{25}{2}\left(\frac{13,469 + 7141 - 23,104}{3967} + 1\right) = 4.6 \text{ mm} < t_c = 25 \text{ mm}$$

Summing all forces about the PNA (Figure 12.5 and Table 12.4), obtain

$$M_p = \sum M_{PNA} = P_{c1}\left(\frac{y_{PNA}}{2}\right) + P_{c2}\left(\frac{t_{fc} - y_{PNA}}{2}\right) + P_s d_s + P_w d_w + P_t d_t$$

$$= \frac{P_c}{2t_c}\left[\bar{Y}^2 + (t_c - \bar{Y})^2\right] + P_s d_s + P_w d_w + P_t d_t$$

$$d_s = \frac{275}{2} + 110 - 25 + 4.0 = 227 \text{ mm}$$

$$d_w = \frac{2440}{2} + 25 - 4.6 = 1240 \text{ mm}$$

$$d_t = \frac{45}{2} + 2440 + 25 - 4.6 = 2483 \text{ mm}$$

$$M_p = \frac{3967}{2(0.025)}\left[0.0046^2 + (0.025 - 0.004)^2\right] + (23,206)(0.227)$$

$$+ (13,469)(1.24) + (7141)(2.483)$$

$$= 39,737 \text{ kN-m}$$

5. **Calculate Nominal Moment**
   a. *Check compactness of steel girder section:*
      - Web slenderness requirement (Table 12.2)

$$\frac{2D_{cp}}{t_w} \leq 3.76\sqrt{\frac{E}{F_{yc}}}$$

Since the PNA is within the top flange, D_{cp} is equal to zero. The web slenderness requirement is satisfied.
      - It is usually assumed that the top flange is adequately braced by the hardened concrete deck; there is, therefore, no requirements for the compression flange slenderness and bracing for compact composite sections at the strength limit state.
      ∴ The section is a compact composite section.
   b. Check ductility requirement (Table 12.2) $D_p/D' \leq 5$:

The purpose of this requirement is to prevent permanent crashing of the concrete slab when the composite section approaches its plastic moment capacity.

D_p = the depth from the top of the concrete deck to the PNA

$$D_p = 275 + 110 - 25 + 4.6 = 364.6 \text{ mm}$$

$$D' = \beta\left(\frac{d + t_s + t_h}{7.5}\right) = 0.7\left(\frac{2485 + 275 + 110}{7.5}\right) = 267.9 \text{ mm}$$

$$\frac{D_p}{D'} = \frac{364.6}{267.9} = 1.36 < 5 \qquad\qquad\qquad \text{OK}$$

c. Check moment of inertia ratio limit (AASHTO Article 6.10.1.1):

The flexural members shall meet the following requirement:

$$0.1 \le \frac{I_{yc}}{I_y} \le 0.9$$

where I_{yc} and I_y are the moments of inertia of the compression flange and steel girder about the vertical axis in the plane of web, respectively. This limit ensures that the lateral torsional bucking formulas are valid.

$$I_{yc} = \frac{(25)(460)^3}{12} = 2.03(10^8) \text{ mm}^4$$

$$I_y = 2.03(10^8) + \frac{(2440)(16)^3}{12} + \frac{(45)(460)^3}{12} = 5.69(10^8) \text{ mm}^4$$

$$0.1 < \frac{I_{yc}}{I_y} = \frac{2.03(10^8)}{5.69(10^8)} = 0.36 < 0.9 \qquad\qquad \text{OK}$$

d. *Nominal flexure resistance M_n* (Table 12.2):

Assume that the adjacent interior pier section is noncompact. For continuous spans with the noncompact interior support section, the nominal flexure resistance of a compact composite section is taken as

$$M_n = 1.3 \, R_h M_y \le M_p$$

with flange stress reduction factor $R_h = 1.0$ for this homogenous girder, we obtain

$$M_n = 1.3(1.0)(21,340) = 27,742 \text{ kN-m} < M_p = 39,712 \text{ kN-m}$$

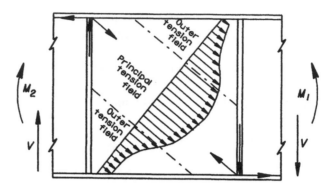

FIGURE 12.11   Tension field action.

12.5   Shear Design

12.5.1   Basic Concept

Similar to the flexural resistance, web shear capacity is also dependent on the slenderness ratio λ in term of width-to-thickness ratio (h/t_w). In calculating shear strength, three failure modes are considered: shear yielding when $\lambda \leq \lambda_p$, inelastic shear buckling when $\lambda_p < \lambda < \lambda_r$, and elastic shear buckling when $\lambda > \lambda_r$. For the web without transverse stiffeners, shear resistance is contributed by the beam action of shearing yield or elastic shear buckling. For interior web panels with transverse stiffeners, shear resistance is contributed by both beam action (the first term of the C_s *equation in Table 12.8) and tension field action* (the second term of the C_s equation in Table 12.8). For end web panels, tension field action cannot be developed because of the discontinuous boundary and the lack of an anchor. It is noted that transverse stiffeners provide a significant inelastic shear buckling strength by tension field action as shown in Figure 12.11. Table 12.8 lists the AASHTO-LRFD [4,5] design formulas for shear strength.

Example 12.2: Shear Strength Design — Strength Limit State I

Given
For the I-girder bridge shown in Example 12.1, factored shear V_u = 2026 and 1495 kN are obtained at the left end of Span 1 and 6.1 m from the left end in Span 1, respectively. Design shear strength for the Strength Limit State I for those two locations.

Solutions

1. **Nominal Shear Resistance V_n**
   a. V_n *for an unstiffened web* (Table 12.8 or AASHTO Article 6.10.7.2):
      For D = 2440 mm and t_w = 16 mm, we have

$$\because \quad \frac{D}{t_w} = \frac{2440}{16} = 152.5 > 3.07\sqrt{\frac{E}{F_{yw}}} = 3.07\sqrt{\frac{200,000}{345}} = 73.9$$

$$\therefore \quad V_n = \frac{4.55 t_w^3 E}{D} = \frac{4.55\,(16)^3(200,000)}{2440} = (1528)10^3\text{ N} = 1528\text{ kN}$$

TABLE 12.8 AASHTO-LRFD Design Formulas of Nominal Shear Resistance at Strength Limit State

Unstiffened homogeneous webs

$$V_n = \begin{cases} V_p = 0.58 F_{yw} D t_w & \text{for } \dfrac{D}{t_w} \leq 2.46 \sqrt{\dfrac{E}{F_{yw}}} & \text{– Shear yielding} \\[3ex] 1.48 t_w^2 \sqrt{E F_{yw}} & \text{for } 2.46 \sqrt{\dfrac{E}{F_{yw}}} < \dfrac{D}{t_w} \leq 3.07 \sqrt{\dfrac{E}{F_{yw}}} & \text{– Inelastic buckling} \\[3ex] \dfrac{4.55 t_w^3 E}{D} & \text{for } \dfrac{D}{t_w} > 3.07 \sqrt{\dfrac{E}{F_{yw}}} & \text{– Elastic buckling} \end{cases}$$

Stiffened interior web panels of compact homogeneous sections

$$V_n = \begin{cases} C_s V_p & \text{for } M_u \leq 0.5 \phi_f M_p \\[2ex] R C_s V_p \geq C V_p & \text{for } M_u > 0.5 \phi_f M_p \end{cases}; \quad C_s = C + \frac{0.87(1-C)}{\sqrt{1+(d_o/D)^2}}$$

$$R = 0.6 + 0.4\left(\frac{M_r - M_u}{M_r - 0.75\phi_f M_y}\right) \leq 1.0$$

Stiffened interior web panels of noncompact homogeneous sections

$$V_n = \begin{cases} C_s V_p & \text{for } f_u \leq 0.5 \phi_f F_y \\[2ex] R C_s V_p \geq C V_p & \text{for } f_u \leq 0.5 \phi_f F_y \end{cases} \qquad R = 0.6 + 0.4\left(\frac{F_r - f_u}{F_r - 0.75\phi_f F_y}\right) \leq 1.0$$

End panels and hybrid sections

$$V_n = C V_p$$

d_o = stiffener spacing (mm)
D = web depth
F_r = factored flexural resistance of the compression flange (MPa)
f_u = factored maximum stress in the compression flange under consideration (MPa)
M_r = factored flexural resistance
M_u = factored maximum moment in the panel under consideration
ϕ_f = resistance factor for flexure = 1.0 for the strength limit state
C = ratio of the shear buckling stress to the shear yield strength

$$C = \begin{cases} 1.0 & \text{For } \dfrac{D}{t_w} \leq 1.1 \sqrt{\dfrac{Ek}{F_{yw}}} \\[3ex] \dfrac{1.1}{D/t_w}\sqrt{\dfrac{Ek}{F_{yw}}} & \text{For } 1.1\sqrt{\dfrac{Ek}{F_{yw}}} \leq \dfrac{D}{t_w} \leq 1.38 \sqrt{\dfrac{Ek}{F_{yw}}} \\[3ex] \dfrac{1.52}{(D/t_w)^2}\sqrt{\dfrac{Ek}{F_{yw}}} & \text{For } \dfrac{D}{t_w} > 1.38 \sqrt{\dfrac{Ek}{F_{yw}}} \end{cases} \qquad (12.8)$$

$$k = 5 + \frac{5}{(d_o/D)^2}$$

b. V_n *for end-stiffened web panel* (Table 12.8 or AASHTO Article 6.10.7.3.3c):
Try the spacing of transverse stiffeners $d_o = 6100$ mm. In order to facilitate handling of web panel sections, the spacing of transverse stiffeners shall meet (AASHTO Article 6.10.7.3.2) the following requirement:

$$d_o \leq D \left[\frac{260}{(D/t_w)} \right]^2$$

$$d_o = 6100 \text{ mm} < D \left[\frac{260}{(D/t_w)} \right]^2 = 2440 \left[\frac{260}{2440/16} \right]^2 = 7090 \text{ mm} \qquad \text{OK}$$

Using formulas in Table 12.8, obtain

$$k = 5 + \frac{5}{(d_o/D)^2} = 5 + \frac{5}{(6100/2440)^2} = 5.80$$

$$\because \frac{D}{t_w} = 152.5 > 1.38 \sqrt{\frac{Ek}{F_{yw}}} = 1.38 \sqrt{\frac{200,000(5.8)}{345}} = 80$$

$$\because C = \frac{1.52}{(152.5)^2} \sqrt{\frac{200,000(5.80)}{345}} = 0.379$$

$$V_p = 0.58 F_{yw} D t_w = 0.58(345)(2440)(16) = 7812(10)^3 \text{ N} = 7812 \text{ kN}$$

$$V_n = CV_p = 0.379(7812) = 2960 \text{ kN}$$

2. Strength Limit State I

AASHTO-LRFD [4] requires that for Strength Limit State I

$$V_u \leq \phi_v V_n$$

where ϕ_v is the shear resistance factor = 1.0.

a. *Left end of Span 1:*

$$\because V_u = 2026 \text{ kN} > \phi_v V_n \text{ (for unstiffened web)} = 1528 \text{ kN}$$

$$\therefore \text{ Stiffeners are needed to increase shear capacity}$$

$$\phi_v V_n = (1.0) 2960 = 2960 \text{ kN} > V_u = 2026 \text{ kN} \qquad \text{OK}$$

b. *Location of the first intermediate stiffeners, 6.1 m from the left end in Span 1:*
Since $V_u = 1459$ kN is less than the shear capacity of the unstiffened web $\phi_v V_n = 1528$ kN the intermediate transverse stiffeners may be omitted after the first intermediate stiffeners.

TABLE 12.9 AASHTO-LRFD Design Formulas of Stiffeners

Location	Stiffener	Required Project Width and Area	Required Moment of Inertia
Compression flange	Longitudinal	$b_l \leq 0.48\, t_s \sqrt{E/F_{yc}}$	$I_s \geq \begin{cases} 0.125k^3 & \text{for } n=1 \\ 0.07k^3n^4 & \text{for } n=2,3,4 \text{ or } 5 \end{cases}$ k see Table 12.5
	Transverse	Same size as longitudinal stiffener; at least one transverse stiffener on compression flange near the dead load contraflexure point	
Web	Longitudinal	$b_l \leq 0.48 t_s \sqrt{E/F_{yc}}$	$I_l \geq D t_w^3 \left[2.4(d_o/D)^2 - 0.13 \right]$ $r \geq 0.234 d_o \sqrt{F_{yc}/E}$
	Transverse intermediate	$50 + d/30 \leq b_l \leq 0.48 t_s \sqrt{E/F_{ys}}$ $16t_p \geq b_l \geq 0.25 b_f$ $A_s \geq \left[0.15 BD t_w \dfrac{(1-C)V_u}{V_r} - 18t_w^2 \right] \dfrac{F_{yw}}{F_{ys}}$ $B = 1$ for stiffener pairs, 1.8 for single angle and 2.4 for single plate	$I_t \geq d_o t_w^3 J$ $J = 2.5\left(D_p/d_o\right)^2 - 2 \geq 0.5$
	Bearing	$b_t \leq 0.48 t_p \sqrt{E/F_{ys}}$ $B_r = \phi_b A_{pn} F_{ys}$	Use effective section (AASHTO-LRFD 6.10.8.2.4) to design axial resistance

b_f = width of compression flange
t_f = thickness of compression flange
f_c = stress in compression flange due to the factored loading
F_{ys} = specified minimum yield strength of the stiffener
ϕ_b = resistance factor of bearing stiffeners = 1.0
A_{pn} = area of the projecting elements of the stiffener outside of the web-to-flange fillet welds, but not beyond the edge of the flange

12.5.2 Stiffeners

For built-up I-sections, the longitudinal stiffeners may be provided to increase bending resistance by preventing local buckling while transverse stiffeners are usually provided to increase shear resistance by the tension field action [10,11]. The following three types of stiffeners are usually used for I-sections:

- *Transverse Intermediate Stiffeners.* These work as anchors for the tension field force so that postbuckling shear resistance can be developed. It should be noted that elastic web shear buckling cannot be prevented by transverse stiffeners. Transverse stiffeners are designed to (1) meet the slenderness requirement of projecting elements to present local buckling, (2) provide stiffness to allow the web to develop its postbuckling capacity, and (3) have strength to resist the vertical components of the diagonal stresses in the web. These requirements are listed in Table 12.9.

- *Bearing Stiffeners.* These work as compression members to support vertical concentrated loads by bearing on the ends of stiffeners (see Figure 12.2). They are transverse stiffeners and connect to the web to provide a vertical boundary for anchoring shear force from tension

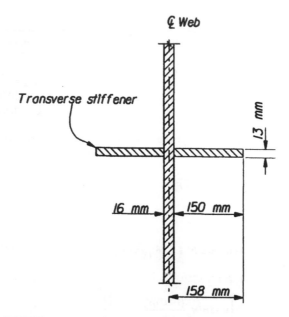

FIGURE 12.12   Cross section of web and transverse stiffener.

field action. They should be placed at all bearing locations and at all locations supporting concentrated loads. For rolled beams, bearing stiffeners may not be needed when factored shear is less than 75% of factored shear capacity. They are designed to satisfy the slenderness, bearing, and axial compression requirements as shown in Table 12.9.

- *Longitudinal Stiffeners:* These work as restraining boundaries for compression elements so that inelastic flexural buckling stress can be developed in a web. It consists of either a plate welded longitudinally to one side of the web, or a bolted angle. It should be located a distance of $2D_c/5$ from the inner surface of the compression flange, where D_c is the depth of web in compression at the maximum moment section to provide optimum design. The slenderness and stiffness need to be considered for sizing the longitudinal stiffeners (Table 12.9).

Example 12.3: Transverse and Bearing Stiffeners Design

Given

For the I-girder bridge shown in Example 12.1, factored shear V_u = 2026 and 1495 kN are obtained at the left end of Span 1 and 6.1 m from the left end in Span 1, respectively. Design the first intermediate transverse stiffeners and the bearing stiffeners at the left support of Span 1 using F_{ys} = 345 MPa for stiffeners.

Solutions

1. **Intermediate Transverse Stiffener Design**

   Try two 150 × 13 mm transverse stiffener plates as shown in Figure 12.12 welded to both sides of the web.

   a. *Projecting width b_t requirements* (Table12.9 or AASHTO Article 6.10.8.1.2):

   To prevent local buckling of the transverse stiffeners, the width of each projecting stiffener shall satisfy these requirements listed in Table 12.9.

   $$b_t = 150 \text{ mm} > \begin{cases} 50 + \dfrac{d}{30} = 50 + \dfrac{2510}{30} = 134 \text{ mm} \\ \\ 0.25\,b_f = 0.25(460) = 115 \text{ mm} \end{cases} \qquad \text{OK}$$

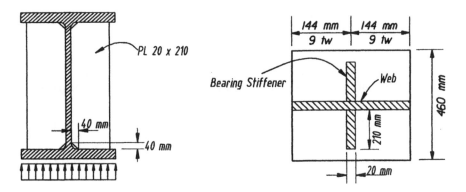

FIGURE 12.13 (a) Bearing stiffeners; (b) effective column area.

$$b_t = 150 \text{ mm} < \begin{cases} 0.48t_p\sqrt{\dfrac{E}{F_{ys}}} = 0.48(13)\sqrt{\dfrac{200,000}{345}} = 150 \text{ mm} & \text{OK} \\ 16t_p = 16(13) = 208 \text{ mm} \end{cases}$$

b. *Moment of inertia requirement* (Table 12.9 AASHTO Article 6.10.8.1.3):
The purpose of this requirement is to ensure sufficient rigidity of transverse stiffeners to develop adequately a tension field in the web.

$$\because J = 2.5\left(\frac{2440}{6100}\right)^2 - 2.0 = -1.6 < 0.5 \quad \therefore \text{ Use } J = 0.5$$

$$I_t = 2\left(\frac{150^3(13)}{3}\right) = 29.3(10)^6 \text{ mm}^4 > d_o t_w^3 J = (6100)(16)^3(0.5)$$

$$= 12.5(10)^6 \text{ mm}^4$$ OK

c. *Area requirement* (Table 12.9 or AASHTO Article 6.10.8.1.4):
This requirement ensures that transverse stiffeners have sufficient area to resist the vertical component of the tension field, and is only applied to transverse stiffeners required to carry the forces imposed by tension field action. From Example 12.2, we have $C = 0.379$; $F_{yw} = 345$ MPa; $V_u = 1460$ kN; $\phi_v V_n = 1495$ kN; $t_w = 16$ mm; $B = 1.0$ for stiffener pairs. The requirement area is

$$A_{sreqd} = \left(0.15(1.0)(2440)(16)(1-0.379)\frac{1459}{1495} - 18(16)^2\right)\left(\frac{345}{345}\right) = -1060 \text{ mm}^2$$

The negative value of A_{sreqd} indicates that the web has sufficient area to resist the vertical component of the tension field.

2. **Bearing Stiffener Design**
Try two 20 × 210 mm stiffness plates welded to each side of the web as shown in Figure 12.13a.
a. *Check local buckling requirement* (Table 12.9 or AASHTO Article 6.19.8.2.2):

$$\frac{b_t}{t_p} = \frac{210}{20} = 10.5 \leq 0.48\sqrt{\frac{E}{F_y}} = 0.48\sqrt{\frac{200,000}{345}} = 11.6 \qquad \text{OK}$$

b. *Check bearing resistance* (Table 12.9 or AASHTO Article 6.10.8.2.3):
Contact area of the stiffeners on the flange $A_{pn} = 2(210 - 40)20 = 6800 \text{ mm}^2$

$$B_r = \phi_b\, A_{pn}F_{ys} = (1.0)(6800)(345) = 2346\,(10)^3\,\text{N} = 2346\text{ kN} > V_u = 2026 \text{ kN} \quad \text{OK}$$

c. *Check axial resistance of effective column section* (Table 12.9 or AASHTO 6.10.8.2.4):
Effective column section area is shown in Figure 12.13b:

$$A_s = 2\big[210(20) + 9(16)(16)\big] = 13,008 \text{ mm}^2$$

$$I = \frac{(20)(420+16)^3}{12} = 138.14(10)^6 \text{ mm}^4$$

$$r_s = \sqrt{\frac{I}{A_s}} = \sqrt{\frac{138.14(10)^6}{13,008}} = 103.1 \text{ mm}$$

$$\lambda = \left(\frac{KL}{r_s\pi}\right)^2 \frac{F_y}{E} = \left(\frac{0.75(2440)}{103.1\pi}\right)^2 \frac{345}{200,000} = 0.055$$

$$P_n = 0.66^\lambda F_y A_s = 0.66^{0.055}(345)(13,008) = 4386\,(10)^3 \text{ N}$$

$$P_r = \phi_c P_n = 0.9(4386) = 3947 \text{ kN} > V_u = 2026 \text{ kN} \qquad \text{OK}$$

Therefore, using two 20 × 210 mm plates are adequate for bearing stiffeners at abutment.

12.5.3 Shear Connectors

To ensure a full composite action, shear connectors must be provided at the interface between the concrete slab and the structural steel to resist interface shear. Shear connectors are usually provided throughout the length of the bridge. If the longitudinal reinforcement in the deck slab is not considered in the composite section, shear connectors are not necessary in negative flexure regions. If the longitudinal reinforcement is included, either additional connectors can be placed in the region of dead load contraflexure points or they can be continued over the negative flexure region at maximum spacing. The two types of shear connectors such as shear studs and channels (see Figure 12.4) are most commonly used in modern bridges. The fatigue and strength limit states must be considered in the shear connector design. The detailed requirements are listed in Table 12.10.

Example 12.4: Shear Connector Design

Given
For the I-girder bridge shown in Example 12.1, design the shear stud connectors for the positive flexure region of Span 1. The shear force ranges V_{sr} are given in Table 12.11 and assume number of cycle $N = 7.844(10)^7$.

TABLE 12.10 AASHTO-LRFD Design Formulas of Shear Connectors

Connector Types	Stud	Channel
Basic requirement	$\dfrac{h_s}{d_s} \geq 4.0$	Fillet welds along the heels and toe shall not smaller than 5 mm
	$6d_s < $ pitch of connector $p = (nZ_r I)/V_{sr}Q < 600$ mm Transverse spacing $\geq 4d_s$ Clear distance between flange edge of nearest connector ≥ 25 mm Concrete cover over the top of the connectors ≥ 50 mm and $d_s \geq 50$ mm	
Special requirement	For noncomposite negative flexure region, additional number of connector: $n_{ac} = (A_r f_{sr})/Z_r$	
Fatigue resistance	$Z_r = \alpha d_s^2 \geq 19 d_s^2$ $\alpha = 238 - 29.5 \log N$	—
Nominal shear resistance	$Q_n = 0.5 A_{sc}\sqrt{f'_c E_c} \leq A_{sc} F_u$	$Q_n = 0.3\left(t_f + 0.5 t_w\right) L_c \sqrt{f'_c E_c}$
Required shear connectors	$n = \dfrac{V_h}{\phi_{sc} Q_n}$; $\quad V_h = $ smaller $\begin{cases} 0.85 f'_c b_{eff} t_s \\ \sum A_{si} F_{yi} \end{cases}$	
	For continuous span between each adjacent zero moment of the centerline of interior support: $V_h = A_r F_{vr}$	

A_{si} = area of component of steel section
b_{eff} = effective flange width
h_s = height of stud
d_s = diameter of stud
n = number of shear connectors in a cross section
E_c = modulus of elasticity of concrete
f_c = stress in compression flange due to the factored loading
f'_c = specified compression strength of concrete
F_{yi} = specified minimum yield strength of the component of steel section
f_{sr} = stress range in longitudinal reinforcement (AASHTO-LRFD 5.5.3.1)
F_u = specified minimum tensile strength of a stud
L_c = length of channel shear connector
Q = first moment of transformed section about the neutral axis of the short-term composite section
I = moment of inertia of short-term composite section
N = number of cycles (AASHTO-LRFD 6.6.1.2.5)
V_{sr} = shear force range at the fatigue limit state
t_s = thickness of concrete slab
t_f = flange thickness of channel shear connector
Z_r = shear fatigue resistance of an individual shear connector

Solutions

1. Stud Size (Table 12.10 AASHTO Article 6.10.7.4.1a)

Stud height should penetrate at least 50 mm into the deck. The clear cover depth of concrete cover over the top of the shear stud should not be less than 50 mm. Try

$$H_s = 180 \text{ mm} > 50 + (110 - 25) = 135 \text{ mm (min)} \qquad \text{OK}$$

$$\text{stud diameter } d_s = 25 \text{ mm} < H_s/4 = 45 \text{ mm} \qquad \text{OK}$$

TABLE 12.11   Shear Connector Design for the Positive Flexure Region in Span 1

Span	Location (x/L)	V_{sr} (kN)	$p_{required}$ (mm)	p_{final} (mm)	$n_{total\text{-}stud}$
	0.0	267.3	253	245	3
	0.1	229.5	295	272	63
	0.2	212.6	318	306	117
	0.3	205.6	329	326	165
1	0.4	203.3	333	326	**210**
	0.4	203.3	333	326	**144**
	0.5	202.3	334	326	99
	0.6	212.6	318	306	51
	0.7	223.7	302	306	3

Notes:

1. $V_{sr} = \left| +(V_{LL+IM})_u \right| + \left| -(V_{LL+IM})_u \right|$.

2. $P_{required} = \dfrac{n_s Z_r I_{com-n}}{V_{sr} Q} = \dfrac{67\,634}{V_{sr}}$.

3. $n_{total\text{-}stud}$ is summation of number of shear studs between the locations of the zero moment and that location.

2. **Pitch of Shear Stud, p, for Fatigue Limit State**

   a. *Fatigue resistance Z_r (Table 12.10 or AASHTO Article 6.10.7.4.2):*

   $$\alpha = 238 - 29.5 \, \log(7.844 \times 10^7) = 5.11$$

   $$Z_r = 19d_s^2 = 19(25)^2 = 11{,}875 \ \text{N}$$

   b. *First moment Q and moment of initial I (Table 12.6):*

   $$Q = \left(\frac{b_{eff} t_s}{8}\right)\left(y_{st-n} + t_h + \frac{t_s}{2}\right)$$

   $$= \left(\frac{3530(275)}{8}\right)\left(380 + 85 + \frac{275}{2}\right) = 73.11 \ (10^6) \ \text{mm}^3$$

   $$I_{com-n} = 138.8(10^9) \ \text{mm}^4$$

   c. *Required pitch for the fatigue limit state:*
   Assume that shear studs are spaced at 150 mm transversely across the top flange of steel section (Figure 12.9) and using $n_s = 3$ for this example and obtain

   $$P_{reqd} = \frac{n_s Z_r I}{V_{sr} Q} = \frac{3(11.875)(138.8)(10)^9}{V_{sr}(73.11)(10)^6} = \frac{67634}{V_{sr}}$$

   The detailed calculations for the positive flexure region of Span 1 are shown in Table 12.11.

3. **Strength Limit State Check**

   a. *Nominal horizontal shear force (AASHTO Article 6.10.7.4.4b):*

$$V_h = \text{the lesser of} \begin{cases} 0.85 \, f'_c \, b_{\text{eff}} t_s \\ F_{yw} D t_w + F_{yt} b_{ft} t_{ft} + F_{yc} b_{fc} t_{fc} \end{cases}$$

$$V_{h-concrete} = 0.5 \, f'_c \, b_{\text{eff}} t_s = 0.85(28)(3530)(275) = 2.31(10)^6 \text{ N}$$

$$V_{h-steel} = F_{yw} D t_w + F_{yt} b_{ft} t_{ft} + F_{yc} b_{fc} t_{fc}$$

$$= 345[(2440)(16) + (460)(45) + (460)(25)] = 2.458(10)^6 \text{ N}$$

$$\therefore \quad V_h = 23\,100 \text{ kN}$$

b. *Nominal shear resistance* (Table 12.10 or AASHTO Article 6.10.7.4.4c):
Use specified minimum tensile strength $F_u = 420$ MPa for stud shear connectors

$$\because \; 0.5\sqrt{f'_c E_c} = 0.5\sqrt{28(25,000)} = 418.3 \text{ MPa} < F_u = 420 \text{ MPa}$$

$$\therefore \; Q_n = 0.5 A_{sc}\sqrt{f'_c E_c} = 418.3\left(\frac{\pi (25)^2}{4}\right) = 205\,332 \text{ N} = 205 \text{ kN}$$

c. *Check resulting number of shear stud connectors* (see Table 12.11):

$$n_{\text{total-stud}} = \begin{cases} 210 & \text{from left end } 0.4\,L_1 \\ 144 & \text{from } 0.4L_1 \text{ to } 0.7L_1 \end{cases} > \frac{V_h}{\phi_{sc} Q_n} = \frac{23\,100}{0.85(205)} = 133 \qquad \text{OK}$$

12.6 Other Design Considerations

12.6.1 Fatigue Resistance

The basic fatigue design requirement limits live-load stress range to fatigue resistance for each connection detail. Special attention should be paid to two types of fatigue: (1) load-induced fatigue for a repetitive net tensile stress at a connection details caused by moving truck and (2) distortion-induced fatigue for connecting plate details of cross frame or diaphragms to girder webs. See Chapter 53 for a detailed discussion.

12.6.2 Diaphragms and Cross Frames

Diaphragms and cross frames, as shown in Figure 12.14, are transverse components to transfer lateral loads such as wind or earthquake loads from the bottom of girder to the deck and from the deck to bearings, to provide lateral stability of a girder bridge, and to distribute vertical loads to the longitudinal main girders. Cross frames usually consist of angles or WT sections and act as a truss, while diaphragms use channels or I-sections as a flexural beam connector. End cross frames or diaphragms at piers and abutments are provided to transmit lateral wind loads and/or earthquake load to the bearings, and intermediate one are designed to provide lateral support to girders.

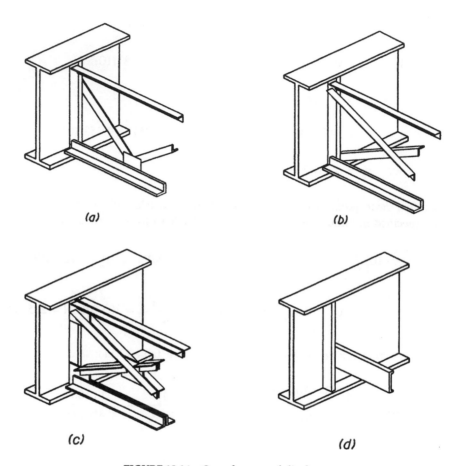

FIGURE 12.14 Cross frames and diaphragms.

The following general guidelines should be followed for diaphragms and cross frames:

- The diaphragm or cross frame shall be as deep as practicable to transfer lateral load and to provide lateral stability. For rolled beam, they shall be at least half of beam depth [AASHTO-LRFD 6.7.4.2].
- Member size is mainly designed to resist lateral wind loads and/or earthquake loads. A rational analysis is preferred to determine actual lateral forces.
- Spacing shall be compatible with the transverse stiffeners.
- Transverse connectors shall be as few as possible to avoid fatigue problems.
- Effective slenderness ratios (KL/r) for compression diagonal shall be less than 140 and for tension member (L/r) less than 240.

Example 12.5: Intermediate Cross-Frame Design

Given

For the I-girder bridge shown in Example 12.1, design the intermediate cross frame as for wind loads using single angles and M270 Grade 250 Steel.

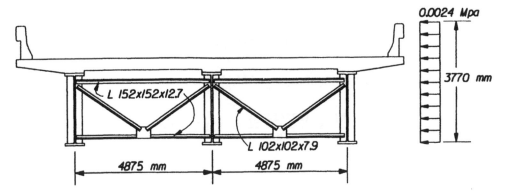

FIGURE 12.15   Wind load distribution.

Solutions:

1. **Calculate Wind Load**

   In this example, we assume that wind load acting on the upper half of girder, deck, and barrier is carried out by the deck slab and wind load on the lower half of girder is carried out by bottom flange. From AASHTO Table 3.8.1.2, wind pressure P_D= 0.0024 MPa, d= depth of structure member = 2,510 mm, and γ = load factor = 1.4 (AASHTO Table 3.4.1-1). The wind load on the structure (Figure 12. 15) is

   $$W = 0.0024(3770) = 9.1 \text{ kN/m} > 4.4 \text{ kN/m}$$

   Factored wind force acting on bottom flange:

   $$W_{bf} = \frac{\gamma P_D d}{2} = \frac{1.4 \,(0.0024)(2510)}{2} = 4.21 \text{ kN / m}$$

   Wind force acting on top flange (neglecting concrete deck diaphragm):

   $$W_{tf} = 1.4(0.0024)\left(3770 - \frac{2510}{2}\right) = 8.45 \text{ kN / m}$$

2. **Calculate forces acting on cross frame**

   For cross frame spacing:

   $$L_b = 6.1 \text{ m}$$

   Factored force acting on bottom strut:

   $$F_{bf} = W_{bf}L_b = 4.21(6.1) = 25.68 \text{ kN}$$

   Force acting on diagonals:

   $$F_d = \frac{F_{tf}}{\cos \phi} = \frac{8.45(6.1)}{\cos 45^o} = 72.89 \text{ kN}$$

3. Design bottom strut

Try $\angle$ 152 × 152 × 12.7; A_s = 3710 mm²; r_{min} = 30 mm; L = 4875 mm

Check member slenderness and section width/thickness ratios.

$$\frac{KL}{r} = \frac{0.75(4875)}{30} = 121.9 < 140 \qquad\qquad \text{OK}$$

$$\frac{b}{t} = \frac{152}{12.7} = 11.97 < 0.45\sqrt{\frac{E}{F_y}} = 0.45\sqrt{\frac{200,000}{250}} = 12.8 \qquad\qquad \text{OK}$$

Check axial load capacity.

$$\lambda = \left(\frac{0.75(4875)}{30.0\pi}\right)^2 \frac{250}{200,000} = 1.88 \ < \ 2.25 \quad \text{(ASSHTO 6.9.4.1-1)}$$

$$P_n = 0.66^\lambda A_s F_y = 0.66^{1.88}(3710)(250) = 424,675 \text{ N} = 425 \text{ kN}$$

$$P_r = \phi_c P_n = 0.9(425) = 382.5 \text{ kN} > F_{bf} = 25.68 \text{ kN} \qquad\qquad \text{OK}$$

4. Design diagonals

Try $\angle$ 102 × 102 × 7.9, A_s = 1550 mm²; r_{min} = 20.1 mm; L = 3450 mm

Check member slenderness and section width/thickness ratios.

$$\frac{KL}{r} = \frac{0.75(3450)}{20.1} = 128.7 < 140 \qquad\qquad \text{OK}$$

$$\frac{b}{t} = \frac{102}{7.9} = 12.9 \approx 0.45\sqrt{\frac{E}{F_y}} = 0.45\sqrt{\frac{200,000}{250}} = 12.8 \qquad\qquad \text{OK}$$

Check axial load capacity.

$$\lambda = \left(\frac{0.75(3450)}{20.1\pi}\right)^2 \frac{250}{200,000} = 2.1 \ < \ 2.25 \quad \text{(ASSHTO 6.9.4.1-1)}$$

$$P_n = 0.66^\lambda A_s F_y = 0.66^{2.1}(1550)(250) = 161,925 \text{ N} = 162 \text{ kN}$$

$$P_r = \phi_c P_n = 0.9(162) = 145.8 \text{ kN} > F_d = 72.89 \text{ kN} \qquad\qquad \text{OK}$$

5. Top strut

The wind force in the top strut is assumed zero because the diagonal will transfer the wind load directly into the deck slab. To provide lateral stability to the top flange during construction, we select angle $\angle$ 152 × 152 × 12.7 for top struts.

12.6.3 Lateral Bracing

The lateral bracing transfers wind loads to bearings and provides lateral stability to compression flange in a horizontal plan. All construction stages should be investigated for the need of lateral bracing. The lateral bracing should be placed as near the plane of the flange being braced as possible. Design of lateral bracing is similar to the cross frame.

12.6.4 Serviceability and Constructibility

The service limit state design is intended to control the permanent deflections, which would affect riding ability. AASHTO-LRFD [AASHTO-LRFD 6.10.3] requires that for Service II (see Chapter 5) load combination, flange stresses in positive and negative bending should meet the following requirements:

$$f_r = \begin{cases} 0.95\,R_h F_{yf} & \text{for both steel flanges of composite section} \\ 0.80\,R_h F_{yf} & \text{for both flanges of noncomposite section} \end{cases} \tag{12.9}$$

where R_h is a hybrid factor, 1.0 for homogeneous sections (AASHTO-LRFD 6.10.5.4), f_f is elastic flange stress caused by the factored loading, and F_{yf} is yield strength of the flange.

An I-girder bridge constructed in unshored conditions shall be investigated for strength and stability for all construction stages, using the appropriate strength load combination discussed in Chapter 5. All calculations should be based on the noncomposite steel section only.

Splice locations should be determined in compliance with both contructibility and structural integrity. The splices for main members should be designed at the strength limit state for not less than (AASHTO 10.13.1) the larger of the following:

- The average of flexural shear due to the factored loads at the splice point and the corresponding resistance of the member;
- 75% of factored resistance of the member.

References

1. Xanthakos, P. P., *Theory and Design of Bridges*, John Wiley & Sons, New York, 1994.
2. Barker, R. M. and Puckett, J. A., *Design of Highway Bridges*, John Wiley & Sons, New York, 1997.
3. Taly, N., *Design of Modern Highway Bridges*, WCB/McGraw-Hill, Burr Ridge, IL, 1997.
4. AASHTO, *AASHTO LRFD Bridge Design Specifications*, American Association of State Highway and Transportation Officials, Washington, D.C., 1994.
5. AASHTO, *AASHTO LRFD Bridge Design Specifications*, 1996 Interim Revisions, American Association of State Highway and Transportation Officials, Washington, D.C., 1996.
6. AISC., *Load and Resistance Factor Design Specification for Structural Steel Buildings*, 2nd ed., American Institute of Steel Construction, Chicago, IL, 1993.
7. AISC, *Manual of Steel Construction — Load and Resistance Factor Design*, 2nd ed., American Institute of Steel Construction, Chicago, IL, 1994.
8. Blodgett, O. W., *Design of Welded Structures*, James F. Lincoln Arc Welding Foundation, Cleveland, OH, 1966.

9. Galambos, T. V., Ed., *Guide to Stability Design Criteria for Metal Structures*, 5th ed., John Wiley & Sons, New York, 1998.

10. Basler, K., Strength of plates girder in shear, *J. Struct. Div.*, *ASCE*, 87(ST7), 1961, 151.

11. Basler, K., Strength of plates girder under combined bending and shear, *J. Struct. Div.*, *ASCE*, 87(ST7), 1961, 181.

13

Steel–Concrete Composite Box Girder Bridges

Yusuf Saleh
California Department of Transportation

Lian Duan
California Department of Transportation

13.1 Introduction

Box girders are used extensively in the construction of urban highway, horizontally curved, and long-span bridges. Box girders have higher flexural capacity and torsional rigidity, and the closed shape reduces the exposed surface, making them less susceptible to corrosion. Box girders also provide smooth, aesthetically pleasing structures.

There are two types of steel box girders: steel–concrete composite box girders (i.e., steel box composite with concrete deck) and steel box girders with orthotropic decks. Composite box girders are generally used in moderate- to medium-span (30 to 60 m) bridges, and steel box girders with orthotropic decks are often used for longer-span bridges.

This chapter will focus on straight steel–concrete composite box-girder bridges. Steel box girders with orthotropic deck and horizontally curved bridges are presented in Chapters 14 and 15.

13.2 Typical Sections

Composite box-girder bridges usually have single or multiple boxes as shown in Figure 13.1. A single cell box girder (Figure 13.1a) is easy to analyze and relies on torsional stiffness to carry eccentric loads. The required flexural stiffness is independent of the torsional stiffness. A single box girder with multiple cells (Figure 13.1b) is economical for very long spans. Multiple webs reduce the flange

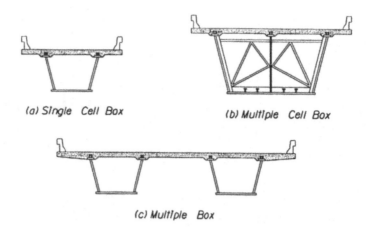

FIGURE 13.1   Typical cross sections of composite box girder.

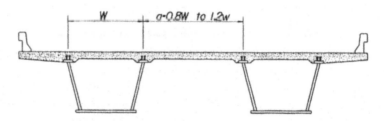

FIGURE 13.2   Flange distance limitation.

shear lag and also share the shear forces. The bottom flange creates more equal deformations and better load distribution between adjacent girders. The boxes in multiple box girders are relatively small and close together, making the flexural and torsional stiffness usually very high. The torsional stiffness of the individual boxes is generally less important than its relative flexural stiffness. For design of a multiple box section (Figure 13.1c), the limitations shown in Figure 13.2. should be satisfied when using the AASHTO-LRFD Specifications [1,2] since the AASHTO formulas were developed from these limitations. The use of fewer and bigger boxes in a given cross section results in greater efficiency in both design and construction [3].

A composite box section usually consists of two webs, a bottom flange, two top flanges and shear connectors welded to the top flange at the interface between concrete deck and the steel section (Figure 13.3). The top flange is commonly assumed to be adequately braced by the hardened concrete deck for the strength limit state, and is checked against local buckling before concrete deck hardening. The flange should be wide enough to provide adequate bearing for the concrete deck and to allow sufficient space for welding of shear connectors to the flange. The bottom flange is designed to resist bending. Since the bottom flange is usually wide, longitudinal stiffeners are often required in the negative bending regions. Web plates are designed primarily to carry shear forces and may be placed perpendicular or inclined to the bottom flange. The inclination of web plates should not exceed 1 to 4. The preliminary determination of top and bottom flange areas can be obtained from the equations (Table 13.1) developed by Heins and Hua [4] and Heins [6].

13.3   General Design Principles

A box-girder highway bridge should be designed to satisfy AASHTO-LRFD specifications to achieve the objectives of constructibility, safety, and serviceability. This section presents briefly basic design principles and guidelines. For more-detailed information, readers are encouraged to refer to several texts [6–14] on the topic.

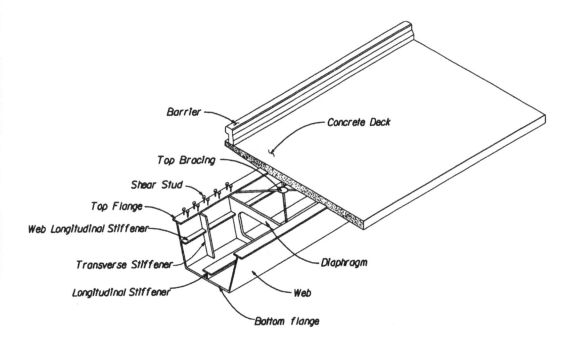

FIGURE 13.3 Typical components of a composite box girder.

In multiple box-girder design, primary consideration should be given to flexure. In single box-girder design, however, both torsion and flexure must be considered. Significant torsion on single box girders may occur during construction and under live loads. Warping stresses due to distortion should be considered for fatigue but may be ignored at the strength limit state. Torsional effects may be neglected when the rigid internal bracings and diaphragms are provided to maintain the box cross section geometry.

13.4 Flexural Resistance

The flexural resistance of a composite box girders depends on the compactness of the cross sectional elements. This is related to compression flange slenderness, lateral bracing, and web slenderness. A "compact" section can reach full plastic flexural capacity. A "noncompact" section can only reach yield at the outer fiber of one flange.

In positive flexure regions, a multiple box section is designed to be compact and a single box section is considered noncompact with the effects of torsion shear stress taken by the bottom flange (Table 13.2). In general, in box girders non-negative flexure regions design formulas of nominal flexure resistance are shown in Table 13.3.

In lieu of a more-refined analysis considering the shear lag phenomena [15] or the nonuniform distribution of bending stresses across wide flanges of a beam section, the concept of effective flange width under a uniform bending stress has been widely used for flanged section design [AASHTO-LRFD 4.6.2.6]. The effective flange width is a function of slab thickness and the effective span length.

13.5 Shear Resistance

For unstiffened webs, the nominal shear resistance V_n is based on shear yield or shear buckling depending on web slenderness. For stiffened interior web panels of homogeneous sections, the postbuckling resistance due to tension-field action [16,17] is considered. For hybrid sections, tension-field action is

TABLE 13.1　Preliminary Selection of Flange Areas of Box-Girder Element

Items	Top Flange		Bottom Flange	
	A_T^+	A_T^-	A_B^+	A_B^-
Single span	$254d\left(1-\dfrac{26}{L}\right)$	—	$328d\left(1-\dfrac{28}{L}\right)$	—
Two span	$0.64A_B^+$	$1.60A_B^+\dfrac{F_y^-}{F_y^+}$	$\dfrac{645}{k}(1.65L^2-0.74L+13)$	$1.17A_B^+\dfrac{F_y^-}{F_y^+}$
Three span	$\dfrac{330n}{k}(L_1-22)$ $0.95A_T^- -$ $\dfrac{A_T^{-2}}{58650}-\dfrac{3484}{k}$	$\dfrac{814n}{k}(L_1-31)$	$\dfrac{423n}{k}(L_1-16)$ $\dfrac{211.67n}{k}(L_2-14.63)$	$\dfrac{645}{kn}(3.16L_2-0.018L_2^2-70)$

A_T^+, A_T^- = the area of top flange (mm²) in positive and negative region, respectively

A_B^+, A_B^- = the area of bottom flange (mm²) in positive and negative region, respectively

d　　　　= depth of girder (mm)

L,L_1, L_2 = length of the span (m);　　for simple span　($27 \le L \le 61$)

　　　　　　　　　　　　　　　for two spans　($30 \le L_2 \le 67$)

　　　　　　　　　　　　　　　for three spans　($27 \le L_1 \le 55$)

W_R　　　= roadway width (m)

N_b　　　= number of boxes

n　　　　= L_2/L_1

k　　　　= $\dfrac{N_B F_y d}{W_R(344,750)}$

F_y　　　= yield strength of the material (MPa)

not permitted and shear yield or elastic shear buckling limits the strength. The detailed AASHTO-LRFD design formulas are shown in Table 12.8 (Chapter 12). For cases of inclined webs, the web depth D shall be measured along the slope and be designed for the projected shear along inclined web.

　　To ensure composite action, shear connectors should be provided at the interface between the concrete slab and the steel section. For single-span bridges, connectors should be provided throughout the span of the bridge. Although it is not necessary to provide shear connectors in negative flexure regions if the longitudinal reinforcement is not considered in a composite section, it is recommended that additional connectors be placed in the region of dead-load contraflexure points [AASHTO-LRFD 1.10.7.4]. The detailed requirements are listed in Table 12.10.

13.6　Stiffeners, Bracings, and Diaphragms

13.6.1　Stiffeners

Stiffeners consist of longitudinal, transverse, and bearing stiffeners as shown in Figure 13.1. They are used to prevent local buckling of plate elements, and to distribute and transfer concentrated loads. Detailed design formulas are listed in Table 12.9.

TABLE 13.2 AASHTO-LRFD Design Formulas of Nominal Flexural Resistance in Negative Flexure Ranges for Composite Box Girders (Strength Limit State)

Compression flange with longitudinal stiffeners	

$$F_n = \begin{cases} R_b R_h F_{yc} & \text{for } \dfrac{w}{t} \le 0.57\sqrt{\dfrac{kE}{F_{yc}}} \\[2ex] 0.592 R_b R_h F_{yc}\left(1+0.687\sin\dfrac{c\pi}{2}\right) & \text{for } 0.57\sqrt{\dfrac{kE}{F_{yc}}} < \dfrac{w}{t} \le 1.23\sqrt{\dfrac{kE}{F_{yc}}} \\[2ex] 181\,000\,R_b R_h k\left(\dfrac{t}{w}\right)^2 & \text{for } \dfrac{w}{t} > 1.23\sqrt{\dfrac{kE}{F_{yc}}} \end{cases}$$

$$c = \frac{1.23 - \dfrac{w}{t}\sqrt{\dfrac{F_{yc}}{kE}}}{0.66}$$

$$k = \text{buckling coefficent} = \begin{cases} \dfrac{8I_s}{wt^3} \le 4.0 & \text{for } n = 1 \\[2ex] \dfrac{14.3I_s}{wt^3 n^4} \le 4.0 & \text{For } n = 2,3,4 \text{ or } 5 \end{cases}$$

Compression flange without longitudinal stiffeners	Use above equations with the substitution of compression flange width between webs, b for w and buckling coefficient k taken as 4

Tension flange	$F_n = R_b R_h F_{yt}$

E = modulus of elasticity of steel
F_n = nominal stress at the flange
F_{yc} = specified minimum yield strength of the compression flange
F_{yt} = specified minimum yield strength of the tension flange
n = number of equally spaced longitudinal compression flange stiffeners
I_s = moment of inertia of a longitudinal stiffener about an axis parallel to the bottom flange and taken at the base of the stiffener
R_b = load shedding factor, R_b = 1.0 — if either a longitudinal stiffener is provided or $2D_c/t_w \le \lambda_b\sqrt{E/f_c}$ is satisfied
R_h = hybrid factor; for homogeneous section, R_h = 1.0, see AASHTO-LRFD (6.10.5.4)
t_h = thickness of concrete haunch above the steel top flange
t = thickness of compression flange
w = larger of width of compression flange between longitudinal stiffeners or the distance from a web to the nearest longitudinal stiffener

13.6.2 Top Lateral Bracings

Steel composite box girders (Figure 13.3) are usually built of three steel sides and a composite concrete deck. Before the hardening of the concrete deck, the top flanges may be subject to lateral torsion buckling. Top lateral bracing shall be designed to resist shear flow and flexure forces in the section prior to curing of concrete deck. The need for top lateral bracing shall be investigated to ensure that deformation of the box is adequately controlled during fabrication, erection, and placement of the concrete deck. The cross-bracing shown in Figure 13.3 is desirable. For 45° bracing, a minimum cross-sectional area (mm²) of bracing of 0.76× (box width, in mm) is required to ensure closed box action [11]. The slenderness ratio (L_b/r) of bracing members should be less than 140.

AASHTO-LRFD [1] requires that for straight box girders with spans less than about 45 m, at least one panel of horizontal bracing should be provided on each side of a lifting point; for spans greater than 45 m, a full-length lateral bracing system may be required.

13.6.3 Internal Diaphragms and Cross Frames

Internal diaphragms or cross frames (Figure 13.1) are usually provided at the end of a span and interior supports within the spans. Internal diaphragms not only provide warping restraint to the box girder, but improve distribution of live loads, depending on their axial stiffness which prevents distortion. Because rigid and widely spaced diaphragms may introduce undesirable large local forces, it is generally good practice to provide a large number of diaphragms with less stiffness than a few very rigid diaphragms. A recent study [18] showed that using only two intermediate diaphragms per span results in 18% redistribution of live-load stresses and additional diaphragms do not significantly improve the live-load redistribution. Inverted K-bracing provides better inspection access than X-bracing. Diaphragms shall be designed to resist wind loads, to brace compression flanges, and to distribute vertical dead and live loads [AASHTO-LRFD 6.7.4].

For straight box girders, the required cross-sectional area of a lateral bracing diagonal member A_b (mm²) should be less than 0.76× (width of bottom flange, in mm) and the slenderness ratio (L_b/r) of the member should be less than 140.

For horizontally curved boxes per lane and radial piers under HS-20 loading, Eq. (13.1) provides diaphragm spacing L_d, which limits normal distortional stresses to about 10% of the bending stress [19]:

$$L_d = \sqrt{\frac{R}{200L-7500}} \le 25 \tag{13.1}$$

where R is bridge radius, ft, and L is simple span length, ft.

To provide the relative distortional resistance per millimeter greater than 40 [13], the required area of cross bracing is as

$$A_b = 750\left[\frac{L_{ds}a}{h}\right]\left[\frac{t^3}{h+a}\right] \tag{13.2}$$

where t is the larger of flange and web thickness; L_{ds} is the diaphragm spacing; h is the box height, and a is the top width of box.

13.7 Other Considerations

13.7.1 Fatigue and Fracture

For steel structures under repeated live loads, fatigue and fracture limit states should be satisfied in accordance with AASHTO 6.6.1. A comprehensive discussion on the issue is presented in Chapter 53.

13.7.2 Torsion

Figure 13.4 shows a single box girder under the combined forces of bending and torsion. For a closed or an open box girder with top lateral bracing, torsional warping stresses are negligible. Research indicates that the parameter ψ determined by Eq. (13.3) provides limits for consideration of different types of torsional stresses.

$$\psi = L\sqrt{GJ/EC_w} \tag{13.3}$$

where G is shear modulus, J is torsional constant, and C_w is warping constant.

For straight box girder (ψ is less than 0.4), pure torsion may be omitted and warping stresses must be considered; when ψ is greater than 10, it is warping stresses that may be omitted and pure

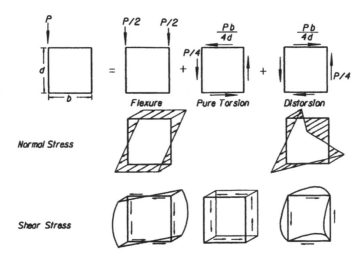

FIGURE 13.4 A box section under eccentric loads.

torsion that must be considered. For a curved box girder, ψ must take the following values if torsional warping is to be neglected:

$$\psi \geq \begin{cases} 10 + 40\theta & \text{for } 0 \leq \theta \leq 0.5 \\ 30 & \text{for } \theta > 0.5 \end{cases} \tag{13.4}$$

where θ is subtended angle (radius) between radial piers.

13.7.3 Constructibility

Box-girder bridges should be checked for strength and stability during various construction stages. It is important to note that the top flange of open-box sections shall be considered braced at locations where internal cross frames or top lateral bracing are attached. Member splices may be needed during construction. At the strength limit state, the splices in main members should be designed for not less than the larger of the following:

- The average of the flexure moment, the shear, or axial force due to the factored loading and corresponding factored resistance of member, and
- 75% of the various factored resistance of the member.

13.7.4 Serviceability

To prevent permanent deflections due to traffic loads, AASHTO-LRFD requires that at positive regions of flange flexure stresses (f_f) at the service limit state shall not exceed $0.95R_h F_{yf}$.

13.8 Design Example

Two-Span Continuous Box-Girder bridge

Given
A two-span continuous composite box-girder bridge that has two equal spans of 45 m. The superstructure is 13.2 m wide. The elevation and a typical cross section are shown in Figure 13.5.

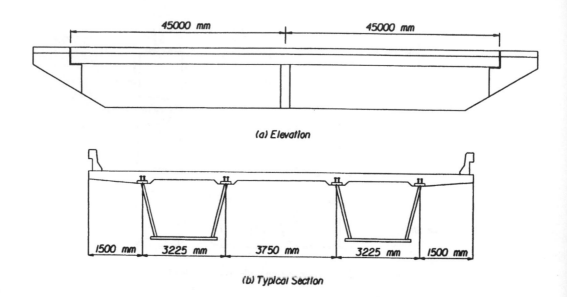

FIGURE 13.5 Two-span continuous box-girder bridge.

Structural steel: AASHTO M270M, Grade 345W (ASTM A709 Grade 345W) uncoated weathering steel with F_y = 345 MPa

Concrete: f'_c = 30.0 MPa; E_c = 22,400 MPa; modular ratio $n = 8$

Loads: Dead load = self weight + barrier rail + future wearing 75 mm AC overlay
Live load = AASHTO Design Vehicular Load + dynamic load allowance
Single-lane average daily truck traffic ADTT in one direction = 3600

Deck: Concrete slabs deck with thickness of 200 mm

Specification: AASHTO-LRFD [1] and 1996 Interim Revision (referred to as AASHTO)

Requirements: Design a box girder for flexure, shear for Strength Limit State I, and check fatigue requirement for web.

Solution

1. **Calculate Loads**

   a. *Component dead load — DC for a box girder:*

   The component dead-load DC includes all structural dead loads with the exception of the future wearing surface and specified utility loads. For design purposes, assume that all dead load is distributed equally to each girder by the tributary area. The tributary width for the box girder is 6.60 m.

   - *DC1*: acting on noncomposite section
     Concrete slab = (6.6)(0.2)(2400)(9.81)   = 31.1 kN/m
     Haunch   = 3.5 kN/m
     Girder (steel-box), cross frame, diaphragm, and stiffener   = 9.8 kN/m
   - *DC2*: acting on the long term composite section
     Weight of each barrier rail   = 5.7 kN/m

   b. *Wearing surface load — DW:*

   A future wearing surface of 75 mm is assumed to be distributed equally to each girder
   - *DW*: acting on the long-term composite section = 10.6 kN/m

2. **Calculate Live-Load Distribution Factors**

   a. *Live-load distribution factors for strength limit state* [AASHTO Table 4.6.2.2.2b-1]:

$$LD_m = 0.05 + 0.85\frac{N_L}{N_b} + \frac{0.425}{N_L} = 0.05 + 0.85\frac{3}{2} + \frac{0.425}{3} = 1.5 \text{ lanes}$$

b. *Live-load distribution factors for fatigue limit state:*

$$LD_m = 0.05 + 0.85\frac{N_L}{N_b} + \frac{0.425}{N_L} = 0.05 + 0.85\frac{1}{2} + \frac{0.425}{1} = 0.9 \text{ lanes}$$

3. Calculate Unfactored Moments and Shear Demands

The unfactored moment and shear demand envelopes are shown in Figures 13.8 to 13.11. Moment, shear demands for the Strength Limit State I and Fatigue Limit State are listed in Table 13.3 to 13.5.

TABLE 13.3 Moment Envelopes for Strength Limit State I

Span	Location (x/L)	M_{DC1} (kN-m); Dead Load-1	M_{DC2} (kN-m); Dead Load-2	M_{DW} (kN-m); Wearing Surface	M_{LL+IM} (kN-m) Positive	M_{LL+IM} (kN-m) Negative	M_u (kN-m) Positive	M_u (kN-m) Negative
	0.0	0	0	0	0	0	0	0
	0.1	3,058	372	681	3338	−442	10,592	4,307
	0.2	5,174	629	1152	5708	−883	18,023	7,064
	0.3	6,350	772	1414	7174	−1326	22,400	8,268
	0.4	6,585	801	1466	7822	−1770	23,864	7,917
1	0.5	5,880	715	1309	7685	−2212	22,473	6,018
	0.6	4,234	515	943	6849	−2653	18,369	2,571
	0.7	1,647	200	367	5308	−3120	11,540	−2,472
	0.8	−1,882	−229	−419	3170	−3822	2,168	−9,457
	0.9	−6,350	−772	−1414	565	−4928	−9,533	−18,745
	1.0	−11,760	−1430	−2618	−1727	−7640	−22,264	−32,095

Notes:
1. Live load distribution factor $LD = 1.467$.
2. Dynamic load allowance $IM = 33\%$.
3. $M_u = 0.95 \left[1.25(M_{DC1} + M_{DC2}) + 1.5 M_{DW} + 1.75 M_{LL+IM}\right]$.

TABLE 13.4 Shear Envelopes for Strength Limit State I

Span	Location (x/L)	V_{DC1} (kN); Dead Load-1	V_{DC2} (kN); Dead Load-2	V_{DW} (kN); Wearing Surface	V_{LL+IM} (kN) Positive	V_{LL+IM} (kN) Negative	V_u (kN) Positive	V_u (kN) Negative
	0.0	784	95	87	877	−38	2626	1104
	0.1	575	70	64	782	−44	2158	784
	0.2	366	44	41	711	−58	1727	449
	0.3	157	19	18	601	−91	1233	83
	0.4	−53	6	−6	482	−138	724	−307
1	0.5	−262	−32	−29	360	−230	208	−773
	0.6	−471	−57	−52	292	−354	−216	−1290
	0.7	−680	−83	−76	219	−482	−648	−1815
	0.8	−889	−108	−99	145	−612	−1083	−2342
	0.9	−1098	−133	−122	67	−750	−1524	−2882
	1.0	−1307	−159	−145	22	−966	−1910	−3553

Notes:
1. Live load distribution factor $LD = 1.467$.
2. Dynamic load allowance $IM = 33\%$.
3. $V_u = 0.95 \left[1.25(V_{DC1} + V_{DC2}) + 1.5 V_{DW} + 1.75 V_{LL+IM}\right]$.

TABLE 13.5   Moment and Shear Envelopes for Fatigue Limit State

Span	Location (x/L)	M_{LL+IM} (kN-m) Positive	M_{LL+IM} (kN-m) Negative	V_{LL+IM} (kN) Positive	V_{LL+IM} (kN) Negative	$(M_{LL+IM})_u$ (kN-m) Positive	$(M_{LL+IM})_u$ (kN-m) Negative	$(V_{LL+IM})_u$ (kN) Positive	$(V_{LL+IM})_u$ (kN) Negative
	0.0	0	0	286	−31	0	0	214	−23
	0.1	1102	−137	245	−31	827	−102	184	−23
	0.2	1846	−274	205	−57	1385	−206	154	−38
	0.3	2312	−412	167	−79	1734	−309	125	−59
	0.4	2467	−550	130	−115	1851	−412	98	−86
1	0.5	2405	−687	97	−153	1804	−515	73	−115
	0.6	2182	−824	67	−190	1636	−618	50	−143
	0.7	1716	−962	45	−226	1287	−721	33	−169
	0.8	1062	−1099	25	−257	1796	−824	19	−193
	0.9	414	−1237	9	−286	311	−928	7	−215
	1.0	0	−1373	0	−309	0	−1030	0	−232

Notes:
1. Live load distribution factor $LD = 0.900$.
2. Dynamic load allowance $IM = 15\%$.
3. $(M_{LL+IM})_u = 0.75(M_{LL+IM})_u$ and $(V_{LL+IM})_u = 0.75(V_{LL+IM})_u$.

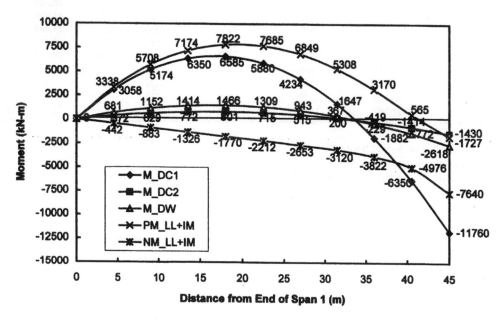

FIGURE 13.6   Unfactored moment envelopes.

4. Determine Load Factors for Strength Limit State I and Fracture Limit State

Load factors and load combinations

   The load factors and combinations are specified as [AASHTO Table 3.4.1-1]:
   Strength Limit State I: $1.25(DC1 + DC2) + 1.5(DW) + 1.75(LL + IM)$
   Fatigue Limit State: $0.75(LL + IM)$
   a. *General design equation* [AASHTO Article 1.3.2]:

$$\eta \sum \gamma_i \, Q_i \leq \phi \, R_n$$

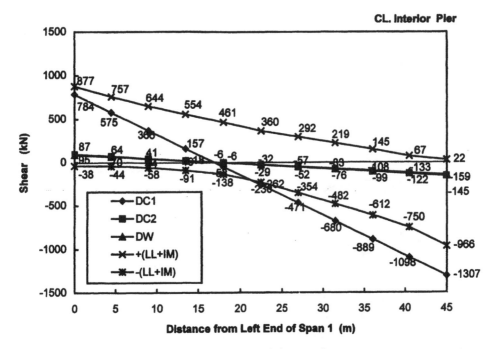

FIGURE 13.7 Unfactored shear envelopes.

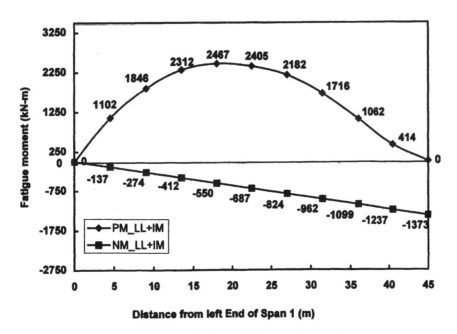

FIGURE 13.8 Unfactored fatigue load moment.

where γ_i is load factor and ϕ resistance factor; Q_i represents force effects or demands; R_n is the nominal resistance; η is a factor related ductility η_D, redundancy η_R, and operational importance η_I of the bridge (see Chapter 5) designed and is defined as:

$$\eta = \eta_D \eta_R \eta_I \geq 0.95$$

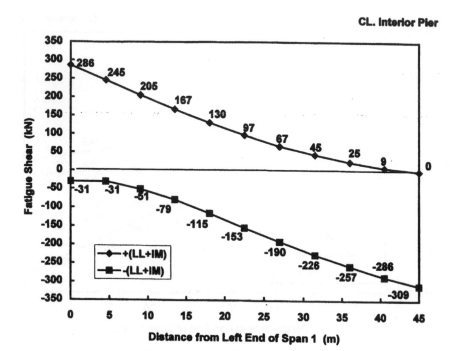

FIGURE 13.9   Unfactored fatigue load shear.

For this example, the following values are assumed:

Limit States	Ductility η_D	Redundancy η_R	Importance η_I	η
Strength limit state	0.95	0.95	1.05	0.95
Fatigue limit state	1.0	1.0	1.0	1.0

5. Calculate Composite Section Properties:

Effective flange width for positive flexure region [AASHTO Article 4.6.2.6]

a. *For an interior web, the effective flange width:*

$$b_{eff} = \text{the lesser of} \begin{cases} \dfrac{L_{eff}}{4} = \dfrac{33750}{4} = 8440 \text{ mm} \\[2mm] 12t_s + \dfrac{b_f}{2} = (12)(200) + \dfrac{450}{2} = 2625 \text{ mm} \quad \text{(controls)} \\[2mm] S = 3750 \text{ mm} \end{cases}$$

b. *For an exterior web, the effective flange width:*

$$b_{eff} = \text{the lesser of} \begin{cases} \dfrac{L_{eff}}{8} = \dfrac{33750}{8} = 4220 \text{ mm} \\[2mm] 6t_s + \dfrac{b_f}{4} = (6)(200) + \dfrac{450}{4} = 1310 \text{ mm} \quad (\text{controls}) \\[2mm] \textit{The width of the overhang} = 1500 \text{ mm} \end{cases}$$

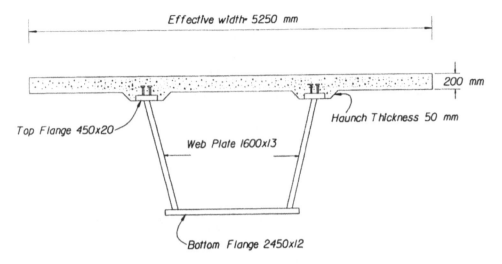

FIGURE 13.10　Typical section for positive flexure region.

Total effective flange width for the box girder $= 1310 + \dfrac{2625}{2} + 2625 = 5250$ mm

where L_{eff} is the effective span length and may be taken as the actual span length for simply supported spans and the distance between points of permanent load inflection for continuous spans; b_f is top flange width of steel girder.

Elastic composite section properties for positive flexure region:
For a typical section (Figure 13.10) in positive flexure region of Span 1, its elastic section properties for the noncomposite, the short-term composite ($n = 8$), and the long-term composite ($3n = 24$) are calculated in Tables 13.6 to 13.8.

TABLE 13.6　Noncomposite Section Properties for Positive Flexure Region

Component	A (mm²)	y_i (mm)	$A_i y_i$ (mm³)	$y_i - y_{sb}$ (mm)	$A_i(y_i - y_{sb})^2$ (mm⁴)	I_o (mm⁴)
2 top flange 450 × 20	18,000	1574.2	28.34 (10⁶)	885	141 (10⁹)	0.60 (10⁶)
2 web 1600 × 13	41,600	788.1	32.79 (10⁶)	99	0.41 (10⁹)	8.35 (10⁹)
Bottom flange 2450 × 12	29,400	6.0	0.17 (10⁶)	−683	13.70 (10⁹)	0.35 (10⁶)
Σ	89,000	—	61.30 (10⁶)	—	28.23 (10⁹)	8.35 (10⁹)

$$y_{sb} = \frac{\sum A_i y_i}{\sum A_i} = \frac{61.30(10^6)}{89,000} = 688.7 \text{ mm} \qquad y_{st} = (12 + 1552.5 + 20) - 688.7 = 895.5 \text{ mm}$$

$$I_{girder} = \sum I_o + \sum A_i(y_i - y_{sb})^2$$

$$= 8.35(10^9) + 28.23(10^9) = 36.58(10^9) \text{ mm}^4$$

$$S_{sb} = \frac{I_{girder}}{y_{sb}} = \frac{36.58(10^9)}{688.7} = 53.11(10^6) \text{ mm}^3 \qquad S_{st} = \frac{I_{girder}}{y_{st}} = \frac{36.58(10^9)}{895.5} = 40.85(10^6) \text{ mm}^3$$

Effective flange width for negative flexure region:
The effective width is computed according to AASHTO 4.6.2.6 (calculations are similar to Step 5a) The total effective of flange width for the negative flexure region is 5450 mm.

TABLE 13.7 Short-Term Composite Section Properties ($n = 8$)

Component	A (mm²)	y_i (mm)	A_iy_i (mm³)	$y_i - y_{sb}$ (mm)	$A_i(y_i - y_{sb})^2$ (mm⁴)	I_o (mm⁴)
Steel section	89,000	688.7	61.30 (10^6)	−611	33.24 (10^9)	36.58 (10^9)
Concrete Slab 5250/8 × 200	131,250	1714.2	225.0 (10^6)	414	22.54 (10^9)	0.43 (10^9)
Σ	220,250	—	386.3 (10^6)	—	55.77 (10^9)	37.02 (10^9)

$$y_{sb} = \frac{\sum A_i y_i}{\sum A_i} = \frac{92.79(10^6)}{220\,250} = 1299.8 \text{ mm} \qquad y_{st} = (12+1552.5+20) - 1299.8 = 284.4 \text{ mm}$$

$$I_{com} = \sum I_o + \sum A_i(y_i - y_{sb})^2$$

$$= 37.02(10^9) + 55.77(10^9) = 92.79(10^9) \text{ mm}^4$$

$$S_{sb} = \frac{I_{com}}{y_{sb}} = \frac{92.79(10^9)}{1299.8} = 71.39(10^6) \text{ mm}^3 \qquad S_{st} = \frac{I_{com}}{y_{st}} = \frac{92.79(10^9)}{284.4} = 326.30(10^6) \text{ mm}^3$$

TABLE 13.8 Long-Term Composite Section Properties ($3n = 24$)

Component	A (mm²)	y_i (mm)	A_iy_i (mm³)	$y_i - y_{sb}$ (mm)	$A_i(y_i - y_{sb})^2$ (mm⁴)	I_o (mm⁴)
Steel section	89,000	688.4	61.3 (10^6)	−338	10.2 (10^9)	36.58 (10^9)
Concrete slab 5250/24 × 200	43,750	1714.2	75.0 (10^6)	688	20.7 (10^9)	5.40 (10^6)
Σ	132,750	—	136.0 (10^6)	—	30.85 (10^9)	36.59 (10^9)

$$y_{sb} = \frac{\sum A_i y_i}{\sum A_i} = \frac{136.0(10^6)}{132\,750} = 1026.7 \text{ mm} \qquad y_{st} = (12+1552.5+20) - 1026.7 = 557.5 \text{ mm}$$

$$I_{com} = \sum I_o + \sum A_i(y_i - y_{sb})^2$$

$$= 36.59(10^9) + 136.0(10^9) = 67.43(10^9) \text{ mm}^4$$

$$S_{sb} = \frac{I_{com}}{y_{sb}} = \frac{67.43(10^9)}{1026.7} = 65.68(10^6) \text{ mm}^3 \qquad S_{st} = \frac{I_{com}}{y_{st}} = \frac{67.43(10^9)}{557.5} = 121.0(10^6) \text{ mm}^3$$

Elastic composite section properties for negative flexure region:
AASHTO (6.10.1.2) requires that for any continuous span the total cross-sectional area of longitudinal reinforcement must not be less than 1% of the total cross-sectional area of the slab. The required reinforcement must be placed in two layers uniformly distributed across the slab width and two thirds must be placed in the top layer. The spacing of the individual bar should not exceed 150 mm in each row.

$$A_{s\,reg} = 0.01(200) = 2.00 \text{ mm}^2 / \text{mm}$$

$$A_{s\,top-layer} = \frac{2}{3}(0.01)(200) = 1.33 \text{ mm}^2 / \text{mm} \quad (\#16 \text{ at } 125 \text{ mm} = 1.59 \text{ mm}^2 / \text{mm})$$

$$A_{s\,bot-layer} = \frac{1}{3}(0.01)(200) = 0.67 \text{ mm}^2 / \text{mm} \quad (\text{alternate } \#10 \text{ and } \#13 \text{ at } 125 \text{ mm} = 0.80 \text{ mm}^2 / \text{mm})$$

Figure 13.11 shows a typical section for the negative flexure region. The elastic properties for the noncomposite and the long-term composite ($3n = 24$) are calculated and shown in Tables 13.9 and 13.10.

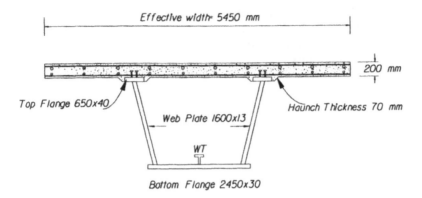

FIGURE 13.11 Typical section for negative flexure region.

TABLE 13.9 Noncomposite Section Properties for Negative Flexure Region

Component	A (mm^2)	y_i (mm)	A_iy_i (mm^3)	$y_i - y_{sb}$ (mm)	$A_i(y_i - y_{sb})^2$ (mm^4)	I_o (mm^4)
2 Top flange 650 × 40	52,000	1602	83.32 (10^6)	911	43.20 (10^9)	6.93 (10^6)
2 Web 1600 × 13	41,600	806	33.53 (10^6)	115	0.55 (10^9)	8.35 (10^9)
Stiffener WT	5,400	224.3	1.21 (10^6)	−466	1.18 (10^9)	37.63 (10^6)
Bottom flange 2450 × 30	73,500	15	1.10 (10^6)	−676	33.57 (10^9)	5.51 (10^6)
Σ	172,500	—	119.2 (10^6)		78.49 (10^9)	8.40 (10^9)

$$y_{sb} = \frac{\sum A_i y_i}{\sum A_i} = \frac{119.2(10^6)}{172500} = 690.8 \text{ mm} \qquad y_{st} = (30+1552.5+40) - 690.8 = 931.4 \text{ mm}$$

$$I_{girder} = \sum I_o + \sum A_i(y_i - y_{sb})^2$$

$$= 8.40(10^9) + 78.50(10^9) = 86.90(10^9) \text{ mm}^4$$

$$S_{sb} = \frac{I_{girder}}{y_{sb}} = \frac{86.90(10^9)}{690.8} = 125.8(10^6) \text{ mm}^3 \qquad S_{st} = \frac{I_{girder}}{y_{st}} = \frac{86.90(10^9)}{931.4} = 93.29(10^6) \text{ mm}^3$$

TABLE 13.10 Composite Section Properties for Negative Flexure Region

Component	A (mm^2)	y_i (mm)	A_iy_i (mm^3)	$y_i - y_{sb}$ (mm)	$A_i(y_i - y_{sb})^2$ (mm^4)	I_o (mm^4)
Steel section	172 500	690.8	119.2 (10^6)	73.2	0.92 (10^9)	86.90 (10^9)
Top reinforcement	8,665	1762.2	15.27 (10^6)	998.2	8.63 (10^9)	—
Bottom reinforcement	4,360	1677.2	7.31 (10^9)	913.2	3.64 (10^9)	—
Σ	185 525	—	141.7 (10^6)	—	13.19 (10^9)	86.90 (10^9)

$$y_{sb} = \frac{\sum A_i y_i}{\sum A_i} = \frac{141.7(10^6)}{185\,525} = 764 \text{ mm} \qquad y_{st} = (30+1552.5+40) - 764 = 858.2 \text{ mm}$$

$$I_{com} = \sum I_o + \sum A_i(y_i - y_{sb})^2$$

$$= 86.90(10^9) + 13.19(10^9) = 100.09(10^9) \text{ mm}^4$$

$$S_{sb} = \frac{I_{com}}{y_{sb}} = \frac{100.09(10^9)}{764} = 131.00(10^6) \text{ mm}^3 \qquad S_{st} = \frac{I_{com}}{y_{st}} = \frac{100.09(10^9)}{858.2} = 116.63(10^6) \text{ mm}^3$$

6. **Calculate Yield Moment M_y and Plastic Moment Capacity M_p**
   a. *Yield moment M_y [AASHTO Article 6.10.5.1.2]:*
      The yield moment M_y corresponds to the first yielding of either steel flange. It is obtained by the following formula

$$M_y = M_{D1} + M_{D2} + M_{AD}$$

where M_{D1}, M_{D2}, and M_{AD} are moments due to the factored loads applied to the steel, the long-term, and the short-term composite section, respectively. M_{AD} can be obtained by solving the equation:

$$F_y = \frac{M_{D1}}{S_s} + \frac{M_{D2}}{S_{3n}} + \frac{M_{AD}}{S_n}$$

$$M_{AD} = S_n \left(F_y - \frac{M_{D1}}{S_s} - \frac{M_{D2}}{S_{3n}} \right)$$

where S_s, S_n and S_{3n} are the section modulus for the noncomposite steel, the short-term, and the long-term composite sections, respectively.

$$M_{D1} = (0.95)(1.25)(M_{DC1}) = (0.95)(1.25)(6585) = 7820 \text{ kN-m}$$

$$M_{D2} = (0.95)(1.25 M_{DC2} + 1.5 M_{DW})$$

$$= (0.95)[1.25(801) + 1.5(1466)] = 3040 \text{ kN-m}$$

For the top flange:

$$M_{AD} = (329.3)10^{-3} \left((345)10^3 - \frac{7.820}{40.85(10)^{-3}} - \frac{3.040}{120(10)^{-3}} \right)$$

$$= 41.912(10)^3 \text{ kN-m}$$

For the bottom flange:

$$M_{AD} = (71.39)10^{-3} \left((345)10^3 - \frac{7.820}{53.11(10)^{-3}} - \frac{3.040}{64.68(10)^{-3}} \right)$$

$$= 10.814(10)^3 \text{ kN-m} \qquad \text{(control)}$$

$$M_y = 7820 + 3040 + 10814 = 21,674 \text{ kN-m}$$

   b. *Plastic moment M_p [AASHTO Article 6.1]:*
      The plastic moment M_p is determined using equilibrium equations. The reinforcement in the concrete slab is neglected in this example.
      • Determine the location of the plastic neutral axis (PNA), $\bar{Y}$
        From the Equation listed in Table 12.4 and Figure 12.7.

$$P_s = 0.85 f_c' b_{\text{eff}} t_s = 0.85(30)(5250)(200) = 26{,}775 \text{ kN}$$

$$P_c = A_{fc} F_{yc} = 2(450)(20)(345) = 6{,}210 \text{ kN}$$

$$P_w = A_w F_{yw} = 2\,(1600)(13)(345) = 14{,}352 \text{ kN}$$

$$P_t = A_{ft} F_{yt} = 2450(12)(345) = 10{,}143 \text{ kN}$$

$$\because P_t + P_w + P_c = 10{,}143 + 14{,}352 + 6{,}210 = 30\ 705 \text{ kN} > P_s = 26{,}755 \text{ kN}$$

$\therefore$ PNA is located within the top flange of steel girder and the distance from the top of compression flange to the PNA, $\overline{Y}$ is

$$\overline{Y} = \frac{t_{fc}}{2}\left(\frac{P_w + P_t - P_s}{P_c} + 1\right)$$

$$\overline{Y} = \frac{20}{2}\left(\frac{14{,}352 + 10{,}143 - 26{,}775}{6{,}210} + 1\right) = 6.3 \text{ mm}$$

- *Calculate M_p:*
  Summing all forces about the PNA, obtain:

$$M_p = \sum M_{\text{PNA}} = \frac{P_c}{2t_c}\left(\overline{Y}^2 + (t_c - \overline{Y})^2\right) + P_s d_s + P_w d_w + P_t d_t$$

where

$$d_s = \frac{200}{2} + 50 - 20 + 6.3 = 136.3 \text{ mm}$$

$$d_w = \frac{1552.5}{2} + 20 - 6.3 = 789.8 \text{ mm}$$

$$d_t = \frac{12}{2} + 1552.5 + 20 - 6.3 = 1571.9 \text{ mm}$$

$$M_p = \frac{6210}{2(20)}\left(6.3^2 + (20 - 6.3)^2\right) + (26{,}775)(136.3) + (14{,}352)(789.8) + (10{,}143)(1571.9)$$

$$M_p = 30{,}964 \text{ kN-m}$$

7. **Flexural Strength Design — Strength Limit State I:**
   a. *Positive flexure region:*
      - *Compactness of steel box girder*
        The compactness of a multiple steel boxes is controlled only by web slenderness. The purpose of the ductility requirement is to prevent permanent crushing of the concrete slab when the composite section approaches its plastic moment capacity. For this example, by referring to Figures 13.2 and 13.4, obtain:

$\dfrac{2D_{cp}}{t_w} \le 3.76 \sqrt{\dfrac{E}{f_c}}$,   PNA is within the top flange $D_{cp} = 0$, the web slenderness require-

ment is satisfied

$D_p = 200 + 50 - 20 + 6.3 = 236.3$ mm

(depth from the top of concrete deck to the PNA)

$D' = \beta \left[\dfrac{d + t_s + t_h}{7.5}\right]$ $\beta = 0.7$ for $F_y = 345$ MPa

$D' = 0.7 \left(\dfrac{1552.5 + 12 + 200 + 50}{7.5}\right) = 169.3$ mm

$\left(\dfrac{D_p}{D'}\right) = \left(\dfrac{236.3}{169.3}\right) = 1.4 \le 5$ OK

- Calculate nominal flexure resistance, M_n (see Table 12.2)

$$1 < \left(\dfrac{D_p}{D'}\right) = 1.4 < 5$$

$$M_n = \dfrac{5M_p - 0.85M_y}{4} + \dfrac{0.85M_y - M_p}{4}\left(\dfrac{D_p}{D'}\right)$$

$$M_n = \dfrac{5(30,964) - 0.85(21,674)}{4} + \dfrac{0.85(21,674) - (30,964)}{4}(1.4)$$

$M_n = 28,960$ kN-m $\ge 1.3(1.0)(21674) = 28,176$ kN-m

$\therefore$ $M_n = 28,176$ kN-m

From Table 13.3, the maximum factored positive moments in Span 1 occurred at the location of $0.4L_1$.

$$\eta \Sigma \gamma_i M_i \le \phi_f M_n$$

$$23,864 \text{ kN-m} < 1.0 \ (28,176) \text{ kN-m} \qquad\qquad \text{OK}$$

b. *Negative flexure region:*
   For multiple and single box sections, the nominal flexure resistance should be designed to meet provision AASHTO 6.11.2.1.3a (see Table 13.2)
   i. Stiffener requirement [AASHTO 6.11.2.1-1]:
      Use one longitudinal stiffener (Figure 13.11), try WT 10.5 × 28.5.
      The projecting width, b_ℓ of the stiffener should satisfy:

$$b_\ell \le 0.48 t_p \sqrt{\dfrac{E}{F_{yc}}}$$

where

$t_p =$ the thickness of stiffener (mm)

$b_\ell =$ the projected width (mm)

$$b_\ell = \frac{267}{2} = 133.5 \, \text{mm}$$

$$I_s = 33.4 \times 10^6 + 4748(190.1)^2 = 20.5 \times 10^6 \, \text{mm}$$

$$133.5 \le 0.48(16.5)\sqrt{\frac{2 \times 10^5}{345}} = 190 \, \text{mm} \qquad\qquad \text{OK}$$

ii. Calculate buckling coefficient, k:
   For $n = 1$

$$k = \left(\frac{8I_s}{Wt^3}\right)^{1/3} = \left(\frac{8(20.5 \times 10^6)}{1225(24)^3}\right)^{1/3} = 2.13 < 4.0$$

iii. Calculate nominal flange stress (see Table 13.2):

$$0.57\sqrt{\frac{Ek}{F_{yc}}} = 0.57\sqrt{\frac{(2.13)(2)10^5}{345}} = 20.03$$

$$1.23\sqrt{\frac{Ek}{F_{yc}}} = 1.23\sqrt{\frac{(2.13)(2)10^5}{345}} = 43.22$$

$$20.03 < \frac{w}{t} = \frac{1225}{30} = 40.83 < 43.22$$

The nominal flexural resistance of compression flange is controlled by inelastic buckling:

$$F_{nc} = 0.592 \, R_b R_h F_{yc}\left(1 + 0.687\sin\frac{c\pi}{2}\right)$$

$$c = \frac{1.23 - \dfrac{w}{t}\sqrt{\dfrac{F_{yc}}{kE}}}{0.66} = \frac{1.23 - 40.8\sqrt{\dfrac{345}{(3.9)2(10^5)}}}{0.66} = 0.56$$

Longitudinal stiffener is provided, $R_b = 1.0$, for homogenous plate girder $R_h = 1.0$:

$$F_{nc} = 0.592.(1.0)(1.0)(345)\left(1 + 0.687\sin\frac{(0.56)\pi}{2}\right) = 313.4 \, \text{MPa}$$

For tension flange:

$$F_{nt} = R_b R_h R_{yt} = (1.0)(1.0)(345) = 345 \text{ MPa}$$

iv. Calculate M_{AD} at Interior Support

$$M_{D1} = (0.95)(1.25)(M_{DC1}) = (0.95)(1.25)(11760) = 13{,}965 \text{ kN-m}$$

$$M_{D2} = (0.95)(1.25 M_{DC2} + 1.5 M_{DW})$$

$$= (0.95)[1.25(1430) + 1.5(2618)] = 5428 \text{ kN-m}$$

$$M_{AD} = S_n \left(F_n - \frac{M_{D1}}{S_s} - \frac{M_{D2}}{S_n} \right)$$

$$M_{AD-\text{comp}} = (0.131)\left(312.4 \times 10^3 - \frac{13\,965}{0.1258} - \frac{5428}{0.131} \right)$$

$$= 20\,954 \text{ kN-m}$$

$$M_{AD-\text{tension}} = (0.1166)\left(345 \times 10^3 - \frac{13\,965}{9.33(10)^{-2}} - \frac{5428}{0.1166} \right)$$

$$= 17\,346 \text{ kN-m} \quad (\text{control})$$

- Calculate nominal flexure resistance, M_n:

$$M_n = 13{,}965 + 5{,}428 + 17{,}346 = 36{,}739 \text{ kN-m}$$

From Table 13.3, maximum factored negative moments occurred at the interior support

$$\eta \Sigma \gamma_i M_i \leq \phi_f M_n$$

$$\underline{32{,}095 \text{ kN-m} < 1.0 \,(36{,}739) \text{ kN-m}} \qquad\qquad\qquad \text{OK}$$

8. Shear Strength Design — Strength Limit State I
a. *End bearing of Span 1*
   - Nominal shear resistance V_n:
   For inclined webs, each web shall be designed for shear, V_{ui} due to factored loads taken as [AASHTO Article 6.11.2.2.1]

$$\therefore V_{ui} = \frac{V_u}{\cos\theta} = \frac{2626}{2\cos(14)} = 1353 \text{ kN} \quad (\text{per web})$$

where θ is the angle of the web to the vertical.

$$\because \quad \frac{D}{t_w} = \frac{1600}{13} = 123.1 > \ 3.07\sqrt{\frac{E}{F_{yw}}} = 3.07\sqrt{\frac{(2.0)10^5}{345}} = 73.9$$

$$\therefore \ V_n = \frac{4.55 t_w^3 E}{D} = \frac{4.55(13)^3(2.0)10^5}{1600} = 1249.5 \ \text{kN}$$

$$\because \ V_{ui} = 1353 \ \text{kN} \ > \ \phi_v V_n = (1.0)(1249.5) \ \text{kN}$$

$$\therefore \ \text{Stiffeners are required}$$

- V_n for end-stiffened web panel [AASHTO 6.10.7.3.3c]

$$V_n = CV_p$$

$$k = 5 + \frac{5}{(d_o / D)^2}$$

in which d_o is the spacing of transverse stiffeners

$$\text{For } d_o = 2400 \ \text{mm} \quad \text{and} \quad k = 5 + \frac{5}{(2400/1600)^2} = 7.22$$

$$\because \ \frac{D}{t_w} = 123.1 > 1.38\sqrt{\frac{Ek}{F_{yw}}} = 1.38\sqrt{\frac{200,000(7.22)}{345}} = 89.3$$

$$\because \ C = \frac{152}{(123.1)^2}\sqrt{\frac{200,000(7.22)}{345}} = 0.65$$

$$V_p = 0.58 F_{yw} D t_w = 0.58(345)(1600)(13) = 4162 \ \text{kN}$$

$$V_n = CV_p = 0.65(4162) = 2705 \ \text{kN} \ > V_{ui} = 1353 \ \text{kN} \qquad \text{OK}$$

b. *Interior support:*
- The maximum shear forces due to factored loads is shown in Table 13.4

$$V_u = \frac{3553}{2} = 1776.5 \ \text{kN (per web)} \ > \ \phi_v V_n = (1.0)(1249.5) \ \text{kN}$$

$\therefore$ Stiffeners are required for the web at the interior support.

c. *Intermediate transverse stiffener design*
The intermediate transverse stiffener consists of a plate welded to one of the web. The design of the first intermediate transverse stiffener is discussed in the following.
- Projecting Width b_t Requirements [AASHTO Article 6.10.8.1.2]
  To prevent local bucking of the transverse stiffeners, the width of each projecting stiffener shall satisfy these requirements:

$$\begin{Bmatrix} 50 + \dfrac{d}{30} \\ 0.25b_f \end{Bmatrix} \le b_t \le \begin{Bmatrix} 0.48t_p\sqrt{\dfrac{E}{F_{ys}}} \\ 16t_p \end{Bmatrix}$$

where b_f is full width of steel flange and F_{ys} is specified minimum yield strength of stiffener. Try stiffener width, $b_t = 180.0$ mm.

$$b_t = 180 > \begin{cases} 50 + \dfrac{d}{30} = 50 + \dfrac{1600}{30} = 103.3 \text{ mm} \\ 0.25b_f = 0.25(450) = 112.5 \text{ mm} \end{cases} \qquad \text{OK}$$

Try $t_p = 16$ mm

$$b_t = 180 < \begin{cases} 0.48t_p\sqrt{\dfrac{E}{F_{ys}}} = 0.48(16)\sqrt{\dfrac{200\,000}{345}} = 185 \text{ mm} \\ 16t_p = 16(14) = 224 \text{ mm} \end{cases} \qquad \text{OK}$$

Use 180 mm × 16 mm transverse stiffener plates.

- Moment of inertia requirement [AASHTO Article 6.10.8.1.3]
  The purpose of this requirement is to ensure sufficient rigidity of transverse stiffeners to develop tension field in the web adequately.

$$I_t \ge d_o t_w^2 J$$

$$J = 2.5\left(\dfrac{D_p}{d_o}\right)^2 - 2.0 \ge 0.5$$

where I_t is the moment of inertia for the transverse stiffener taken about the edge in contact with the web for single stiffeners and about the midthickness of the web for stiffener pairs and D_p is the web depth for webs without longitudinal stiffeners.

$$\because J = 2.5\left(\dfrac{1600}{2400}\right)^2 - 2.0 = -0.89 < 0.5 \qquad \because \underline{\text{Use } J = 0.5}$$

$$I_t = \dfrac{(180)^3(16)}{3} = 31.1(10)^6 \text{ mm}^4 > \quad d_o t_w^2 J = (2400)(16)^3(0.5) = 4.9(10)^6 \text{ mm}^4 \quad \text{OK}$$

- Area Requirement [AASHTO Article 6.10.8.1.4]:
  This requirement ensures that transverse stiffeners have sufficient area to resist the vertical component of the tension field, and is only applied to transverse stiffeners required to carry the forces imposed by tension-field action.

$$A_s \ge A_{s\,\text{min}} = \left(0.15BDt_w(1-C)\dfrac{V_u}{\phi_v V_n} - 18t_w^2\right)\left(\dfrac{F_{yw}}{F_{ys}}\right)$$

where $B = 1.0$ for stiffener pairs. From the previous calculation:

$C = 0.65$ $F_{yw} = 345$ MPa $F_{ys} = 345$ MPa

$V_u = 1313$ kN (per web) $\phi_f V_n = 1249.5$ kN $t_w = 13$ mm

$A_s = (180)(16) = 2880$ in.^{2}

$$> A_{s\,min} = \left(0.15(2.4)(1600)(13)(1-0.65)\frac{1313}{1249.5} - 18(13)^2\right)\left(\frac{345}{345}\right)$$

$$= -288\,\text{mm}^2$$

The negative value of $A_{s\,min}$ indicates that the web has sufficient area to resist the vertical component of the tension field.

9. Fatigue Design — Fatigue and Fracture Limit State

a. *Fatigue requirements for web in positive flexure region* [AASHTO Article 6.10.4]:

The purpose of these requirements is to control out-of-plane flexing of the web due to flexure and shear under repeated live loadings. The repeated live load is taken as twice the factored fatigue load.

$$D_c = \frac{f_{DC1} + f_{DC2} + f_{DW} + f_{LL+IM}}{\dfrac{f_{DC1}}{y_{st}} + \dfrac{f_{DC2} + f_{DW}}{y_{st-3n}} + \dfrac{f_{LL+IM}}{y_{st-n}}} - t_{fc}$$

$$= \frac{\dfrac{M_{DC1}}{S_{st}} + \dfrac{M_{DC2} + M_{DW}}{S_{st-3n}} + \dfrac{2(M_{LL+IM})_u}{S_{st-n}}}{\dfrac{M_{DC1}}{I_{girder}} + \dfrac{M_{DC2} + M_{DW}}{I_{com-3n}} + \dfrac{2(M_{LL+IM})_u}{I_{com-n}}} - t_{fc}$$

$$D_c = \frac{\dfrac{6585}{40.9(10)^6} + \dfrac{(801+1466)}{121(10)^6} + \dfrac{2(2467)}{326.3(10)^6}}{\dfrac{6585}{36.58(10)^9} + \dfrac{(801+1466)}{67.43(10)^9} + \dfrac{2(2467)}{92.79(10)^9}} - 20$$

$$D_c = 710\text{ mm}$$

$$\frac{2D_c}{t_w} = \frac{2(710)}{13(\cos 14)} = 113 < 5.76\sqrt{\frac{E}{F_{yc}}} = 5.76\sqrt{\frac{2(10)^5}{345}} = 137.2$$

$$\therefore \quad f_{cf} = F_{yw}$$

f_{cf}= maximum compression flexure stress in the flange due to unfactored permanent loads and twice the fatigue loading

$$f_{cf} = \frac{M_{DC1}}{S_{st}} + \frac{M_{DC2} + M_{DW}}{S_{st-3n}} + \frac{2(M_{LL+IM})_4}{S_{st-n}} = 161 + 18.7 + 15.3$$

$$= 195\text{ Mpa} < F_{yw} = 435\text{ MPa} \qquad\qquad \textbf{OK}$$

References

1. AASHTO, *AASHTO LRFD Bridge Design Specifications*, American Association of State Highway and Transportation Officials, Washington, D.C., 1994.
2. AASHTO, *AASHTO LRFD Bridge Design Specifications*, 1996 Interim Revisions, American Association of State Highway and Transportation Officials, Washington, D.C., 1996.
3. Price, K. D., Big Steel Boxes, in *National Symposium on Steel Bridge Construction*, Atlanta, 1993, 15-3.
4. Heins, C. P. and Hua, L. J., Proportioning of box girder bridges girder, *J. Struct. Div. ASCE*, 106(ST11), 2345, 1980.
5. Subcommittee on Box Girders of the ASCE-AASHTO Task Committee on Flexural Members, Progress report on steel box girder bridges, *J. Struct. Div. ASCE*, 97(ST4), 1971.
6. Heins, C. P., Box girder bridge design — state of the art, *AISC Eng. J.*, 15(4), 126, 1978.
7. Heins, C. P., Steel box girder bridges — design guides and methods, *AISC Eng. J.*, 20(3), 121, 1983.
8. Wolchuck, R., Proposed specifications for steel box girder bridges, *J. Struct. Div. ASCE*, 117(ST12), 2463, 1980.
9. Wolchuck, R., Design rules for steel box girder bridges, in *Proc. Int. Assoc, Bridge Struct. Eng.*, Zurich, 1981, 41.
10. Wolchuck, R., 1982. Proposed specifications for steel box girder bridges, Discussion, *J. Struct. Div. ASCE*, 108(ST8), 1933, 1982.
11. Seim, C. and Thoman, S., Proposed specifications for steel box girder bridges, discussion, *J. Struct. Div. ASCE*, 118(ST12), 2457, 1981.
12. AISC, *Highway Structures Design Handbook*, Vol. II, AISC Marketing, Inc., 1986.
13. Heins, C. P. and Hall, D. H., *Designer's Guide to Steel Box Girder Bridges*, Bethlehem Steel Corporation, Bethlehem, PA, 1981.
14. Wolchuck, R., Steel-plate deck bridges, in *Structural Engineering Handbook*, 3rd ed., Gaylord, E. H., Jr. and Gaylord, C. N., Eds., McGraw-Hill, New York, 1990, Sect. 113.
15. Kuzmanovic, B. and Graham, H. J., Shear lag in box girder, *J. Struct. Div. ASCE*, 107(ST9), 1701, 1981.
16. Balser, Strength of plate girders under combined bending and shear, *J. Struct. Div. ASCE*, 87(ST7), 181, 1971.
17. Balser, Strength of plate girders in shear, *J. Struct Div. ASCE*, 87(ST7), 151, 1971.
18. Foinquinos, R., Kuzmanovic, B., and Vargas, L. M., Influence of diaphragms on live load distribution in straight multiple steel box girder bridges, in *Building to Last, Proceedings of Structural Congress XV*, Kempner, L. and Brown, C. B., Eds., American Society of Civil Engineers, 1997, Vol. I., 89–93.
19. Olenik, J. C. and Heins, C. P., Diaphragms for curved box beam bridges, *J. Struct. Div.*, ASCE, 101(ST10), 1975.

14

Orthotropic Deck Bridges

Alfred R. Mangus
*California Department
of Transportation*

Shawn Sun
*California Department
of Transportation*

14.1 Introduction

This chapter will discuss the major design issues of orthotropic steel-deck systems. Emphasis will be given to the design of the closed-rib system, which is practically the only system selected for orthotropic steel deck by the engineers around the world. Examples of short spans to some of the world's long-span bridges utilizing trapezoidal ribs will be presented. The subject of fabrication detailing and fatigue resistant details necessary to prepare a set of contract bridge plans for construction is beyond the scope of this chapter. However, the basic issues of fatigue and detailing are presented. For more detailed discussion, the best references are four comprehensive books on orthotropic steel deck systems by Wolchuk [1], Troitsky [2], and the British Institution of Civil Engineers [3,4].

14.2 Conceptual Decisions

14.2.1 Typical Sections

Modern orthotropic welded steel-deck bridge rib systems were developed by German engineers in the 1950s [1,2]. They created the word *orthotropic* which is from *orthogonal* for *ortho* and *anisotropic* for *tropic*. Therefore, an orthotropic deck has anisotropic structural properties at 90°. Structural steel is used by most engineers although other metals such as aluminum can be used, as well as advanced composite (fiberglass) materials.

The open (torsionally soft) and closed (torsionally stiff) rib-framing system for orthotropic deck bridges developed by the Germans is shown in Figure 14.1. The open-rib and closed-rib systems are the two basic types of ribs that are parallel to the main span of the bridge. These ribs are also used to stiffen other plate components of the bridge. Flat plates, angles, split Ts, or half beams are types of open ribs that are always welded to the deck plate at only one location. A bent or rolled piece of steel plate is welded to the deck plate to form a closed space. The common steel angle can either be used as an open or closed rib depending on how it is welded to the steel deck plate. If the angle is welded at only one leg, then it is an open rib. However, if the angle is rotated to 45° and both legs are welded to the deck plate forming a triangular space or rib, it is a closed rib. Engineers have experimented with a variety of concepts to shape, roll, or bend a flat plate of steel into the optimum closed rib. The trapezoidal rib has been found to be the most practicable by engineers and the worldwide steel industry. Recently, the Japanese built the record span suspension bridge plus the record span cable-stayed bridge with trapezoidal rib construction (see Chapter 65).

The ribs are normally connected by welding to transverse floor beams, which can be a steel hot-rolled shape, small plate girder, box girder, or full-depth diaphragm plate. In Figure 14.1 small welded plate girders are used as the transverse floor beams. The deck plate is welded to the web(s) of the transverse floor beam. When full-depth diaphragms are used, access openings are needed for bridge maintenance purposes. The holes also reduce dead weight and provide a passageway for mechanical or electrical utilities. Since the deck plate is welded to every component, the deck plate is the top flange for the ribs, the transverse floor beam, and the longitudinal plate girders or box girders. All these various choices for the ribs, floor beam, and main girders can be interchanged, resulting in a great variety of orthotropic deck bridge superstructures.

14.2.2 Open Ribs vs. Closed Ribs

A closed rib is torsionally stiff and is essentially a miniature box girder [6]. The closed-rib deck is more effective for lateral distribution of the individual wheel load than the open-rib system. An open rib has essentially no torsional capacity. The open-rib types were initially very popular in the precomputer period because of simpler analysis and details. Once the engineer, fabricator, and contractor became familiar with the flat plate rib system shown in Figure 14.1 and Table 14.1, the switch to closed ribs occurred to reduce the dead weight of the superstructure, plus 50% less rib surface area to protect from corrosion. Engineers discovered these advantages as more orthotropic decks were built. The shortage and expense of steel after the World War II forced the adoption of closed ribs in Europe. The structural detailing of bolted splices for closed ribs requires handholds located in the bottom flange of the trapezoidal ribs to allow workers access to install the nut to the bolt. For a more-detailed discussion on handhold geometry and case histories for solutions to field-bolted splicing, refer to the four comprehensive books [1-4].

Compression stress occurs over support piers when the rib is used as a longitudinal interior stiffener for the bottom flanges of continuous box girders and can be graphically explained [6]. Ribs are usually placed only on the inside face of the box to achieve superior aesthetics and to minimize exterior corrosion surface area that must also be painted or protected. Compression also occurs when the rib is used as a longitudinal interior stiffener for columns, tower struts, and other components. The trapezoidal rib system quite often is field-welded completely around the super-structure cross section to achieve full structural continuity, rather than field bolted.

Table 14.2 [5] shows the greater bending efficiency in load-carrying capacity and stiffness achieved by the trapezoidal (closed) rib. It is readily apparent that a series of miniature box girders placed side by side is much more efficient that a series of miniature T-girders placed side by side. In the tension zones, the shape of the rib can be open or closed depending on the designers' preferences.

A trapezoidal rib can be quickly bent from a piece of steel as shown in Figure 14.2. A brake press is used to bend the shape in a jig in a few minutes. Rollers can also be used to form these trapezoidal ribs.

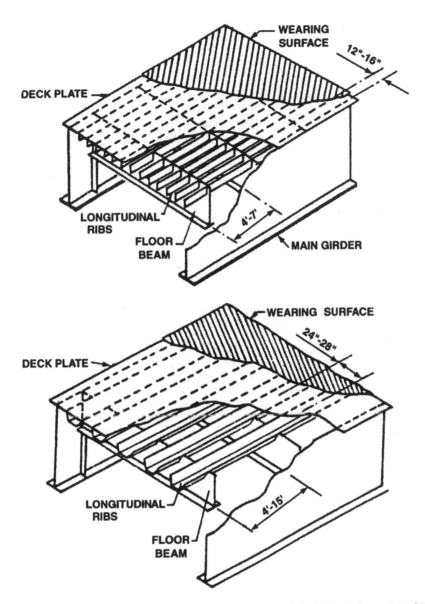

FIGURE 14.1 Typical components of orthotropic deck bridges. (From Troitsky, M. S., *Orthotropic Steel Deck Bridges*, 2nd ed., JFL Arch Welding Foundation, Cleveland, OH, 1987. Courtesy of The James F. Lincoln Arc Welding Foundation.)

A trapezoidal rib can be quickly bent from a piece of steel as shown in Figure 14.2. A brake press is used to bend the shape in a jig in a few minutes. Rollers can also be used to form these trapezoidal ribs. One American steel company developed Table 14.3 to encourage the utilization of orthotropic deck construction. This design aid was developed using main-frame computers in 1970, but due to lack of interest in orthotropic deck by bridge engineers this design aid eventually went out of print; nor was it updated to reflect changes in the AASHTO Bridge Code [5]. Tables 14.4 and 14.5 are excerpts from this booklet intended to assist an engineer quickly to design an orthotropic deck system and comply with minimum deck plate thickness; maximum rib span; and rib-spacing requirements of AASHTO [5]. AASHTO standardization of ribs has yet to occur, but many bridges built in the United States using ribs from Table 14.3 are identified throughout this chapter. The German and Japanese steel companies have developed standard ribs (see Table 14.3)

TABLE 14.1 Limiting Slenderness for Various Types of Ribs

Flanges and web Stiffeners

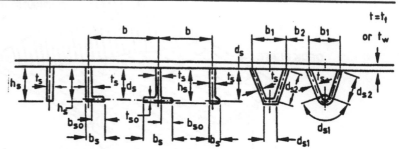

d, h = stiffener depth
b_s = width of angle
t_o, t', t_s = stiffener thickness
t = plate thickness
w, b = spacing of stiffeners
l_s = span of stiffener between supporting members
r_y = radius of gyration of stiffener (without plate) about axis normal to plate
F_y = yield stress of plate, N/mm²
F_{ys} = yield stress of stiffener, N/mm²
F_{max} = maximum factored compression stress, N/mm²

Draft U.S. rules

Effective slenderness coefficient C_s shall meet requirement

$$C_s = \begin{cases} \dfrac{d}{15t_o} + \dfrac{w}{12t} & \text{for flats} \\[3mm] \dfrac{d}{1.35t_o + 0.56r_y} + \dfrac{w}{12t} & \text{for Ts or angles} \end{cases} \leq \begin{cases} \dfrac{0.4}{\sqrt{F_y/E}} & \text{for } f_{max} \geq 0.5\,F_y \\[3mm] \dfrac{0.65}{\sqrt{F_y/E}} & \text{for } f_{max} \leq 0.5\,F_y \end{cases}$$

For any outstand of a stiffener $\dfrac{b'}{t'} \leq \dfrac{0.48}{\sqrt{F_{ys}/E}}$

British Standard 5400

For flats: $\dfrac{h_s}{t_s}\sqrt{\dfrac{F_{ys}}{355}} \leq 10$

For angles: $b_s \leq h_s$; $\dfrac{b_s}{t_s}\sqrt{\dfrac{F_{ys}}{355}} \leq 11$; $\dfrac{h_s}{t_s}\sqrt{\dfrac{F_{ys}}{355}} \leq 7$

Source: Galambos, T. V., Ed., *Guide to Stability Design Criteria for Metal Structures,* 4th ed., John Wiley & Sons, New York, 1988. With permission.

TABLE 14.2 AASHTO Effective Width of Deck Plate Acting with Rib

Calculation of		
Rib section properties for calculation of deck rigidity and flexural effects due to dead loads	$a_o = a$	$a_o + e_o = a + e$
Rib section properties for calculation of flexural effects due to wheel loads	$a_o = 1.1a$	$a_o + e_o = 1.3(a + e)$

Source: American Association of State Highway and Transportation Officials, *LRFD Bridge Design Specifications,* Washington, D.C., 1994. With permission.

FIGURE 14.2 Press brake forming rib stiffener sections. (Photo by Lawrence Lowe and courtesy of Universal Structural, Inc.)

14.2.3 Economics

Orthotropic deck bridges become an economical alternative when the following issues are important: lower mass, ductility, thinner or shallower sections, rapid bridge installation, and cold-weather construction.

Lower superstructure mass is the primary reason for the use of orthotropic decks in long-span bridges. Table 14.6 shows the mass achieved by abandoning the existing reinforced concrete deck and switching to a replacement orthotropic deck system relationship. The mass was reduced from 18 to 25% for long-span bridges, such as suspension bridges. This is extremely important since dead load causes 60 to 70% of the stresses in the cables and towers [7,8]. The mass is also important for bridge responses during an earthquake. The greater the mass, the greater the seismic forces. The Golden Gate Bridge, San Francisco, California, was retrofitted from a reinforced concrete deck built in 1937 to an orthotropic deck built in 1985 (see Figure 14.3). This retrofit reduced seismic forces in the suspension bridge towers and other bridge components. The engineering statistics of redecking are shown in Table 14.6. The Lions Gate Bridge of Vancouver, Canada was retrofitted in 1975 from a reinforced concrete deck to an orthotropic deck, which increased its seismic durability. Economics or cost of materials can be multiplied against the material saved to calculate money saved by reducing the weight.

A very thin deck structure can be built using this structural system, as shown by the Creitz Road Grade Separation in Figure 14.4 or German Railroad Bridge in Figure 14.5. An orthotropic deck may be the most expensive deck system per square meter in a short-span bridge. So why would the most expensive deck be a standard for the German railroads? The key component in obtaining the thinnest superstructure is the deck thickness. An orthotropic deck is thin because the ribs nest between the floor beams. Concrete decks are poured on top of steel beams. Thin superstructures can be very important for a grade-crossing situation because of the savings to a total project. The two components are bridge costs plus roadway or site costs. High-speed trains require minimal grade changes. Therefore, the money spent on highway or railway approach backfill can far exceed the cost of a small-span bridge. A more expensive superstructure will greatly reduce the backfill work and cost. In urban situations, approach fills may not be possible. The local street may need to be excavated below the railway bridge; therefore a more expensive thin orthotropic deck–floor system may result in the lowest total cost for the entire project.

TABLE 14.3 Properties of Trapezoidal Ribs

| American Rib (English Units) | | | | Japanese Rib (Metric) | | |

Depth of Rib d (in.)	Width at Top, a (in.)	Rib Wall Thickness, t_f (in.)	Weight per Foot, w (lb)	Moment of Inertia, I_{xx} (in^4)	Neutral Axis Location, Y_{xx} (in.)	Sloping Face Length, h' (in.)
8.0	11.50	5/16	23.43	46.3	3.09	8.382
		3/8	27.95	54.6	3.12	
		7/16	32.40	62.7	3.14	
9.0	12.12	5/16	25.64	63.8	3.56	9.428
		3/8	30.60	75.5	3.59	
		7/16	35.53	86.8	3.61	
10.0	12.75	5/16	27.88	85.1	4.04	10.477
		3/8	33.29	100.8	4.06	
		7/16	38.66	116.1	4.09	
11.0	13.38	5/16	30.09	110.4	4.52	11.525
		3/8	35.94	131.0	4.54	
		7/16	41.57	151.0	4.57	
12.0	14.00	5/16	32.33	140.2	5.00	12.572
		3/8	38.62	166.4	5.02	
		7/16	44.88	192.1	5.05	
13.0	14.63	5/16	34.53	174.7	5.48	13.621
		3/8	41.31	207.6	5.51	
		7/16	48.01	239.7	5.53	
14.0	15.25	5/16	36.75	214.4	5.97	14.668
		3/8	43.96	254.8	5.99	
		7/16	51.10	294.4	6.02	

Depth of Rib d (mm)	Width at Top, a (mm)	Rib Wall Thickness, t_f (mm)	Weight per Foot, w (Kg/m)	Moment of Inertia, I_{xx} (cm^4)	Neutral Axis Location, Y_{xx} (mm)	Sloping Face Length, h' (mm)
240	320	6	31.6	2460	88.6	246
260	320	6	33.1	3011	99.1	266
242	324	8	42.3	3315	89.9	248
262	324	8	44.3	4055	100.3	268

14.3 Applications

Some of the most notable world bridges were built using an orthotropic steel deck with trapezoidal rib construction. There are about only 50 bridges in North America using orthotropic decks, and eight are built and two more being designed in California. However, there is a vast array of bridge types utilizing the orthotropic deck from very small to some of the longest clear-span bridges of the world. Some orthotropic deck bridges have unique framing systems. Bridges featured and discussed in the following sections were selected to demonstrate the breath of reasons for selecting orthotropic deck superstructure [11-15]. All of these bridges utilize trapezoidal ribs in the deck area or compression zone of the superstructure. These types are: simple span with two plate or box girders, multiple plate girder, single-cell box girder, multicell box girder, wide bridges that have cantilever floor beams supported by struts, a monoarch bridge, a dual-arch bridge, a through-truss

TABLE 14.4 Orthotropic Deck Design Properties — Rigid Floor Beams

H = effective torsional rigidity of orthotropic plate (kip-in.²/in.)	
D_y = flexural rigidity of orthotropic plate in y direction (kip-in.²/in.)	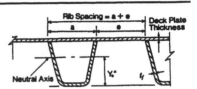
H/D_y = rigidity ratio (unitless)	
I_r = moment of inertia (in.⁴)	
Y_b = centroid (in.)	
t_p = deck plate thickness (in.)	

Deck Plate t_p (in.)	$a+e$ (in.)	Rib Wall (in.)	Value	Span (ft)	8 in.	Rib Depth 9 in.	10 in.	Span (ft)	11 in.	Rib Depth 12 in.	13 in.	14 in.
9/16	22	5/16	H/D_y	7	0.039	0.034	0.030	10	0.048	0.045	0.042	0.040
			I_r		165	217	278		351	431	520	620
			Y_b		6.45	7.14	7.81		8.54	9.20	9.85	10.49
9/16	26	3/8	H/D_y	11	0.057	0.049	0.043	14	0.056	0.051	0.047	0.044
			I_r		197	259	331		417	512	620	740
			Y_b		6.48	7.18	7.86		8.56	9.23	9.88	10.53
9/16	30	7/16	H/D_y	15	0.066	0.056	0.049	18	0.057	0.051	0.047	0.043
			I_r		226	298	382		480	591	716	855
			Y_b		6.50	7.19	7.88		8.57	9.24	9.89	10.54
5/8	22	5/16	H/D_y	7	0.044	0.038	0.033	10	0.053	0.049	0.046	0.043
			I_r		171	225	288		364	446	539	643
			Y_b		6.59	7.30	7.99		8.73	9.41	10.07	10.72
5/8	30	7/16	H/D_y	15	0.079	0.067	0.058	18	0.067	0.061	0.055	0.051
			I_r		234	309	396		498	612	742	886
			Y_b		6.64	7.35	8.05		8.76	9.44	10.11	10.77
11/16	22	5/16	H/D_y	7	0.048	0.041	0.036	10	0.056	0.052	0.048	0.045
			I_r		177	232	297		375	460	557	664
			Y_b		6.72	7.44	8.14		8.90	9.59	10.27	10.93
11/16	30	7/16	H/D_y	15	0.090	0.078	0.068	18	0.077	0.070	0.064	0.059
			I_r		242	318	408		513	632	765	915
			Y_b		6.76	7.49	8.21		8.93	9.62	10.31	10.98
3/4	22	5/16	H/D_y	7	0.052	0.044	0.038	10	0.059	0.054	0.050	0.047
			I_r		182	239	305		386	474	573	683
			Y_b		6.84	7.57	8.29		9.06	9.76	10.45	11.13
3/4	26	3/8	H/D_y	11	0.084	0.073	0.064	14	0.079	0.072	0.067	0.063
			I_r		216	284	364		458	563	682	815
			Y_b		6.87	7.61	8.34		9.08	9.78	10.48	11.16
3/4	30	7/16	H/D_y	15	101	0.087	0.077	18	0.086	0.078	0.072	0.067
			I_r		248	327	420		528	650	787	941
			Y_b		6.88	7.62	8.35		9.08	9.79	10.49	11.17

(Excerpts from out of print booklet. Permission to reprint granted. Anonymous source as requested.)

bridge, a deck-truss bridge, a monoplane cable-stayed bridge, a dual-plane cable-stayed bridge, a monocable suspension bridge, and a dual-cable suspension bridge.

14.3.1 Plate-Girder Bridges

In the 1960s small orthotropic steel-deck bridges were built in California, Michigan, and for the Poplar Street Bridge as prototypes to examine steel construction systems as well as various wearing surface materials. Each bridge used trapezoidal ribs with a split-beam section as floor beam and two plate girders as the main girders. The California Department of Transportation (Caltrans) built the I-680 over U.S. 580 bridge as their test structure [9,13] in 1968. This bridge has two totally different rib/deck systems including two different wearing surfaces. The two-lane cross section of the bridge and a very similar one were built by the Michigan Department of Transportation. The Creitz Road Bridge is a typical grade crossing built over I-496 near Lansing, Michigan (see

TABLE 14.5 Orthotropic Deck Design Properties — Flexible Floor Beams

Wt. = weight (PSF)
I_r = moment of inertia (in⁴)
Y_b = centroid (in.)
t_p = deck plate thickness (in.)

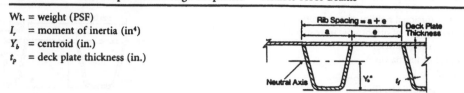

Deck Plate t_p (in.)	$a + e$ (in.)	Rib Wall (in.)	Value	Rib Depth						
				8 in.	9 in.	10 in.	11 in.	12 in.	13 in.	14 in.
9/16	22	5/16	Wt.	35.7	36.9	38.1	39.4	40.6	41.8	43.0
			I_r	169	222	284	355	437	528	630
			Y_b	6.54	7.24	7.94	8.62	9.29	9.95	10.60
9/16	26	3/8	Wt.	35.8	37.1	38.3	39.5	40.8	42.0	43.2
			I_r	199	262	335	420	517	625	747
			Y_b	6.54	7.24	7.93	8.61	9.28	9.94	10.59
9/16	30	7/16	Wt.	35.9	37.2	38.4	39.7	40.9	42.1	43.4
			I_r	228	301	386	484	595	721	861
			Y_b	6.54	7.24	7.93	8.61	9.28	9.93	10.58
5/8	22	5/16	Wt.	38.3	39.5	0.033	41.9	43.1	44.3	45.6
			I_r	175	230	288	368	452	547	653
			Y_b	6.68	7.40	8.11	8.80	9.49	10.17	10.83
5/8	30	7/16	Wt.	38.5	39.7	41.0	42.2	43.5	44.7	45.9
			I_r	236	311	399	501	616	745	892
			Y_b	6.68	7.40	8.10	8.80	9.48	10.15	10.82
11/16	22	5/16	Wt.	40.8	42.0	43.2	44.5	45.7	46.9	48.1
			I_r	180	237	303	379	466	564	674
			Y_b	6.81	7.54	8.26	8.97	9.67	10.36	11.04
11/16	30	7/16	Wt.	41.0	42.3	43.5	44.8	46.0	47.2	48.5
			I_r	244	321	412	516	636	770	920
			Y_b	6.81	7.54	8.26	8.96	9.66	10.35	11.03
3/4	22	5/16	Wt.	43.4	44.6	45.8	47.0	48.2	49.4	50.7
			I_r	185	243	311	390	479	580	693
			Y_b	6.92	7.67	8.40	9.13	9.84	10.54	11.24
3/4	26	3/8	Wt.	43.5	44.7	46.0	47.2	48.4	49.7	50.9
			I_r	218	287	368	461	567	687	821
			Y_b	6.92	7.66	8.40	9.12	9.83	10.53	11.22
3/4	30	7/16	Wt.	43.6	44.8	46.1	47.3	48.6	49.8	51.0
			I_r	250	330	423	531	653	792	947
			Y_b	6.92	7.67	8.40	9.12	9.83	10.53	11.22

Excerpts from out of print booklet. Permission to reprint granted. Anonymous source as requested.

Figure 14.4). It is a typical two-lane bridge that carries local traffic over an interstate freeway, with two symmetrical spans of 29 m. The bridge uses the 5/16-in.-thick, 9-in.-deep, 25.64-plf rib as shown in Table 14.3. The rigid steel bent comprises three welded steel box members aesthetically shaped [14,15]. Caltrans built a weigh station as an orthotropic deck prototype for the Hayward–San Mateo Bridge [17]. All of these short-span orthotropic deck bridges are still in use after 30 years of service, but the wearing surface has been replaced on many of these bridges.

Single-track railroad bridges through steel plate are the most common type for short spans. A two-girder bridge shown in Figure 14.5 is AASHTO fracture critical because, if one girder fractures, the bridge will collapse [18]. Both versions of the current AASHTO codes require the designer to label fracture-critical components, which have more stringent fabrication requirements. Orthotropic bridges can be erected quickly when the entire superstructure is fabricated as a full-width component. Many railroads prefer weathering steel since maintenance painting is not required. The German Federal Railroads have a standard, classic two plate girder with orthotropic deck system

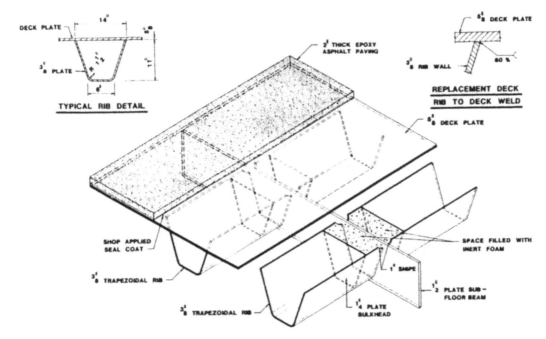

FIGURE 14.3   Golden Gate orthotropic deck details. (From Troitsky, M. S., *Orthotropic Steel Deck Bridges*, 2nd ed., JFL Arc Welding Foundation, Cleveland, OH, 1987. Courtesy of the James F. Lincoln Arc Welding Foundation.)

for their common short-span railroad bridges. Edge plates are used to keep the gravel ballast in place on top of the superstructure (Figure 14.5).

14.3.2 Box-Girder Bridges

Box-girder bridges can be subdivided into three basic categories: the single-cell box, the multicell box, and the box with struts supporting a cantilevered deck. In Figure 14.6, the typical cross section of the Valdez Floating City Dock Transfer Bridge is shown. The entire bridge, only about 3 to 6 m above the waterline, was completed in 1981 using the orthotropic deck with trapezoidal ribs. The two identical bridges were built at each end of the floating dock. Each bridge has only two box girders and has a simple span of 61 m. The transfer bridge provides traffic access to and from the floating dock and serves as the primary mooring tie for dock forces perpendicular to the shoreline. Trapezoidal rubber marine fenders absorb kinetic energy as the floating dock moves with the waves. These bumpers are at each end of the bridge. Box girders are more efficient in transmitting compression forces than plate girders [19]. ASTM A-36 steel was used to meet charpy impact requirements of 15 foot-pounds at −15°F. Automatic flux cored welding was used, and all full-penetration welds were either radiographically or ultrasonically inspected. The bridge uses the ⅜-in.-thick, 12-in.-deep, 38.62-plf rib as shown in Table 14.3. The floor beams are 2 ft deep by 1 ft wide ⅜-in.-thick plate bent in a U shape pattern. The ribs pass through the floor beam.

The typical cross sections of the Yukon River or "E. L. Patton" Bridge are shown in Figure 14.7. The 671-m-long bridge, with spans of 128 m, crosses over the Yukon River and was completed in 1976. The haul road is a gravel road originally built to transport supplies for the pipeline and oil field facilities at Prudhoe Bay, Alaska. The bridge was field-bolted together in cold weather, since the Alaskan winter lasts 6 months. It was important to keep the construction on schedule since the bridge was built to carry the 1.46-m-diameter trans-Alaskan crude oil pipeline. The bridge is the first built in Alaska across the Yukon River [2,20]. It is still the only bridge in Alaska across the

TABLE 14.6 Orthotropic Redecking Statistics Table (weight, deck area, etc.) Metric

Bridge	Lions Gate, Vancouver, BC, Canada	George Washington, New York, NY, USA	Golden Gate, San Francisco, CA, USA	Throngs Neck Viaduct, New York, NY, USA	Ben Franklin, Philadelphia, PA, USA	Champlain, Montreal, PQ, Canada
Bridge type	Girder	Suspension	Suspension + Approaches	Girder	Suspension + Approaches	Trusses
Redecking	Finished 1975	Finished 1978	Finished 1985	Finished 1986	Finished 1987	Finished 1992
Main Spans	13 to 38	186 1,067 186	343 1,280 343	42 to 58	219 534 219	118 215 118
Redecked area (m²)	8600	40,320	52,680	45,800	55,740	18,620
Rib type	Closed	Open (Ts)	Closed	Closed	Open (bulb section)	Closed
Rib spans (m)	4.12	1.60	7.62	6.10 to 8.50	5.80 to 6.70	6.40 to 9.80
Wearing surface	40 mm epoxy asphalt	40 mm bitum. asphalt	50 mm epoxy asphalt	40 mm bitum. asphalt	32 mm epoxy asphalt + 32 mm bitum. asphalt	50 mm epoxy asphalt
Original concrete deck weight	N/A	517 (kg/m²)	508 (kg/m²)	522 (kg/m²)	601 (kg/m²)	N/A
Total weight of new deck[a]	300 (kg/m²)	293 (kg/m²)	386 (kg/m²)	406 (kg/m²)	435 (kg/m²)	402 (kg/m₂)
Weight savings	N/A	224 (kg/m²)	122 (kg/m²)	116 (kg/m²)	166 (kg/m²)	N/A
New deck + main members	Yes, integral	No, integral	No, integral	No, integral	Yes, integral	Yes, integral
Cost/m²[b]	U.S. $500	U.S. $460	U.S. $1070	U.S. $770	U.S. $1010	U.S. $402
Redeck	Nighttime	Nighttime	Nighttime	Nighttime	Daytime	Nighttime

[a] Including surfacing, parapets, shear connectors;
[b] Total bid price/deck area. Note that bid prices reflect such variable factors as specific project characteristics, contractors profit margins, etc.
Source: Wolchuk, R., Structural Engineering International, IABSE, Zurich, 2(2), 125, 1992. With permission.

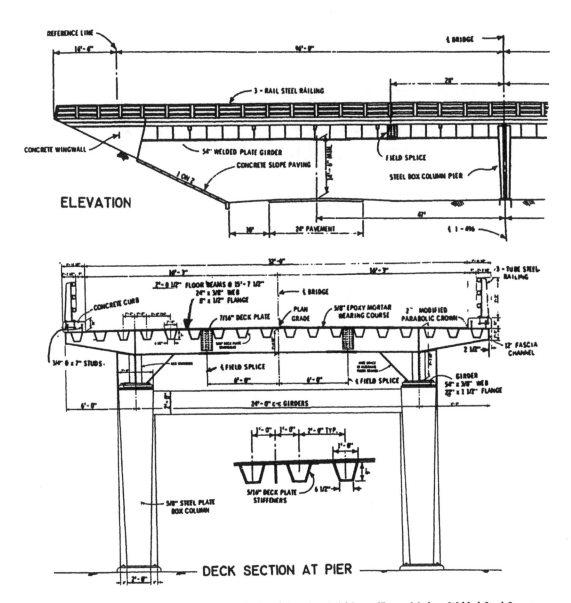

FIGURE 14.4 Typical grade separation — Creitz Road, Lansing, Michigan. (From *Modern Welded Steel Structure,* III, JFL Arc Welding Foundation, Cleveland, OH, 1970, B-10. Courtesy of the James F. Lincoln Arc Welding Foundation.)

Yukon, which has river ice 2 m thick. The superstructure consists of constant-depth twin rectangular box girders which have unique cantilevering brackets that support the trans-Alaskan crude oil pipeline on one side. The future trans-Alaska natural gas line, yet to be built, can be supported on the opposite side of the bridge with these specially designed cantilever support brackets. The bridge, which was fabricated in Japan, uses ribs as shown in Table 14.3. The ⅜-in.-thick, 11-in.-deep, 35.94-plf rib for the deck and the 5/16-in.-thick, 8-in.-deep, 23.43-plf rib in other locations are shown in Figure 14.7. Concrete deck construction requires curing temperature of a minimum of 40°F, otherwise, the water freezes during hydration. This was another factor in the selection of the orthotropic deck system, which was erected in temperatures as cold as –60°F [20]. The main bridge components, such as tower columns, tower cross frames, box girders and orthotropic steel deck were stiffened by trapezoidal ribs. The goal of maximizing the number of locations of trapezoidal rib was to reduce

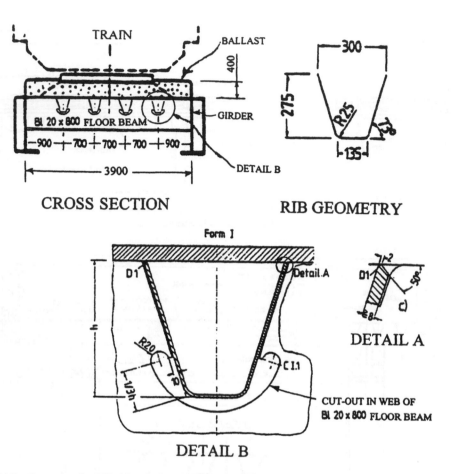

CROSS SECTION   **RIB GEOMETRY**

DETAIL A

DETAIL B

FIGURE 14.5 German railroad bridge. (From Haibach E. and Plasil, I., *Der Stahlbau*, 269, Ernst & Sohn, Berlin, Germany, 1983 [in German]. With permission.) (Metric units).

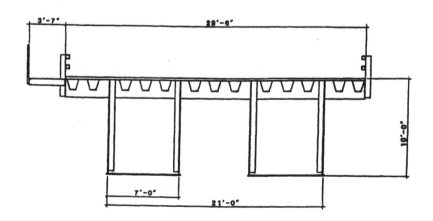

CROSS-SECTION OF RAMP

FIGURE 14.6 Valdez Floating City Dock Transfer Bridge. (Courtesy of Berger/ABAM Engineers, Federal Way, WA.) (English units)

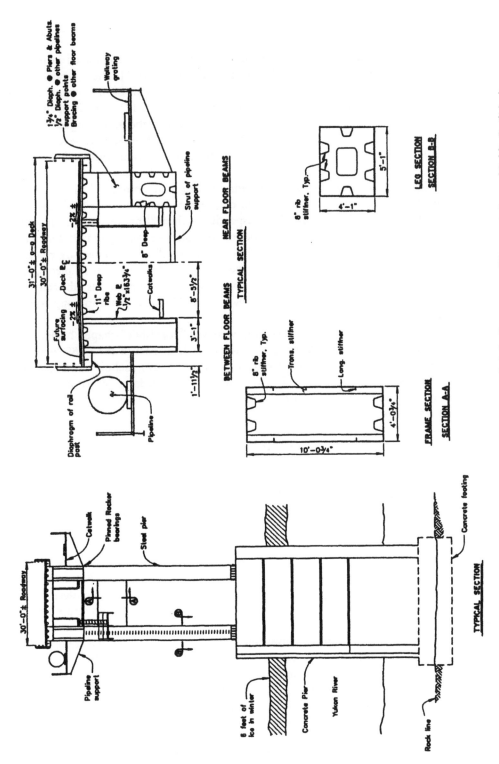

FIGURE 14.7 Yukon River Bridge — orthotropic deck and columns. (Courtesy of Alaska Department of Transportation and Public Facilities.)

the fabricator's setup costs to make a rib. The bridge tower columns and cross frames were built of shop-welded steel and field-bolted splices. The wearing surface of treated timber boards was bolted to the steel bridge because this could be built during cold weather, plus the haul road remains unpaved. The bridge utilized cold-weather steel (ASTM A537 and A514) with high charpy impact test characteristics for subzero temperatures [2, 20].

Shown in Figure 14.8a is the San Diego–Coronado Bridge, which was completed in June 1969. This California toll bridge sweeps around the harbor area of San Diego [3]. The Caltrans engineers selected single-cell box-girder orthotropic steel deck (continuous length of orthotropic portion = 573 m) because a constant-depth box could be used for the 201, -201, 171 m main spans over the shipping channel. Steel plate girders with concrete deck were used on the remaining length of 1690 m. The bridge was erected in these large pieces with a barge crane. The sections were field-bolted together. The bridge is painted on the inside and outside to resist corrosion and carries six lanes of traffic [21–23].

The Queens Way Bridges, identical three span twin bridges, were completed in June 1970 and are near the tourist attraction of the decommissioned Queen Mary ocean liner [25]. Each orthotropic bridge has a main drop-in span of 88 m suspended with steel hanger bars from cantilever side spans of 32 m. Thus, the center to center of the concrete piers or clear span is 152 m, with two side spans of 107 m. Each superstructure cross section is a single-cell box with deck overhangs with components similar to the San Diego–Coronado. The bridge was fabricated in 14 pieces, and the superstructure was erected in 11 days. The 88-m suspended or drop-in span was fabricated in one piece weighing 618 U.S. tons in Richmond, California, floated 700 miles south to Long Beach, and lifted up 15.2 m. by the same barge crane [26].

Shown in Figure 14.8b is the Maritime Off-Ramp a curved "horseshoe"-shaped bridge crossing over I-80 in Oakland, California, which was completed in 1997 as part of the I-880 Replacement Project. This superstructure has a very high radius of 76 m and a very shallow web depth of only 2.13 m for 58 m spans. The bridge has two lanes of traffic that create large centrifugal forces. The box-girder superstructure is divided into three separate cells to resist the very high torsional forces. To reduce the fabrication costs, the trapezoidal ribs were used in the top and bottom box-girder flanges since this was a continuous structure. The bridge sections were erected over busy I-80 on two Saturday nights creating an instant superstructure. The bridge was fabricated in 13 segments weighing as much as 350 U.S. tons and erected with two special hydraulic jacks supported by special multiwheeled trailers [27-29]. The orthotropic superstructure has a wider top deck plate with a 16-mm thickness and narrower flange plate of 19-mm thickness. In addition, each of three cells has four ribs for the top deck and two in the bottom flange. There are two exterior inclined webs and two interior vertical webs.

When a bridge gets very wide in relationship to the depth of the box girder superstructure, the German solutions shown in Figure 14.9 become the most economical. Shown in Figure 14.9a is the Jagst Viaduct, Widdern in the German Interstate system or Heilbronn–Wurzburg Autobaun carrying eight lanes of traffic over a deep valley. The superstructure is 30 m wide and 5.25 m deep. The most economical solution was to brace the ends of the cantilevering floor beams with struts attached to the bottom flange of the box girder. The box system remains constant depth so the strut remains a constant length. This keeps fabrication costs lower since the struts are all identical components [30].

Shown in Figure 14.9b is the Moselle Viaduct, Winningen, which is near Koblenz, Germany. The engineers decided to utilize a bottom soffit or flange following a parabolic curve. The superstructure is 30.5 m wide and 6 m deep at midspan and 8.5 m deep at the concrete piers. Therefore, to keep the struts a constant length, additional interior framing or cross-bracing members was devised. The struts are bolted to the side of the superstructure at a constant depth. At the inside face of the box girder, web cross-bracing members were attached and aligned with the exterior struts. This also produces a more pleasant architectural appearance [30]. The top deck utilizes "martini-glass"-shaped ribs, which consist of two standard shapes welded together. First a V-shaped rib is welded to the deck, second, a split-T is welded to the bottom of a V-shaped rib. It is a hybrid rib because

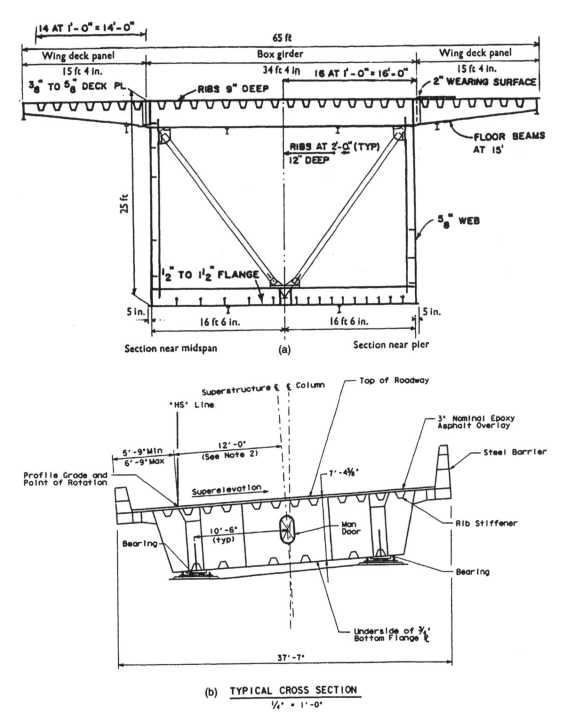

FIGURE 14.8 Steel box-girder bridges (a) San Diego–Coronado Bridge, California. (From Institute of Civil Engineers, Steel box-girder bridges, in *Proceedings of the International Conference,* Thomas Telford Publishing, London, 1973. With permission. (b) Maritime Off-Ramp Bridge, California. (Courtesy of the ICF KAISER Engineers.)

it has characteristics of both the open and closed rib. Some references have categorized it with closed ribs. The top portion provides good torsional stability to the deck, but the lower split-T portion has the buckling and corrosion disadvantages of an open rib. The split-T provides much greater bending strength, but no torsional stiffness to the deck. An open rib has one weld, a closed rib has

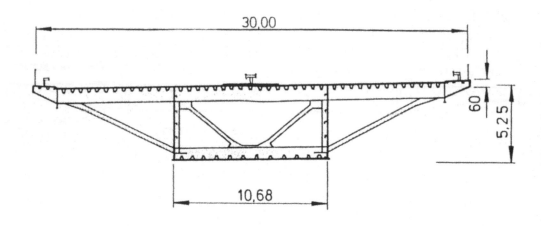

(a) Wurzburg Viaduct

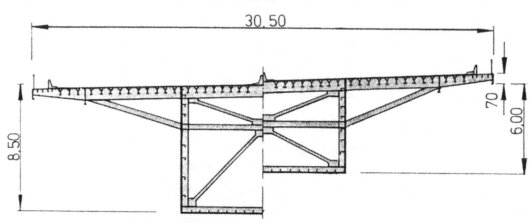

(b) Moseltal Bridge Winninzen (Autobahn A61 Koblenz)

FIGURE 14.9 Steel box girder with strutted deck bridges. (From Leonhardt, F., *Bridge Aesthetics and Design*, MIT Press, Cambridge, MA, 1984. Deutsche Verlags-Anstalt, Stuttgart, Germany. With permission.)

two welds, and a hybrid rib has three welds. This third weld is another possible source for fatigue cracking (see Section 14.4.5). The martini-glass-shaped ribs have only been used by the Germans for about 30 bridges, and ignored by engineers in the United States (See Table 14.1).

14.3.3 Arch Bridges

Arch superstructures also utilize orthotropic deck systems. The arch can either support the deck in one line or two lines of support. The Barqueta Bridge (Figure 14.10a) is a unique signature bridge with high aesthetic appearance built for the Expo 90 Fair held in Seville, Spain [31,32]. This bridge utilizes an aerodynamic system with torsional rigidity because it is supported by a single line of suspender cables from a single arch located at the centerline of the bridge. The center portion of the superstructure cross section is reinforced for the high stress concentrations from the suspender cables. Trapezoidal ribs are also used for the bottom flange or soffit. The bridge was erected on one side of the river and floated in a rotating pattern into a permanent orientation.

The Fremont Bridge of Portland Oregon has an orthotropic deck with trapezoidal ribs for the upper deck with a conventional reinforced concrete slab on steel girders for the lower deck

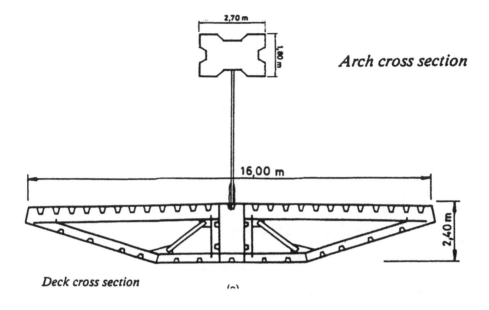

Arch cross section

Deck cross section

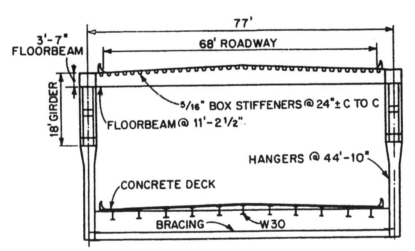

FIGURE 14.10   Arch bridges. (a) Barqueta Bridge, Seville, Spain. (From Arenas, J. J., and Pantaleon, M. J., *Structural Engineering International*, IABSE, 2(4), 251, 1992. With permission.) (b) Fremont Bridge, Portland, Oregon. (From Hedelfine, A. and Merritt, F. S., Ed., *Structural Steel Engineering Handbook*, McGraw Hill, New York, NY., 1972. With permission.)

[Figure 14.10b]. This bridge is a tied arch with a 383-m main span and was the fourth longest arch in the world upon completion in 1973. The deck provides lateral stability to the truss. The truss has two levels of traffic and is painted. The bridge was erected in large pieces with an oceangoing barge crane from the river dividing downtown Portland [33].

14.3.4   Movable Bridges

The orthotropic deck system has the lowest weight for movable bridges, so it is surprising how few examples there are of this excellent system. The swing-bridge design across a 42-m-wide navigation channel near Naestvad, Denmark was selected over 14 proposals from five competing consulting firms (Figure 14.11a). The superstructure is divided into two symmetrical components [34]. The

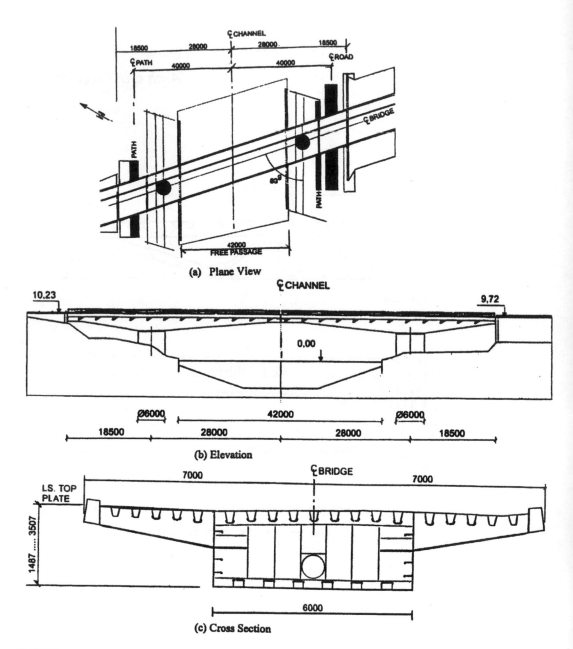

FIGURE 14.11 Swing bridge near Naestved, Denmark. (a) Plane view; (b) elevation; (c) cross section. (Thomsen, K. and Pedersen, K. E, *Structural Engineering International,* IABSE, 8(3), 201, 1998. With permission.)

main component is a 6-m-wide variable-depth box girder with tapered cantilevered floor beams (Figure 14.11b). The top deck plate is stiffened by a cold-rolled trapezoidal rib (Figure 14.11c). The bottom flange plate of the box girder is stiffened by a cold-rolled rectangular-shaped rib. The two exterior web plates of the box girder are stiffened by a bulb-shaped rib. The two sections were fabricated in a shop and barged to the bridge site utilizing the navigation channel. The exterior is painted, and the interior uses dehumidification equipment (see Section 14.4.6) in each of the two sections. The bridge opened to traffic in 1997.

There is an orthotropic swing bridge at the southern mouth of a small slough east of the main channel of the St. Clair River on the Walpole Island Indian Reservation, Ontario, Canada built in

1970. This movable bridge allows small pleasure craft traveling between Lake Huron and Lake St. Clair to pass through. Vehicles can travel to and from the island across this bridge, which was the first movable bridge built with an orthotropic deck in North America. The advantages of this solution are discussed in detail in Reference [35].

The Danziger Vertical Lift Bridge, completed in 1988, is the world's widest vertical-lift bridge and carries seven lanes of traffic on U.S. 90 through downtown New Orleans across the Industrial Canal. The orthotropic deck that is lifted is a 33 m wide × 97 m span supported by three steel 4.26 m deep × 1.82 m wide box beams. The spacing of the box beams is 11.5 m with split-T shaped tapered floor beams at 4.42 m on center. The cantilever on the floor beams is 3.31 m from the face of the box girder. The rectangular boxes are fabricated of ASTM A572 and A588 steel for the main plate and A36 steel for secondary members including all steel median barriers. The ASTM A572 ribs are 5⁄16-in.-thick, 10-in.-deep, 27.88-plf rib as shown in Figure 14.7.

The world's largest double-leaf bascule bridge was opened to traffic in 1969 in Cadiz, Spain [2]. The main girders cantilever 48.3 m, providing a channel between Puerto de Santa Maria and Cadiz of 96.7 m. The orthotropic deck spans 2 m from between split-T-shaped transverse floor beams, which cantilever 2.6 m. At each side of the 12-m deck plate are sidewalks. The two main plate girders are tapered, with maximum depth of 5 m, and are 6 m on center. Sway struts are between floor beam and midspan of plate girders. The signature bridge Erasmus of Rotterdam, the Netherlands utilizes trapezoidal orthotropic deck on both the cable-stayed portion and bascule span. This 33-m-wide by 50-m-long bascule span is skewed at 22°, with an opening of 56 m. The bridge has a very thin, 8-mm wearing surface and was opened in 1996. The Miller Sweeny bridge of Alameda Island California is the only orthotropic bascule bridge in North America.

A unique concept is to have an entire 11.6-m-wide by 33-m-long midsection removed by two cranes to allow ship traffic to pass through once every 2 to 3 years [36]. A conventional concrete box-girder bridge supports a drop-in orthotropic box-girder component that has a much smaller mass than concrete. This allows two smaller cranes to move the drop-in unit. The ribs were fabricated from 610-mm-wide plate, a standard plate dimension in the United States. Four of these plates would be cut without waste from warehoused stock plate received directly from the factory. Apparently, there have been no plate optimization studies performed by the steel industry. Also 254-mm-thick urethane foam insulation was sprayed on the bottom face of the steel deck to reduce the tendency of the steel deck to change temperatures more quickly.

A much more dramatic system is planned to move an entire 410-m superstructure in Japan. An all-steel superstructure is planned for the Yumeshima–Maishima Bridge in the "Tech Port Osaka" to be completed in the year 2000. Each end of the all-steel bridge is supported by a 58 × 58 m steel pontoon and will be moved by tugboats. This would be the world's largest movable bridge, with a deck area of 12,000 m² [37]. A scale model has been built and the estimated cost of the completed bridge is U.S. $400 million.

A unique civil engineering structure is the curved tidal surge gates of Rotterdam, The Netherlands. The two floating gates, each about the size of the Eiffel Tower, are made of orthotropic deck with trapezoidal ribs. Each gate has 20,500 tons of steel, with 14 mm deck plate. A seawater ballast system is used to adjust this structure to various tidal surge freeboard heights.

14.3.5 Truss Bridges

The German Federal Railways uses its standard orthotropic deck system, shown in Figure 14.5, for one-track bridges using a steel truss superstructure. The lateral bracing for the through truss is provided by the stiffness of the orthotropic deck. The standardization of their steel bridge deck plus floor beams keeps fabrication cost to a minimum [18].

A steel truss is used in both the transverse and longitudinal directions for the double-wall steel pontoon for the port of Iquitos on the Amazon River, in Peru (Figure 14.12a). The orthotropic steel deck with trapezoidal rib is used for the top deck that supports vehicular traffic. The side walls that

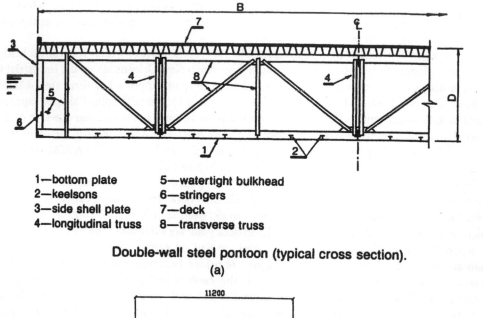

1—bottom plate   5—watertight bulkhead
2—keelsons   6—stringers
3—side shell plate   7—deck
4—longitudinal truss   8—transverse truss

Double-wall steel pontoon (typical cross section).
(a)

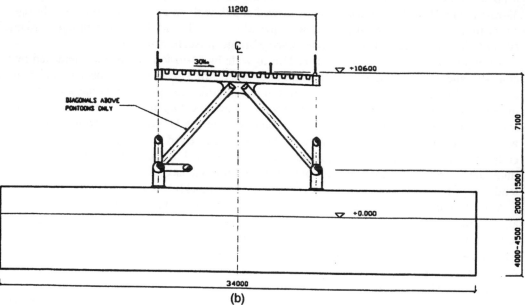

(b)

FIGURE 14.12 Truss bridges. (a) Floating dock on Amazon River. (From Tsinker, G. P., *Floating Ports Design and Construction Practices*, Gulf Publishing Company, Houston, TX, 1986. With permission.) (b) Bergsoysund Floating Bridge. (From Solland, G., Stein, H., and Gustavsen, J. H., *Structural Engineering International*, IABSE, 3(3), 142, 1993. With permission.)

have lower water pressure utilize small angles for the ribs. The bottom plate ribs, or keelsons (nautical terminology), are larger split-Ts because of the higher water pressure. This watertight floating pontoon is completely built of welded steel [38]. This framing system is also known as a "space truss" commonly used for building roofs.

The Bergsoysund Floating Bridge comprises floating concrete pontoons with a painted steel truss superstructure (Figure 14.12b). Floating orthotropic bridges become very economical for Norwegian fjords, which are actually deeper than the adjacent Atlantic Ocean floor. Lateral stability of the entire bridge is provided by an arch shape (in plan view) rather than cables with anchors in the 300-m-deep

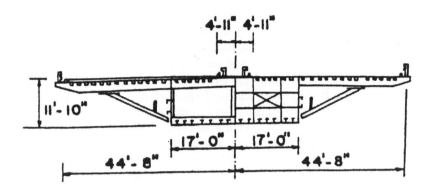

Cross section.

FIGURE 14.13 Papineau–Leblanc cable-stayed bridges. (From Troitsky, M. S., *Orthotropic Steel Deck Bridges*, 2nd ed., JFL Arc Welding Foundation, Cleveland, OH, 1987. Courtesy of the James F. Lincoln Arc Welding Foundation.)

fjord. The lateral stability of the top chords of the trusses is assisted by transverse stiffness of the orthotropic deck. The three-dimensional space truss is built of hollow steel pipe tubular joints, which have the minimum exposed area to resist corrosion. Detailing and design of these joints were based on experience developed for tubular offshore structures built in the North Sea. The closed trapezoidal rib was used, since the bridge is totally exposed to corrosive saltwater spray. Also, it was very important to minimize total bridge weight to reduce the size of the concrete pontoons. To avoid future painting in the ocean water, concrete was selected over orthotropic pontoons similar to Figure 14.12a. This bridge is a state-of-the-art solution utilizing offshore oil platform technology combined with floating bridge design technology [39].

14.3.6 Cable-Stayed Bridges

In cable-stayed bridges, the superstructures can be supported by one or two planes or lines of cables. Additional compression stresses occur in a cable-stayed bridge superstructure where the orthotropic deck is the compression component since the cables are the tension component. The Papineau–Leblanc Bridge, completed in 1969, linking the city of Montreal to Laval Islands, Canada, is a strutted deck box girder supported by a single line of cable stays at the centerline of the superstructure as shown in Figure 14.13. The bridge has spans of 90, 241, and 90 m with a superstructure width of 27.44 m [2,40]. Extra diaphragm plates were located at the cable support locations to transfer the loads from the deck into the cable stay as shown in right side of the split cross section of Figure 14.13. Closed U-shaped ribs were used with the top 7/16-in.-thick top deck plate and open ribs for the lower flange plate. The two bridge piers are cone-shaped. The pier face is at a 23° slope to bend the river ice, thus breaking the ice into pieces.

The Bratislava Bridge, completed in 1972 with a main span of 303 m crosses the Danube River, a major transportation river for barges, in Czechoslovakia. The orthotropic superstructure is a double-deck cellular box girder supported by a single line of cable stays at the centerline of the superstructure. The bridge has an anchor span of 75 m, and a superstructure width of 21 m. The feature that makes this bridge a unique signature span is the circular coffeehouse on top of the 85-m-high A-frame tower. Tourists can ride up an elevator in one tower leg to reach the sight-seeing windows. An emergency staircase is located in the other steel tower leg. Another nice feature is the protected pedestrian walkways on each side of the lower orthotropic deck. This feature gives wind and rain protection. The interior of the cross section contains utilities that cross the Danube [40, 41]. The coffeehouse is "saucer shaped," probably inspired by the Seattle Space Needle. The framing consists of steel bowstring trusses for the roof and floor.

The Luling or Luling–Destrehan or Hale Boggs Bridge near New Orleans is a weathering steel bridge that spans the Mississippi River and was completed in 1983. Its superstructure has twin trapezoidal box girders. The floor beams and deck have four bolted splice points in the longitudinal direction and are supported by two planes of cable stays. Trapezoidal ribs are used for the deck system [42]. The main span is 383 m and has an aerodynamic shape to withstand hurricanes. The bridge uses the ⁵⁄₁₆-in.-thick, 9-in.-deep, 25.64-plf rib as shown in Table 14.3. The center barrier and exterior barriers are welded steel plate bolted to the deck with welded studs. This bridge was fabricated in Japan and shipped to the United States. The world's longest clear-span cable-stayed bridge is a steel orthotropic deck bridge in Japan, and the second longest is the Normandie Bridge in France. Both bridges have two planes of cable stays.

Shown in Figure 14.14 is half of the proposed orthotropic superstructure option B for the cable-stayed replacement bridge for the East Span of the San Francisco–Oakland Bay Bridge. A family of solutions with single and dual towers has reached the 30% design development. The concept illustrated has a divided or separated superstructure connected by steel stiffening trusses to be built of steel tubes. Each half is planned to carry five lanes of traffic and be supported by a single tower. The separated superstructure allows the wind to flow around each side, as well as through the center, reducing wind forces [44].

14.3.7 Suspension Bridges

Suspension bridge superstructures may be supported by one or two planes of cables. Shown in Figure 14.15 is the orthotropic aerodynamic superstructure option for the suspension replacement bridge for the East Span of the San Francisco–Oakland Bay Bridge. A family of solutions with single and dual towers has reached the 30% design development. The concept illustrated has a separated superstructure, and each bridge is planned to carry five lanes of traffic in one direction. Each half is actually an independent bridge, and this superstructure solution is based on the British Severn Bridge completed in 1966. Note how each rib has a different cutout hole to eliminate fatigue cracks depending on the rib shape [44]. The San Francisco–Oakland Bay Bridge East replacement Spans design is currently evolving, and final approved plans have not yet been completed.

The Konohana Bridge in Japan has a main span of 300 m with side spans of 120 m [45]. This is a monocable self-anchoring suspension bridge. Its superstructure is supported only at the centerline of the bridge (Figure 14.16a). This signature span is supported by suspender cables attached to the hanger connection plate. An isometric of the main stiffening plates and diaphragms of the super-structure is shown in Figure 14.16a. Note the concentration of plates needed to distribute cable forces throughout the superstructure. The cantilevering sides of the superstructure mandate an aerodynamic box-girder superstructure. The Japanese rib 242 mm × 324 mm × 8 mm was used (Table 14.3).

Shown in Figure 14.16b is the suspension bridge with two planes of cables and a 762 m main span currently under final design for the Third Carquinez Strait Bridge between San Francisco and Sacramento, California. It is an aerodynamic superstructure with trapezoidal ribs for the deck and bottom soffit [46]. This bridge design has been heavily influenced by the successful bridges built in Europe and Japan.

Shown in Figure 14.17 is a drawing of the world's longest concept suspension bridge, proposed to span the straits of Messina [47, 48]. This clear span between Sicily and Italy has been a historical challenge to engineers. This concept has been given technical approval for the final design by the various Italian authorities and political approval is in progress. European engineers have already performed wind tunnel studies in order to develop contract plans and specifications. The super-structure is supported by two pairs of cables and a 3000-m main span is composed of the middle of welded steel framing system that is 52 m wide. A separated superstructure concept allows wind to flow through the superstructure. The Messina and the option B for the cable-stayed replacement bridge for the East Span of the San Francisco–Oakland Bay Bridge are similar in principle. The

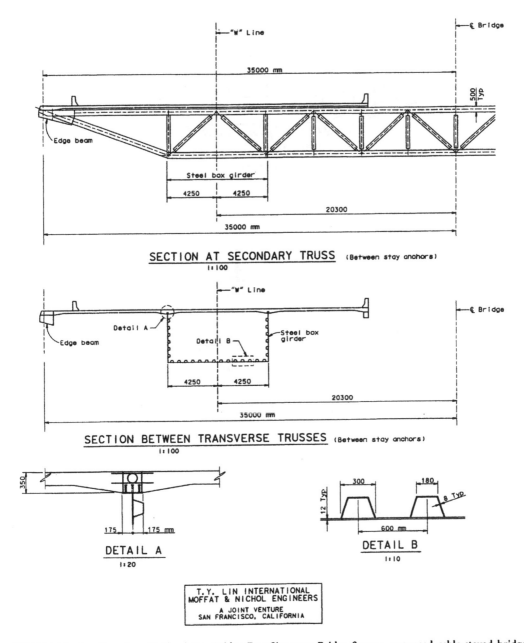

FIGURE 14.14 San Francisco–Oakland Bay Bridge East Signature Bridge Span — proposed cable-stayed bridge option. (Courtesy of T. Y. Lin International and Moffat & Nichols.)

open spaces, while strongly increasing the aeraelastic stability (e.g. flutter stability), reduce wind loading and have a grillage or grating that allow lanes for emergency stop and maintenance vehicle traffic. Three longitudinal independent wing-shaped box girders are linked transversely with very large welded steel box-shaped cross girders. The orthotropic deck is stiffened by trapezoidal ribs for the top deck and open ribs for the bottom soffit. All the barriers have aerodynamic shape and grating to reduce wind forces. New suspension bridges with double-decker superstructures with an upper orthotropic deck for vehicles and a lower orthotropic deck for commuter trains are in use in Asia. Examples are the Tsing Ma Bridge of Hong Kong, China and Yong Jong Grand self-anchoring suspension bridge of Seoul, Korea.

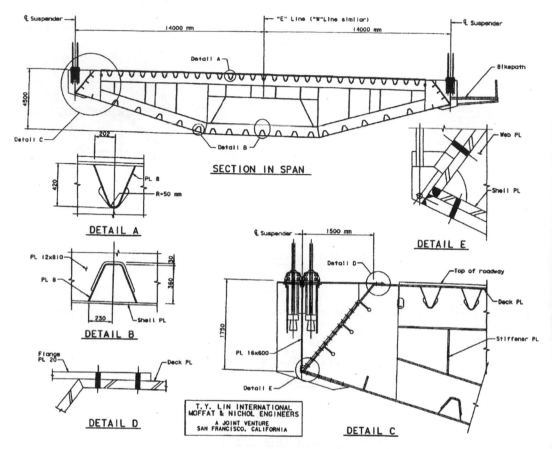

FIGURE 14.15 San Francisco–Oakland Bay Bridge East Signature Bridge Span — selected suspension bridge option. (Courtesy of T. Y. Lin International and Moffat & Nichols.)

Some older suspension bridges have been retrofitted with installation of small orthotropic panels to accommodate higher traffic loading and to extend the useful or fatigue life of bridges. Currently, there are more retrofitted North American suspension bridges than new bridges with orthotropic decks. Small deck panels have been trucked onto various bridges to replace the portion of reinforced concrete deck that was removed (see Table 14.6 and Figure 14.3). The Golden Gate Bridge uses essentially the same 3/8-in.-thick, 11-in.-deep, 35.94-plf rib as shown in Table 14.3. The Wakato Bridge in Kitakusyu, Japan was redecked, very wide pedestrian sidewalks with only two traffic lanes were eliminated, and the bridge was converted to four lanes of vehicular traffic without pedestrian access. The historical Williamsburg Suspension Bridge, built in 1903, is the most recent redecking project and, essentially, uses this rib shape again [49]. Extensive testing of a full-size mockup of the designer's concept was performed to verify its durability or fatigue life. An extra internal plate or miniature diaphragm or rib bulkhead aligns with the floor beam web. Welding detail options and cutout holes or scallops were tested and showed that the selected system should have a long fatigue life.

14.4 Design Considerations

14.4.1 General

In contrast to the conventionally designed bridge, where the individual structural elements (stringers, floor beams, and main girders) are assumed to perform separately, an orthotropic steel deck bridge is a complex structural system in which the component members are closely interrelated.

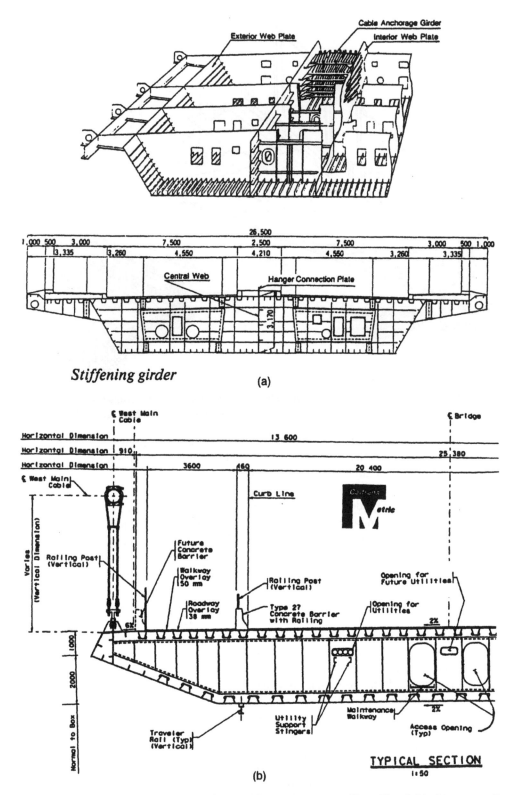

FIGURE 14.16 Suspension bridges (a) Konohana Bridge superstructure. (From Kamei, M., Maruyama, T. and Tanaka H., *Structural Engineering International*, IABSE, 2(1), 4, 1992. With permission.) (b) Third Carquinez Bridge stiffening girder. (Courtesy of De Leuw-OPAC-Steinman.)

FIGURE 14.17   World's biggest single-span bridge. The deck arrangement of the Messina Bridge can be clearly seen from this drawing; grillages will help with the aerodynamics. (Courtesy of Brown, W. at Brown Beech & Associates.)

The stress in the deck plate is the combination of the effects of the various functions performed by the deck. For structural analysis under dead and live loads, it is necessary for design convenience to treat the following structural members separately.

In a typical orthotropic deck system as shown in Figure 14.18, Member I is defined as the deck, a flat plate supported by welded ribs as shown in Figure 14.1 and Table 14.1. The deck plate acts locally as a continuous member directly supporting the concentrated wheel loads placed between the ribs and transmitting the reactions to the ribs. The design of the deck plate is discussed in Section 14.4.2.

Member II is defined as a rib spanning from a floor beam to a floor beam (normally a continuous element of at least two spans) as shown in Figure 14.18. The stiffened steel plate deck (acting as a bridge floor between the floor beams) consists of the ribs plus the deck plate that is the common upper flange. A detailed discussion of rib design is given in Section 14.4.3.

Member III is the floor beam that spans between the main girders. Member IV is defined as a girder spanning from a column (or cable) to column (or cable) as shown in Figure 14.18 and is normally a continuous element of at least two spans to be economically viable. The deck also acts as part of this member. In the computation of stresses, the effective cross-sectional area of the deck plus the inclusion of all longitudinal ribs is considered as the flange. The determination of the effective width of the deck and design stresses will be discussed in Section 14.4.4. The orthotropic deck plate receives stresses under multiple loading combinations as shown in Figure 14.19. This is because the deck plate is the top flange of the Member II, Member III, and Member IV. An orthotropic steel deck should be considered an integral part of the bridge superstructure. The deck plate acts as a common flange of the ribs, the floor beams, and the main longitudinal components of the bridge. Any structural arrangement in which the deck plate is made to act independently from the main components is undesirable.

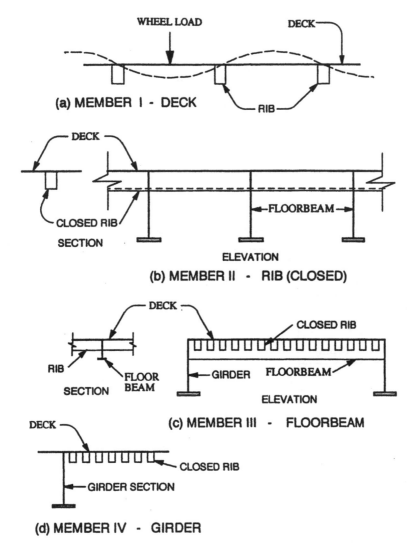

FIGURE 14.18 Four members to be analyzed in an orthotropic deck. (a) Member I — deck; (b) Member II — rib (closed); Member III — floorbeam; (d) Member IV — girder.

When redecking the bridge, if the orthotropic deck is supported by existing floor beams, the connection between the deck and the floor beam should be designed for full composite action, even if the effect of composite action is neglected in the design of floor beams. Where practical, connection suitable to develop composite action between the deck and the main longitudinal components should be provided.

The effects due to global tension and compression should be considered and combined with local effects. When decks are in global tension, the factored resistance of decks subject to global tension, P_u, due to the factored loads with simultaneous global shear combined with local flexural must satisfy [5]:

$$\frac{P_u}{P_r} + \frac{M_{lr}}{M_{rr}} \le 1.33 \tag{14.1}$$

$$P_u = A_{d,\text{eff}}\sqrt{f_g^2 + 3f_{vg}^2} \tag{14.2}$$

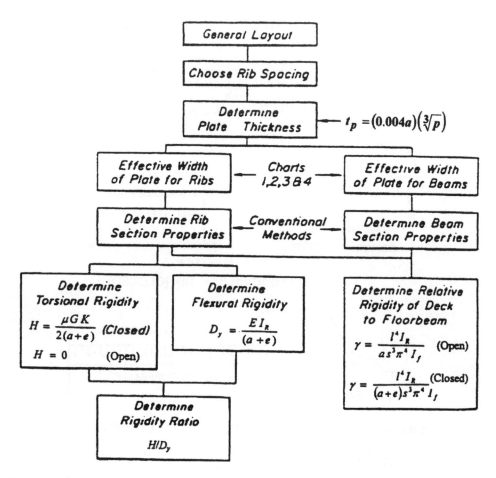

FIGURE 14.19 Determination of required section properties flow chart (metric formula). (From Milek, W. A., Jr., *Eng. J. AISC*, 40, 1974. With permission.)

where

f_g = axial global stress in deck (MPa)

f_{vg} = simultaneous global shear in the deck (MPa)

$A_{d,\,eff}$ = effective cross section area of the deck, including longitudinal ribs (mm²)

P_r = nominal tensile resistance of the deck with consideration of effective deck width (N)

M_{lr} = local flexural moment of longitudinal rib due to the factored loads (N-mm)

M_{rr} = Flexural resistance of longitudinal rib, governed by yielding in extreme fiber (N-mm)

The effect of simultaneous shear is usually not significant in orthotropic decks of girder or truss bridges, but may be important in decks used as tension ties in arches or in compression for cable-stayed bridges.

When decks are under global compression, longitudinal ribs, including effective width of deck plate, should be designed as individual columns assumed to be simply supported at transverse beams. Buckling formulas for steel decks can be found in the AISC *Design Manual for Orthotropic Steel Plate Deck Bridges* [1].

Diaphragms or cross frames should be provided at each support and should have sufficient stiffness and strength to transmit lateral forces to the bearings and to resist transverse rotation, displacement, and distortion. Intermediate diaphragms or cross frames should be provided at locations consistent with the analysis of the girders and should have sufficient stiffness and strength to resist transverse distortions.

14.4.2 Deck Design

The primary function of the steel deck (Member I) is to directly support the traffic loads and to transmit the reactions to the longitudinal ribs. An important characteristic of the steel deck is its capacity for carrying concentrated loads. When loads approach the ultimate load the deck plate practically acts as a membrane and can carry on the order of 15 to 20 times the ultimate load computed in accordance with the ordinary flexural theory. Thus, the bridge deck plate possesses an ample local overload-carrying capacity.

The minimum thickness of the deck plate may be determined by allowable deflection of the deck plate under a wheel load, which should not exceed $1/300$ of the spacing of the deck supports. Based on this criteria, the plate thickness, t_p, may be determined by Kloeppel's formula:

$$t_p \geq (0.004\,a)\left(\sqrt[3]{p}\right) \tag{14.3}$$

where:

a = spacing of the open ribs, or the maximum spacing of the walls of the closed ribs, in mm
p = wheel load unit pressure, under the AASHTO LRFD design tandem wheel load 55 kN, including 33% dynamic load allowance, in kPa. For 50 mm wearing surface, p is 449 kPa.

The distribution of the wheel load is assumed in a 45° footprint from the top of the wearing surface to the top of the deck plate. The AASHTO Specifications, 16th edition [50] tabulates wheel loads and contact area:

Wheel Load (kN)	Width Perpendicular to Traffic (mm)	Length Direction of Traffic (mm)
36	508 + 2t	203 + 2t
54	508 + 2t	203 + 2t
72	610 + 2t	203 + 2t

t = the thickness of the wearing surface in mm.

Using the AASHTO-LRFD [5], the wheel loads and contact area can be tabulated:

Wheel Load (kN)	Width Perpendicular to Traffic (mm)	Length Direction of Traffic (mm)
17.5 (truck)	510 + 2t	53 + 2t
55 (tandem)	510 + 2t	167 + 2t
72.5 (truck)	510 + 2t	220 + 2t

t = the thickness of the wearing surface in mm.

The current AASHTO-LRFD [5] requires that the minimum deck plate thickness, t_p, shall not be less than either 14 mm or 4% of the largest rib spacing. Experience from the durability of previously built bridges shows that this requirement is advisable for both constructibility and long-term bridge life.

For a rib spacing of 300 mm, a 14-mm plate is required per the AASHTO-LRFD, while a 9-mm plate can be derived using Eq. 14.3. For a rib spacing of 380 mm, a 15-mm plate is required per the AASHTO-LRFD, while a 12-mm can be derived using Eq. 14.3.

14.4.3 Rib Design

Table 14.2 gives the effective width of the deck plate acting with a rib. Since most of the ribs used in the current practice are closed type, this section will only discuss closed ribs. The ribs span between and are continuous at floor beams (Figure 14.18). Spacing of ribs depends on the deck plate thickness and usually for closed ribs $(a + e)$ varies between 610 and 760 mm. Rib spans of approximately 4500 mm have been common in North American practice. But, spans up to 8500 mm, required by spacing of existing floor beams, have been used in bridge redeckings. Long rib spans are feasible and may be economical.

The minimum thickness of closed ribs should not be less than 4.75 mm per AASHTO-LRFD. Fatigue tests concluded that local out-of-plane flexural stress in the rib web at the junction with the deck plate should be minimized. It is necessary to limit the stress in the rib web caused by the rotation of the rib–deck plate junction by making the rib webs relatively slender compared with the deck plate. To achieve this, AASHTO-LRFD [5] specifies that the cross-sectional dimensions of an orthotropic steel deck shall satisfy:

$$\frac{t_r a^3}{t_{d,\text{eff}}^3 h'} \leq 400 \tag{14.4}$$

where
t_f = thickness of rib web (mm)
$t_{d,\text{eff}}$ = effective thickness of the deck plate, with consideration of the stiffening effect of the wearing surfacing (mm)
a = largest spacing between the rib webs (mm)
h' = length of the inclined portion of the rib web (mm)

To prevent overall buckling of the deck under compression induced by the bending of the girder, the slenderness, L/r, of a longitudinal rib shall not exceed the value given by the following equation in the AASHTO Standard Specification [50]:

$$\left(\frac{L}{r}\right)_{\text{max}} = 83\sqrt{\frac{1500}{F_y} - \frac{2700F}{F_y^2}} \tag{14.5}$$

where
L = distance between transverse floor beams
r = radius of gyration about the horizontal centroidal axis of the rib including an effective width of the deck plate
F = maximum compressive stress in MPa in the deck plate as a result of the deck acting as top flange of the girder; this stress shall be taken as positive
F_y = yield strength of rib material in MPa

Orthotropic analysis furnishes distribution of loads to ribs and stresses in the member. Despite many simplifying assumptions, orthotropic plate theories that are available and reasonably in accordance with testing results and behaviors of existing structures require long, tedious computations. Computer modeling and analysis may be used to speed up the design. The AISC manual [1] was used when only expensive main-frame computers were available. The flowchart for this process is shown in Figure 14.19. Tables 14.3 to 14.5 were later created to assist those without computers. Today, engineers can write their own software or create the appropriate spreadsheet using a personal computer to expedite the iterative computations. Published tables provide useful databases to check against "bugs" in software. The following method, known as the Pelikan–Esslinger method, has been used in design of orthotropic plate bridges. In this method, the closed-rib decks are analyzed in two stages.

First, the deck with closed ribs is assumed on rigid supports. Only the longitudinal flexural rigidity and the torsional rigidity of the ribs are considered. The transverse flexural rigidity can be negligible. A good approximation of the deflection w may be presented in the form of Huber's differential equation:

$$D_y \frac{\partial^4 w}{\partial y^4} + 2H \frac{\partial^4 w}{\partial x^2 \partial y^2} = p(x, y) \tag{14.6}$$

where
D_y = flexural rigidity of the substitute orthotropic plate in the direction of the rib
H = equivalent torsional rigidity of the substitute orthotropic plate
$p(x,y)$ = load expressed as function of coordinates x and y

In the computation of D_y and H, the contribution of the plate to these parameters must be included. The flexural rigidity in the longitudinal direction usually is calculated as the rigidity of one rib with effective deck width divided by the rib spacing:

$$D_y = \frac{EI_y}{a+e} \tag{14.7}$$

where I_r = moment of inertia, including a rib and effective plate width.

Because of the flexibility of the orthotropic plate in the transverse direction, the full cross section is not completely effective in resisting torsion. Therefore, the formula for computing H includes a reduction factor u.

$$H = \frac{\mu GK}{2(a+e)} \tag{14.8}$$

where
G = shearing modulus of elasticity of steel
K = torsional factor, a function of the cross section

In general, for hollow closed ribs, the torsional factor may be obtained from

$$K = \frac{4A_y^2}{\dfrac{p_e}{t_r} + \dfrac{a}{t_p}} \tag{14.9}$$

where
A_r = mean of area enclosed by inner and outer boundaries of ribs
p_r = perimeter of rib, exclusive of top flange
t_r = rib thickness
t_p = plate thickness

The reduction factor u for a trapezoidal rib may be closely approximated by

$$\frac{1}{u} = 1 + \frac{GK}{EI_p} \frac{a^3}{12(a+e)^2} \left(\frac{\pi}{S_e}\right)^2 \left[\left(\frac{e}{a}\right)^3 + \left(\frac{e-b}{a+b} + \frac{b}{a}\right)^2\right] \tag{14.10}$$

where
I_p = moment of inertia = $Et_p^3 / 12(1-u^2)$
s_2 = effective rib span for torsion = $0.81s$
s = rib span

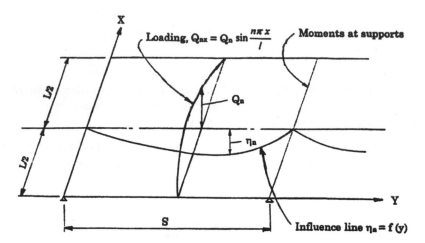

FIGURE 14.20   Computation of bending moments in orthotropic plate.

The values of D_y and H are combined in the relative rigidity coefficient H/D_y, which is a parameter characterizing the load-distributing capacity of the deck in the direction perpendicular to the ribs. For any given rib size, spacing, and deck plate thickness, H does not remain constant but increases with the rib span. Therefore, the parameter H/D_y is also a function of span, and the transverse load distribution of the deck structure improves as the span of the ribs is increased.

The general solution of Eq. (14.4) can only be given as an infinite series:

$$w_n = \left(C_{1n}\sinh a_n y + C_{2n}\cosh a_n y + C_{3n}a_n y + C_{4n}\right)\sin\frac{n\pi x}{l} \tag{14.11}$$

where
n = integer ranging from 1 to ∞ (odd numbers for symmetrical loads)
l = floor beam span
C_{in} = integration constant, determined by boundary conditions

$$a_n = \frac{n\pi}{l}\sqrt{\frac{2H}{D_y}} \tag{14.12}$$

The plate parameter H/D_y can be obtained from Table 14.4 or calculated.

Bending moments in the substitute orthotropic plate due to the given loading can be computed by formulas derived from the above solution. Since the solution can only be given as an infinite series, values of the influence ordinates, bending moments, etc. must be expressed as sums of the component values for each term n of the series of component loads Q_{nx}. The values needed for the design of the ribs are the bending moments in the orthotropic plate over the *floor beams* and at the midspan of the ribs. These moments are obtained by multiplying the values of the component loads Q_{nx} by corresponding influence-ordinate components η_n.

$$M = \sum Q_{nx}\eta_n \tag{14.13}$$

This is shown in Figure 14.20. Formulas and design charts for Q_{nx} and η_n for the various AASHTO loading cases are given in Ref. [1].

The bending moments M_y in the direction of the ribs are obtained in the substitute orthotropic plate system of unit width of the deck. Usually only the maximum moment ordinate M_{ymax} at the center of the loaded rib is computed and the moment acting on one rib is then obtained, conservatively, as

$$M_R = M_{ymax}(a+e) \tag{14.14}$$

In the second stage of the design, the effect of the floor beam flexibility is considered. This effect will result in an increase of bending moment in the middle span and reduction of bending moment at the floor beam. The magnitude of the effect is determined based on the relative rigidity between ribs and floor beam. The effective width of the plate used for the first stage calculations generally can be used for the second stage with small error. Modifications of bending moments and shears in the ribs due the floor beam flexibility may be computed in the formulas and design charts in Ref. [1].

14.4.4 Floor Beam and Girder Design

Dead-load bending moments and shears in a *floor beam* (Member III) are calculated based on its own weight and the weights of tributary area of the deck. Live-load bending moments and shears in a floor beam are computed in two stages. In the first, the floor beams are assumed to act as rigid supports for the continuous ribs. The bending moments and shears are determined using a conventional method.

In the second stage, the effect of the floor beam flexibility is calculated, which tends to distribute the load on a directly loaded floor beam to the adjacent floor beams. The magnitude of this effect is a function of the relative rigidities between ribs and floor beams. The floor beam flexibility will reduce the bending moments and shears calculated in the first stage. The bending moment and shear corrections in the floor beams may be computed by the formulas and charts in Ref. [1].

The floor beam design per AASHTO-LRFD [5] for transverse flexure is

$$\frac{M_{fb}}{M_{rb}} + \frac{M_{ft}}{M_{rt}} \le 1.0 \tag{14.15}$$

where
M_{fb} = applied moment due to the factored loads in transverse beam (N-mm)
M_{rb} = factored moment resistance of transverse beam (N-mm)
M_{ft} = applied transverse moment in the deck plate due to the factored loads as a result of the plate-carrying wheel loads to adjacent longitudinal ribs (N-mm)
M_{rt} = factored moment resistance of deck plate in carrying wheel loads to adjacent ribs (N-mm)

For deck configurations with the spacing of the transverse beams larger than at least three times the spacing of longitudinal rib webs, the second term in Eq. (14.15) may be ignored. It should be noted that the applied moment M_{fb} should be obtained and based on the superstructure configuration.

The methods of analysis and design of the girders (Member IV) are described in other chapters of this book. The effective width of an orthotropic deck acting as the top flange of a girder or a floor beam is a function of the ratio of the span length to the girder or floor beam spacing, the cross-sectional area of the stress-carrying stiffeners, and the type and position of loading. Values of effective widths for the case of uniform loading based on a study by Moffatt and Dowling[58] are given in Figure 14.21 [5]. This figure was originally developed to determine the effective width of deck to be considered active with each web of a box girder, but is also applicable to other types of girders. The cumulative effect of the stresses on a specific orthotropic deck needs to be carefully reviewed by the design engineer and is generalized by Figure 14.22.

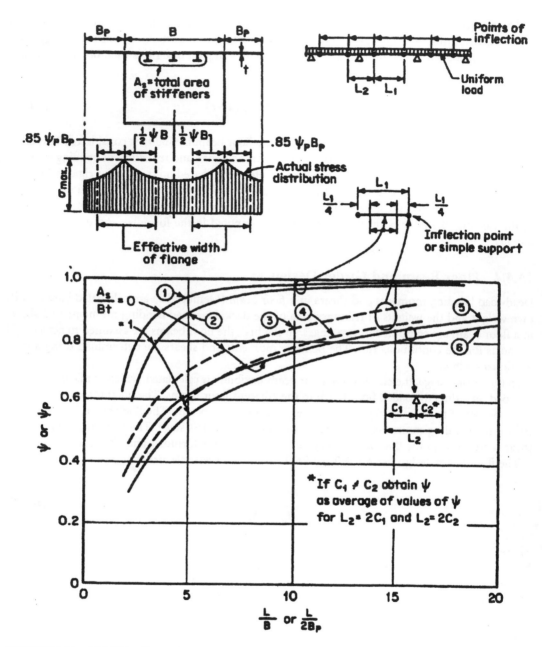

FIGURE 14.21 AASHTO effective deck width. (From American Association of State Highway and Transportation Officials, *LRFD Bridge Design Specifications*, Washington, D.C., 1994.)

14.4.5 Fatigue Considerations

Detailing of an orthotropic bridge is more involved because of all the numerous plates (see Figure 14.16 isometric). In-fill plates and complex geometric detailing make many orthotropic deck bridges one of a kind with unique details. The fatigue strength of the deck plate is very high, but thin deck plates may cause fatigue failures in other components. Fatigue of low-alloy deck plate of usual proportions subject to the AASHTO wheel loads is not considered a critical design factor. Reference [51] shows sketches of several European bridges with steel cracking caused by fatigue stress, and Ref. [52] shows how one bridge with wine-glass ribs was repaired. The original design

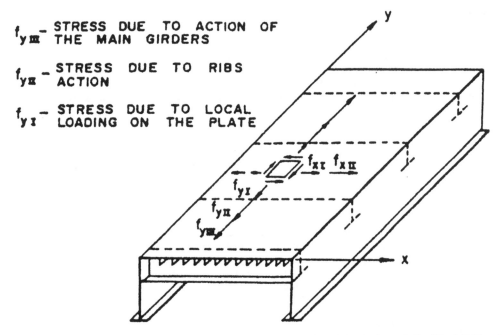

FIGURE 14.22 Resulting stresses in an orthotropic deck. (From Trotsky, M. S., *Orthotropic Steel Deck Bridges,* 2nd ed., JFL Arch Welding Foundation, Cleveland, OH, 1987. Courtesy of the James F. Lincoln Arc Welding Foundation.)

engineers did not fully understand long-term fatigue stress and CAFL (constant-amplitude fatigue life) issues. The AASHTO LRFD commentary [5] states: "Fatigue stress tests indicate that local out-of-plane flexural stress in the rib web at the junction with the deck plate should be minimized. One way to achieve this is to limit the stress in the rib web caused by the rotation of the rib–deck plate junction by making the rib welds relatively slender compared to the deck plate." Fatigue issues add to the complexity of orthotropic box-girder bridge design.

Excessive welding causes shrinkage of the weld metal which may cause additional locked-in stresses. In addition there are more components that are cut to fit between other components. If a component is cut too short this may also cause problems during welding. If additional filler welding is installed to span a larger incorrectly fabricated gap or root opening, then more weld width can increase the probability of shrinkage stresses. Detailing, imperfections in the alignment of a components, and quality control in fabrication have a direct result on the end product. Some designers prefer to let the fabricator select the welding processes (submerged arc welding, gas welding, etc.). Therefore, the welders and management of the fabrication plant are only required to achieve the strength and performance. Both AASHTO specifications require the use of 80% partial penetration welds of the ribs to deck plate (see Figure 14.3).

The AASHTO LRFD [5] maximum allowable stress limit for various details is shown in Table 14.7. The floor beam web cutouts around the rib were developed by the Germans. When ribs are designed to be continuous, it has been found to be best to make the rib go through the floor beam or diaphragm plate. Coping near the bottom of the rib and below greatly reduces the chances of fatigue cracking. There is still research going on about the actual shape of the holes. The cutout pattern per AASHTO is shown in Figure 14.23. However, it should be pointed out that in Figure 14.5 the German Railways has experimented with a different shape. AASHTO does not provide guidance on V-shaped ribs as shown in Figure 14.15. When components align on opposite sides of a plate, AASHTO Detail A must be followed for stress limits. Misaligned plates normally contribute to long-term fatigue failures. Full-scale fatigue testing of the Williamsburg Suspension Bridge has shown that code-acceptable details have a shorter fatigue life than what experts are able to develop with

TABLE 14.7 AASHTO Fatigue Detailing

Illustrative Example	Detail	Description of Condition	Detail Category
	Transverse or longitudinal deck plate splice or rib splice	1. Ceramic backing bar. Weld ground flush parallel to stress.	B
		1. Ceramic backing bar.	C
		3. Permanent backing bar. Backing bar fillet welds shall be continuous if outside of groove or may be intermittent if inside of groove.	D
	Bolted deck plate or rib splice	4. In unsymmetrical splices, effects of eccentricity shall be considered in calculating stress.	B
	Deck plate or rib splice Double groove welds	5. Plates of similar cross-section with welds ground flush. Weld run-off tabs shall be used and subsequently removed, plate edges to be ground flush in direction of stress. 6. The height of weld convexity shall not exceed 20% or weld width. Run-off tabs as for 5.	B
	Welded rib "window" field splice Single groove butt weld	7. Permanent backing bar—deck plate weld made with ceramic backing bar only. Welding gap > rib wall thickness f = axial stress range in bottom of rib	D
	Rib wall at rib/floorbeam intersection Fillet welds between rib and floorbeam web	8. Closed rib with internal diaphragm inside the rib or open rib. Welding gap > rib wall thickness f = axial stress range in rib wall 9. Closed rib, no internal diaphragm inside of rib $f = f_1 = f_2$ f2 = local bending stress range in rib wall due to out-of-plane bending caused by rib–floorbeam interaction, obtained from a rational analysis	C C

Source: American Association of State Highway and Transportation Officials, *AASHTO LRFD Bridge Design Specifications,* 2nd ed. Washington, D.C., 1998. With permission.

proper funding, as shown in Figure 14.24. This solution was selected as the optimal detail [49]. Termination of a weld is a very common place for a fatigue crack to begin. A runoff tab plate allows the welder to terminate the weld in this piece of steel. The tab plate is cut off and tossed in the recycle bin. The source of a potential flaw is now no longer part of the final structure. The weld tab plate system was introduced to the orthotropic deck detail. This tab plate is cut off with the 1/2-in. radius as shown in Figure 14.24 after welding. This extra step was used on one test panel to compare against test panel without tab plates. These researchers believe that deck plates of 8 to 12 mm are too thin for more than 2 million cycles. For more-detailed discussion of fatigue, see Chapter 53.

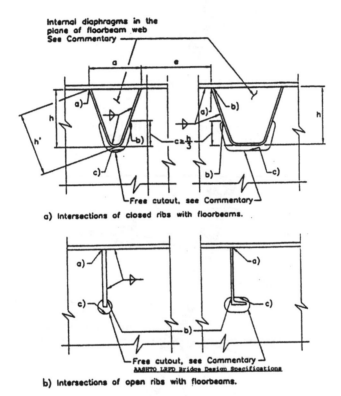

FIGURE 14.23 Detailing requirements for orthotropic deck. (From American Association of State Highway and Transportation Officials, *LRFD Bridge Design Specifications*, 2nd ed., Washington, D.C., 1998.)

14.4.6 Bridge Failures

Unfortunately, from 1969 to 1971 there were four steel box-girder collapses in four different countries, which caused the bridge engineering industry to reevaluate its different design code formulas and methods of erections. The Rhine River Bridge at Koblenz, Germany; the fourth Danube Bridge in Vienna, Austria; the Milford Haven Bridge in Wales, Great Britain; and the Yarra River Bridge in Australia were the four bridges that collapsed, making it a global bridge design issue [53-55]. It is a sobering thought to realize that 35 construction workers died at the Yarra River Bridge collapse [56,57]. This bridge was redesigned with an orthotropic steel deck to reduce its original dead weight. In Great Britain, a Board of Inquiry under the chairmanship of A. W. Merrison produced a list of new design rules and code details for fatigue issues. Extensive testing, research, and symposiums were held in Great Britain [3,4]. The British Institution of Civil Engineers [1979] held a symposium in London to have engineers from around the world share their experiences and ideas. It is very important to remember that codes are imperfect and the long-term fatigue details are still evolving and research continuing [58-64]. When the West Virginia bridge collapsed due to fracture critical failure of a single steel member, 50 individuals died. The FHWA responded by initiating a mandatory bridge maintenance program.

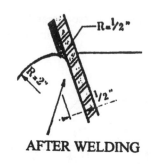

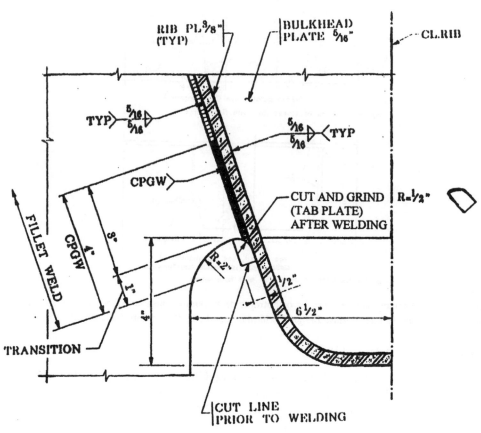

FIGURE 14.24 Fatigue-resistant detail Williamsburg Suspension Bridge. (From Khazen, D., *Civil Eng. ASCE*, June, 1998. With permission.)

14.4.7 Corrosion Protection

The AASHTO LRFD commentary [5] states: " The interior of the closed ribs cannot be inspected and/or repaired. It is essential to hermetically seal them against the ingress of moisture and air." (see Figure 14.3 for a solution). Atmospheric corrosion of steel requires water and a continuous supply of fresh air. Abrasion will speed up the process.

The three different methods that can be used to protect corrosion for new bridges with orthotropic decks are painting, weathering steel, or dehumidification. Painting is the most common method (see Table 14.8 for one Japanese Standard) reference [64]. Weathering steel was invented by the steel industry to eliminate painting. Corrosion can continue if the rusted layer is abraded away exposing bare steel, thus allowing the continuation of corrosion or rusting. Therefore, some designers normally provide an extra thickness of steel with weathering steel in case abrasion occurs. The steel

TABLE 14.8 Japanese Painting Specifications for Steel Bridges

Surface	N	Painting in Workshop	Painting in Field	Remarks
A	6	1 Etching primer 2,3 Lead anticorrosive paint[a] 4 Phenolic MIO[b] paint	5,6 Long oil phethalic resin coating 5,6 Chlorinated rubber paint	General location Near sea and over sea
B	3	1 Etching primer 2,3 Epoxy coal-tar paint		
C	5	1 Inorganic zinc-rich primer 2 Organic zinc-rich paint 3 Epoxy MIO paint	4,5 Urethane resin paint 4,5 Chlorinated rubber paint	Painting before surfacing with asphalt Painting after surfacing
D	3	1 Inorganic zinc-rich primer 2,3 Epoxy coal-tar paint	No painting	Inner surface of box girder with steel deck
E	2	1 Inorganic zinc-rich primer 2 Organic zinc-rich paint		

[a] For example, lead suboxide anticorrosive paint (red lead).
[b] MIO = micaceous iron oxide
N = number of painting film
Source: Nakai, H. and Yoo, C.H., *Analysis and Design of Curved Steel Bridges*, MacGraw-Hill, New York, 1968. With permission.

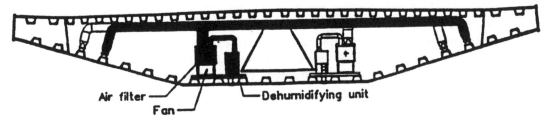

FIGURE 14.25 Dehumidification plant. (Courtesy of Monberg and Thorsen.)

towers for the Luling Bridge are also fabricated of weathering steel; for a color photo of its aesthetic appearance, see Ref. [42]. Another type of steel has been developed for use for contact with salt water or mariner steel.

Inspecting bridges with small access holes is facilitated with an electric-driven inspection cart; the individual rides inside long shallow orthotropic superstructures looking for fatigue cracks or corrosion [65]. The Normandie Bridge in France utilizes a mechanical dehumidification system (see Figure 14.25). The segments were shop-fabricated and then were full butt field-welded at the bridge site. This produced full structural continuity plus completely sealed the bridge superstructure. Exterior access doors for maintenance personnel have gasket seals to eliminate infiltration of air into the superstructure. Dehumidification is popular in European bridges (see Figure 14.11). An air supply system or aspirators are needed for maintenance personnel during bridge inspections. Some engineers are skeptical that dehumidification can actually remove all the moisture. Proponents of dehumidification point out that an air supply system or aspirators are needed for maintenance personnel during future maintenance painting inside the confined space of the box girder. State-of-the-art European technology went into the detailing of this bridge. Rather than paint the interior of the bridge, a mechanical dehumidification process is utilized to prevent corrosion of the interior of the superstructure. A central bridge maintenance walkway gallery was created in a triangular opening. The sections were prefabricated in an assembly-line process, then full butt-welded together at the bridge site.

Hot-dip galvanizing may be utilized for corrosion protection of smaller field-bolted orthotropic deck panels, which are used on deck retrofits and temporary bridges. The fabricated steel component is dipped into a molten zone in a tub. The maximum size of a component is limited by the tub size. A disadvantage is that warping of fabricated components can occur from the hot zinc. Hot-dip galvanizing can be cost-effective for limited-size components, especially near highly corrosive salt water. Drain holes would be needed in any closed ribs to allow molten zinc to drip out. Field welding of hot-dipped galvanizing creates toxic fumes.

14.3.8 Wearing Surface

An orthotropic steel plate deck must be paved with a wearing surface to provide a durable and skid-resistant surface for vehicular traffic. AASHTO [5] states: "The wearing surface should be regarded as an integral part of the total orthotropic deck system and shall be specified to be bonded to the top of the deck plate. For the purpose of designing the wearing surface itself, and its adhesion to the deck plate, the wearing surface shall be assumed to be composite with the deck plate, regardless of whether or not the deck plate is designed on that basis." AREA (American Railroad Engineers Association) specifications should be used for railroad bridges as shown in Figure 14.5. Some materials used in wearing surfaces are proprietary or patented. Aesthetic issues may control for pedestrian bridges or sidewalk areas as shown in Figure 14.16. The surfacing for vehicular traffic performs several functions:

- Exhibits high skid resistance throughout the life of the surfacing;
- Provides a smooth riding surface for the comfort of the drivers using the bridge;
- Helps waterproof the steel deck;
- Resists cracking, delaminating, and displacing for a long service life with low maintenance cost and disruption to traffic for repairs;
- A stiff and well bonded surfacing can also provide some reduction in fatigue stresses in the steel deck, ribs, and welds by dynamic composite action with the deck plate. In addition, surfacing with thickness of about 50 mm provides some distribution of the wheel loads and damping of the steel plate.

The number of wheel load applications during the life of the surfacing can be enormous as each passage of a truck wheel stresses the surfacing and the steel deck. For example, assume a bridge carries 10,000 vehicles a day with 5% being three-axle trucks. If half of these are loaded trucks, the annual full-load applications are about a quarter of a million or 4.5 million for a 20-year service life. A busy bridge carrying a high percentage of trucks can exceed this figure. The wheel load applications for the design of a surfacing should be calculated for the specific bridge site. Vehicle tires also wear and polish the aggregates on the surface of the surfacing causing a reduction in skid resistance. Hard, durable, and polish-resistant aggregates should be selected, preferably with small asperities projecting from the surface. The asperities help to increase the dry skid resistance and also provide a rough surface for improved bonding with the aggregate within the matrix of the surfacing. A long life is required for the surfacing of a heavily traveled bridge as replacement of the surfacing, whether for fatigue cracking, debonding, or for lack of skid resistance, is costly. The surfacing is costly to remove and relay and is also costly to the user as it disrupts and delays traffic.

Treated timber plank wearing surfaces have been used for a few bridges, as shown in Figure 14.7 and described in Section (14.3). Timber planks are not watertight so the steel deck must be painted or other corrosion protection provided. Welded threaded studs or economy head bolts can hold the planks in place. There are three basic classifications of surfacing for orthotropic decks:

1. Thin surfacings from about 4 to 8 mm thick;
2. Surfacings composed of mastics with binders of asphalt or polymer resins usually laid from about 12 to 75 mm thick;
3. Surfacings composed of concertos with binders of asphalt or polymer resins usually laid from about 18 to 60 mm thick.

All three types require a bond coat on the steel deck to hold the surfacing in place against the forces of braking truck wheels and to provide intimate contact with the steel deck for dynamic composite action for each passage of a wheel. The strength of the bond must last for the life of the surfacing.

Thin surfacings are usually laid by flooding the deck with a thermosetting polymer resin and broadcasting a hard aggregate that is locked into the binder to provide skid resistance. Thin surfacings are not appropriate for decks carrying high truck traffic. The repetitive wheel loads will wear away the aggregate exposing the slick resin surface producing a very low skid resistance. The repetitive wheel loads can also wear away the resin and expose the steel deck. However, a thin surfacing can be used as a temporary wearing course for bridge deck replacements. The new orthotropic deck panels can be paved in the shop and installed a panel at a time during replacement of an existing concrete deck. Traffic can run on the temporary surfacing up to 2 years, as was done for the deck of the Golden Gate Bridge. After all the panels are installed, the permanent pavement can be laid as a continuous operation for a smooth riding surface.

Mastics are usually a mixture of aggregates and binder of asphalt, polymer-modified asphalt, or polymer resins. The binder is proportioned in excess of that required to fill the air voids. The strength of mastic surfacing is dependent on the strength of the binder material rather than on the interlock of the aggregates. During placement, the mix is flowed onto the deck and leveled, usually by a vibrating screed, before the binder sets. Hard, durable stones can be broadcast over the surface to improve skid resistance. The high-binder-content polymer resins may cause high shrinkage strains and high bond stress on the deck plate. Gussasphalt (German word for poured asphalt) is a mastic using a low penetration asphalt as a binder heated to a high temperature of about 200°C and applied usually by poring and leveling to a thickness of about 75 mm or more by hand labor. It has been used apparently successfully in Europe and Japan.

Concrete is usually a mixture of polymer resins or polymer-extended asphalt with aggregates with some air voids up to 4% remaining unfilled. Concrete surfacing for steel decks should not be confused with portland cement concrete, which is not suitable for steel-deck surfacing. The strength of concrete surfacings is dependent on both the strength of the interlock of the aggregates and the strength of the binder material. They require compaction usually by a steel roller or vibrating screed. They have the advantage of being mixed, placed, and compacted using conventional paving equipment. Ordinary or modified asphalt concrete surfacings have low first cost, but have not given long, trouble-free service life. If the binder is a thermoset-resin-extended asphalt, such as epoxy asphalt, the cost is increased somewhat but the added strength imparted by the thermoset resin greatly improves the performance and life of the surfacing.

Failure of the wearing surface is common, but the Hayward-San Mateo Bridge still has its original wearing surface after 30 years [24]. Prior to the construction of this bridge a small test panel (truck weight scale) was used to test the durability of the wearing surface under actual truck traffic. The San Diego–Coronado wearing surface was replaced after 25 years. The Caltrans test bridge used two types of wearing surfaces (thin and thick) [13]. The failure of a wearing surface can be caused by the deck plate thickness (stiffness); poor construction practices; installation quality control; bridge deck splice details (bolt heads and splice plates); and/or the temperature range (freeze–thaw action) plus humidity conditions expected at the bridge site. Due to all these factors wearing surfaces can fail very quickly, and few last over 20 years. Flexible orthotropic decks can cause a stiffer wearing surface to pop off. This is one reason AASHTO wants the designer to think of the wearing surface as a composite material that can "structurally" fail, when not deflecting in synchronization with the deck. For case history details on the wearing surfaces of many bridges, the reader can refer to References [1, 2]. The Miller Sweeney bascule bridge's wearing surface failed by creep while the movable span was in the vertical position.

A wearing surface of a sacrificial material is placed on top of the deck since vehicles' wheels cause abrasion. Asphalt and epoxy concrete are the most commonly used materials for wearing surfaces. Timber wearing surfaces have been used as described in Section 14.3. For details on the many possible types of wearing surfaces, the reader may refer to Ref. [1–4]. The proposed test system has been developed for Caltrans [44].

14.3.9 Future Developments

The second generation of orthotropic deck bridges will be better based on lessons learned from the first group of bridges built. Many orthotropic bridges were never built and have remained only a dream. The most interesting is the Ruck-A-Chucky cable-stayed bridge designed by Prof. T. Y. Lin to be built across the flooded American River Canyon after the completion of the Auburn Dam near Sacramento, California [66]. This horseshoe-shaped superstructure was planned to be supported by cable stays anchored into the sloping canyon rock walls. Therefore, no towers would be necessary. Scale models were built for wind tunnel and earthquake shake-table testing. The 1977 orthotropic design featured trapezoidal ribs for a five-cell, 14.61-m-wide superstructure. The bridge plans used a 396-m clear span on a 457-m radius for two lanes of traffic and an equestrian trail.

A promising concept, patent pending to a Redding, California firm, is to have the bridge deck comprise only nested trapezoidal ribs, which are welded together. The three-sided "rectangular" ribs are placed at 90° to the driving surface, so that the sides of the rib become the top and bottom flanges. The result is to achieve a 12-m span that is about 275 mm deep or wide rib is required. This system would compete economically against concrete slab bridges and has been marketed at bridge conventions. Another unique rib system for floating steel bridges or pontoons is the "biserrated-rib" developed by Dr. Arsham Amirikian, consultant to Naval Facilities Engineering Command, Department of the Navy. Portions of the sides of the trapezoidal ribs are removed in a repetitive scallop pattern to reduce the weight of the rib. The structural strength is almost identical to a full rib. The disadvantage would be increased (double) surface area exposed to corrosion.

A panelized orthotropic deck system was developed and built for trapezoidal ribs for a temporary detour bridge in New York [67]. A cross section of this bridge built in 1991 is shown in Figure 14.26a. This system is the engineering evolution of a previous concept for open ribs. A system was developed and engineered for a steel grating company utilizing open ribs for a bridge system [68]. A cross section of the bridge proposed in 1961 is shown in Figure 14.26b. This system would use similar materials stored in the steel grating company warehouse, and allow them to market another product.

The future of orthotropic deck cost reduction lies in the standardization of ribs and details by AASHTO or the steel industry. Such standardization has led to the popularization of precast prestressed concrete girders.

Acknowledgments

The writing of this chapter included the support and suggestions of our co-workers at Caltrans especially Dr. Lian Duan PE, and other individuals plus their employers as identified in the credit lines and references. Mr. Scott Whitaker P.E., Engineer of the Bethlehem Steel Corporation, provided written suggestions for the text, as well as a wealth of articles, many which are included in the references. Special thanks to Mr. Chuck Seim P.E. of T. Y. Lin International for sharing his ideas on wearing surface issues.

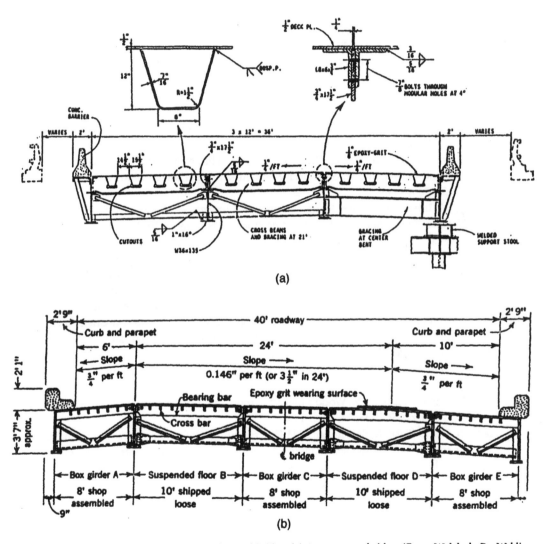

(a)

(b)

FIGURE 14.26 Prefabricated orthotropic deck panel bridge. (a) A temporary bridge. (From Wolchuk, R., *Welding Innovation Q.*, IX(2), 19, 1992. Courtesy of the James F. Lincoln Arc Welding Foundation.) (b) A short-span bridge. (From Chang, J.C.L., *Civil Eng. ASCE*, Dec. 1961. With permission.)

References

1. Wolchuk, R., *Design Manual for Orthotropic Steel Plate Deck Bridges*, American Institute of Steel Construction, Chicago, 1963.
2. Troitsky, M.S., *Orthotropic Bridges — Theory and Design*, 2nd ed., James F. Lincoln Arc Welding Foundation, Cleveland, OH, 1987.

3. ICE, Steel box girder bridges, in Proceedings of the International Conference, the Institution of Civil Engineers, Thomas Telford Publishing, London, 1972.

4. Cartledge, P., Ed., *Proceedings of the International Conference on Steel Box Girder Bridges,* the Institution of Civil Engineers, Thomas Telford Publishing, London, 1973.

5. AASHTO, *LRFD Bridge Design Specifications,* American Association of State Highway and Transportation Officials, Washington, D.C., 1994.

6. Galambos, T. V. Ed., *Guide to Stability Design Criteria for Metal Structures,* 4th ed., John Wiley & Sons, New York, 1988, Chap. 7.

7. Wolchuk, R., Orthotropic redecking of bridges on the North American continent, *Struct. Eng. Int.,* IABSE, 2(2), 125, 1992.

8. Wolchuk, R., Applications of orthotropic decks in bridge rehabilitation, *Eng. J. AISC,* 24(3), 113, 1987.

9. JFL, *Modern Welded Structure Volume Selection,* Vols. I and II, James F. Lincoln Arc Welding Foundation, Cleveland, OH, 1968, C-1.

10. Guadalajara Bridge, Mexico, *Eng. News Rec.,* McGraw-Hill, New York, Aug 7. page 58, 1969.

11. Popov, O. and Seliverstov, V., Steel bridges on Ankara's Perimeter Motorway, *Struct. Eng. Int.,* IABSE, 8(3), 205, 1998.

12. Ramsay, W., Innovative bridge design concepts steel bridges — the European way, in *Proceedings of the National Symposium on Steel Bridge Construction,* November 10–12 Atlanta, AASHTO, AISC, FHWA, 1993.

13. Davis R. E., *Field Testing of an Orthotropic Steel Deck Bridge,*Vol. 1 and 2, California Department of Public Works, Division of Highways, Bridge Department, Sacramento, CA, 1968.

14. JFL, Orthotropic bridge designed as solution to concrete deterioration (Crietz Road), in *Modern Welded Steel Structure,* III, James F. Lincoln Arc Welding Foundation, Cleveland, OH, 1970, B-10.

15. Risch, J. E., Final Report of Experimental Orthotropic Bridge S05 of 23081 A Crietz Road Crossing over I-496 Three Miles West of the City Limits of Lansing, Michigan, DOT project 67 G-157; FHWA Project I 496-7(21), 1971.

16. Heins, C. and Firmage, D. A., *Design of Modern Steel Highway Bridges,* John Wiley & Sons, New York, 1979.

17. Foley, E. R. and Murphy J. P., World's longest orthotropic section feature of San Mateo–Hayward Bridge, *Civil Eng. ASCE,* 38, 54, April 1968.

18. Haibach E. and Plasil, I., September 1983 Untersuchungen zur Betriebsfestigkeit von Stahlleichtfahrbahnen mit Trapezholsteifen im Eisenbahnbruckenbau [The fatigue strength of an orthotropic steel deck plate with trapezoidal closed longitudinal ribs intended for use in railway bridges] in *Der Stahlbau* [The Steel-builder], 269, Ernst & Sohn, Berlin, 1983 [in German].

19. Ozolin, E., Wilson, W., and Hutchison, B., Valdez floating dock mooring system, in *The Ocean Structural Dynamic Symposium,* Oregon State University, Corvallis, Sept., 1982, 381.

20. Carlson, L. A., Platzke, R., and Dreyer, R. C. J., First bridge across the Yukon River, *Civil Eng. ASCE,* 47, August 1976.

21. Hedefine, A., Orthotropic-plate girder bridges, in *Structural Steel Engineering Handbook,* Merritt, F. S., Ed., McGraw Hill, New York, 1972, chap. 11.

22. Merritt, F. S. and Geschwindner, L. F. Analysis of special structures, in *Structural Steel Engineering Handbook,* 2nd ed., Merritt, F. S., Ed., McGraw Hill, New York, 1994, chap. 4.

23. Bouwkamp, J. G., Analysis of the Orthtropic Steel Deck of the San Diego–Coronado Bridge, Report No. 67-20, Structural Engineering Laboratory, University of California, Berekely, 1969.

24. Balala, B., First orthotropic bridge deck paved with epoxy asphalt, (San Mateo-Haywood) *Civil Eng. ASCE,* page 59, April 1968.

25. Curtis, G. N., Design of the Queens Way Bridge, *Modern Welded Structure,* IV, James F. Lincoln Arc Welding Foundation, Cleveland, OH, 1980, A-34.

26. ENR, Prefab steel bridge girders are biggest ever lifted (Queens Way Bridge), *Engineering News Record,* McGraw Hill, New York, August 27, 1970 p. 34.

27. Construction Marketing of Bethlehem Steel Corporation in cooperation with Universal Structural Inc. Steel Bridge Report BG-502 — Cypress Reconstruction — Contract E (Maritime Off- Ramp), Bethlehem Steel Corporation, Bethlehem, PA, 1997.

28. Marquez, T., Huang, C., Beauvoir, C., Benoit, M., and Mangus, A., California's 2356 foot long orthotropic bridge for I-880 Cypress Replacement Project (Maritime Off-Ramp), in *Proceedings of 15th International Bridge Conference*, Pittsburgh, 1998. (IBC 98-22).

29. Marquez, T., Williams, J., Huang, C., Benoit, M., and Mangus, A., Unique steel curved orthotropic bridge for I-880 Cypress Replacement Project (Maritime Off-Ramp), in *Proceedings of International Steel Bridge Symposium*, the National Steel Bridge Alliance, Chicago, IL, 1998.

30. Leonhardt, F., *Bridges Aesthetics and Design*, MIT Press, Cambridge, MA, 1984.

31. Arenas, J. J. and Pantaleon, M. J., Barqueta Bridge Seville, Spain, *Struct. Eng. Int.*, IABSE, 2(4), 251, 1992.

32. Cerver, F. A., Ed., *The Architecture of Bridges*, Barcelona, Spain, 1992, 186.

33. Tokola, A. J. and Wortman, E. J., Erecting the center span of the Fremont Bridge, *Civil Eng. ASCE*, 62, July 1973.

34. Thomsen, K. and Pedersen, K. E, Swing bridge across a navigational channel, Denmark, *Struct. Eng. Int.*, IABSE, 8(3), 201, 1998.

35. Bowen, G. J. and Smith, K. N., Walpole swing span has orthotropic deck, *Heavy Constr. News*, Canada, Feb page 6, 1970.

36. Bender O., Removable section — Sacramento River Bridge at Colusa, in *Arc Welded in Manufacturing and Construction*, II, James F. Lincoln Arc Welding, Cleveland, OH, 1984, page (D-15).

37. Maruyama, T., Watanabe, E., and Tanaka, H., Floating swing bridge with a 280 m span, Osaka, *Struct. Eng. Int.*, IABSE, 8(3), 174, 1998.

38. Tsinker, G. P., *Floating Ports Design and Construction Practices*, Gulf Publishing Company, Houston, TX, 1986.

39. Solland, G., Stein, H., and Gustavsen, J. H., The Bergsoysund Floating Bridge, *Struct. Eng. Int.*, IABSE, 3(3), 142, 1993.

40. Troitsky, M. S., *Cable-Stayed Bridges an Approach to Modern Bridge Design*, 2nd ed., Van Nostrand Reinhold, New York, 1988.

41. Podolny, W., Jr. and Scalzi, J. B., *Construction & Design of Cable-Stayed Bridges*, John Wiley & Sons, New York, 1976.

42. ASCE, Bridge [discusses the Luling Bridge], *Civil Eng. ASCE*, 31, July 1984.

43. ENR, Stayed-girders reaches record [discusses the Luling Bridge], *Engineering News Record*, McGraw-Hill, New York, April 8, 1992, page 33.

44. T. Y. Lin International–Moffat & Nichcols, a joint venture, San Francisco–Oakland Bay Bridge Structure Type Selection Report to Caltrans, San Francisco, CA, May 1998.

45. Kamei, M., Maruyama, T., and Tanaka H., Konohana Bridge Japan, *Struct. Eng. Int.*, 2(1), 4, 1992.

46. DeLeuw Cather–OPAC–Steinman, Third Carquinez Strait Bridge Structure Type Selection Report — Caltrans Contract No. 59A0007, San Francisco, CA, 1997.

47. BD&E, Monster of Messina, in *Bridge Design & Engineering*, Route One Publishing, London, 9, 7, 1997.

48. Gimsing, N., *Cable-Supported Bridges*, John Wiley & Sons, New York, 1997.

49. Kaczinski, M. R., Stokes, F. E., Lugger, P., and Fisher J. W., Williamsburg Bridge Orthotropic Deck Fatigue Test, ATLSS Report No. 97-04, Lehigh University, Bethlehem, PA, 1997.

50. AASHTO, *Standard Specifications for Highway Bridges*, 16th ed., American Association of State Highway and Transportation Officials, Washington, D.C., 1996.

51. Wolchuk, R., Lessons from weld cracks in orthotropic decks on three European bridges, *Welding Innovation Q.*, II(I), 1990.

52. Nather, F., Rehabilitation and strengthening of steel road bridges, *Struct. Eng. Int. IABSE*, 1(2), 24, 1991.

53. *ENR*, Rhine River Bridge Collapse, Koblenz Germany, *Engineering News Record*, McGraw Hill, New York, Nov., 18, 1971, p. 17; Nov., 23, 1972, p. 10; Dec. 20, 1973, p. 26).

54. *ENR*, 4th Danube Bridge Collapse, Austria, *Engineering News Record*, McGraw Hill, New York, Nov. 13 p. 11, and Dec. 4, P. 15, 1969.

55. *ENR*, Milford Haven Bridge Collapse, Great Britain, *Engineering News Record*, McGraw Hill, New York, June 13, p. 11; Dec 4, 69 p. 15, 1970.

56. Kozak, J. and Seim, C., Structural design brings West Gate Bridge failure (Yarra River Melbourne, Australia), *Civil Eng. ASCE*, June 1972, pp 47-50.

57. Wolfram, H. G. and Toakley, A. R., Design modifications to West Gate Bridge, Melbourne, Institution of Engineers, *Civil Engineering Transactions* CE16, Australia, 143, 1974.

58. Moffatt, K. R. and Dowling, D. J., *Parametric Study on the Shear Lag Phenomenon in Steel Box-Girder Bridges*, Engineering Structures Laboratory, Imperial College, London, 1972.

59. Wolchuk, R. and Mayrbourl, R. M. Proposed Design Specification for Steel Box Girder Bridges, RN FHWA-TS 80-205, U.S. Department of Transportation, Federal Highway Administration Washington, D.C., 1980.

60. Milek, W. A., Jr., How to use the AISC Orthotropic Plate Design Manual, *AISC Eng. J.*, 40, April 1964 page 40.

61. Wolchuk, R., Steel-plate-deck bridges and steel box girder bridges, *Structural Engineering Handbook*, 4th ed., Gaylord, E. H., Gaylord C. N., and Stallmeyer, J. E., Eds., McGraw Hill, New York, 1997, chap 19.

62. Xanthakos, P. P., *Theory and Design of Bridges*, John Wiley & Son, New York, 1994.

63. AISC, Orthotropic plate deck bridges, in *Highway Structures Design Handbook*, Vol. 1, American Institute of Steel Construction Marketing, Inc., Chicago, IL, 1992.

64. Nakai, H. and Yoo, C. H., *Analysis and Design of Curved Steel Bridges*, McGraw-Hill, New York, 1988.

65. Thorsen, N. E. and Rouvillain F., *The Design of Steel Parts*, draft of Normandie Bridge France personal correspondence with Mondberg & Thorsen A/S Copenhagen, Denmark, 1997.

66. Lin, T.Y., Kulka, F., Chow P., and Firmage A., Design of Ruck-A-Chucky Bridge, American River, California USA, in *48TH Annual Convention of the Structural Engineers Association of California*, Sacramento, 1979, 133–146.

67. Wolchuk, R., Temporary bridge with orthotropic deck, *Welding Innovation Q.*, IX(2), 19, 1992.

68. Chang, J. C. L., Orthotropic-plate construction for short-span bridges, *Civil Eng. ASCE*, Dec. 1961.

69. Wolchuk, R., Steel Orthotropic Decks — Development in the 1990s, Transportation Research Board 1999 Annual Meeting.

70. Siem, C. and Ferwerda, Fatigue Study of Orthotropic Bridge Deck Welds (for proposed southern crossing of San Francisco Bay), 1972. California Dept. of Public Works Division of Highways and Division of Bay Toll Crossings.

15
Horizontally Curved Bridges

Ahmad M. Itani
University of Nevada at Reno

Mark L. Reno
California Department of Transportation

15.1 Introduction

As a result of complicated geometrics, limited rights of way, and traffic mitigation, horizontally curved bridges are becoming the norm of highway interchanges and urban expressways. This type of superstructure has gained popularity since the early 1960s because it addresses the needs of transportation engineering. Figure 15.1 shows the 20th Street HOV in Denver, Colorado. The structure is composed of curved I-girders that are interconnected to each other by cross frames and are bolted to the bent cap. Cross frames are bolted to the bottom flange while the concrete deck is supported on a permanent metal form deck as shown in Figure 15.2. Figure 15.3 shows the elevation of the bridge and the connection of the plate girders into an integral bent cap. Figure 15.4 shows the U.S. Naval Academy Bridge in Annapolis, Maryland which is a twin steel box-girder bridge that is haunched at the interior support. Figure 15.5 shows Ramp Y at I-95 Davies Blvd. Interchange in Broward County, Florida. The structure is a single steel box girder with an integral bent cap. Figure 15.6 shows a photo of Route 92/101 Interchange in San Mateo, California. The structure is composed of several cast-in-place curved P/S box-girder bridges.

The American Association of Highway and Transportation Officials (AASHTO) governs the structural design of horizontally curved bridges through *Guide Specifications for Horizontally Curved Highway Bridges* [1]. This guide was developed by Consortium of University Research Teams (CURT) in 1976 [2] and was first published by AASHTO in 1980. In its first edition the guide specification included allowable stress design (ASD) provisions that was developed by CURT and load factor design (LFD) provisions that were developed by American Iron and Steel Institute under project 190 [15]. Several changes have been made to the guide specifications since 1981. In 1993 a new version of the guide specifications was released by AASHTO. However, these new specifications did not include the latest extensive research in this area nor the important changes that affected the design of straight I-girder steel bridges.

FIGURE 15.1   Curved I-girder bridge under construction — 20th St. HOV, Denver, Colorado.

FIGURE 15.2   Bottom view of curved I-girder bridge.

FIGURE 15.3 Curved I-girder bridge with integral bent cap.

The guide specifications for horizontally curved bridges under Project 12-38 of the National Cooperative Highway Research Program (NCHRP) [3] have been modified to reflect the current state-of-the-art knowledge. The findings of this project are fully documented in NCHRP interim reports: "I Girder Curvature Study" and "Curved Girder Design and Construction, Current Practice" [3]. The new "Guide Specifications for Horizontally Curved Steel Girder Highway Bridges" [18] proposed by Hall and Yeo was adopted as AASHTO Guide specifications in May, 1999. In addition to these significant changes, the Federal Highway Administration (FHWA) sponsored extensive theoretical and experimental research programs on curved girder bridges. It is anticipated that these programs will further improve the current curved girder specifications. Currently, the NCHRP 12–50 is developing "LRFD Specifications for Horizontally Curved Steel Girder Bridges" [19].

The guidelines of curved bridges are mainly geared toward structural steel bridges. Limited information can be found in the literature regarding the structural design of curved structural concrete (R/C and P/S) bridges. Curved structural concrete bridges have a box shape, which makes the torsional stiffness very high and thus reduces the effect of curvature on the structural design.

The objective of this chapter is to present guidelines for the design of curved highway bridges. Structural design of steel I-girder, steel, and P/S box-girder bridges is the main thrust of this chapter.

15.2 Structural Analysis for Curved Bridges

The accuracy of structural analysis depends on the analysis method selected. The main purpose of structural analysis is to determine the member actions due to applied loads. In order to achieve reliable structural analysis, the following items should be properly considered:

- Mathematical model and boundary conditions
- Application of loads

FIGURE 15.4   Twin box-girder bridge — U.S. Naval Academy Bridge, Annapolis, Maryland.

FIGURE 15.5   Single box girder bridge with integral bent cap — Ramp Y, I-95 Davies Blvd., Broward County, Florida.

FIGURE 15.6   Curved concrete box-girder bridges — Route 92/101 Interchange, San Mateo, California.

The mathematical model should reflect the structural stiffness properly. The deck of the super-structure should be modeled in such a way that is represented as a beam in a grid system or as a continuum. The boundary conditions in the mathematical model must be represented properly. Lateral bearing restraint is one of the most important conditions in curved bridges because it affects the design of the superstructure. The deck overhang, which carries a rail, provides a significant torsion resistance. Moreover, the curved bottom flange would participate in resisting vertical load. This participation increases the applied stresses beyond those determined by using simple structural mechanics procedures [3].

Due to geometric complexities, the gravity load will induce torsional shear stresses, warping normal stresses, and flexural stresses to the structural components of horizontally curved bridges. To determine these stresses, special analysis accounting for torsion is required. Various methods were developed for the analysis of horizontally curved bridges, which include simplified and refined analysis methods. The simplified methods such as the *V-Load* method [4] for I-girders and the *M/R* method for box girders are normally used with "regular" curved bridges. However, refined analysis will be required whenever the curved bridges include skews and lateral or rotational restraint. Most refined methods are forms of finite-element analysis. Grillage analysis as well as three-dimensional (3-D) models have been used successfully to analyze curved bridges. The grillage method assumes that the member can be represented in a series of beam elements. Loads are normally applied through a combination of vertical and torsion loads. The 3-D models that represent the actual depth of the superstructure will capture the torsion responses by combining the responses of several bridge elements.

15.2.1   Simplified Method: V-Load

In 1984, AISC Marketing, Inc. published "V-Load Analysis" for curved steel bridges [4]. This report presented an approximate simplified analysis method to determine moments and shears for horizontally

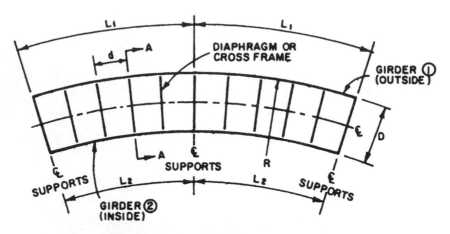

FIGURE 15.7   Plan view of two-span curved bridge.

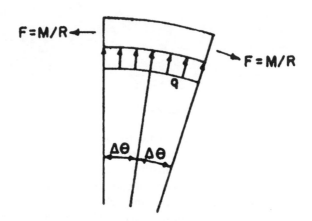

FIGURE 15.8   Plan view of curved bridge top flange.

curved open-framed highway bridges. This method is known as the V-Load method because a large part of the torsion load on the girders is approximated by sets of vertical shears known as "V-Loads." The V-Load method is a two-step process. First, the bridge is straightened out so that the applied vertical load is assumed to induce only flexural stresses. Second, additional fictitious forces are applied to result in final stresses similar to the ones in a curved bridge. The additional fictitious forces are determined so that they result in no net vertical, longitudinal, or transverse forces on the bridge.

Figure 15.7 shows two prismatic girders continuous over one interior support with two equal spans, L_1. Girder 1 has a radius of R and the distance between the girders is D. The cross frames are uniformly spaced at distance equal to d. As shown later, the cross frames in curved bridges are primary members since they are required to resist the radial forces applied on the girder due to bridge curvature.

When the gravity load is applied, the flanges of the plate girder will be subjected to axial forces $F = M/R$, as shown in Figure 15.8. However, due to the curvature of the girder, laterally distributed load q will be applied to flanges of the plate girder in order to achieve equilibrium. By assuming that the flanges resist most of the bending moment, the longitudinal forces in the flanges at any point will be equal to the moment, M, divided by the section height, h. Due to the curvature of the bridge, these forces are not collinear along any given segment of the flange. Thus, radial forces must be developed along the girder in order to maintain equilibrium. The forces cause lateral bending of the girder flanges resulting in warping stresses. The magnitude of the radial forces is equal to M/hR and has the same shape of the bending moment diagram as shown in Figure 15.9.

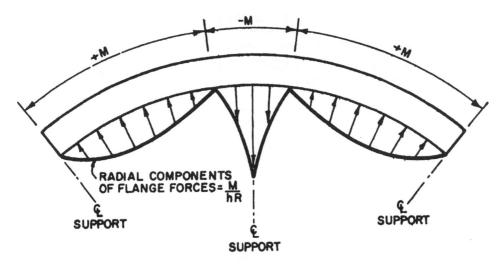

FIGURE 15.9   Lateral forces on curved girder flange.

This distributed load creates equal and opposite reaction forces at every cross frame as shown in Figure 15.10. By assuming the spacing between the cross frames is equal to d, the reaction force at the cross frame is equal to H, which is equal to Md/hR.

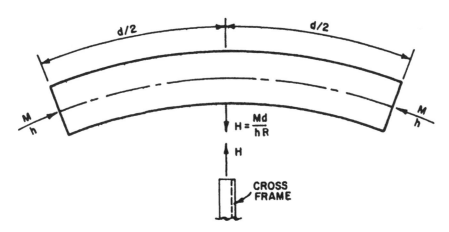

FIGURE 15.10   Reaction at cross frame location.

To maintain equilibrium of the cross frame forces, vertical shear forces must develop at the end of the cross frames as a result of cross frame rigidity and end fixity as shown in Figure 15.11.

15.3   Curved Steel I-Girder Bridges

15.3.1   Geometric Parameters

According to the current AASHTO specifications [13], the effect of curvature may be neglected in determining the primary bending moment in longitudinal members when the central angle of each span in a two or more span bridge is less than 5° for five longitudinal girders. The framing system

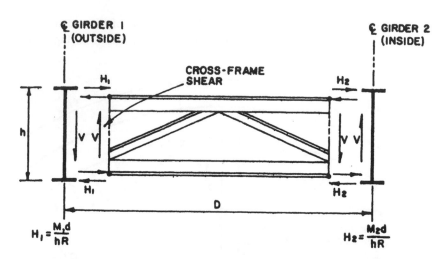

FIGURE 15.11   Equilibrium at cross frame location and the formation of V-loads.

for curved I-girder bridges may follow the preliminary design of straight bridges in terms of span arrangement, girder spacing, girder depth, and cross frame types. The choice of the exterior span length is normally set to give relatively equal positive dead-load moments in the exterior and interior spans. The arrangement results in the largest possible negative moment, which reduces both positive moments and related deflections. Normally, the depth of the superstructure is the same for all spans. Previous successful design showed a depth-to-span ratio equal to 25 for the exterior girder to be adequate. This ratio has been based on vibration and stiffness needed to construct the plate girders. Also, this ratio helps to ensure that the girders do not experience excessive vertical deflections. The uplift of the exterior girder should be prevented as much by extending the span length of the exterior girder rather than dealing with the use of tie-down devices.

Girder spacing plays a significant role in the deck design and the determination of the number of girders. Wider spacing tends to increase the dead load on the girders, while closer spacing requires additional girders, which increases the fabrication and erections costs. For curved steel I-girder bridges, the girder spacing varies between 3.05 m (10 ft) and 4.87 m (16 ft). Wider spacing, common in Europe and Japan, requires a post-tensioned concrete deck, which is not common practice in the United States. The overhang length should not exceed 1.22 m (4 ft) because it tends to increase the load on the exterior girders by adding more dead load and permitting truckload to be applied on the cantilever. The flanges of the plate girder should have a minimum width to avoid out-of-plane buckling during construction. Many steel erectors limit the length of girder shipping pieces to 85 times the flange width [5]. Based on that, many bridge engineers tend to limit the width of the flange to 40.6 mm (16 in) based on a maximum shipping length equal to 36.6 m (120 ft). It is also recommended that the minimum web thickness be limited to 11.1 mm (7/16 in) because of weld distortion problems. The thickness of the web depends on its depth and the spacing of the transverse stiffeners. This represents a trade-off between having extra material or adding more stiffeners. Many bridge engineers use the ratio of $D/t=150$ to choose the thickness of the web.

The spacing of the cross frame plays an important factor in the amount of force carried out by it and the value of flange lateral bending. Normally, cross-frame spacing is held between 4.57 m (15 ft) and 7.62 m (25 ft).

15.3.2 Design Criteria

The design guidelines, according to the Recommended Specifications for Steel Curved Girder Bridges [3], are established based on the following principles:

- Statics
- Stability

- Strength of materials
- Inelastic behavior

External and internal static equilibrium should be maintained under every expected loading condition. Stability of curved steel girder bridges is a very important issue especially during construction. By their nature, curved girders experience lateral deflection when subjected to gravity loading. Therefore, these girders should be braced at specified intervals to prevent lateral torsional buckling. The compactness ratio of the web and the flanges of curved I-girders are similar to the straight girders. The linear strain distribution is normally assumed in the design of curved girder bridges. The design specification recognizes that compact steel sections can undergo inelastic deformations; however, current U.S. practice does not utilize a compact steel section in the design of curved I-girder bridges.

The design criteria for curved girder bridges can be divided into two main sections.

- Strength
- Serviceability

Limit state design procedures are normally used for the strength design, which includes flexure and shear. Service load design procedures are used for fatigue design and deflection control. The primary members should be designed to be such that their applied stress ranges are below the allowable fatigue stress ranges according to AASHTO fatigue provisions [6]. The deflection check is used to ensure the serviceability of the bridge. According to the recommended specifications for the design of curved steel bridges [3], the superstructure should be first analyzed to determine the first mode of flexural vibration. The frequency of this mode is used to check the allowable deflection of the bridge as indicated in the *Ontario Bridge Code* [7].

15.3.3 Design Example

Following the 1994 Northridge Earthquake in California, the California Department of Transportation (Caltrans) embarked on a task of rebuilding damaged freeways as soon as possible. At the SR 14–I-5 interchange in the San Fernando Valley, several spans of cast-in-place prestressed concrete box girders have collapsed [9]. These were the same ramps that were previously damaged during the 1971 San Fernando Earthquake [8]. Because of the urgency of completion and the restrictions on geometry, steel plate girders were considered a viable replacement alternative. The idea was that the girders could be fabricated while the substructure was being constructed. Once the footings and columns were completed, the finished girders would be delivered to the job site. Therefore, in a period of 5 weeks Caltrans designed two different alternatives for two ramps approximately 396 m (1300 ft) and 457 m (1500 ft) in length. The South Connector Ramp will be discussed in this section. The "As-Built" South Connector was approximately 397 m (1302 ft) in length set on a horizontal curve with a radius of 198 m (650 ft) producing a superelevation of 11%. This ramp was designed utilizing Bridge Software Development International (BSDI) curved girder software package [10] as one frame with expansion joints at the abutments. This computer program is considered one of the most-advanced programs for the analysis and design of curved girder bridges. The program analyses the curved girders based on 3-D finite-element analysis and utilizes the influence surface for live-load analysis. The program has also an interactive postprocessor for performing designs and code checking. The design part of the program follows the 15th edition of AASHTO [13] and the Curved Girder Guide Specifications [1]. The ramp was then checked using DESCUS I [14], another software package, and spot-checked with in-house programs developed by Caltrans. A cross-sectional width of 11.43 m (37.5 ft) was selected for two lanes of traffic (3.66 m, 12 ft), two shoulders (1.52 m, 5 ft), and two concrete barriers (0.533 m, 1.75 ft). This ramp has a 212.7 mm (8⅜ inch) concrete deck, which was composite with four continuous welded plate girders with bolted field splices for erection. The material selected was A709 Grade 50W. The spans ranged from

35.97 m (118 ft) up to 66.44 m (218 ft) in length, which meant the girder depths alone were around 2.2 m (7.25 ft) deep and the composite section was 2.44 m (8 ft) deep. The cross frames were a mixture of inverted K frames and plate diaphragms at the bents. The K frames were inverted so as to place the catwalks between the girders, and the braces were changed to plate sections at the bents to help handle the large seismic forces that are transmitted from the superstructure to the "hammerhead" bent caps both longitudinally and transversely. The bracing was designed for both live-load and seismic-load conditions. Figure 15.12 shows the elevation of intermediate cross frames. The bracing was held to a spacing of less than 6.1 m (20 ft).

The BSDI program works by placing unit loads on a defined geometry pattern of the deck. Then an influence surface is developed so that application of loads for maximum and minimum stresses becomes a simple numerical solution. This program was thoroughly checked utilizing the V-Load method and using an SC-Bridge package that utilizes GT Strudl [11] for the moving load generator. Good correlation was seen by all methods with the exception of the V-Load, which consistently gave more conservative results. As is frequently the case with curved girders, the outside girder ends up being designed heavier than the remaining sections. This difference can be as little as 15%, but as great at 40%, depending on location. It should also be understood that by designing a stiffer girder for the outside, there is the tendency to attract more loads, thereby requiring more material. This is a similar phenomenon to that seen in seismic design. The BSDI system allows the designer to check for construction loads and sequencing. This was absolutely critical on a project like this as the girder sections were often controlled by the sequence of construction load application. Limits on concrete pours were set around limiting stresses on the girders.

Girder plate sizes were optimized both for the design and for the fabrication. A typical span would have five different sections in it. There were two sections at either end over the bents. The top and bottom flanges were very similar at point of maximum negative moment. Then on either side a transition section would be utilized until the inflection point. Finally, a maximum positive section where there is usually a significant difference in the top and bottom flanges was designed. The elevation of the plate girder that shows the different flange dimensions is shown in Figure 15.13. The five different flange dimensions were justified by considering the material costs vs. the welded splice costs. In addition, the "transition" sections were often sized such that the top flange width was the same as the negative moment sections. This way the plates could be welded end to end and then all four girders could be cut on one bed with one operation, saving handling costs. Plate sections were also set based on erection and shipping capabilities.

Steel was a good choice of structure type for this project because of the seismic risk, which exists in this location. Several faults pass in the vicinity of this interchange, and the structure would be subjected to "near-fault" phenomenon. This structure was designed with vertical acceleration. The plate girder with concrete deck superstructure weighs one third as much as the traditional cast-in-place box structure. Some ductile steel details were developed for this project [12]. Since the girders rest on a hammerhead bent cap, the load transfer mechanism is through the bearings and the shear can be as much as the plastic shear of the column. To make this load transfer possible, plate diaphragms were designed at the bent caps. With the plates in place, a concrete diaphragm could be poured that would not only add stiffness, but strength to handle these large seismic forces. The diaphragms were approximately 0.91 m (3 ft) wide by the depth of the girder. The plates were covered with shear studs and reinforcing was placed prior to the concrete. In addition, pipe shear keys were installed in the top of the bent cap on either side of the diaphragm. This structure was redundant in that if the displacements were excessive, the pipes would be engaged.

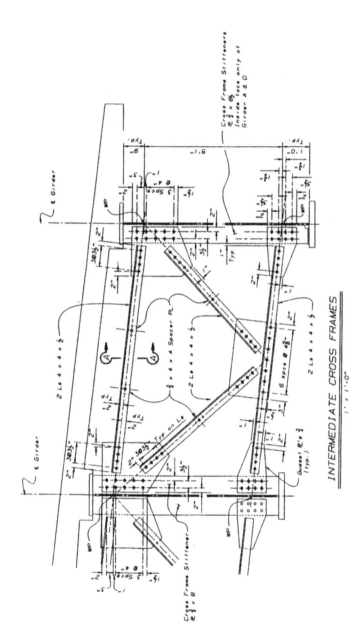

INTERMEDIATE CROSS FRAMES

FIGURE 15.12   Elevation of intermediate cross frames.

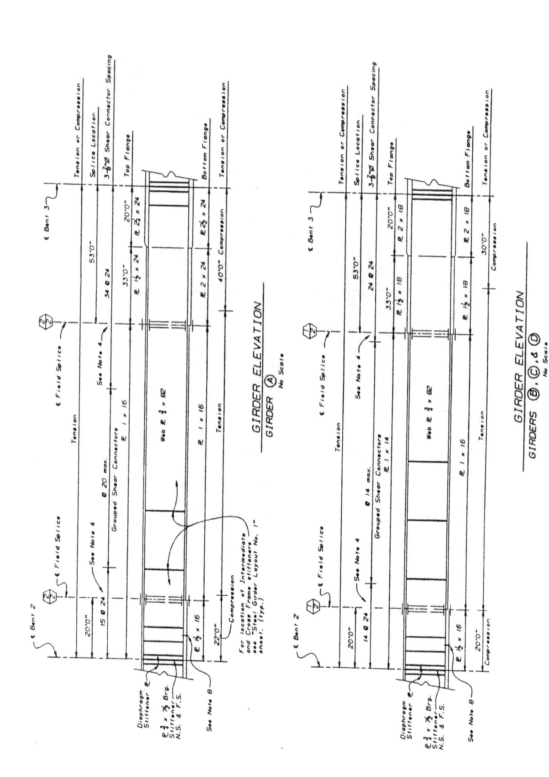

FIGURE 15.13   Elevation of interior and exterior curved plate girder.

15.4 Curved Steel Box-Girder Bridges

The most common type of curved steel box girder bridges are tub girders that consist of independent top flanges and cast-in-place reinforced concrete decks. The design guidelines are covered in the "Recommended Specifications for Steel Curved Girder Bridges"[3]. Normally the tub girder is composed of a bottom plate flange, two web plates, and an independent top flange attached to each web. The top flanges should be braced to become capable of resisting loads until the girder acts in a composite manner. The tub girders require internal bracing because of the distortion of the box due to the bending stresses. Finite-element analysis, which accounts for the distortion, is normally utilized to calculate the stresses and displacement of the box.

The webs of the box girder may be inclined with a ratio of one-to-four, width-to-depth. The AASHTO provisions for straight box girders apply for curved boxes regarding the shear capacity of the web and the ultimate capacity of the tub girders. The maximum bending stresses are determined according to the factored loads with the considerations of composite and noncomposite actions. Bending stresses should be checked at critical sections during erection and deck placement. The bending stresses may be assumed uniform across the width of the box. Prior to curing of concrete, the top flanges of tub girders are to be assumed laterally supported at top flange lateral bracing. The longitudinal warping stresses in the bottom flange are computed based on the stiffness and spacing of internal bracing. It is recommended that the warping stresses should not exceed 15% of the maximum bending stresses.

As mentioned earlier, the *M/R* method is usually used to analyze curved box girder bridges. The basic concept behind this method is the conjugate beam analogy. The method loads a conjugate simple span beam with a distributed loading, which is equal to the moment in the real simple or continuous span induced by the applied load divided by the radius of curvature of the girder. The reactions of the supports are obtained and thus the shear diagram can be constructed representing the internal torque diagram of the curved girder. After the concentrated torque at the ends of the floor beam is known, the end shears are computed from statics. These shears are applied as vertical concentrated loads at each cross frame location to determine the moment of the developed girder. This procedure constitutes a convergence process whereby the *M/R* values are applied until convergence is attained.

15.5 Curved Concrete Box-Girder Bridges

Current curved bridge specifications in the United States do not have any guidelines regarding curved concrete box-girder bridges. It is generally believed that the concrete monolithic box girders have high torsional rigidity, which significantly reduces the effect of curvature. However, during the last 15 years a problem has occurred with small-radius horizontally curved, post-tensioned box-girder bridges. The problem has occurred at two known sites during the construction [16]. The problem can be summarized as, during the prestressing of tendons in a curved box girder, they break away from the web tearing all the reinforcement in the web along the profile of the tendon. Immediate inspection of the failure indicated that the tendons exerted radial horizontal pressure along the wall of the outermost web.

In recognition of this problem, Caltrans has prepared and implemented design guidelines since the early 1980s [17]. Charts and reinforcement details were developed to check girder webs for containment of tendons and adequate stirrup reinforcement to resist flexural bending. Caltrans' Memo-to-Designers 11-31 specifies that designers of curved post-tensioned bridges should consider the lateral prestress force, F, for each girder. This force F is equal to the jacking force, P_j, of each girder divided by the horizontal radius of the girder. If the ratio of $P_j/R > 100$ kN/m per girder or the horizontal radius is equal to 250 m or less, Detail A, as shown in Figure 15.14 should be used. Charts for No. 16 and No. 19 stirrups were developed to be used with the ratio of P_j/R in order to get minimum web thickness and spacing between the No. 16 stirrups, as shown in Figure 15.15.

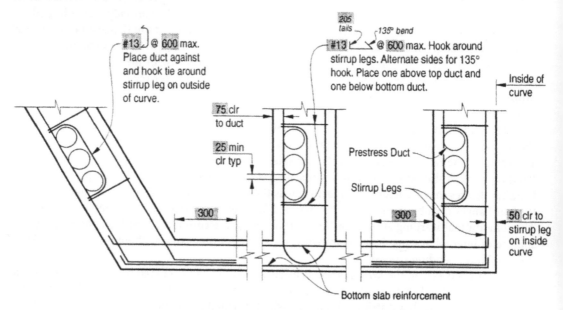

FIGURE 15.14 Caltrans duct detail in curved concrete bridges.

The first step is to enter the chart with the value of F on the vertical axis of the chart and travel horizontally until the height of the web h_c is reached. The chart then indicates the minimum web thickness and the spacing of the No. 16 stirrups.

These charts were developed assuming that the girder web is a beam with a length equal to the clear distance between top and bottom slabs. The lateral force, F, is acting at the center point of the web creating a bending moment in the web. This moment is calculated by the simple beam formula reduced by 20% for continuity between the web and slabs. The value of this bending moment is equal to

$$M_u = 0.8 \ \frac{P_j}{4 \ R} \ h_c \tag{15.1}$$

In the commentary of this memo, Caltrans considered the stirrups to be capable of handling the bending and shear stresses for the following reasons:

- M_u is calculated for the maximum conditions of F acting at $h_c/2$. This occurs at only two points in a span due to tendon drape.
- The jacking force, P_p, is used in the calculation of M_u and, at the time P_j is applied, the structure is supported on falsework. When the falsework is removed and vertical shear forces act, the prestressing forces will be reduced by the losses.

In addition, for curve box girders with an inside radius of under 243.8 m (800 ft), intermediate diaphragms are required at a maximum spacing of 24.4 m (80 ft) unless shown otherwise by tests or structural analysis. The code goes further to say that if the inside radius is less than 121.9 m (400 ft), the diaphragm spacing must not exceed 12.2 m (40 ft).

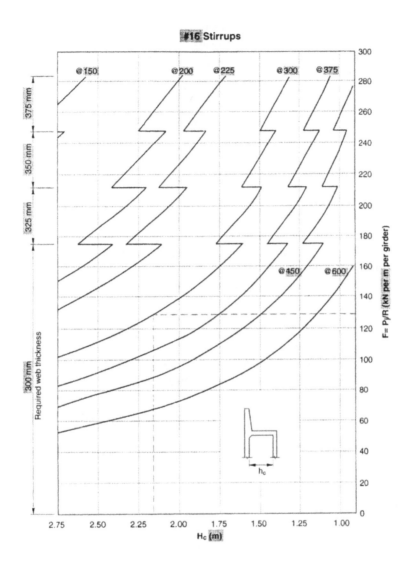

FIGURE 15.15 Caltrans chart design for web thickness and reinforcement.

Acknowledgments

The authors would like to thank Dr. Duan and Prof. Chen for selecting them to participate in this *Bridge Engineering Handbook*. The National Steel Bridge provided the photographs of curved bridges in this document for which the authors are sincerely grateful. The support and the cooperation of Mr. Dan Hall of BSDI are appreciated. Finally, the two authors warmly appreciate the continued support of Caltrans.

References

1. AASHTO, *Guide Specifications for Horizontally Curved Highway Bridges*, American Association of State Highway and Transportation Officials, Washington, D.C., 1993, 1–111.

2. Mozar, J., Cook, J., and Culver, C., Horizontally Curved Highway Bridges-Stability of Curved Plate Girders, Report No. P1, Carnegie-Mellon University, CURT Program, Pittsburgh, PA, Sept. 1971.
3. Hall, D. H. and Yoo, C. H., Curved Girder Design and Construction, Current Practice, NCHRP Project 12-38, 1995, 1–136.
4. V-Load analysis, in *USS Highway Structures Design Handbook*, Vol. 1, AISC Marketing, Inc., Chicago, IL, 1984, chap. 12, 1–56
5. AISC, *Highway Structure Design Handbook*, Newsletter Issue No. 2, AISC Marketing, Chicago, IL, 1991
6. AASHTO, *LRFD Bridge Design Specifications*, American Association of State Highway and Transportation Officials, Washington, D.C, 1994.
7. *Ontario Highway Bridge Design Code*, 3rd ed., Ministry of Transportation and Communications, Highway Engineering Division, Toronto, Ontario, 1991.
8. The San Fernando Earthquake — Field Investigation of Bridge Damage, State of California, Caltrans, Division of Structures, Sacramento, 1991.
9. Northridge Earthquake — Field Investigation of Bridge Damage, State of California, Caltrans, Division of Structures, Sacramento, 1994.
10. Hall, D. H., BSDI 3D System, Internal Document, Bridge Software Development International, Ltd., Coopersburg, PA, 1994.
11. *SC-Bridge, User's Guide*, Version 2.1, SC Solutions, Mountain View, CA, 1994.
12. Itani, A. and Reno, M., Seismic design of modern steel highway connectors, in *ASCE Structures Congress*, Vol. 2, 1995, 1528–1531.
13. AASHTO, *Standard Specifications for Highway Bridges*, 15th ed. with interim's, American Association of State Highway and Transportation Officials, Washington, D.C., 1994.
14. *DESCUS I and II*, Opti-Mate, Inc., Bethlehem, PA.
15. Analysis and Design of Horizontally Curved Steel Bridges, U.S. Steel Structural Report ADUCO 91063, May 1963.
16. Podolny, W., The cause of cracking in post-tensioned concrete box girder bridges and retrofit procedures, *PCI J.*, March 1985.
17. Caltrans, Bridge Memo-to-Designers, Vol. 1, California Department of Transportation, Sacramento, 1996.
18. Hall, D.H. and Yoo, C.H., Recommended Specifications for Steel Curved-Girder Bridges, NCHRP Project 12–38. BSDI, Ltd., Coopersburg, PA, Dec. 1998.
19. NCHRP (12–52), LRFD Specifications for Horizontally Curved Steel Girder Bridges, Transportation Research Board, Washington, D.C.

16

Highway Truss Bridges

John M. Kulicki
Modjeski and Masters, Inc.

16.1 Truss Configurations

16.1.1 Historical

During the 1800s, truss geometries proliferated. *The Historic American Engineering Record* illustrates 32 separate bridge truss geometries in its 1976 print shown in Figure 16.1 [1]. These range from the very short King Post and Queen Post Trusses and Waddell "A" trusses to very complex indeterminate systems, including the Town Lattice and Burr Arch truss. Over a period of years following Squire Whipple's breakthrough treatise on the analysis of trusses as pin-connected assemblies, i.e., two force members, a number of the more complex and less functional truss types gradually disappeared and the well-known Pratt, Howe, Baltimore, Pennsylvania, K truss and Warren configurations came into dominance. By the mid-20th century, the Warren truss with verticals was a dominant form of truss configuration for highway bridges, and the Warren and K trusses were dominant in railroad bridges. *The Historic American Engineering Record* indicates that the Warren truss without verticals may have appeared as early as the mid-1880s, but was soon supplanted by the Warren truss with verticals, as this provided a very convenient way to brace compression chords, reduce stringer lengths, and frame sway frames into the relatively simple geometry of the vertical members.

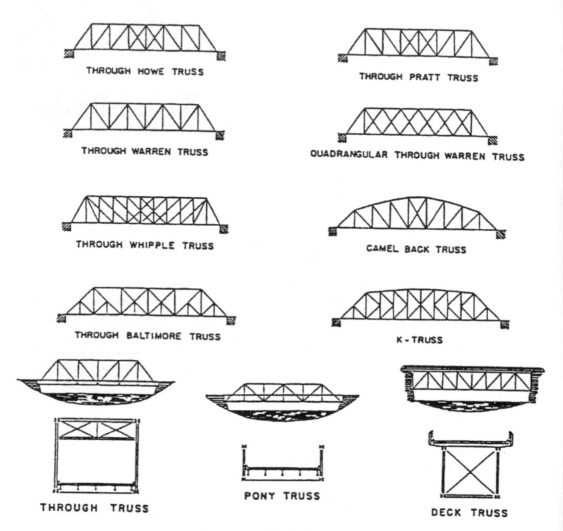

FIGURE 16.1   Historic trusses.

16.1.2 Modern

Few single span trusses are used as highway bridges today, although they are still used for railroad bridges. Modern highway trusses are usually either continuous or cantilever bridges and are typically Warren trusses with or without verticals. Some typical configurations are shown in Figure 16.2.

Throughout the 1980s and 1990s, the Warren truss without verticals has resurfaced as a more aesthetically pleasing truss configuration, especially in the parallel chord configuration, and this has led to a significant simplification in truss detailing, because sway frames are typically omitted in this form of truss, except for portals. The Warren truss without verticals received a great deal of use on the Japanese Railroad System and, more recently, in U.S. highway practice as exemplified in the Cooper River Bridge near Charleston, South Carolina, and the Kanawha River Bridge near Charleston, West Virginia; such a bridge configuration is shown in Figure 16.3 as it was considered one option for the Second Blue Water Bridge (281 m main span) between Port Huron, Michigan, and Point Edward, Ontario, Canada.

The truss bridge behaves much like a closed box structure when it has four planes capable of resisting shear and end portals sufficient to transmit shear back into vertical loads at the bearings. Given the need for a box configuration to resist vertical and lateral loads, it is possible that the configuration could be either rectangular, i.e., four-sided, or triangular, if that geometry is able to

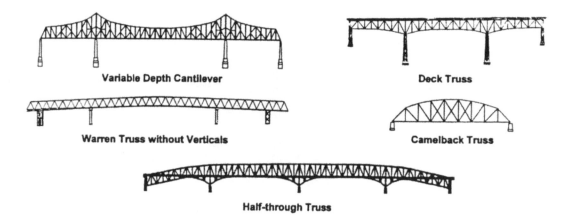

Variable Depth Cantilever

Deck Truss

Warren Truss without Verticals

Camelback Truss

Half-through Truss

FIGURE 16.2   Typical modern highway truss configuration.

FIGURE 16.3   Second Blue Water Bridge — parallel chord truss study option (on-site photo).

accommodate the roadway clearances. Issues of redundancy should be addressed, either by supplementary load paths, e.g., prestressing, or by sufficiently improved material properties, primarily toughness, to make a triangular configuration acceptable to owners, but it is certainly within the technical realm of reason.

16.2   Typical Components, Nomenclature, and Materials

Components and Nomenclature

The truss bridge is usually characterized by a plethora of bracing and wind-carrying members in addition to those members seen in front elevation. Typical members of a simple single span through-truss are identified in Figure 16.4, taken from Ref. [2].

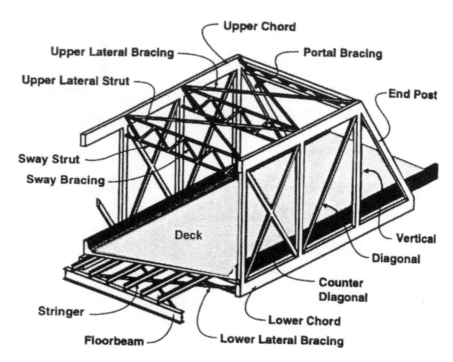

FIGURE 16.4 Typical truss members.

The lateral members in the planes of the top and bottom chords resist wind loads and brace the compression chords. Sway frames are thought to square the truss and increase its torsional rigidity. End portals carry torsional loads resulting from uneven vertical loads and wind loads into the bearings.

It is the visual impact of the various members, especially bracing members, which contribute to aesthetic opposition to many truss designs. However, if unforeseen events cause damage to a main truss member, these bracing members can serve as additional load paths to carry member load around a damaged area.

Truss Members

Some of the cross sections used as modern truss members are shown in Figure 16.5. Truss members have evolved from rods, bars, and eyebars to box and H-shaped members. Generally speaking, the box members are more structurally efficient and resist the tendency for wind-induced vibration better than H-shapes, whereas H-shapes are perceived as being more economical in terms of fabrication for a given tonnage of steel, generally easier to connect to the gusset plates because of open access to bolts, and easier to maintain because all surfaces are accessible for painting. The use of weathering steel offsets these advantages.

Even in the late 1990s, box members are widely used and, in some cases, the apparent efficiency of the H-shape is offset by the need to make the members aerodynamically stable. The choice is clearly project specific, although the H-shaped sections have a relatively clear advantage in the case of tension members because they are easier to connect to gusset plates and easier to paint as indicated above, without the stability design requirements needed for compression members. They are, however, more susceptible to wind-induced vibrations than box shapes. Box shapes have an advantage in the case of compression members because they usually have lower slenderness ratios about the weak axis than a corresponding H-shaped member.

The sealing of box shapes to prevent corrosion on the inside of the members has been approached from many directions. In some cases, box shapes may be fully welded, except at access locations at

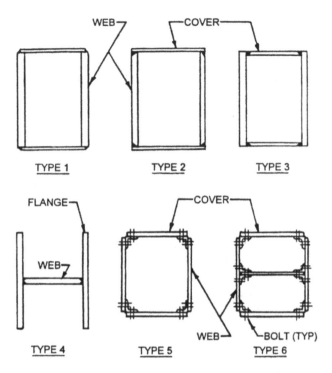

FIGURE 16.5   Cross sections of modern truss members.

the ends used to facilitate connection to gusset plates. Sealing of box members has met with mixed success. In some instances, even box members which have been welded on all four sides and have had welded internal squaring and sealing diaphragms have been observed to collect moisture. The issue is that the member need not be simply watertight to prevent the infiltration of water in the liquid form, it must also be airtight to prevent the natural tendency for the member to "breath" when subjected to temperature fluctuations, which tends to draw air into the member through even the smallest cracks or pinholes in the sealing system. This air invariably contains moisture and can be a recurring source of condensation, leading to a collection of water within the member. In some cases, box members have been equipped with drainage holes, even though nominally sealed, in order to allow this condensate to escape. In some instances, box members have been sealed and pressurized with an inert gas, typically nitrogen, in order to establish that adequate seals have been developed, as well as to eliminate oxygen from the inside of the member, thus discouraging corrosion. Box members have been built with valve stems in order to monitor the internal pressure, as well as to purge and refill the inert gas corrosion protection. Various types of caulking have been used to try to seal bolted joints with mixed success.

Due to the increased interest in redundancy of truss bridges, stitch-bolted members have been used in some cases. Because a bolt does not completely fill a hole, this does leave a path for water ingress, making adequate ventilation and drainage of the member important.

The future of truss member configurations is somewhat dependent on the evolution of new materials as discussed below.

Materials

Early trusses were made of timber. Over a period of time, the combination of timber compression members and wrought iron tension members was evolved and eventually the timber components were replaced by cast iron. After the construction of the Eads Bridge in St. Louis, steel became more widely used and remains today the predominant, and almost exclusive, material for truss construction.

As truss bridges reached longer and longer spans, they became the natural platform for the introduction of new steels, including the silicon steels in the earlier part of this century, weathering steels, copper-bearing steels, low-alloy steels, and even the high-yield-strength quenched and tempered alloy steels, such as ASTM A514/A517. An earlier version of weathering-type steels was used in trusses during the first quarter of the 20th century and was little recognized for their weathering capabilities until relatively recently.

By the current era (1998), 345 MPa yield steel dominated, with some use of higher strength materials, especially for very long-span bridges. The recent high-performance steel (HPS) initiative on the part of the Federal Highway Administration, the steel industry, and the U.S. Navy has led to the development of steels which hold the promise for relatively high strength, e.g., 485 MPa yield and higher, relatively favorable yield-to-tensile ratios on the order of 0.85, extraordinary toughness which could eliminate fracture as a consideration and with it fracture-critical members designation, reduced interpass temperature controls, and reduced preheat. This new material not only holds the promise for increased efficiency and cost-effectiveness using conventional box and I-shaped members in truss design, but may also lead to the use of tubular members and cast joints. This sounds like a return to the old Phoenix bridge system, but the increased efficiency of new materials, as well as the advances in the casting industry, may make field welding of tubular shapes to nodes possible and efficient. The use of high-performance concrete (HPC) truss members may evolve in which concrete serves as a composite stiffener to relatively thin steel plates in compression members.

Advanced composite materials may lead to further truss efficiencies. A pultruded shape as a compression member, butting into a metallic, concrete, or composite node may be a near-term possibility. Composite tension members may require a bonding agent, e.g., glue, before it becomes possible to make a fatigue-resistant joint. The problems associated with mechanical fastener-type connection will probably be solved sooner or later facilitating the use of advanced composites.

16.3   Methods of Analysis

16.3.1   Two-Force Member Methods — Pin-Connected Truss

In the 1840s, a method of analyzing trusses as pin-connected assemblages was developed and is still in wide use today. This method is based on assuming that the truss joints are frictionless pins. This assumption means that as long as loads are applied to the joints and not along the member length, the only bending is caused by self-weight. Thus, the major force in the member is assumed to act along its length. This is often called a "two-force member." The two forces are the axial load at each end of the member.

Throughout the 19th century and even into the early part of the 20th century, it was common to use physical pins in truss joints in order to facilitate the interconnection of components of members, and also to replicate the mathematical assumptions. As a truss deflects under loads, the joints rotate through what are typically very small angles. If the pins truly were frictionless, the truss members would rotate relative to each other and no end moments would be developed on the members. The physical pins never really were friction-free, so some moments developed at the ends in truss members and these were typically regarded as secondary forces. When pin-ended construction gave way to riveted joints and then to bolted or welded joints, the truss joints were detailed so that the working lines of the members intersected either at a common point, so as to reduce eccentricities or to utilize eccentricities to compensate for the bending caused by the dead weight of the members. In either event, it was widely regarded that the pin-connected analysis model was applicable. As will be discussed later, as long as a bridge is properly cambered, it often is an accurate analysis tool.

Two variations of the pin-connected truss model are in common usage; the method of joints and the method of sections. Each of these are illustrated below.

16.3.1.1 Method of Joints

As the name implies, the method of joints is based on analysis of free-body diagrams of each of the truss joints. As long as the truss is determinate, there will be enough joints and equations of equilibrium to find the force in all the members. Consider the simple example shown in Figure 16.6. This six-panel truss supports a load P at Joint L3. By taking the summation of the moments about each end of the bridge, it is possible to determine that the left-hand reaction is ⅔ P and the right-hand reaction is ⅓ P.

Isolating Joint L0, it can be seen that there are two unknowns, the force in Member L0-U1 and the force in Member L0-L1. For this small truss, and as typically illustrated in most textbooks, the truss is assumed to be in a horizontal position, so that it is convenient to take one reference axis through Member L0-L1 and establish an orthogonal axis through L0. These are commonly called the horizontal and vertical axes. The forces parallel to each must be in equilibrium. In this case, that means that the vertical component of the force in Member L0-U1 is equal the reaction RL. By considering the forces in the horizontal direction, the force in Member L0-L1 is equal to the horizontal component of the force in Member L0-U1. Thus, all of the member forces at L0 can be determined.

If we proceed to Joint L1, at which there is no applied load, it is clear that vertical equilibrium of the joint requires that the force in L1-U1 be equal to 0, and that the force in L1-L2 be equal to L0-L1.

Proceeding to Joint U1, it can be seen that, although four members frame into that joint, the force in two of the members are now known, and the force in the other two members can be found from the equation of equilibrium of forces along the two axes.

The analysis continues in this way from joint to joint.

16.3.1.2 Method of Sections

The method of sections proceeds by identifying free-body diagrams which contain only two unknowns, so that equilibrium of the sum of the moment about one joint and the equilibrium of the sum of the shears through a panel are sufficient to determine the two unknown truss forces. Consider Section AA in Figure 16.7 which shows a portion of the same truss shown in Figure 16.6. If we consider the free-body diagram to the left of Section AA, it is clear that the shear in the panel is equal to the reaction R_L and that this can be reacted only by the force L0-U1. Similarly, since the section and hence the free-body diagram is taken just to the left of the Joints L1 and U1, summing moments about Joint U1 or, more accurately, the end of Member L0-U1 an infinitesimally small distance to the left of the section line enables us to compute the force in L0-U1.

If we consider Section BB, it can be seen that the sum of the moments about the lower chord joint enables us to find the force in the top chord, and the shear in this panel enables us to find the force in the diagonal directly.

The analysis then proceeds from section to section along the truss. As a practical matter, a combination of the method of section and the method of joints usually results in the most expeditious calculations.

16.3.1.3 Influence Lines for a Truss

An influence line is a graphical presentation of the force in a truss member as the load moves along the length of the structure.

Influence lines for forces in the members are usually found by applying a unit load at each of the affected chord joints. This information is then shown pictorially, as indicated in Figure 16.8, which shows the influence line for a Chord Force U1-U2 (or U2-U3) and Diagonal Force L2-U3. If the truss is statically determinate, the influence line is a series of straight line segments. Since panel point loading is usually used in a truss, the influence lines for diagonals typically pass through a truss panel, as shown in Figure 16.8. If the truss is statically indeterminate, then the influence lines will be a series of chords to a curve, not a straight line.

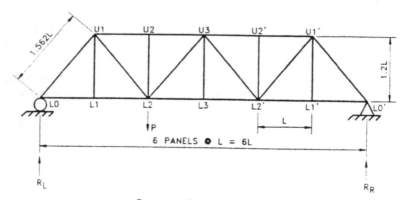

Computer Reactions

Summing Moments About Right End:

$$4PL = 6R_L L$$

$$R_L = \frac{2}{3}P$$

Summing Moments About Left End:

$$2PL = 6R_R L$$

$$R_R = \frac{1}{3}P$$

L0

From $\Sigma V = 0$:

$$\frac{1.2\ L0U1}{1.562} = R_L = \frac{2}{3}P$$

$$L0U1 = 0.868\ P$$

From $\Sigma H = 0$:

$$\frac{L0U1}{1.562} = L0L1$$

$$L0L1 = 0.555\ P$$

L1

From $\Sigma V = 0$:

$$L1U1 = 0$$

From $\Sigma H = 0$:

$$L1L2 = L0L1 = 0.555\ P$$

U1

From $\Sigma V = 0$:

$$\frac{1.2\ L0U1}{1.562} = \frac{1.2\ U1L2}{1.562}$$

$$U1L2 = L0U1 = 0.868\ P$$

From $\Sigma H = 0$:

$$U1U2 = \frac{L0U1}{1.562} + \frac{U1L2}{1.562} = \frac{2 \times 0.868\ P}{1.562}$$

$$U1U2 = 1.111\ P$$

U2

From $\Sigma V = 0$:

$$L2U2 = 0$$

From $\Sigma H = 0$:

$$U2U3 = U1U2 = 1.111\ P$$

Determined So Far:

$$L2U2 = 0$$

$$U1L2 = 0.858\ P$$

$$L1L2 = 0.555\ P$$

L2

From $\Sigma V = 0$

$$\frac{1.2\ U1L2}{1.562} + \frac{1.2\ U3L2}{1.562} = P$$

$$U3L2 = 0.434\ P$$

From $\Sigma H = 0$

$$L2L3 = L1L2 + \frac{U1L2}{1.562} = \frac{U3L2}{1.562}$$

$$L2L3 = 0.833\ P$$

FIGURE 16.6 Method of joints.

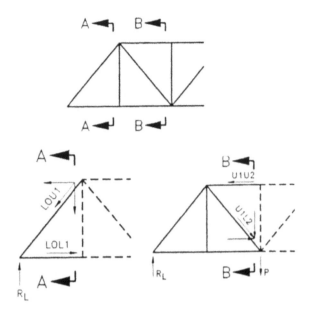

FIGURE 16.7 Method of sections.

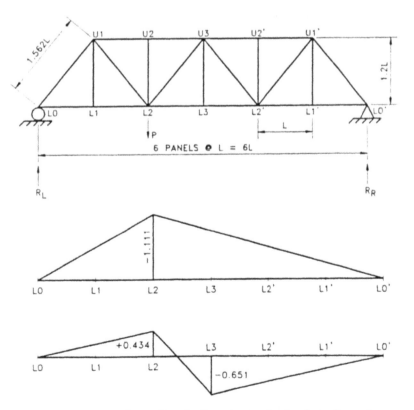

FIGURE 16.8 Influence lines for forces in one chord and one diagonal.

16.3.2 Computer Methods

The method of joints and the method of sections identified above appear to be very simple as long as the geometry of the truss is also simple and the structure is statically determinate. This is particularly true if one or both chords are horizontal. On most modern trusses, the span is sufficiently long that the change in vertical geometry can be significant. In fact, most larger trusses are on vertical curves if they cross a waterway. The chord joints are usually parallel to the deck profile. Thus, in many practical truss bridges, one or both truss chords is a series of chord segments representing a parabolic curve over at least part of the length of the bridge. This significantly complicates the geometry with respect to the use of either the method of joints or the method of sections. It does not negate the use of either of these methods, but certainly makes them less attractive.

There are many software packages for computers which permit the analysis of trusses, as can be done typically as either the pin-connected assemblage or as a frame with moment-resisting joints.

If the bridge is determinate in the plane of a truss, and if the truss is analyzed with two force members, then the cross-sectional area of the members does not affect the analysis. Assuming unit area for all members will give the proper forces, but not necessarily the proper displacements. If the truss is indeterminate in a plane, then it will be necessary to use realistic areas for the truss members and may be important to include the camber of the members in order to get realistic results in some cases. This will be true of the so-called "geometric case" which is usually taken as the state of the bridge under all dead load, at which time it is supposed to have the proper grades and profile. An analysis for a subsequent load, such as unit loads for the assembly of influence lines, or a transient load, does not require inclusion of the member camber. In fact, inclusion of the camber for other than the loads acting in the geometric condition would yield erroneous results for the indeterminate truss.

Where software contains the ability to put in a unit length change within a member and an analysis similar to this will be required to account properly for camber of the members, then it is often found efficient to calculate the influence lines for truss members using the Mueller–Bresslau principle as found in any text on structural mechanics.

When a truss is analyzed as a three-dimensional assemblage with moment-resisting joints, then the method of camber becomes even more important. It is common practice for some of the members to be cambered to a "no-load position" and in order for these members to have no load in them as analyzed, all the other members in the truss will have to be properly cambered in the computer model. With the usual fabrication techniques and adequate care for camber in both primary and secondary member (which may have no camber), a three-dimensional computer analysis of a roadway truss will typically result in determining truss member forces which are very close to those obtained by the pin-connected truss analogy. The secondary stresses from joint rotation resulting from transient loads will be determined directly from the computer analysis.

16.4 Floor Systems and Framing Details

16.4.1 Conventional Deck Systems Not Integral with Truss Chords

Initially, floor system framing was intended to be as structurally simple as possible. In the past, floor beams were often hung from truss pins with yokes, the simple-span stringers framing between floor beams often supported by saddle brackets on floor beam webs. As time went on, the advantages of continuous stringers, particularly in highway bridges, became very evident, and framing involving stringers-over-floor-beams developed, as did improved details for framing simply supported stringers between floor beams. Composite design of stringers and/or stringers and floor beams continued to add strength, stiffness, and robustness to trusses, while simultaneously eliminating many of the

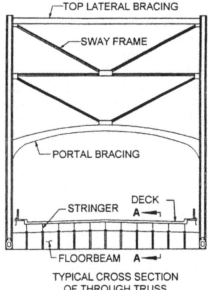

TYPICAL CROSS SECTION
OF THROUGH TRUSS

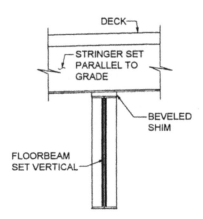

SECTION A-A THROUGH THE FLOORBEAM

FIGURE 16.9 Typical truss cross-section.

sources of uncontrolled drainage, and hence, corrosion, which have been the perceived source of excessive maintenance in trusses. Currently, floor beams are either vertical or set normal to roadway grade, and stringers are usually normal to crown and parallel to grade. If they are vertical, some sort of beveled fill will be necessary between the floor beam and the stringers. A typical through-truss cross-section is shown in Figure 16.9.

Most modern truss designs continue to use concrete decks, as well as filled grid, or grid and concrete composite systems as efficient durable decks. Relatively little use has been made of orthotropic decks in conjunction with original design (as opposed to rehabilitation) of trusses in the United States but this is certainly a feasible alternative. The use of newer lightweight deck systems, such as the proprietary Aluma-Deck or possibly advanced composite orthotropic deck systems can lead to further reduction in weight and, hence, savings in a competitive environment, as well as holding the potential for significantly reduced maintenance in future trusses.

16.4.2 Decks Integral with Truss Chords

So far, deck systems have almost always been designed to be structurally separate from the main supporting truss systems. As the need for efficiency and reduced cost, as well as increased redundancy, continues, a possible merging of the deck and truss system is a technical possibility. An orthotropic deck has been used as part of the bottom or top chord on some foreign bridges. Redundancy issues should be thoroughly considered as more traditional load paths are reduced. The available computer capabilities allow modeling of damage scenarios and the emerging knowledge on the computation of reliability indexes for damaged structures can provide designs with high levels of confidence, but such sophisticated calculations will have to be justified by cost savings and/or other benefits. Merging chords and deck has the potential to eliminate more joints within the deck system, perhaps at the expense of accommodating certain differential temperature features. Generally, as in all types of bridge structures, the elimination of joints is perceived as a favorable development. If the use of orthotropic decks as part of the chord system, and the lateral system for that matter, were to evolve, designers would have to consider the possibility of using either reinforced or prestressed concrete in a similar manner. This would, of course, tend to lead toward loading the chord at other than the panel points, but this situation has been handled in the past where, in some situations, deck chord members directly supported the roadway deck over their full length, not simply at the panel points.

16.5 Special Details

16.5.1 Hangers and Dummy Chords for Cantilever Bridges

The cantilevered truss has been used effectively on long-span structures since the Firth of Forth Bridge was built in Scotland in the late 1800s. This structural system was developed to provide most of the economy of continuous construction, as well as the longer spans possible with continuity, while simultaneously providing the simplicity of a statically determinate structural system. Consider the system shown in Figure 16.10. Figure 16.10a shows what appears to be a Warren truss configuration for a three-span continuous unit. The parallel diagonal configuration shown in the detail in Figure 16.10a is an indication that the framing system for the standard Warren truss has been interrupted. The statical system for the cantilever truss, indicated by the parallel diagonals, is shown in Figure 16.10b. The continuity has been interrupted by providing two points along the structure where the chords carry no axial force, resulting in a "shear only" connection. This is, by definition, a structural hinge. The unit between the two hinges is commonly referred to as a "suspended span." The remaining portions of the structure are called the "cantilever arms" and the "anchor spans," as indicated in Figure 16.10a. The mechanism for supporting suspended span is shown in concept in Figure 16.10c, which indicates that two chords are missing and hinges have been placed in the strap, or hanger, carrying the load of the suspended span into the anchor arm. The configuration with the link and two hinges allows the portions of the structure to expand and contract relative to each other.

In practice, the unnecessary top and bottom chords are added to the structure to allay public concerns, and are articulated in a manner which prevents them from carrying any axial load. These elements are typically called "false chords, or "dummy chords." A typical top chord joint at the hanger point is shown in Figure 16.11. Figure 16.11a is a plan view of the top chord element, and the corresponding elevation view is shown in Figure 16.11b. The false chord in this case is supported by the anchor arm, utilizing a pin. The false chord is slotted, so that it may move back and forth with expansion and contraction without carrying any load. It simply moves back and forth relative to the pin in the slot provided. The pin carries the vertical weight of the member to the top chord joint. Also shown in Figure 16.11b, and extending further into Figure 16.11c, are details of the hanger assembly and the top pin of the pins in the hanger used to allow it to swing back and forth. Hangers are potentially fracture-critical members, as a failure of this member would almost certainly

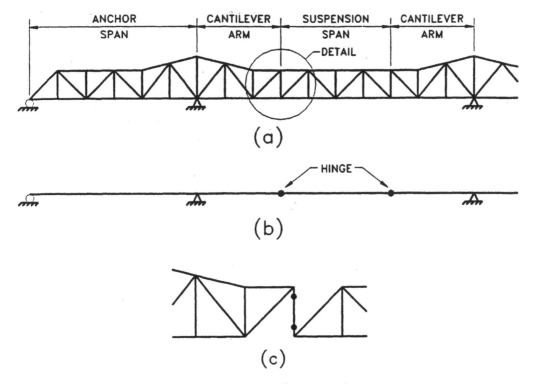

FIGURE 16.10 Cantilever suspension.

result in a collapse of at least the suspended span portion of the structure. These members are usually built of multiple components to add redundancy. The particular assembly shown has multiple plates bolted together to compensate for the hole occupied by the pin. In recent years, many truss bridges have been retrofitted with redundancy-adding assemblies usually consisting of rods or cables parallel to the hanger and attaching to the top and bottom chord. These assemblies are intended to pick up the load if the hanger or pin were to fail. Some of the details for the hanger pin are also shown in Figure 16.11c.

The corresponding portions of the structure at the bottom chord are shown in Figure 16.12. An elevation view of the lower chord joint is shown in Figure 16.12a, and a partial plan view is shown in Figure 16.12b. The concepts are very similar to those utilized in the upper joints, in that there are pins and slots to allow the false chord to move without picking up the axial load and pins and gusset plates to transfer the load from the suspended span into the hanger.

After completion of erection of the anchor span and the cantilever arm, the suspended span may be erected component by component, often referred to as "stick erection," or the entire suspended span may be assembled off site and hoisted into position until it can be brought into bearing at the hanger pins. If stick erection is used, the bridge will sag toward the middle as the cantilevers reach midspan. The bridge will be in the sag position because, once assembled to midspan, the cantilevers will be much shorter than they are at the midspan closure. It will thus be necessary to raise or lower portions of the bridge in order to get the closure members in and to transfer loads to the intended statical system. Also during this time, the false chords have to carry loads to support the cantilevering. With this type of erection, the false chords may be temporarily fixed, and one or both of the chords may have mechanical or hydraulic jacks for transferring load and for repositioning the two cantilevers for closure. Provisions for this type of assembly are often made in the false chord, at least to the point of being certain that the required space is available for jacks of sufficient capacity and that bearing plates to transmit the load are either in place or can be added by the contractor. A

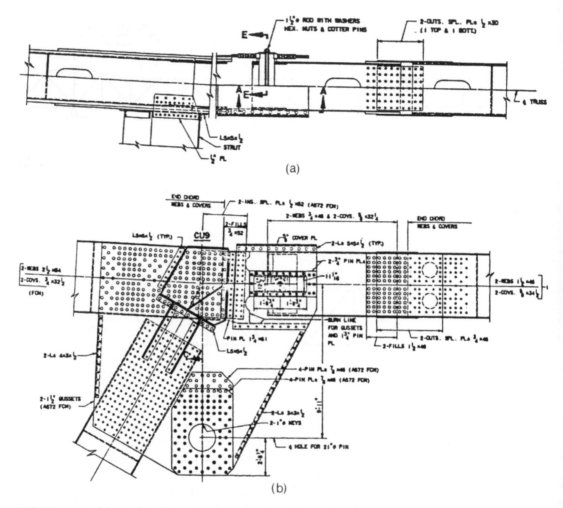

FIGURE 16.11 (a) Top false chord details — plan view; (b) top false chord details — elevation view; (c) hanger details.

typical detail for providing for jack assemblies is shown in Figure 16.13, which indicates how jacks would fit in the bottom false chord shown in Figure 16.12, and bear against the rest of the structure, so as to swing the cantilevered portion of the suspended span upward to facilitate closure.

16.5.2 Bearings

From the viewpoint of bearings, the cantilever form of erection offers several other advantages. The bearings used on the main piers can be fixed to the pier tops, and the chords framing into that point can be pinned into gusset plates. This provides a very simple and relatively maintenance-free connection to carry the major reaction of the bridge. The bearings on the end piers at the ends of the anchor spans are sometimes unique. Depending upon the requirements of the site or to reduce costs, the end spans are sometimes quite short, so that even under dead load, the reaction on the end piers is negative, which is to say an uplift condition exists under the dead load. Under some patterns of live load, this uplift will increase. When this condition exists, hanger assemblies similar to those described in Section 16.5.1 may be used to connect to a bearing fixed to the pier, and connected to an embedded steel grillage or similar device used to engage the weight of the pier to hold down the superstructure. Such a bearing is shown in Figure 16.14. The link accommodates

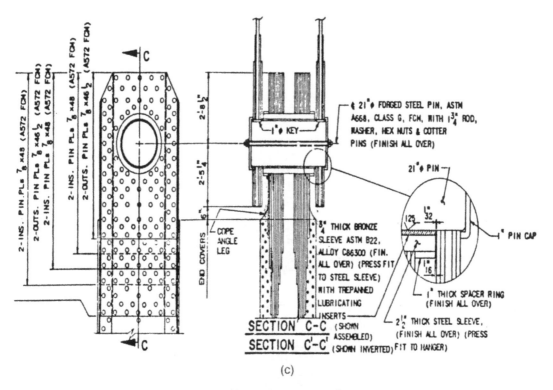

(c)

FIGURE 16.11 (continued)

the movement of the superstructure relative to the substructure required by expansion and contraction. The length of the arch swing of the link is designed so that the vertical displacement associated with the swing of the link can be accounted for and accommodated.

Where positive reactions are possible under all loadings, the bearing on the back span pier may be a rocker, roller nest, such as that shown in Figure 16.15, roller and gear assembly, or low-profile modern bearings, such as a pot bearing shown in Figure 16.16 as applied to a girder bridge or the disk bearing shown in Figure 16.17. It will usually be necessary for this bearing to provide for expansion and contraction while minimizing the forces put on the piers and to allow for rotation about the major bending axis of the bridge. Depending upon the designer's preferences, these bearings may or may not also carry the horizontal forces on the structure, such as wind loads, into the piers in the transverse direction. In some instances, the chord bearings serve this function through guide bars or pintles and, in some cases, a separate wind bearing, such as that shown in Figure 16.18, is provided to carry the transverse loads into piers separate from the main chord bearings.

Where structures are continuous, as opposed to cantilevered, it will usually be necessary for three of the four span bearings supporting the typical three-span truss to move. Individual movement will be greater in these bearings than they would be for a comparable-length cantilevered truss, and additional requirements will be placed on the bearing at one of the two main piers which moves, because of the large vertical reaction that will be transmitted at that point. Additionally, the continuous bridge will have two expansion joints, instead of four on the cantilever bridge, which is both an advantage and a disadvantage. These joints will have to be larger for the continuous bridge than for the cantilevered bridge and, therefore, more expensive. On the other hand, the tendency toward minimizing the number of joints in structures in order to reduce damage from deck drainage favors the continuous structure. Generally speaking, the extra points of expansion and contraction, associated deck joints and articulation hardware in the cantilevered bridge have required above-average maintenance.

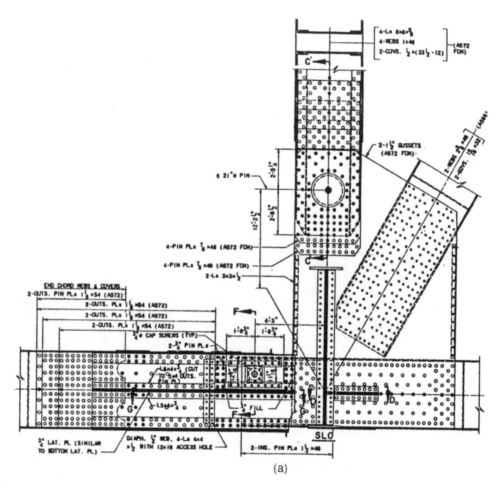

FIGURE 16.12   (a) Bottom false chord details — elevation view; (b) bottom false chord details — plan view.

16.5.3 Wind Tongues and Bearings for Transverse Forces

Wind loads carried by the suspended span in the cantilevered bridge have to be carried to the bearings on the piers. Thus, it is necessary for the wind loads to be carried through the panels framed with the false chords. Typically, in a through-truss, all of the wind loads on the suspended span are reacted by the hangers and by a special-purpose mechanism used to transmit horizontal forces at the lower chord level from the suspended span into the anchor arms. Wind load tributary to the upper lateral truss system in the plane of the top chord joints is carried into the anchor arms as a shear at the lower chord joint and the torque necessary to react to the transfer of loads from the top chord to the lower chord is carried as equal and opposite vertical reactions on the hangers. The horizontal forces at the lower chord level are then transmitted from the suspended span to the cantilever arm by a device called a "wind tongue," shown schematically in Figure 16.19. Because of the offset in chord joints at the suspended span, the horizontal force creates a torque in the plane of the bottom lateral system as the shear is transmitted across the expansion joint. Additionally, because expansion and contraction movements are accommodated at this point, allowance has to be made for some of the lateral members to swing along with that expansion and contraction. Thus, in Figure 16.19, there are four pin assemblies shown in the detail. The two horizontal links thus swing back and forth to accommodate the relative movement occurring at the open joint. The torque caused by the offset shear is reacted by the members framing from the open joint back into

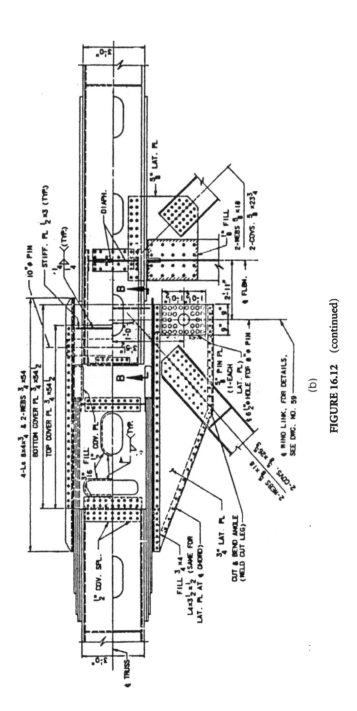

FIGURE 16.12 (continued)

(b)

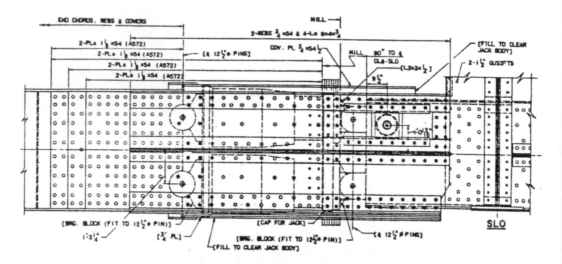

JACKING ARRANGEMENT IN BOTTOM CHORD

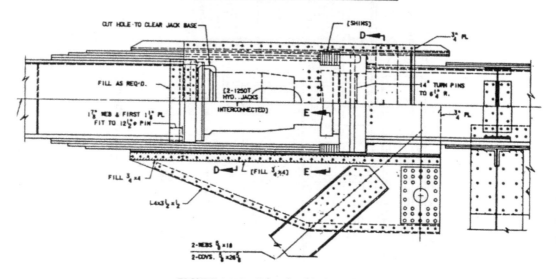

FIGURE 16.13   False chord jacking details.

the next panel point of the suspended span. These members form a lever to react to torque and prevent significant rotation of the wind tongue.

The typical details for accomplishing this wind transfer are shown in Figure 16.20a and b. Figure 16.20a shows the assembly that spans the open joint between the suspended span and the anchor arm. Also shown is one of the horizontal link members. The reacting members that form the lever to react the torque are also shown in this view, and are shown again in Figure 16.20b as they converge back to a common work point in the lateral truss of the suspended span. A typical pin assembly is shown in Figure 16.21.

In the case of the continuous truss, since the open joint does not exist, no assembly similar to the wind tongue, described above, is necessary. As can be seen in a plan view, the colinear force system can be developed to transmit wind and other transverse forces into the piers without creating torque in the bottom lateral system. Despite this, designers will often support the bearing point in order to accommodate accidental eccentricies that might exist. As seen in a vertical plane, there will almost certainly be an eccentricity between the center of transverse forces and the bearing. This will also be typically framed into a triangular system to carry this eccentricity through truss action,

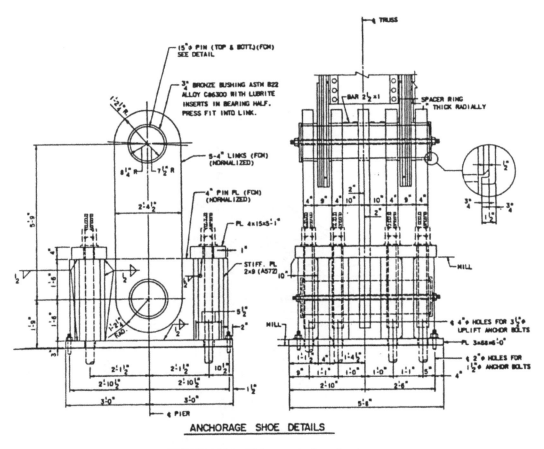

FIGURE 16.14 Link-type tie-down bearing.

rather than bending. These details are usually much simpler than the wind tongue at the suspended span, because it is not necessary to account simultaneously for expansion and contraction in the lateral truss system. This is usually handled by allowing the reaction points to move relative to the bearing while they are stationary relative to the lateral truss.

16.6 Camber and Erection

16.6.1 Camber for Vertical Geometry

It is obvious that all bridges have a theoretical geometric location as determined by the final design drawings. Every member has a theoretical length and location in space. One goal of the designer, fabricator, and erector is to produce a bridge as close to the theoretical position as possible, thereby ensuring actual stresses similar to design stresses.

To accomplish this, main members are usually cambered. Tensioned members that stretch under load are fabricated, such that their unstressed length is shorter than their length under the effect of dead load of the structure. The opposite is true for compression members. The cambered lengths are then accounted for during the erection stress and geometry studies. The state of the bridge when the camber "comes out" is called the geometric position. In this state the loads on the bridge are sufficient to return all of the members to their theoretical length. At any other state, the bridge will be out of shape and additional forces may result from the difference between its shape at any time and the final shape. This is relatively easy to see in the case of a continuous truss because it is clearly statically indeterminant and the shears and moments producing member forces are dependent on

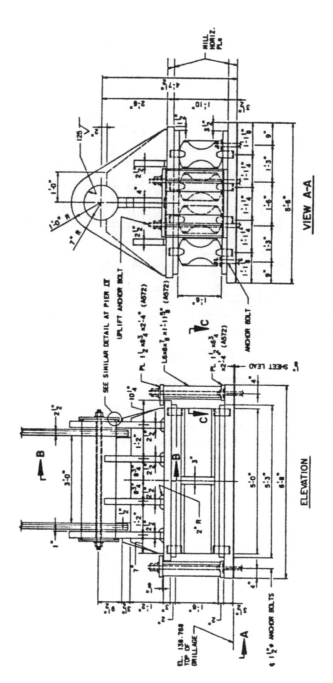

FIGURE 16.15   Roller bearing.

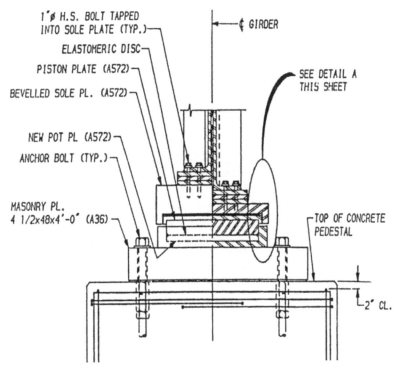

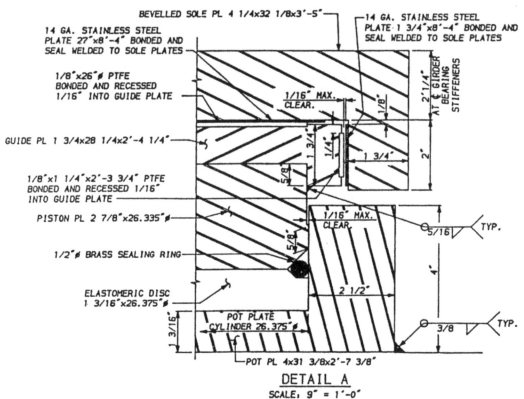

FIGURE 16.16 (a) Typical pot bearing; (b) typical pot bearing details.

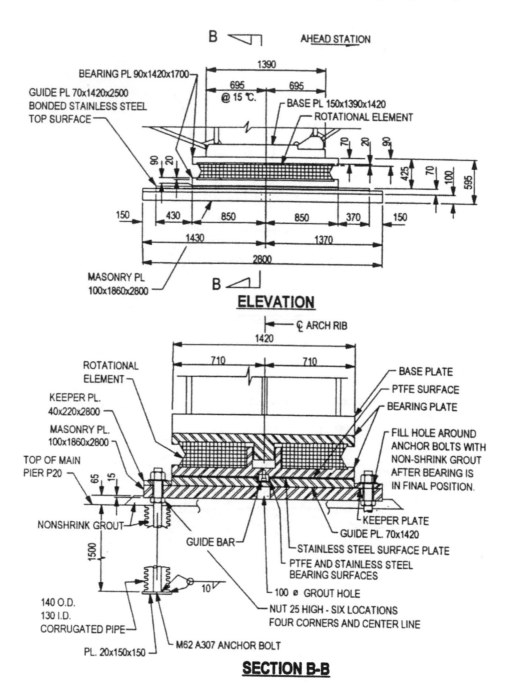

FIGURE 16.17 Typical disc bearing.

its shape. However, this is also true for a simple span truss because the joints are not frictionless pins as may have been assumed in the analysis. Because of this, even the simplest form of truss can have significant temporary member forces and moments until it reaches the geometric position. As will be discussed in the next section, there may be secondary moments in the geometric position, depending on the positioning of connector patterns on member ends in the shop.

Secondary members such as laterals and sway frame members are usually not cambered. They are usually intended to be stress free in the geometric position. Thus, at intermediate stages of erection they may also be subject to temporary forces.

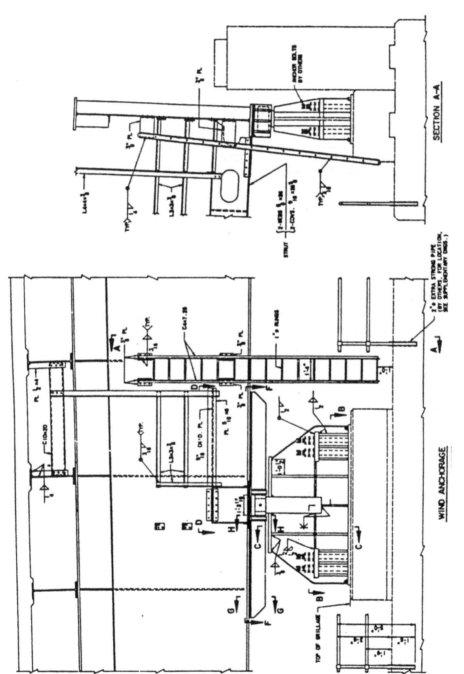

FIGURE 16.18 Truss wind bearing.

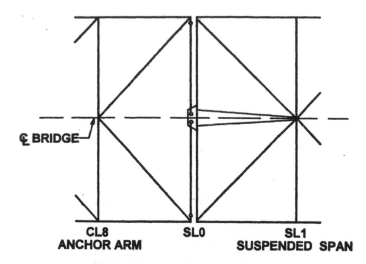

FIGURE 16.19   Schematic of wind tongue.

The importance of camber in achieving the designer's intent for the structure will be shown in the following discussion. In a determinant structure, the forces in components are uniquely determined by the geometry, loading, and support condition through the equations of equilibrium. The designer cannot alter the structural actions.

In a redundant structure, the designer can alter the forces associated with any one loading case. After this, the distribution of forces is again uniquely determined by all the conditions above, plus the relative stiffness of the various structural components.

Consider a simple case. If a two-span simply supported beam supporting a uniform load is intended to be horizontal under that load, the natural order is to have the reactions and negative moment at the pier equal to the value shown below at Piers 1, 2, and 3, respectively.

$$R_1 = \frac{3w\ell}{8} = R_3 \tag{16.1}$$

$$R_2 = \frac{10w\ell}{8} \tag{16.2}$$

$$M_2 = \frac{w\ell^2}{8} \tag{16.3}$$

Given the reactions, the load and length of the beam, the deflection at any point of the beam can be determined. Given that the example started with the condition that the beam is to be horizontal, calculation of deflections would reveal that the deflection is zero at all three support points and that the beam deflects downward between the support points. In order to put the beam on a horizontal position under a load, the web plates must be cut to the reverse of the deflected shape.

An alternative set of reactions and a corresponding moment diagram can be determined for this beam by cambering it so that not all three support points are on a straight line. For example, suppose it was desired to reduce the negative moment at the center pier. This could be achieved by cutting the web plate such that, in the horizontal layout position, the girder would rest such that a straight line between the two external reactions would be below the middle of the beam (with gravity acting downward). Thus, when the beam is assembled in the field, a certain amount of bending will have to take place as a simply supported beam of span equal to $2L$ until such time as the girder touches the middle bearing. After that, the remainder of the loads will be carried as a two-span unit. Unless a loading is sufficient in magnitude and opposite in sense so as to create uplift at the center bearing, all loads will then be handled as a two-span continuous unit. In this way, one might determine that an optimum condition is to have equal dead-load moments in the two spans and over the support. All that is necessary to achieve this is to determine the offset distance off the bearing point in the laydown position. Then cut the beam to that configuration using the cambered shape and erect it.

In truss construction, this adjustment of natural forces is seldom actually done. On occasion, camber to produce a determinant structure at steel closure has been used to facilitate erection. Camber for force control is very common in other types of bridges.

16.6.2 Camber of Joints

When the principal operations on a main member, such as punching, drilling, and cutting are completed, and when the detail pieces connecting to it are fabricated, all the components are brought together to be fitted up, i.e., temporarily assembled with fit-up bolts, clamps, or tack welds. At this time, the member is inspected for dimensional accuracy, squareness, and, in general, conformance with shop detail drawings. Misalignment in holes in mating parts should be detected then and holes reamed, if necessary, for insertion of bolts. When fit-up is completed, the member is bolted or welded with final shop connections.

The foregoing type of shop preassembly or fit-up is an ordinary shop practice, routinely performed on virtually all work. There is another class of fit-up, however, mainly associated with highway and railroad bridges that may be required by project specifications. These may specify that the holes in bolted field connections and splices be reamed while the members are assembled in the shop. Such requirements should be reviewed carefully before they are specified. The steps of subpunching (or subdrilling), shop assembly and reaming for field connections add significant costs. Modern computer-controlled drilling equipment can provide full-size holes located with a high degree of accuracy. AASHTO Specifications, for example, include provisions for reduced shop assembly procedures when computer-controlled drilling operations are used.

16.6.3 Common Erection Methods

The most common construction methods for trusses include cantilever construction, falsework, float-ins, and tiebacks. It is common for more than one method to be used in the construction of any single bridge. The methods selected to erect a bridge may depend on several factors, including the type of bridge, bridge length and height, type and amount of river traffic, water depth, adjacent geographic conditions, cost, and weight, availability, and cost of erection equipment.

Regardless of the method of erection that is used, an erection schedule should be prepared prior to starting the erection of any long-span bridge. The study should include bridge geometry ,member stress, and stability at all stages of erection. Bridges under construction often work completely differently than they do in their finished or final condition, and the character of the stress is changed, as from tension to compression. Stresses induced by erection equipment must also be checked, and, it goes without saying, large bridges under construction must be checked for wind stresses and sometimes wind-induced vibrations. An erection schedule should generally include a fit-up schedule for bolting major joints and a closing procedure to join portions of a bridge coming from opposite directions. Occasionally, permanent bridge members must be strengthened to withstand temporary erection loads. Prior to the erection of any bridge, proper controls for bridge line and elevation must be established and then maintained for the duration of the construction period.

Most long-span bridge construction projects have a formal closing procedure prepared by bridge engineers. The bridge member or assembled section erected to complete a span must fit the longitudinal

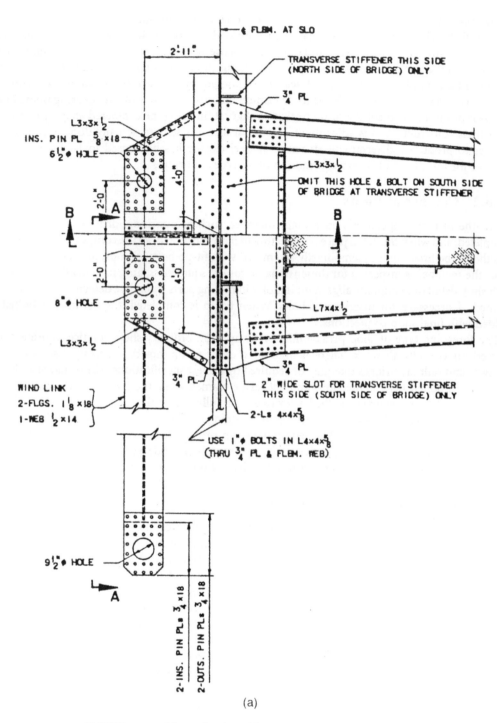

(a)

FIGURE 16.20   (a) Details of wind bearing; (b) details of wind tongue.

opening for it, and it must properly align with both adjoining sections of the bridge. Proper alignment of the closing piece is generally obtained by vertical jacking of falsework, lifting the existing bridge with a tieback system, or horizontal jacking of truss chords. At the Greater New Orleans Bridge No. 2, a scissors jack was inserted in place of a dummy top chord member, and it was used to pivot the main span truss for proper alignment.

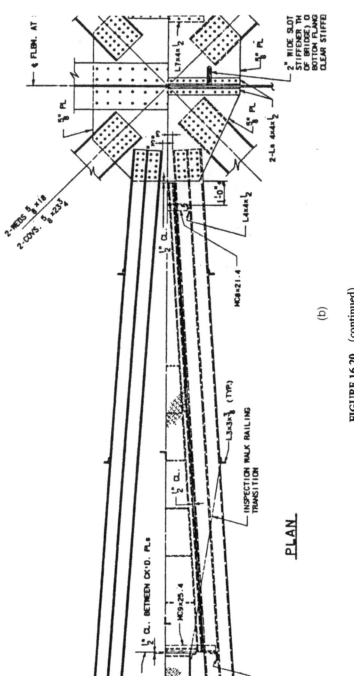

(b)

FIGURE 16.20 (continued)

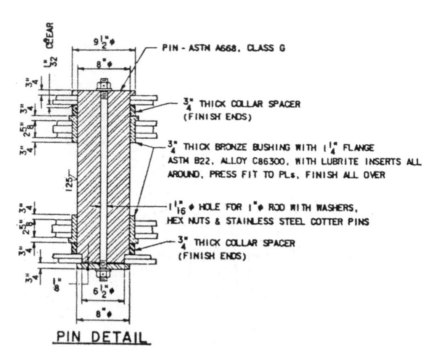

FIGURE 16.21   Typical wind tongue pin details.

FIGURE 16.22   Cantilever erection (on-site photo).

FIGURE 16.23 Early stage.

16.6.3.1 Cantilever Construction

In balanced cantilever construction, a bridge cantilevers in both directions from a single pier, as shown in Figure 16.22. The loads on each side of a pier must be kept reasonably in balance. In this type of construction, the first horizontal member erected in each direction may have to be temporarily supported by a brace, as shown in Figure 16.23.

In other types of cantilever construction, one span of a truss may be complete, and the bridge then cantilevers into an adjacent span, as shown in Figure 16.24. This type of construction is quite common.

16.6.3.2 Falsework

Falsework is commonly used when building medium-span trusses, especially for bridge approaches over land, as shown in Figures 16.25 and 16.26. It is also used in long-span bridge construction, but heavy river traffic, deep water, or poor foundation material may restrict its use. Falsework is sometimes used in the anchor or side span of cantilever trusses.

16.6.3.3 Float-In

Float-in is another commonly used construction method for long-span bridges. In this method, a portion of the bridge is generally assembled on a barge. The barge is then moved to the construction site, and the bridge is then either set into place off the barge or pulled vertically into place from the barge and connected to the part of the bridge already constructed. This method was used on the second Newburgh–Beacon Bridge across the Hudson River as discussed below.

First, a short section of the main span truss was erected on a barge; then sections of the approach spans over the river were assembled on top of the main span truss, floated into position, and erected, as shown in Figures 16.27 and 16.28. The section of bridge floated in is generally higher than its final position, and it is lowered into position with jacks. The two anchor spans were then assembled on a barge and erected the same way, as shown in Figure 16.29.

FIGURE 16.24   Advanced stage of cantilevering (on-site photo).

FIGRE 16.25   Use of falsework — 1 (on-site photo).

The cantilever portion of the bridge was then erected member by member by the cantilever erection method out into the main span to the pin hangers, as shown in Figure 16.30, which also shows the arrival of the suspended span of the main span. The suspended span was then barged into position and hoisted in place as shown in Figure 16.31.

FIGURE 16.26   Use of falsework — 2.

FIGURE 16.27   Float-in — 1.

The forces obtained during various stages of construction may be entirely different from those applicable to the final condition, and, in fact, the entire mechanism for resisting forces within the structure may change. During the erection of a truss, falsework bents may be used to support a portion of the structure. The gravity and lateral loads still act in the sense that they do on the final

FIGURE 16.28 Float-in — 2.

FIGURE 16.29 Float-in — 3.

condition, and the basic internal mechanism of resisting forces, as outlined in Chapter 10 of the Standard Specifications or Section 6 of the AASHTO-LRFD Specifications, remains unchanged, i.e., the primary load path involves axial load in the chords and diagonals, with the vertical components of those forces adding up to equal the applied shear within the panel, and the bending moment accounted for by the horizontal component of those forces. The camber of the truss will not pertain

FIGURE 16.30 Float-in of suspended span.

FIGURE 16.31 Lift of suspended span.

to an intermediate construction case, and, therefore, there is apt to be more joint rotation and, hence, more secondary bending moments within the truss members. Nonetheless, the primary load-carrying mechanism is that of axial forces in members. As the truss is erected, it is entirely possible that members that are in tension in the permanent condition will be under compression during erection, and vice versa. In fact, the state of stress may reverse several times during the erection of the bridge. Clearly, this has to be taken into account, not only in the design of the members, but also in the design of the connections. Compression members are apt to have been designed to transmit part of the forces in bearing. This will not be applicable when the member is in tension during erection. Similarly, a compression member that has tension during erection has to be reviewed for net section provisions and shear lag provisions.

16.7   Summary

Truss bridges have been an effective and efficient force of long-span bridges for over 150 years. As plate girder bridges have been utilized for spans of about 550 ft, box girders for spans of up to 750 ft, segmental concrete box girders for spans of up to about 800 ft, and cable-stayed bridges for spans of about 500 feet to 2000 ft, the use of trusses has declined over the last 25 years. Nonetheless, they remain a cost-effective bridge form, one with which many fabricators and erectors are experienced. Emerging materials and the use of computer analysis to treat the bridge as a three-dimensional structure will keep the truss form viable for the foreseeable future.

References

1. *Historic American Engineering Record,* National Park Service, Washington, D.C., 1976
2. Hartle, R. A., Amrheim, W. J., Willson, K. E., Baughman, D. R., and Tkacs, J. J., Bridge Inspector's Training Manual/90, FHWA-PD-91-015, Federal Highway Administration, Washington, D.C., 1990.

17

Arch Bridges

Gerard F. Fox
HNTB Corporation (Retired)

17.1   Introduction

17.1.1   Definition of Arch

An arch is sometimes defined as a curved structural member spanning an opening and serving as a support for the loads above the opening. This definition omits a description of what type of structural element, a moment and axial force element, makes up the arch. Nomenclature used to describe arch bridges is outlined on Figure 17.1. The true or perfect arch, theoretically, is one in which only a compressive force acts at the centroid of each element of the arch. The shape of the true arch, as shown in Figure 17.2, can be thought of as the inverse of a hanging chain between abutments. It is practically impossible to have a true arch bridge except for one loading condition. The arch bridge is usually subject to multiple loadings (dead load, live load, temperature, etc.) which will produce bending moment stresses in the arch rib that are generally small compared with the axial compressive stress.

17.1.2   Comparison of Arch Bridge with Other Bridge Types

The arch bridge is very competitive with truss bridges in spans up to about 275 m. If the cost is the same or only slightly higher for the arch bridge, then from aesthetic considerations the arch bridge would be selected instead of the truss bridge.

For longer spans, usually over water, the cable-stayed bridge has been able to be more economical than tied arch spans. The arch bridge has a big disadvantage in that the tie girder has to be constructed before the arch ribs can function. The cable-stayed bridge does not have this disadvantage, because deck elements and cables are erected simultaneously during the construction process. The true arch bridge will continue to be built of long spans over deep valleys where appropriate.

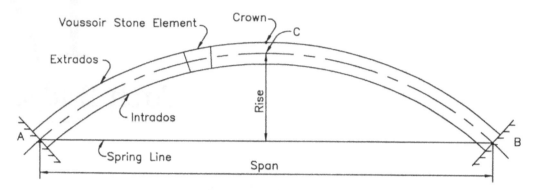

FIGURE 17.1   Arch nomenclature.

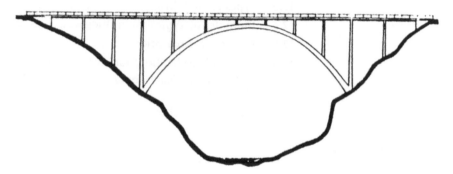

FIGURE 17.2   Concrete true arch.

17.1.3 Aesthetics of Arch Bridges

There is no question that arch bridges are beautiful, functional, and a pleasure for the motorist to drive over. Long-span arch bridges over deep valleys have no competitors as far as aesthetics is concerned. Arch bridges are better looking than truss bridges.

Many of the masonry arch bridges built for the last 2000 years are in the middle of cities whose residents consider these bridges not only necessary for commerce but also for their beautiful appearance. It is regrettable that masonry arches, in general, will not be constructed in today's environment since stonework has become too expensive.

17.2   Short History of Arch Bridges

Even before the beginning of recorded history, humans had a need to cross fast-moving streams and other natural obstacles. Initially, this was accomplished by means of stepping-stones and later by using felled tree trunks that either were supported by stones in the stream or spanned the entire distance between shores. Humans soon discovered that a vine attached to a treetop enabled them to swing across a wide river. This led to the construction of primitive suspension bridges with cables of vines or bamboo strips twisted into ropes. Post and lintel construction utilizing timber and stone slabs, as exemplified by the monument at Stonehenge, soon followed.

The arch came much later as applied to bridge building. The Sumerians, a community that lived in the Tigris-Euphrates Valley, made sunbaked bricks that were used as their main building material. To span an opening, they relied on corbel construction techniques. Around 4000 B.C. they discovered the advantages of the arch shape and its construction and they began to construct arch entranceways and small arch bridges with their sunbaked bricks [1].

Other communities with access to stone soon began to build arches with stone elements. By the time of the Romans most bridges were constructed as stone arches, also known as masonry or Voussoir arches. Empirical rules for dimensioning the shape of the arch and the wedge-shaped stones were developed. The Romans were magnificent builders and many of their masonry bridges are still standing. Probably the most famous is the Pont du Gard at Nimes in France, which is not only a bridge but an aqueduct as well. It stands as a monument to its builders. Excellent descriptions of other great Roman bridges can be found in Reference [1]. Stone arch bridges are very beautiful and much admired. A number have become centerpieces of their cities. There were few failures since stone is able to support very large compressive forces and is resistant to corrosive elements. Also, the arch is stable as long as the thrust line is contained within the cross-sectional area. Masonry arch bridges are very durable and most difficult to destroy [2].

Today, arch bridges are generally constructed of concrete or steel. However, there is still a great deal of research on stone arches directed toward determining their ultimate load capacity, their remaining life, their stability, their maintenance requirements, and also to determine the best methods to retrofit the structures. The reason for this great interest is, of course, that there are thousands of these stone arch bridges all over the world that are still carrying traffic and it would be an enormous cost to replace them all, especially since many of them are national monuments.

In 1779, the first Cast Iron Bridge was constructed at Coalbrookdale, England to span the Severn River. It is a semicircular arch spanning 43 m. By the year 1800, there were very few long-span masonry bridges built because they were not competitive with this new material. The 19th century was really the century of iron/steel bridges, suspension bridges, trusses, large cantilever bridges, viaducts, etc. Gustave Eiffel designed two notable steel arch bridges, a 160-m span at Oporto, Portugal and a 165-m span over the Truyeres River at St. Flour, France. Another notable arch bridge is the Eads steel bridge at St. Louis which has three spans of 155 m. In addition, a very beautiful shallow steel arch, the Pont Alexandre II, was constructed in Paris over the Seine River.

Concrete bridges began to be constructed at the end of the 19th century, and the arch designs of Robert Maillart should be noted since they are original and so beautiful [3].

17.3 Types of Arch Bridges

There are many different types and arrangements of arch bridges. A deck arch is one where the bridge deck which includes the structure that directly supports the traffic loads is located above the crown of the arch. The deck arch is also known as a true or perfect arch. A through-arch is one where the bridge deck is located at the springline of the arch. A half-through arch is where the bridge deck is located at an elevation between a deck arch and a through arch.

A further classification refers to the articulation of the arch. A fixed arch is depicted in Figure 17.1 and implies no rotation possible at the supports, A and B. A fixed arch is indeterminate to the third degree. A three-hinged arch that allow rotation at A, B, and C is statically determinate. A two-hinged arch allows rotation at A and B and is indeterminate to one degree.

A tied arch is shown in Figure 17.3 and is one where the reactive horizontal forces acting on the arch ribs are supplied by a tension tie at deck level of a through or half-through arch. The tension tie is usually a steel plate girder or a steel box girder and, depending on its stiffness, is capable of carrying a portion of the live loads. A weak tie girder, however, requires a deep arch rib and a thin arch rib requires a stiff deep tie girder. Since they are dependent on each other, it is possible to optimize the size of each according to the goal established for aesthetics and/or cost.

While most through or half-through arch bridges are constructed with two planes of vertical arch ribs there have been a few constructed with only one rib with the roadways cantilevered on each side of the rib.

Hangers usually consist of wire ropes or rolled sections. The hangers are usually vertical but truss-like diagonal hangers have also been used as shown in Figure 17.4. Diagonal hangers result in smaller deflections and a reduction in the bending moments in the arch rib and deck.

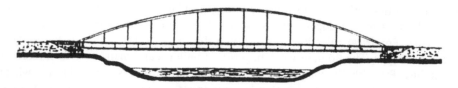

FIGURE 17.3   Steel tied-arch bridge.

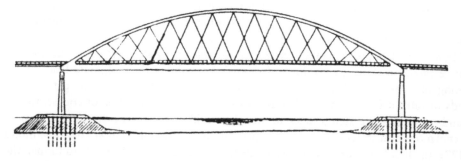

FIGURE 17.4   Arch with diagonal hangers.

There have also been arch bridges constructed with the arch ribs tilted so they can be connected at the crown. This is done for aesthetic reasons but it does add to the lateral stiffness of the arch bridge and could result in reduced bracing requirements.

17.4 Examples of Typical Arch Bridges

The Cowlitz River Bridge in the State of Washington, shown in Figure 17.5, is a typical true concrete arch that consists of a four-rib box section that spans 159 m with a rise of 45 m. Practically all concrete arches are of this type.

Multiple concrete arch spans have also been constructed such as the bridge, shown in Figure 17.6, across the Mississippi River at Minneapolis. The bridge consists of two 168-m shallow arch spans.

The longest reinforced concrete arch span in the world is the Wanxian Yangtze Bridge located in China with a span of 425 m and a rise of 85 m which gives a rise-to-span ratio of 1:5 [4]. It is unusual in that a stiff three-dimensional arch steel truss frame consisting of longitudinal steel tubes filled with concrete as the upper and lower chords was erected to span the 425 m. It served as the steel reinforcing of the arch and supported the cast-in-place concrete that was deposited in stages. This bridge is really a steel–concrete composite structure.

The Chinese have built many long-span concrete arches utilizing steel tubes filled with concrete. They have also constructed one half of an arch rib on each bank of a river, parallel to the river flow, and when completed, rotated both into their final position [5].

The longest steel true arch bridge has a span of 518 m and crosses the New River Gorge at Fayetteville, West Virginia. The arch rib consists of a steel truss. The deck, which is also a steel truss, is supported by transverse braced steel frames that are very slender longitudinally. The arch has rise-to-span ratio of 1:4.6 and was opened to traffic in 1977.

Most tied arch superstructures are of steel construction. A typical tied arch through structure is the Interstate 65 Twin Bridges over the Mobile River in Alabama as shown in Figure 17.7. The bridges are constructed of weathering steel. Also note the good appearance of the Vierendeel bracing between the arch ribs.

The Milwaukee Harbor Bridge is a three-span half-through steel-tied arch as shown in Figure 17.8. The steel tie at deck level is deep and very stiff while the arch rib is very thin. The main span length is 183 m with 82 m flanking spans. Again, note the appearance of the bracing between ribs.

FIGURE 17.5   Cowlitz River concrete arch.

FIGURE 17.6   Two shallow concrete arch spans.

Another three-span half-through steel-tied arch is the Fremont Bridge across the Willamette River in Portland, Oregon, as shown in Figure 17.9. It is a double-deck structure with an orthotropic top deck and a concrete lower deck. The main span of the arch is 383 m between spring points which makes it one of the longest tied-arch spans, if not the longest, in the world.

FIGURE 17.7   Steel twin-tied arches over the Mobile River.

FIGURE 17.8   Milwaukee Harbor half-through arch.

An unusual long-span half-through arch, shown in Figure 17.10, is the Roosevelt Lake Bridge in Arizona. The level of the lake is to be raised, and above the 200-year level of the lake the arch ribs and bracing are constructed of steel and below this level in concrete. To preserve structural continuity at the junction of the steel and concrete, the steel ribs are prestressed into the concrete rib. The arch is not a tied arch and spans 335 m with a rise-to-span ratio of about 1:5. Extensive wind tunnel testing was performed on models of this arch structure during the design phase.

17.5   Analysis of Arch Bridges

The dead load, live load, and temperature loads are covered in Chapter 6 of this Handbook. Wind effects are covered in Chapter 57 and seismic effects in Part IV.

FIGURE 17.9 Fremont Bridge across the Willamette River.

FIGURE 17.10 Roosevelt Lake Bridge, Arizona.

Before the age of structural analysis by computer methods, analysis of arches was not too difficult to accomplish with the help of a slide rule. In general, the analysis was a force method approach. For example, a two-hinged arch is statically indeterminate to the first degree and therefore has one redundant reaction. In the force method, the structure is made determinate, say, by freeing the right horizontal support and letting it move horizontally. The horizontal deflection ($d1$) at the support is then calculated for the applied loads. Next the horizontal deflection ($d2$) at the support is calculated for a horizontal force of 1 N acting at the support. Since the sum of these two deflections must vanish, then the total horizontal reaction (H) at the support for the applied loading must be

$$H = -\frac{(d1)}{(d2)} \tag{17.1}$$

Having the horizontal reaction the moments and axial forces can be calculated for the arch. A good example of early design procedures for a tied two-hinged arch is contained in a paper on the "Design of St. Georges Tied Arch Span" [6]. This arch, completed in 1941, spans the Chesapeake and Delaware Canal and was the first of its type in the United States to have a very stiff tie and shallow rib. The designer of the bridge was Professor Jewell M. Garrelts who was for many years head of the Civil Engineering Department at Columbia University. While this design procedure may be crude compared with modern methods, many fine arches were constructed using such methods.

Modern methods of analysis, of course, utilize a three-dimensional nonlinear finite-element computer program. For information on the finite-element method refer in this Handbook to Chapter 7 on Structural Theory and Chapter 36 on Nonlinear Analysis. Additional information on nonlinear analysis with accompanying computer programs on disk is available in Reference [7]. To use the finite element program it is necessary to have a preliminary design whose properties could then serve as the initial input to the computer program. In a published discussion of the St. Georges paper referred to above, Jacob Karol derived an approximate formula for calculating influence values for the horizontal force in the arch which depends only on the rise-to-span ratio. He also in the same discussion paper gave an approximate formula for the division of the total moment between the tie and the rib depending only on the depths of the rib and the tie girder [8]. These formulas are very useful in obtaining a preliminary design of a tied arch for input to the finite-element program.

17.6 Design Considerations for an Arch Bridge

17.6.1 Arch Bridge Design

Many chapters of this Handbook in Part II, Superstructure Design, have information concerning the design of decks that also apply to the design of decks of arches. By deck is meant the roadway concrete slab or orthotropic steel plate and its structural supports.

The rise-to-span ratio for arches may vary widely because an arch can be very shallow or, at the other extreme, could be a half-circle. Most arches would have rise-to-span ratios within the range of 1:4.5 to 1:6.

After the moments and axial forces become available from the three-dimensional finite-element nonlinear analysis the arch elements, such as the deck, ribs, ties, hangers, and columns can be proportioned. Steel arch ribs are usually made up of plates in the shape of a rectangular box. The ties are usually either welded steel box girders or plate girders.

In the 1970s there were problems in several arch bridges in that cracks appeared in welded tie girders. Repairs were made, some at great cost. However, there were no complete failures of any of the tie girders. Nevertheless, it caused the engineering community to take a new look at the need for redundancy. One proposal for arch bridges is not to weld the plates of the steel tie girders together but rather to use angles to connect them secured by bolts. Another proposal is to prestress the tie girder with post-tensioning cables. Another is to have the deck participate with the tie girder.

17.6.2 Vortex Shedding

Chapter 57 in this Handbook covers Wind Effects on Long Span Bridges. However, it seems appropriate to discuss briefly some problems in arches that are caused by vortex shedding.

Every now and then an arch is identified that is having problems with hanger vibrations especially those with I-section hangers. The vibrations are a result of vortex shedding. The usual retrofit is to

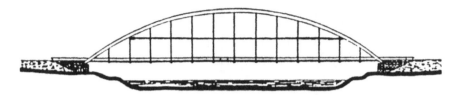

FIGURE 17.11 Horizontal cable connecting hangers.

connect the hangers as shown in Figure 17.11, which effectively reduces the length of the hangers and changes the natural frequency of the hangers. Another method is to add spoiler devices on the hangers [9]. In addition to the hangers, there have also been vortex shedding problems on very long steel columns that carry loads from the arch deck down to the arch rib.

17.6.3 Buckling of Arch Rib

Since the curved rib of the arch bridge is subject to a high axial force, the chance of a failure due to buckling of the rib cannot be ignored and must be accounted for. The subject of stability of arches is very well handled in Reference [10]. Values to use in formulas for critical buckling loads are listed in tables for many different cases of loading and various arch configurations.

As an example, the buckling critical load for a two-hinged parabolic arch rib supporting a uniform vertical deck load distributed on a horizontal projection will be calculated:

Arch span: $L = 120$ m
Arch rise: $S = 24$ m
Rise-to-span ratio = $24/120 = 0.2$
Rib moment of inertia: $I = 7.6 \times 10^9$ mm^4
Modulus of elasticity: $E = 20 \times 10^4$ N/mm^2
Horizontal buckling force:

$$H = C_1 \frac{EI}{L^2} \tag{17.2}$$

Uniform load causing buckling:

$$q = C_2 \frac{EI}{L^3} \tag{17.3}$$

$C_1 = 28.8$ and $C_2 = 46.1$ are from Table 17.1 of Reference [10]

$$H = \frac{28.8 \times 20 \times 10^4 \times 7.6 \times 10^9}{\left(120 \times 10^3\right)^2} = 3.04 \text{ MN}$$

$$q = \frac{46.1 \times 3.04 \times 10^3}{28.8 \times 120} = 40.55 \text{ kN/m}$$

The above calculation of critical loads is for buckling in the plane of the arch which assumes very good bracing between ribs to prevent out-of-plane buckling. The bracing types that are generally used include K type bracing shown in Figure 17.12, diamond-shaped bracing shown in Figure 17.13 and Vierendeel type bracing shown in Figure 17.14.

Also the above calculation is for an arch rib without any restraint from the deck. If the deck is taken into account, the buckling critical load will increase.

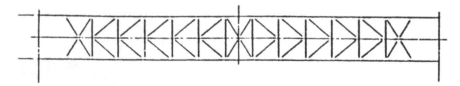

FIGURE 17.12   K-type of bracing.

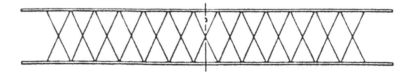

FIGURE 17.13   Diamond type of bracing.

FIGURE 17.14   Vierendeel type of bracing.

17.7 Erection of Arches

Most concrete deck arches have been constructed with the concrete forms and wet concrete being completely supported by falsework. They have also been constructed by means of tieback cables from a tower that is supported by cables anchored into the ground. Each tieback cable would support the forms for a segment length of concrete rib. For multiple arches, the tiebacks would support the forms and wet concrete in balanced cantilever segments off a tower erected on a common pier.

Segmental concrete arches have been erected by cranes that are supported on the ground or on barges. They pick up a concrete segment and connect it to a previously erected segment. Another way of delivering and erecting the concrete segments is by means of a cableway spanning between tower bents.

Steel deck arches have also been erected in segment lengths by means of tieback cables. Another erection method is to have steel rib segments span the distance between temporary erection bents. The New River Gorge steel arch bridge was erected by means of both a cableway and tiebacks. Good examples of steel arch bridge construction are presented in Chapter 45 on Steel Bridge Construction.

For the usual tied-arch span the deck and steel tie can be erected on temporary erection bents. When this operation has been completed, the ribs of the arch, bracing, and hangers can be constructed directly off the deck. An alternative erection scheme that has been used is one in which the deck, steel ties, and ribs are erected simultaneously by means of tieback cables. A more spectacular erection scheme, that is economical when it can be used, involves constructing the tied-arch span on the shore or on piles adjacent and parallel to the shore. When completed the tied arch is floated on barges to the bridge site and then pulled up vertically to its final position in the bridge. For example, Figure 17.15 shows the Fremont Bridge 275 m center-tied arch span being lifted up vertically to connect to the steel cantilevers.

Some arches have had a rib erected on each shoreline in the vertical position as a column. When completed, the ribs are then rotated down to meet at the center between the shorelines.

FIGURE 17.15   Fremont tied arch being lifted into place.

Acknowledgments

My thanks to Lou Silano of Parsons Brinckerhoff for furnishing the two pictures of the Fremont Bridge and thanks to Ray McCabe of HNTB Corp. for the rest of the photos. I appreciate the help I received from Dr. Lian Duan of Caltrans, Professor Wai-Fah Chen of Purdue University, and the people at CRC Press. They are all very patient, for which I am very grateful.

References

1. Steinman, D. B. and Watson, S. R., *Bridges and Their Builders*, G. P. Putnam's Sons, New York, 1941.
2. Heyman, J., *Structural Analysis – A Historical Approach*, Cambridge University Press, Cambridge, 1998.
3. Billington, D. P., *The Tower and the Bridge*, Princeton University Press, Princeton, NJ, 1985.
4. Yan, G. and Yang, Z.-H., Wanxian Yangtze Bridge, China, *Struct. Eng. Int.*, 7, 164, 1997.
5. Zhou, P. and Zhu, Z., Concrete-filled tubular arch bridges in China, *Struct. Eng. Int.*, 7, 161, 1997.
6. Garrelts, J. M., Design of St. Georges tied arch span, *Proc. ASCE*, Dec., 1801, 1941.
7. Levy, R. and Spillers, W. R., *Analysis of Geometrically Nonlinear Structures*, Chapman & Hall, New York, 1995.
8. Karol, J., Discussion of St. Georges tied arch paper, *Proc. ASCE*, April, 593, 1942.
9. Simiu, E. and Scanlan, R. H., *Wind Effects on Structures*, 2nd ed., John Wiley & Sons, New York, 1986.
10. Galambos, T. V., *Stability Design Criteria for Metal Structures*, 5th ed., John Wiley & Sons, New York, 1998.
11. O'Connor, C., *Design of Bridge Superstructures*, John Wiley & Sons, New York, 1971.

18

Suspension Bridges

Atsushi Okukawa
Shuichi Suzuki
Ikuo Harazaki
Honshu–Shikoku Bridge Authority
Japan

18.1 Introduction

18.1.1 Origins

The origins of the suspension bridge go back a long way in history. Primitive suspension bridges, or simple crossing devices, were the forebears to today's modern suspension bridge structures. Suspension bridges were constructed with iron chain cables over 2000 years ago in China and a similar record has been left in India. The iron suspension bridge, assumed to have originated in the Orient, appeared in Europe in the 16th century and was developed in the 18th century. Although wrought iron chain was used as the main cables in the middle of the 18th century, a rapid expansion of the center span length took place in the latter half of the 19th century triggered by the invention of steel. Today, the suspension bridge is most suitable type for very long-span bridge and actually represents 20 or more of all the longest span bridges in the world.

18.1.2 Evolution of Modern Suspension Bridges

Beginning of the Modern Suspension Bridge

The modern suspension bridge originated in the 18th century when the development of the bridge structure and the production of iron started on a full-scale basis. Jacobs Creek Bridge was constructed by Finley in the United States in 1801, which had a center span of 21.3 m. The bridge's distinguishing feature was the adoption of a truss stiffening girder which gave rigidity to the bridge to distribute the load through the hanger ropes and thus prevent excessive deformation of the cable. The construction of the Clifton Bridge with a center span of 214 m, the oldest suspension bridge now in service for cars, began in 1831 and was completed in 1864 in the United Kingdom using wrought iron chains.

Progress of the Center Span Length in the First Half of the 20th Century in the United States

The aerial spinning method (AS method) used for constructing parallel wire cables was invented by Roebling during the construction of the Niagara Falls Bridge, which was completed in 1855 with a center span of 246 m. The technology was established in the Brooklyn Bridge, completed in 1883 with a center span of 486 m, where steel wires were first used. The Brooklyn Bridge, which is hailed as the first modern suspension bridge, was constructed across New York's East River through the self-sacrificing efforts of the Roebling family — father, son, and the daughter-in-law — over a period of 14 years.

In 1903, the Manhattan Bridge, with a center span of 448 m, and in 1909 the Williamsburg Bridge, with a center span of 488 m, were constructed on the upper reaches of the river. The first center span longer than 1000 m was the George Washington Bridge across the Hudson River in New York. It was completed in 1931 with a center span of 1067 m. In 1936, the San Francisco–Oakland Bay Bridge, which was twin suspension bridge with a center span of 704 m, and in 1937, the Golden Gate Bridge with a center span of 1280 m were constructed in the San Francisco Bay area.

In 1940, the Tacoma Narrows Bridge, with a center span of 853 m, the third longest in the world at that time, exhibited bending mode oscillations of up to 8.5 m with subsequent torsional mode vibrations. It finally collapsed under a 19 m/s wind just 4 months after its completion. After the accident, wind-resistant design became crucial for suspension bridges. The Tacoma Narrows Bridge, which was originally stiffened with I-girder, was reconstructed in 1950 with the same span length while using a truss-type stiffening girder.

The Mackinac Straits Bridge with a center span of 1158 m was constructed as a large suspension bridge comparable to the Golden Gate Bridge in 1956 and the Verrazano Narrows Bridge with a center span of 1298 m, which updated the world record after an interval of 17 years, was constructed in 1964.

New Trends in Structures in Europe from the End of World War II to the 1960s

Remarkable suspension bridges were being constructed in Europe even though their center span lengths were not outstandingly long.

In the United Kingdom, though the Forth Road Bridge, with a center span of 1006 m, was constructed using a truss stiffening girder, the Severn Bridge, with a center span of 988 m, was simultaneously constructed with a box girder and diagonal hanger ropes in 1966. This unique design revolutionized suspension bridge technology. The Humber Bridge, with a center span of 1410 m, which was the longest in the world before 1997, was constructed using technology similar as the Severn Bridge. In Portugal, the 25 de Abril Bridge was designed to carry railway traffic and future vehicular traffic and was completed in 1966 with a center span of 1013 m.

In 1998, the Great Belt East Bridge with the second longest center span of 1624 m was completed in Denmark using a box girder.

Developments in Asia since the 1970s

In Japan, research for the construction of the Honshu–Shikoku Bridges was begun by the Japan Society of Civil Engineers in 1961. The technology developed for long-span suspension bridges as part of the Honshu–Shikoku Bridge Project contributed first to the construction of the Kanmon Bridge, completed in 1973 with a center span of 712 m, then the Namhae Bridge, completed in 1973 in the Republic of Korea with a center span of 400 m, and finally the Hirado Bridge, completed in 1977 with a center span of 465 m.

The Innoshima Bridge, with a center span of 770 m, was constructed in 1983 as the first suspension bridge of the Honshu–Shikoku Bridge Project, followed by the Ohnaruto Bridge, which was designed to carry future railway traffic in addition to vehicular loads and was completed in 1985 with a center span of 876 m. The center route of the Honshu–Shikoku Bridge Project, opened to traffic in 1988, incorporates superior technology enabling the bridges to carry high-speed trains. This route includes long-span suspension bridges such as the Minami Bisan–Seto Bridge, with a center span of 1100 m, the Kita Bisan–Seto Bridge, with a center span of 990 m, and the Shimotsui–Seto Bridge with a center span of 910 m. The Akashi Kaikyo Bridge, completed in 1998 with the world longest center span of 1991 m, represents the accumulation of bridge construction technology to this day.

In Turkey, the Bosporus Bridge, with a center span of 1074 m, was constructed in 1973 with a bridge type similar to the Severn Bridge, while the Second Bosporus Bridge with a center span of 1090 m, called the Fatih Sultan Mehmet Bridge now, was completed in 1988 using vertical instead of diagonal hanger ropes.

In China, the Tsing Ma Bridge (Hong Kong), a combined railway and roadway bridge with a center span of 1377 m, was completed in 1997. The construction of long-span suspension bridges of 1000 m is currently considered remarkable, the Xi Ling Yangtze River Bridge with a center span of 900 m and the Jing Yin Yangtze River Bridge with a center span of 1385 m are now under construction [1]. Both suspension bridges have a box stiffening girder and concrete main towers. Besides these bridges, additional long-span suspension bridges are planned.

18.1.3 Dimensions of Suspension Bridges in the World

Major dimensions of long-span suspension bridges in the world are shown in Table 18.1.

18.2 Structural System

18.2.1 Structural Components

The basic structural components of a suspension bridge system are shown in Figure 18.1.

1. Stiffening girders/trusses: Longitudinal structures which support and distribute moving vehicle loads, act as chords for the lateral system and secure the aerodynamic stability of the structure.
2. Main cables: A group of parallel-wire bundled cables which support stiffening girders/trusses by hanger ropes and transfer loads to towers.
3. Main towers: Intermediate vertical structures which support main cables and transfer bridge loads to foundations.
4. Anchorages: Massive concrete blocks which anchor main cables and act as end supports of a bridge.

18.2.2 Types of Suspension Bridges

Suspension bridges can be classified by number of spans, continuity of stiffening girders, types of suspenders, and types of cable anchoring.

Number of Spans

Bridges are classified into single-span, two-span, or three-span suspension bridges with two towers, and multispan suspension bridges which have three or more towers (Figure 18.2). Three-span suspension bridges are the most commonly used. In multispan suspension bridges, the horizontal displacement of the tower tops might increase due to the load conditions, and countermeasures to control such displacement may become necessary.

TABLE 18.1 Dimensions of Long-Span Suspension Bridges

No.	Bridge	Country	Year of Completion	Span Lengths (m)	Type	Remarks
1	Akashi Kaikyo	Japan	1998	960+1991+960	3-span, 2-hinged	
2	Great Belt East	Denmark	1998	535+1624+535	Continuous	
3	Humber	U.K.	1981	280+1410+530	3-span, 2-hinged	
4	Jing Yin Yangtze River	China[a]	(1999)	(336.5)+1385+(309.34)	Single-span	
5	Tsing Ma	China[a]	1997	455+1377 (+300)	Continuous	Highway+Railway
6	Verrazano Narrows	U.S.	1964	370.3+1298.5+370.3	3-span, 2-hinged	
7	Golden Gate	U.S.	1937	342.9+1280.2+342.9	3-span, 2-hinged	
8	Höga Kusten	Sweden	1997	310+1210+280	3-span, 2-hinged	
9	Mackinac Straits	U.S.	1957	548.6+1158.2+548.6	3-span, 2-hinged	
10	Minami Bisan–Seto	Japan	1988	274+1100+274	Continuous	Highway+Railway
11	Fatih Sultan Mehmet	Turkey	1988	(210+) 1090 (+210)	Single-span	
12	Bosphorus	Turkey	1973	(231+) 1074 (+255)	Single-span	
13	George Washington	U.S.	1931	185.9+1066.8+198.1	3-span, 2-hinged	
14	3rd Kurushima Kaikyo	Japan	1999	(260+) 1030 (+280)	Single-span	
15	2nd Kurushima Kaikyo	Japan	1999	250+1020 (+245)	2-span, 2-hinged	
16	25 de Abril	Portugal	1966	483.4+1012.9+483.4	Continuous	Highway+Railway
17	Forth Road	U.K.	1964	408.4+1005.8+408.4	3-span, 2-hinged	
18	Kita Bisan–Seto	Japan	1988	274+990+274	Continuous	Highway+Railway
19	Severn	U.K.	1966	304.8+987.6+304.8	3-span, 2-hinged	
20	Shimotsui–Seto	Japan	1988	230+940+230	Single-span with cantilever	Highway+Railway
21	Xi Ling Yangtze River	China[a]	1997	225+900+255	Single-span	
22	Hu Men Zhu Jiang	China[a]	1997	302+888+348.5	Single-span	
23	Ohnaruto	Japan	1985	93+330+876+330	3-span, 2-hinged	Highway+Railway
24	Second Tacoma Narrows	U.S.	1950	335.3+853.4+335.3	3-span, 2-hinged	
25	Askøy	Norway	1992	(173+) 850 (+173)	Single-span	
26	Innoshima	Japan	1983	250+770+250	3-span, 2-hinged	
27	Akinada	Japan	(2000)	255+750+170	3-span, 2-hinged	
28	Hakucho	Japan	1998	330+720+330	3-span, 2-hinged	
29	Angostura	Venezuela	1967	280+712+280	3-span, 2-hinged	
29	Kanmon	Japan	1973	178+712+178	3-span, 2-hinged	
31	San Francisco–Oakland Bay	U.S.	1936	356.9+704.1+353.6 353.6+704.1+353.6	3-span, 2-hinged	

[a] The People's Republic of China.

Continuity of Stiffening Girders

Stiffening girders are typically classified into two-hinge or continuous types (Figure 18.3). Two-hinge stiffening girders are commonly used for highway bridges. For combined highway–railway bridges, the continuous girder is often adopted to ensure train runnability.

Types of Suspenders

Suspenders, or hanger ropes, are either vertical or diagonal (Figure 18.4). Generally, suspenders of most suspension bridges are vertical. Diagonal hangers have been used, such as in the Severn Bridge, to increase the damping of the suspended structures. Occasionally, vertical and diagonal hangers are combined for more stiffness.

Types of Cable Anchoring

These are classified into externally anchored or self-anchored types (Figure 18.5). External anchorage is most common. Self-anchored main cables are fixed to the stiffening girders instead of the anchorage; the axial compression is carried into the girders.

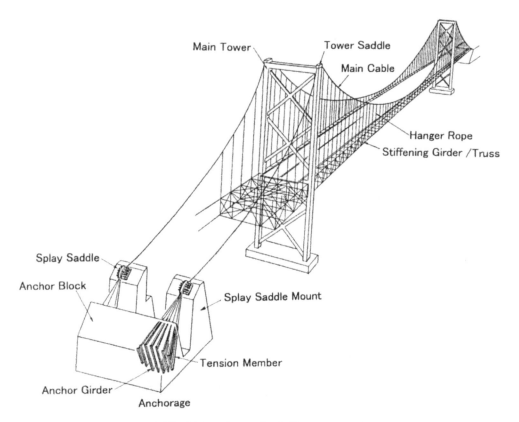

FIGURE 18.1 Suspension bridge components.

18.2.3 Main Towers

Longitudinal Direction

Towers are classified into rigid, flexible, or locking types (Figure 18.6). Flexible towers are commonly used in long-span suspension bridges, rigid towers for multispan suspension bridges to provide enough stiffness to the bridge, and locking towers occasionally for relatively short-span suspension bridges.

Transverse Direction

Towers are classified into portal or diagonally braced types (Table 18.2). Moreover, the tower shafts can either be vertical or inclined. Typically, the center axis of inclined shafts coincides with the centerline of the cable at the top of the tower. Careful examination of the tower configuration is important, in that towers dominate the bridge aesthetics.

18.2.4 Cables

In early suspension bridges, chains, eye-bar chains, or other material was used for the main cables. Wire cables were used for the first time in suspension bridges in the first half of the 19th century, and parallel-wire cables were adopted for the first time in the Niagara Falls Bridge in 1854. Cold-drawn and galvanized steel wires were adopted for the first time in the Brooklyn Bridge in 1883. This type has been used in almost all modern long-span suspension bridges. The types of parallel wire strands and stranded wire ropes that typically comprise cables are shown in Table 18.3. Generally, strands are bundled into a circle to form one cable. Hanger ropes might be steel bars, steel

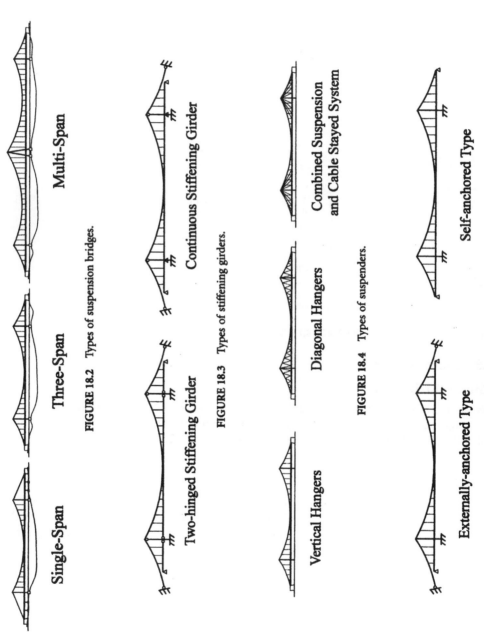

Single-Span   Three-Span   Multi-Span

FIGURE 18.2 Types of suspension bridges.

Two-hinged Stiffening Girder   Continuous Stiffening Girder

FIGURE 18.3 Types of stiffening girders.

Vertical Hangers   Diagonal Hangers   Combined Suspension and Cable Stayed System

FIGURE 18.4 Types of suspenders.

Externally-anchored Type   Self-anchored Type

FIGURE 18.5 Types of cable anchoring

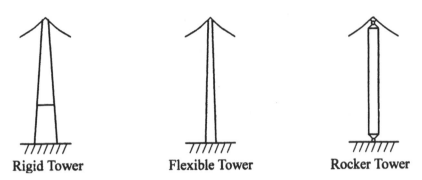

FIGURE 18.6 Main tower structural types.

TABLE 18.2 Types of Main Tower Skeletons

	Truss	Portal	Combined Truss and Portal
Shape			
Bridge	Akashi Kaikyo Forth Road	Great Belt East Humber	Golden Gate Second Tacoma Narrows

TABLE 18.3 Suspension Bridge Cable Types

Name	Shape of section	Structure	Bridge
Parallel Wire Strand		Wires are hexagonally bundled in parallel.	Brooklyn Humber Great Belt East Akashi Kaikyo
Strand Rope		Six strands made of several wires are closed around a core strand.	St.Johns
Spiral Rope		Wires are stranded in several layers mainly in opposite lay directions.	Little Belt Tancarville Wakato
Locked Coil Rope		Deformed wires are used for the outside layers of Spiral Rope.	Kvalsund Emmerich Älbsborg New Köln Rodenkirchen

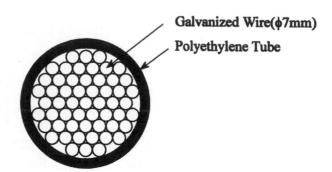

Galvanized Wire(ϕ7mm)

Polyethylene Tube

FIGURE 18.7 Parallel wire strands covered with polyethylene tubing.

rods, stranded wire ropes, parallel wire strands, and others. Stranded wire rope is most often used in modern suspension bridges. In the Akashi Kaikyo Bridge and the Kurushima Kaikyo Bridge, parallel wire strands covered with polyethylene tubing were used (Figure 18.7).

18.2.5 Suspended Structures

Stiffening girders may be I-girders, trusses, and box girders (Figure 18.8). In some short-span suspension bridges, the girders do not have enough stiffness themselves and are usually stiffened by storm ropes. In long-span suspension bridges, trusses or box girders are typically adopted. I-girders become disadvantageous due to aerodynamic stability. There are both advantages and disadvantages to trusses and box girders, involving trade-offs in aerodynamic stability, ease of construction, maintenance, and so on (details are in Section 18.3.8).

18.2.6 Anchorages

In general, anchorage structure includes the foundation, anchor block, bent block, cable anchor frames, and protective housing. Anchorages are classified into gravity or tunnel anchorage system as shown in Figure 18.9. Gravity anchorage relies on the mass of the anchorage itself to resist the tension of the main cables. This type is commonplace in many suspension bridges. Tunnel anchorage takes the tension of the main cables directly into the ground. Adequate geotechnical conditions are required.

18.3 Design

18.3.1 General

Naveir [2] was the first to consider a calculation theory of an unstiffened suspension bridge in 1823. Highly rigid girders were adopted for the suspended structure in the latter half of the 19th century because the unstiffened girders which had been used previously bent and shook under not much load. As a result, Rankine in 1858 [3] attempted to analyze suspension bridges with a highly rigid truss, followed by Melan, who helped complete the elastic theory, in which the stiffening truss was regarded as an elastic body. Ritter in 1877 [4], Lévy in 1886 [5], and Melan in 1888 [6] presented the deflection theory as an improved alternative to the elastic theory. Moisseiff realized that the actual behavior of a suspension bridge could not be explained by the elasticity theory in studies of the Brooklyn Bridge in 1901, and confirmed that the deflection theory was able to evaluate the deflection of that bridge more accurately. Moisseiff designed the Manhattan Bridge using the deflection theory in 1909. This theory became a useful design technique with which other long-span suspension bridges were successfully built [7]. Moreover, together with increasing the span length of the suspension bridge, horizontal loads such as wind load and vertical loads came to govern the design of the stiffening

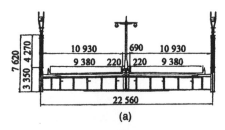

I-girder
(Bronx-Whitestone Bridge)

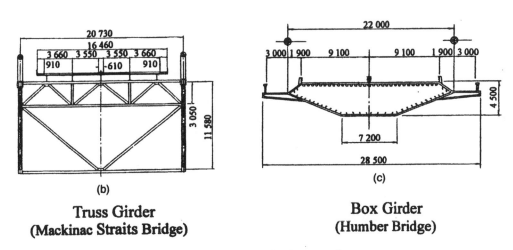

Truss Girder
(Mackinac Straits Bridge)

Box Girder
(Humber Bridge)

FIGURE 18.8 Types of stiffening girders.

girder. Moisseiff was among the first to establish the out-of-plane analysis method for suspension bridges [8].

Currently, thanks to rapid computer developments and the accumulation of matrix analysis studies on nonlinear problems, the finite deformation theory with a discrete frame model is generally used for the analysis of suspension bridges. Brotton [9,10] was the first to analyze the suspension bridge to be a plane structure in the matrix analysis and applied his findings to the analysis at erection stage for the Severn Bridge with good results. Saafan [11] and Tezcan's [12] thesis, which applied the general matrix deformation theory to the vertical in-plane analysis of a suspension bridge was published almost at the same time in 1966. The Newton–Raphson's method or original iteration calculation method may be used in these nonlinear matrix displacement analyses for a suspension bridge.

18.3.2 Analytical Methods

Classical Theory

Elastic Theory and Deflection Theory
The elastic theory and the deflection theory are in-plane analyses for the global suspension bridge system. In the theories, the entire suspension bridge is assumed a continuous body and the hanger ropes are closely spaced. Both of these analytical methods assume:

- The cable is completely flexible.
- The stiffening girder is horizontal and straight. The geometric moment of inertia is constant.

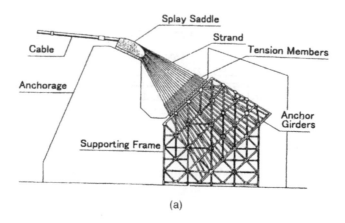

(a)

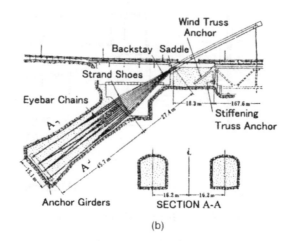

(b)

FIGURE 18.9   Types of anchorages. (a) Gravity, Akashi Kaikyo Bridge; (b) tunnel, George Washington Bridge.

• The dead load of the stiffening girder and the cables is uniform. The coordinates of the cable are parabolic.
• All dead loads are taken into the cables.

The difference between the two theories is whether cable deflection resulting from live load is considered. Figure 18.10 shows forces and deflections due to load in a suspension bridge. The bending moment, $M(x)$, of the stiffening girder after loading the live load is shown as follows:

Elastic Theory:

$$M(x) = M_0(x) - H_p y(x) \tag{18.1}$$

Deflection Theory:

$$M(x) = M_0(x) - H_p y(x) - (H_w + H_p)\eta(x) \tag{18.2}$$

where
$M_0(x)$ = bending moment resulting from the live load applied to a simple beam of the same span
                   length as the stiffening girder
$y(x)$ = longitudinal position of the cable
$\eta(x)$ = deflection of the cable and the stiffening girder due to live load
H_w, H_p = cable horizontal tension due to dead load and live load, respectively

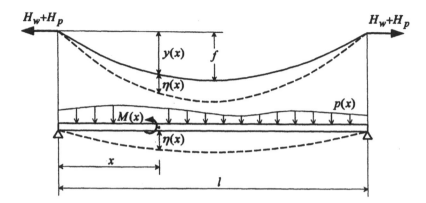

FIGURE 18.10 Deformations and forces of a suspension bridge.

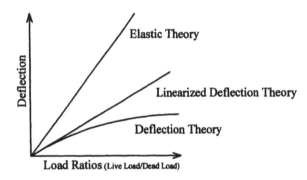

FIGURE 18.11 Deflection–load ratios relations among theories. (*Source:* Bleich, F. et al., *The Mathematical Theory of Vibrations in Suspension Bridges*, Bureau of Public Roads, Washington, D.C., 1950.)

It is understood that the bending moment of the stiffening girder is reduced because the deflection induced due to live load is considered in the last product of Eq. (18.2). Since the deflection theory is a nonlinear analysis, the principle of superposition using influence lines cannot be applied. However, because the intensity of live loads is smaller than that of dead loads for long-span suspension bridges, sufficient accuracy can be obtained even if it is assumed that $H_w + H_p$ is constant under the condition of $H_w \gg H_p$. On that condition, because the analysis becomes linear, the influence line can be used. Figure 18.11 shows the deflection–load ratio relations among the elastic, deflection, and linearized deflection theories [13]. When the ratio of live load to dead load is small, linearized theory is especially effective for analysis. In the deflection theory, the bending rigidity of towers can be neglected because it has no significance for behavior of the entire bridge.

Out-of-Plane Analysis Due to Horizontal Loads
Lateral force caused by wind or earthquake tends to be transmitted from the stiffening girder to the main cables, because the girder has larger lateral deformation than the main cables due to difference of the horizontal loads and their stiffness. Moisseiff [8] first established the out-of-plane analysis method considering this effect.

Out-of-Plane Analysis of the Main Tower
Birdsall [14] proposed a theory on behavior of the main tower in the longitudinal direction. Birdsall's theory utilizes an equilibrium equation for the tower due to vertical and horizontal forces from the cable acting on the tower top. The tower shaft is considered a cantilevered beam with variable cross section, as shown in Figure 18.12. The horizontal load (F) is obtained on the condition that the vertical load (R), acting on the tower top, and the horizontal displacement (Δ) are calculated by using Steinman's generalized deflection theory method [15].

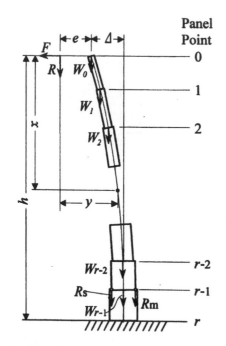

FIGURE 18.12 Analytical model of the main tower. (*Source:* Birdsall, B., *Trans. ASCE,* 1942. With permission.)

F :desired horizontal tower-top load
R :vertical external load on tower top
e :eccentricity of R with respect to the center line of the top of tower
Δ :required deflection of tower top
$W_0, W_1, \cdots W_{r-1}$:parts of tower weight assumed to be concentrated at the panel points indicated by the subscripts
R_s, R_m :reactions on tower at roadway level

Modern Design Method

Finite Deformation Method

With the development of the computer in recent years, finite displacement method on framed structures has come to be used as a more accurate analytical method. This method is used for plane analysis or space frame analysis of the entire suspension bridge structure. The frame analysis according to the finite displacement theory is performed by obtaining the relation between the force and the displacement at the ends of each element of the entire structural system. In this analytical method, the actual behavior of the bridge such as elongation of the hanger ropes, which is disregarded in the deflection theory, can be considered. The suspension bridges with inclined hanger ropes, such as the Severn Bridge, and bridges in the erection stage are also analyzed by the theory. While the relation between force and displacement at the ends of the element is nonlinear in the finite displacement theory, the linearized finite deformation theory is used in the analysis of the eccentric vertical load and the out-of-plane analysis; because the geometric nonlinearity can be considered to be relatively small in those cases.

Elastic Buckling and Vibration Analyses

Elastic buckling analysis is used to determine an effective buckling length that is needed in the design of the compression members, such as the main tower shafts. Vibration analysis is needed to determine the natural frequency and vibrational modes of the entire suspension bridge as part of the design of wind and seismic resistance. Both of these analyses are eigenvalue problems in the linearized finite deformation method for framed structures.

18.3.3 Design Criteria

Design Procedure

A general design procedure for a suspension bridge superstructure is shown in Figure 18.13. Most rational structure for a particular site is selected from the result of preliminary design over various alternatives. Then final detailed design proceeds.

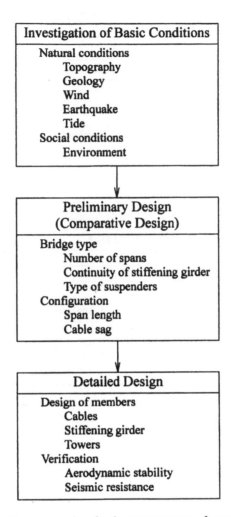

FIGURE 18.13 Design procedure for the superstructure of a suspension bridge.

Design Load

Design loads for a suspension bridge must take into consideration the natural conditions of the construction site, the importance of a bridge, its span length, and its function (vehicular or railway traffic). It is important in the design of suspension bridges to determine the dead load accurately because the dead load typically dominates the forces on the main components of the bridge. Securing structural safety against strong winds and earthquakes is also an important issue for long-span suspension bridges.

1. In the case of wind, consideration of the vibrational and aerodynamic characteristics is extremely important.
2. In the case of earthquake, assumption of earthquake magnitude and evaluation of energy content are crucial for bridges in regions prone to large-scale events.

Other design loads include effects due to errors in fabrication and erection of members, temperature change, and possible movement of the supports.

Analysis Procedure

General procedure used for the design of a modern suspension bridge is as follows (Figure 18.14):

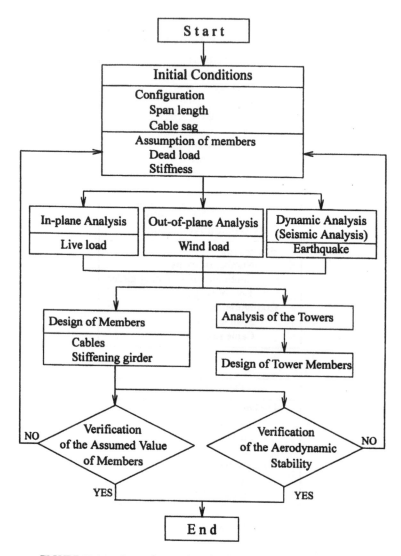

FIGURE 18.14 General procedure for designing a suspension bridge.

1. *Select Initial Configuration:* Span length and cable sag are determined, and dead load and stiffness are assumed.
2. *Analysis of the Structural Model:* In the case of in-plane analysis, the forces on and deformations of members under live load are obtained by using finite deformation theory or linear finite deformation theory with a two-dimensional model. In the case of out-of-plane analysis, wind forces on and deformations of members are calculated by using linear finite deformation theory with a three-dimensional model.
3. *Dynamic Response Analysis:* The responses of earthquakes are calculated by using response spectrum analysis or time-history analysis.
4. *Member Design:* The cables and girders are designed using forces obtained from previous analyses.
5. *Tower Analysis:* The tower is analyzed using loads and deflection, which are determined from the global structure analysis previously described.
6. *Verification of Assumed Values and Aerodynamic Stability:* The initial values assumed for dead load and stiffness are verified to be sufficiently close to those obtained from the detailed analysis. Aerodynamic stability is to be investigated through analyses and/or wind tunnel tests using dimensions obtained from the dynamic analysis.

18.3.4 Wind-Resistant Design

General

In the first half of the 19th century, suspension bridges occasionally collapsed under wind loads because girders tended to have insufficient rigidity. In the latter half of the 19th century, such collapses decreased because the importance of making girders sufficiently stiff was recognized.

In the beginning of the 20th century, stiffening girders with less rigidity reappeared as the deflection theory was applied to long-span suspension bridges. The Tacoma Narrows Bridge collapsed 4 months after its completion in 1940 under a wind velocity of only 19 m/s. The deck of the bridge was stiffened with I-girders formed from built-up plates. The I-girders had low rigidity and aerodynamic stability was very inferior as shown in recent wind-resistant design. After this accident, wind tunnel tests for stiffening girders became routine in the investigation of aerodynamic stability. Truss-type stiffening girders, which give sufficient rigidity and combined partially with open deck grating, have dominated the design of modern suspension bridges in the United States.

A new type of stiffening girder, however, a streamlined box girder with sufficient aerodynamic stability was adopted for the Severn Bridge in the United Kingdom in 1966 [16,17]. In the 1980s, it was confirmed that a box girder, with big fairings (stabilizers) on each side and longitudinal openings on upper and lower decks, had excellent aerodynamic stability. This concept was adopted for the Tsing Ma Bridge, completed in 1997 [18]. The Akashi Kaikyo Bridge has a vertical stabilizer in the center span located along the centerline of the truss-type stiffening girder just below the deck to improve aerodynamic stability [19].

In the 1990s, in Italy, a new girder type has been proposed for the Messina Straits Bridge, which would have a center span of 3300 m [20]. The 60-m-wide girder would be made up of three oval box girders which support the highway and railway traffic. Aerodynamic dampers combined with wind screens would also be installed at both edges of the girder. Stiffening girders in recent suspension bridges are shown in Figure 18.15.

Design Standard

Figure 18.16 shows the wind-resistant design procedure specified in the Honshu–Shikoku Bridge Standard [21]. In the design procedure, wind tunnel testing is required for two purposes: one is to verify the airflow drag, lift, and moment coefficients which strongly influences the static design; and the other is to verify that harmful vibrations would not occur.

Analysis

Gust response analysis is an analytical method to ascertain the forced vibration of the structure by wind gusts. The results are used to calculate structural deformations and stress in addition to those caused by mean wind. Divergence, one type of static instability, is analyzed by using finite displacement analysis to examine the relationship between wind force and deformation. Flutter is the most critical phenomenon in considering the dynamic stability of suspension bridges, because of the possibility of collapse. Flutter analysis usually involves solving the motion equation of the bridge as a complex eigenvalue problem where unsteady aerodynamic forces from wind tunnel tests are applied.

Wind Tunnel Testing

In general, the following wind tunnel tests are conducted to investigate the aerodynamic stability of the stiffening girder.

1. Two-Dimensional Test of Rigid Model with Spring Support: The aerodynamic characteristics of a specific mode can be studied. The scale of the model is generally higher than $1/100$.
2. Three-Dimensional Global Model Test: Test used to examine the coupling effects of different modes.

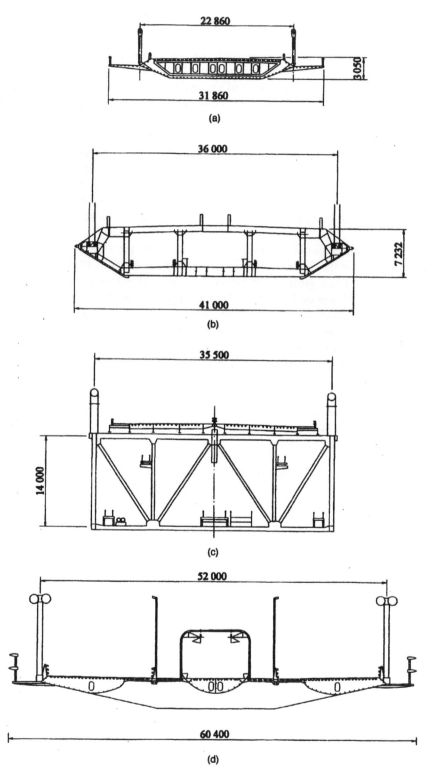

FIGURE 18.15   Cross sections through stiffening girders. (a) Severn Bridge, (b) Tsing Ma Bridge; (c) Akashi Kaikyo Bridge, (d) Messina Straits Bridge.

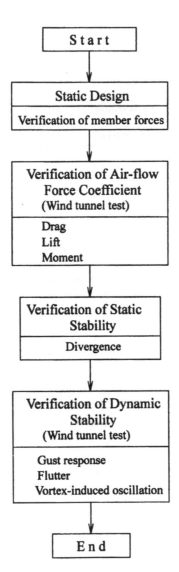

FIGURE 18.16 Procedure for wind–resistant design. (*Source:* Honshu–Shikoku Bridge Authority, Wind–Resistant Design Standard for the Akashi Kaikyo Bridge, HSBA, Japan, 1990. With permission.)

For the Akashi Kaikyo Bridge, a global ¹⁄₁₀₀ model about 40 m in total length, was tested in a boundary layer wind tunnel laboratory. Together with the verification of the aerodynamic stability of the Akashi Kaikyo Bridge, new findings in flutter analysis and gust response analysis were established from the test results.

Countermeasures against Vibration

Countermeasures against vibration due to wind are classified as shown in Table 18.4.

1. *Increase Structural Damping.* Damping, a countermeasure based on structural mechanics, is effective in decreasing the amplitude of vortex-induced oscillations which are often observed during the construction of the main towers and so on. Tuned mass dampers (TMD) and tuned liquid dampers (TLD) have also been used to counter this phenomenon in recent years. Active mass dampers (AMD), which can suppress vibration amplitudes over a wider frequency band, have also been introduced.

TABLE 18. 4   Vibration Countermeasures

Category	Item	Countermeasures
Structural mechanics	Increase damping	TMD,[a] TLD,[b] AMD[c]
	Increase rigidity	Increase cross-sectional area of girder
	Increase mass	
Aerodynamic mechanics	Cross section	Streamlined box girder
		Open deck
	Supplements	Spoiler, Flap

[a] Tuned mass damper.
[b] Tuned liquid damper.
[c] Active mass damper.

2. *Increase Rigidity*: One way to increase rigidity is to increase the girder height. This is an effective measure for suppressing flutter.
3. *Aerodynamic Mechanics*: It may also be necessary to adopt aerodynamic countermeasures, such as providing openings in the deck, and supplements for stabilization in the stiffening girder.

18.3.5   Seismic Design
General

In recent years, there are no cases of suspension bridges collapsing or even being seriously damaged due to earthquakes. During construction of the Akashi Kaikyo Bridge, the relative location of four foundations changed slightly due to crustal movements in the 1995 Hyogo-ken Nanbu Earthquake. Fortunately, the earthquake caused no critical damage to the structures. Although the shear forces in the superstructure generated by a seismic load are relatively small due to the natural frequency of the superstructure being generally low, it is necessary to consider possible large displacements of the girders and great forces transferring to the supports.

Design Method

The superstructure of a suspension bridge should take into account long-period motion in the seismic design. A typical example of a seismic design is as follows. The superstructure of the Akashi Kaikyo Bridge was designed with consideration given to large ground motions including the long-period contribution. The acceleration response spectrum from the design standard is shown in Figure 18.17 [22]. Time-history analysis was conducted on a three-dimensional global bridge model including substructures and ground springs.

18.3.6   Main Towers
General

Flexible-type towers have predominated among main towers in recent long-span suspension bridges. This type of tower maintains structural equilibrium while accommodating displacement and the downward force from the main cable. Both steel and concrete are feasible material. Major bridges like the Golden Gate Bridge and the Verrazano Narrows Bridge in the United States as well as the Akashi Kaikyo Bridge in Japan consist of steel towers. Examples of concrete towers include the Humber and Great Belt East Bridges in Europe and the Tsing Ma Bridge in China. Because boundary conditions and loading of main towers are straightforward in suspension bridge systems, the main tower can be analyzed as an independent structural system.

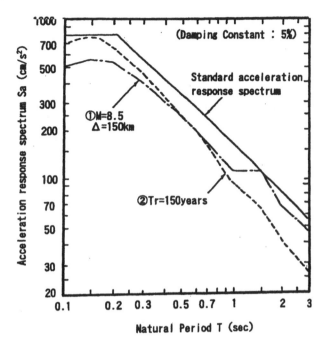

FIGURE 18.17 Design acceleration response spectrum. (*Source:* Honshu–Shikoku Bridge Authority, Seismic Design Standard for the Akashi Kaikyo Bridge, Japan, 1988. With permission.)

Design Method

The design method for steel towers follows. The basic concepts for design of concrete towers are similar. For the transverse direction, main towers are analyzed using small deformation theory. This is permissible because the effect of cable restraint is negligible and the flexural rigidity of the tower is high. For the longitudinal direction, Birdsall's analysis method, discussed in Section 18.3.2, is generally used. However, more rigorous methods, such as finite displacement analysis with a three-dimensional model which allows analysis of both the transverse and longitudinal directions, can be used, as was done in the Akashi Kaikyo Bridge. An example of the design procedure for main towers is shown in Figure 18.18 [23].

Tower Structure

The tower shaft cross section may be T-shaped, rectangular, or cross-shaped, as shown in Figure 18.19. Although the multicell made up of small box sections has been used for some time, cells and single cells have become noticeable in more recent suspension bridges.

The details of the tower base that transmits the axial force, lateral force, and bending moment into the foundation, are either of grillage (bearing transmission type) or embedded types (shearing transmission type), as shown in Figure 18.20. Field connections for the tower shaft are typically bolted joints. Large compressive forces from the cable act along the tower shafts. Tight contact between two metal surfaces acts together with bolted joint to transmit the compressive force across joints with the bearing stresses spread through the walls and the longitudinal stiffeners inside the tower shaft. This method can secure very high accuracy of tower configuration. Another type of connection detail for steel towers using tension bolts was used in the Forth Road Bridge, the Severn Bridge, the Bosporus Bridge, and the first Kurushima Kaikyo Bridge (Figure 18.21).

18.3.7 Cables

General

Parallel wire cable has been used exclusively as the main cable in long-span suspension bridges. Parallel wire has the advantage of high strength and high modulus of elasticity compared with

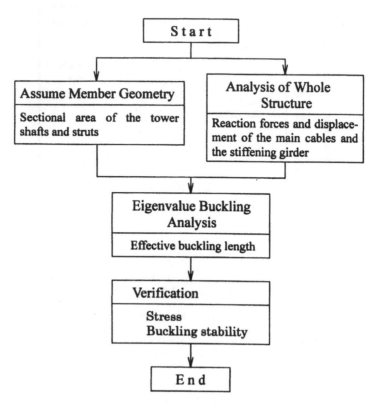

FIGURE 18.18 Design procedure for the main towers. (*Source:* Honshu–Shikoku Bridge Authority, Design Standard of the Main Tower for a Suspension Bridge, HSBA, Japan, 1984. With permission.)

stranded wire rope. The design of the parallel wire cable is discussed next, along with structures supplemental to the main cable.

Design Procedure

Alignment of the main cable must be decided first (Figure 18.22). The sag–span ratios should be determined in order to minimize the construction costs of the bridge. In general, this sag–span ratio is around 1:10. However, the vibration characteristics of the entire suspension bridge change occasionally with changes in the sag–span ratios, so the influence on the aerodynamic stability of the bridge should be also considered. After structural analyses are executed according to the design process shown in Figure 18.14, the sectional area of the main cable is determined based on the maximum cable tension, which usually occurs at the side span face of the tower top.

Design of Cable Section

The tensile strength of cable wire has been about 1570 N/mm² (160 kgf/mm²) in recent years. For a safety factor, 2.5 was used for the Verrazano Narrows Bridge and 2.2 for the Humber Bridge, respectively. In the design of the Akashi Kaikyo Bridge, a safety factor of 2.2 was used using the allowable stress method considering the predominant stress of the dead load. The main cables used a newly developed high-strength steel wire whose tensile strength is 1770 N/mm² (180 kgf/mm²) and the allowable stress was 804 N/mm² (82 kgf/mm²) which led to this discussion. Increase in the strength of cable wire over the years is shown in Figure 18.23. In the design of the Great Belt East Bridge which was done using limit state design methods, a safety factor of 2.0 was applied for the critical limit state [24]. Cable statistics of major suspension bridges are shown in Table 18.5.

Supplemental Components

Figure 18.24 shows the supplemental components of the main cable.

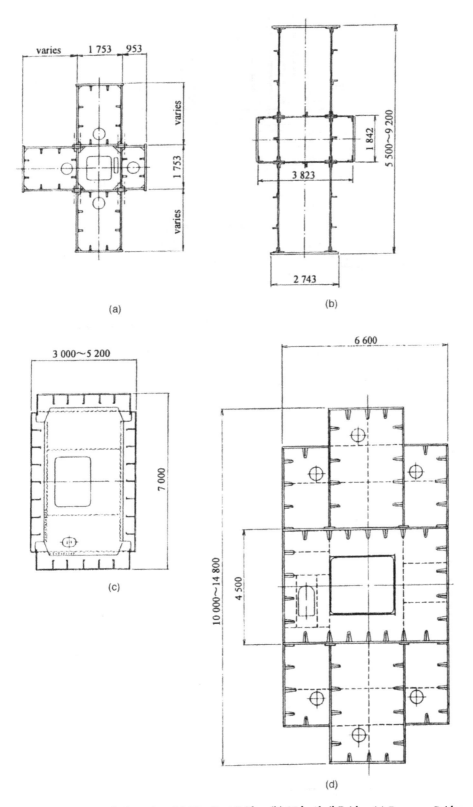

FIGURE 18.19 Tower shaft section. (a) New Port Bridge, (b) 25de Abril Bridge, (c) Bosporus Bridge, (d) Akashi Kaikyo Bridge.

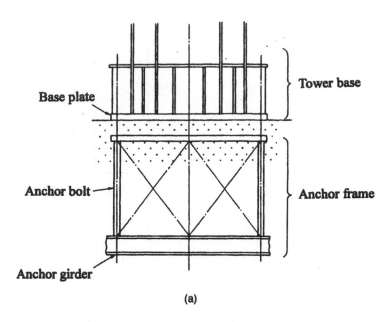

(a)

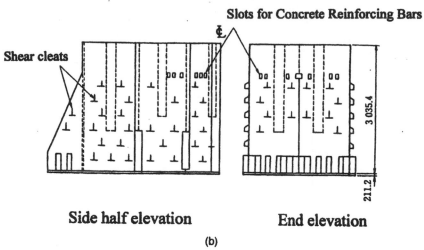

Side half elevation   **End elevation**

(b)

FIGURE 18.20   Tower base. (a) Grillage structure (bearing — transmission type), Akashi Kaikyo Bridge; (b) embedded base (shearing transmission type), Bosporus Bridge.

1. Cable strands are anchored in the cable anchor frame which is embedded into the concrete anchorage.
2. Hanger ropes are fixed to the main cable with the cable bands.
3. Cable saddles support the main cable at the towers and at the splay bents in the anchorages; the former is called the tower saddle and the latter is called the splay saddle.

18.3.8 Suspended Structures

General

The suspended structure of a suspension bridge can be classified as a truss stiffening girder or a box stiffening girder, as described in Section 18.3.4. Basic considerations in selecting girder types are shown in Table 18.6. The length of the bridge and the surrounding natural conditions are also factors.

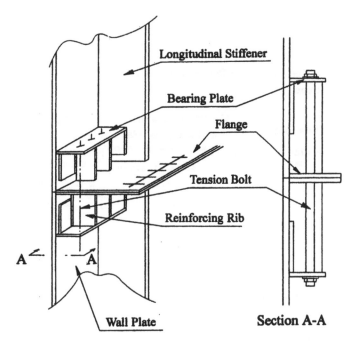

FIGURE 18.21 Connection using tension bolts. (First Kurushima Kaikyo Bridge, Bosporus Bridge.)

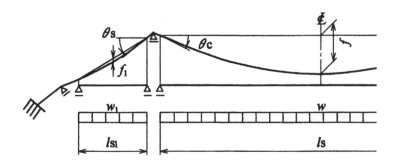

f : Center span sag	θc : Tangential angle of cable (Center span)
f_1 : Side span sag	θs : Tangential angle of cable (Side span)
w : Uniform dead load (Center span)	ls : Center span length
w_1 : Uniform dead load (Side span)	ls_1 : Side span length

FIGURE 18.22 Configuration of suspension bridge.

Design of the Stiffening Girder

Basic Dimensions

The width of the stiffening girder is determined to accommodate carriageway width and shoulders. The depth of the stiffening girder, which affects its flexural and torsional rigidity, is decided so as to ensure aerodynamic stability. After examining alternative stiffening girder configurations, wind tunnel tests are conducted to verify the aerodynamic stability of the girders.

In judging the aerodynamic stability, in particular the flutter, of the bridge design, a bending–torsional frequency ratio of 2.0 or more is recommended. However, it is not always necessary to satisfy this condition if the aerodynamic characteristics of the stiffening girder are satisfactory.

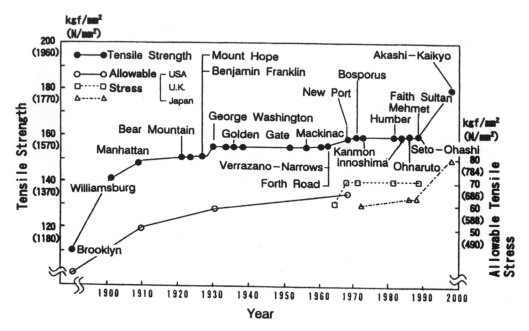

FIGURE 18.23   Increase in strength of cable wire. (*Source:* Honshu–Shikoku Bridge Authority, Akashi Kaikyo Bridge — Engineering Note, Japan, 1992. With permission.)

TABLE 18.5   Main Cable of Long-Span Suspension Bridges

No.	Bridge	Country	Year of Completion	Center Span Length (m)	Erection Method[c]	Composition of Main Cable[d]
1	Akashi Kaikyo	Japan	1998	1991	P.S.	127 × 290
2	Great Belt East	Denmark	1998	1624	A.S.	504 × 37
3	Humber	U.K.	1981	1410	A.S.	404 × 37
4	Jing Yin Yangtze River	China[a]	(1999)[b]	1385	P.S.	127 × 169(c/s), 177(s/s)
5	Tsing Ma	China[a]	1997	1377	A.S.	368 × 80 + 360 × 11 (c/s, Tsing Yi s/s)
						368 × 80+360 × 11 + 304 × 6(Ma Wan s/s)
6	Verrazano Narrows	U.S.	1964	1298.5	A.S.	428 × 61 × 2 cables
7	Golden Gate	U.S.	1937	1280.2	A.S.	452 × 61
8	Höga Kusten	Sweden	1997	1210	A.S.	304 × 37(c/s)
						304 × 37 + 120 × 4(s/s)
9	Mackinac Straits	U.S.	1957	1158.2	A.S.	340 × 37
10	Minami Bisan-Seto	Japan	1988	1100	P.S.	127 × 271
11	Fatih Sultan Mehmet	Turkey	1988	1090	A.S.	504 × 32(c/s), 36(s/s)
12	Bosphorus	Turkey	1973	1074	A.S.	550 × 19
13	George Washington	U.S.	1931	1066.8	A.S.	434 × 61 × 2 cables
14	3rd Kurushima Kaikyo	Japan	1999[b]	1030	P.S.	127 × 102
15	2nd Kurushima Kaikyo	Japan	1999[b]	1020	P.S.	127 × 102
16	25 de Abril	Portugal	1966	1012.9	A.S.	304 × 37
17	Forth Road	UK	1964	1005.8	A.S.	(304~328) × 37
18	Kita Bisan-Seto	Japan	1988	990	P.S.	127 × 234
19	Severn	UK	1966	987.6	A.S.	438 × 19
20	Shimotsui-Seto	Japan	1988	940	A.S.	552 × 44

[a] The People's Republic of China.

[b] Under construction

[c] P.S.: prefabricated parallel wire strand method A.S.: aerial spinning erection method.

[d] Wire/strand × strand/cable.

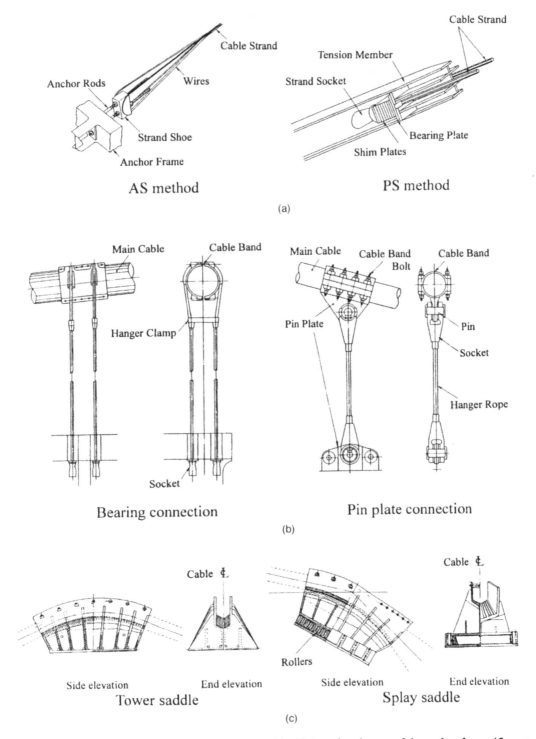

FIGURE 18.24 Supplemental components of the main cable. (a) Strand anchorage of the anchor frame. (*Source:* Japan Society of Civil Engineers, *Suspension Bridge,* Japan, 1996. With permission.) (b) Hanger ropes. (*Source:* Japan Society of Civil Engineers, *Suspension Bridge,* Japan, 1996. With permission.) (c) Cable Saddles. (*Source:* Honshu–Shikoku Bridge Authority, *Design of a Suspension Bridge,* Japan, 1990. With permission.)

TABLE 18.6 Basic Considerations in Selecting Stiffening Structure Types

Item	Truss Girder	Box Girder
Girder height	High	Low
Aerodynamic stability	Flutter should be verified	Vortex-induced oscillation tends to occur
		Flutter should be verified
Maintenance	Coating area is large	Coating area is small
Construction	Both plane section and section erection methods can be used	Only section erection method is permissible

Truss Girders

The design of the sectional properties of the stiffening girder is generally governed by the live load or the wind load. Linear finite deformation theory is commonly applied to determine reactions due to live loads in the longitudinal direction, in which theory the influence line of the live load can be used. The reactions due to wind loads, however, are decided using finite deformation analysis with a three-dimensional model given that the stiffening girder and the cables are loaded with a homogeneous part of the wind load. Linearized finite deformation theory is used to calculate the out-of-plane reactions due to wind load because the change in cable tension is negligible.

Box Girders

The basic dimensions of a box girder for relatively small suspension bridges are determined only by the requirements of fabrication, erection, and maintenance. Aerodynamic stability of the bridge is not generally a serious problem. The longer the center span becomes, however, the stiffer the girder needs to be to secure aerodynamic stability. The girder height is determined to satisfy the rigidity requirement. For the Second and Third Kurushima Kaikyo Bridges, the girder height required was set at 4.3 m based on wind tunnel tests. Fatigue due to live loads needs to be especially considered for the upper flange of the box girder because it directly supports the bridge traffic. The diaphragms support the floor system and transmit the reaction force from the floor system to the hanger ropes.

Supplemental Components

Figure 18.25 shows supplemental components of the stiffening girder.

1. The stay ropes fix the main cable and the girder to restrict longitudinal displacement of the girder due to wind, earthquake, and temperature changes.
2. The tower links and end links support the stiffening girder at the main tower and the anchorages.
3. The wind bearings, which are installed in horizontal members of the towers and anchorages, prevent transverse displacement of the girders due to wind and earthquakes.
4. Expansion joints are installed at the main towers of two-hinged bridges and at the anchorages to absorb longitudinal displacement of the girder.

18.4 Construction

18.4.1 Main Towers

Suspension bridge tower supports the main cable and the suspended structure. Controlling erection accuracy to ensure that the tower shafts are perpendicular is particularly important. During construction, because the tower is cantilevered and thus easily vibrates due to wind, countermeasures for vibration are necessary. Recent examples taken from constructing steel towers of the Akashi Kaikyo Bridge and concrete towers of the Tsing Ma Bridge are described below.

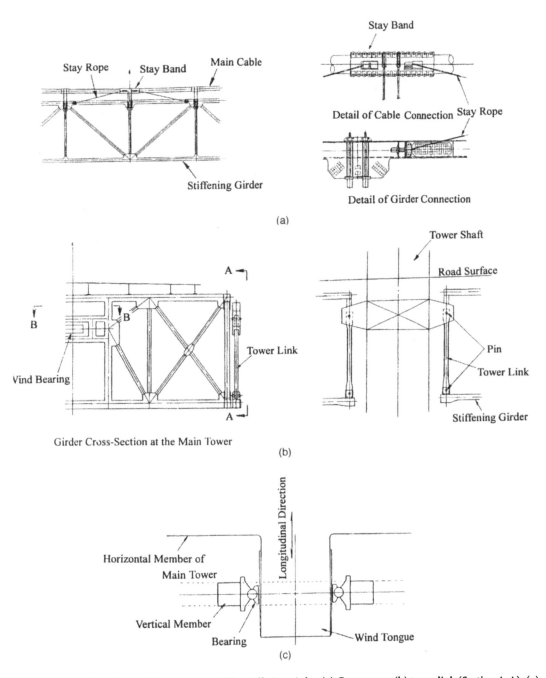

FIGURE 18.25 Supplemental components of the stiffening girder. (a) Center stay; (b) tower link (Section A-A); (c) wind bearing (Section B-B). (*Source:* Honshu–Shikoku Bridge Authority, Design of a Suspension Bridge, Japan, 1990. With permission.)

Steel Towers

Steel towers are typically either composed of cells or have box sections with rib stiffening plates. The first was used in the Forth Road Bridge, the 25 de Abril Bridge, the Kanmon Bridge, and most of the Honshu–Shikoku Bridges. The latter was applied in the Severn Bridge, the Bosporus Bridge, the Fatih Sultan Mehmet Bridge, and the Kurushima Kaikyo Bridges. For the erection of steel towers,

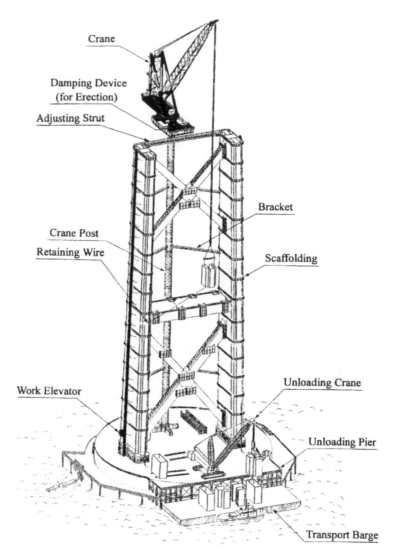

FIGURE 18.26 Overview of main tower construction. (*Source:* Honshu–Shikoku Bridge Authority, Akashi Kaikyo Bridge — Engineering Note, Japan, 1992. With permission.)

floating, tower, and creeper traveler cranes are used. Figure 18.26 shows the tower erection method used for the Akashi Kaikyo Bridge. The tower of the Akashi Kaikyo Bridge is 297 m high. The cross section consists of three cells with clipped corners (see Figure 18.19). The shaft is vertically divided into 30 sections. The sections were prefabricated and barged to the site. The base plate and the first section was erected using a floating crane. The remainder was erected using a tower crane supported on the tower pier. To control harmful wind-induced oscillations, TMD and AMD were installed in the tower shafts and the crane.

Concrete Towers

The tower of the Tsing Ma Bridge is 206 m high, 6.0 m in width transversely, and tapered from 18.0 m at the bottom to 9.0 m at the top longitudinally. The tower shafts are hollow. Each main tower was slip-formed in a continuous around-the-clock operation, using two tower cranes and concrete buckets (Figure 18.27).

FIGURE 18.27   Tower erection for the Tsing Ma Bridge. (Courtesy of Mitsui Engineering & Shipbuilding Co., Ltd.)

18.4.2 Cables

Aerial Spinning Method

The aerial spinning method (AS method) of parallel wire cables was invented by John A. Roebling and used for the first time in the Niagara Falls Bridge which was completed in 1855 with a center span of 246 m (Figure 18.28). He established this technology in the Brooklyn Bridge where steel wire was first used. Most suspension bridges built in the United States since Roebling's development of the AS method have used parallel wire cables. In contrast, in Europe, the stranded rope cable was used until the Forth Road Bridge was built in 1966.

In the conventional AS method, individual wires were spanned in free-hang condition, and the sag of each wire had to be individually adjusted to ensure all were of equal length. In this so-called sag-control method, the quality of the cables and the erection duration are apt to be affected by site working conditions, including wind conditions and the available cable-spinning equipment. It also requires a lot of workers to adjust the sag of the wires.

A new method, called the tension-control method, was developed in Japan (Figure 18.29). The idea is to keep the tension in the wire constant during cable spinning to obtain uniform wire lengths. This method was used on the Hirado, Shimotsui–Seto, Second Bosporus, and Great Belt East Bridges (Figure 18.30). it does require adjustment of the individual strands even in this method.

Prefabricated Parallel Wire Strand Method

Around 1965, a method of prefabricating parallel wire cables was developed to cut the on-site work intensity required for the cable spinning in the AS method. The prefabricated parallel wire strand method (PS method) was first used in the New Port Bridge. That was the first step toward further progress achieved in Japan in enlarging strand sections, developing high-tensile wire, and lengthening the strand.

18.4.3 Suspended Structures

There are various methods of erecting suspended structures. Typically, they have evolved out of the structural type and local natural and social conditions.

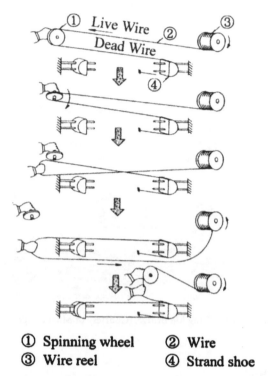

① Spinning wheel   ② Wire
③ Wire reel   ④ Strand shoe

FIGURE 18.28 Operating principle of aerial spinning. (*Source:* Honshu–Shikoku Bridge Authority, Technology of Seto–Ohashi Bridge, Japan, 1989. With permission.)

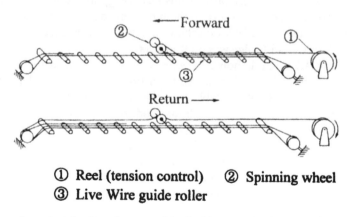

① Reel (tension control)   ② Spinning wheel
③ Live Wire guide roller

FIGURE 18.29 Operating principle of tension control method.(*Source:* Honshu–Shikoku Bridge Authority, Technology of Set–Ohashi Bridge, Japan, 1989. With permission.)

Girder Block Connection Methods

The connections between stiffening girder section may be classified as one of two methods.

All Hinge Method

In this method the joints are loosely connected until all girder sections are in place in general. This method enables simple and easy analysis of the behavior of the girders during construction. Any temporary reinforcement of members is usually unnecessary. However, it is difficult to obtain enough aerodynamic stability unless structures to resist wind force are given to the joints which were used in the Kurushima Kaikyo Bridges, for example.

FIGURE 18.30   Aerial spinning for the Shimotsui–Seto bridge. (Courtesy of Honshu–Shikoku Bridge Authority.)

Rigid Connection Method
In this method full-splice joints are immediately completed as each girder block is erected into place. This keeps the stiffening girder smooth and rigid, providing good aerodynamic stability and high construction accuracy. However, temporary reinforcement of the girders and hanger ropes to resist transient excessive stresses or controlled operation to avoid overstress are sometimes required.

Girder Erection Methods

Stiffening girders are typically put in place using either the girder-section method or cantilevering from the towers or the anchorages.

Girder-Section Method
The state of the art for the girder-section method with hinged connections is shown in Figure 18.31. At the Kurushima Kaikyo Bridges construction sites, the fast and complex tidal current of up to 5 m/s made it difficult for the deck barges and tugboats to maintain their desired position for a long time. As a result, a self-controlled barge, able to maintain its position using computer monitoring, and a quick joint system, which can shorten the actual erection period, were developed and fully utilized.

Cantilevering Method
A recent example of the cantilevering method of girders on the Akashi Kaikyo Bridge is shown in Figure 18.32. Preassembled panels of the stiffening girder truss were erected by extending the stiffening girders as a cantilever from the towers and anchorages. This avoided disrupting marine traffic, which would have been required for the girder-section method.

18.5   Field Measurement and Coatings

18.5.1   Loading Test

The purpose of loading tests is chiefly to confirm the safety of a bridge for both static and dynamic behavior. Static loading tests were performed on the Wakato, the Kanmon, and the President Mobutu

FIGURE 18.31　　Block erection method on the Kurushima Kaikyo Bridge. (Courtesy of Honshu–Shikoku Bridge Authority.)

FIGURE 18.32　　Cantilevering method in the Akashi Kaikyo Bridge. (Courtesy of Honshu–Shikoku Bridge Authority.)

Sese–Seko Bridges by loading heavy vehicles on the bridges. Methods to verify dynamic behavior include vibration tests and the measurement of micro-oscillations caused by slight winds. The former test is based on the measured response to a forced vibration. The latter is described in Section 18.5.2. Dynamic characteristics of the bridge, such as structural damping, natural frequency, and mode of vibration, are ascertained using the vibration test. As the real value of structural damping is difficult to estimate theoretically, the assumed value should be verified by an actual measurement. Examples of measured data on structural damping obtained through vibration tests are shown in Table 18.7.

TABLE 18.7 Structural Damping Obtained from Vibration Tests

Bridge	Center Span Length (m)	Logarithmic Decrement[a]
Minami Bisan–Seto	1100	0.020 ~ 0.096
Ohnaruto	876	0.033 ~ 0.112
Kanmon	712	0.016 ~ 0.062
Ohshima	560	0.017 ~ 0.180

[a] Structural damping.

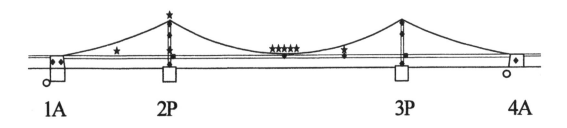

Symbol	Name of instruments
★	wind vane and anemometer
◆	accelerometer, displacement speedmeter
■	deckend displacement gauge
○	seismometer

FIGURE 18.33 Placement of measuring instruments in the Akash Kaikyo Bridge. (*Source:* Abe, K. and Amano, K., Monitoring system of the Akashi Kaikyo Bridge, *Honshi Tech. Rep.*, 86, 29, 1998. With permission.)

18.5.2 Field Observations

Field observations are undertaken to verify such characteristics of bridge behavior as aerodynamic stability and seismic resistance, and to confirm the safety of the bridge. To collect the necessary data for certification, various measuring instruments are installed on the suspension bridge. Examples of measuring instruments used are given in Figure 18.33 [25]. A wind vane and anemometer, which measure local wind conditions, and a seismometer, to monitor seismic activity, gather data on natural conditions. An accelerometer and a displacement speedometer are installed to measure the dynamic response of the structure to wind and earthquake loads. A deck end displacement gauge tracks the response to traffic loads. The accumulated data from these measuring instruments will contribute to the design of yet-longer-span bridges in the future.

18.5.3 Coating Specification

Steel bridges usually get a coating regimen which includes a rust-preventive paint for the base coat, and a long oil-base alkyd resin paint or chlorinated rubber resin paint for the intermediate and top coats. This painting regimen needs to be repeated at several-year intervals. Because long-span suspension bridges are generally constructed in a marine environment, which is severely corrosive, and have enormous painting surfaces, which need to be regularly redone, a heavy-duty coating method with long-term durability is required. The latest coating technology adopted for major suspension bridges is shown in Table 18.8. Previous painting methods relied on oil-base anticorrosive paints or red lead anticorrosive paints for base coats with phthalic resin or aluminum paints as intermediate and top coats. The latest coating specification aimed at long-term durability calls for an inorganic zinc-enriched base

TABLE 18.8 Coating Systems of Major Suspension Bridges

Country	Bridge	Year of Completion	Coating Specification
U.S.	George Washington	1931	Base: oil–based anticorrosive paint Top: phthalic resin paint
	San Francisco–Oakland Bay Golden Gate	1936 1937	Base: red lead anticorrosive paint To: oil–modified phenolic resin aluminum paint
	Mackinac Straits Verrazano Narrows	1957 1965	Base: oil–based anticorrosive paint Top: phthalic resin paint
Canada	Pierre La Porte	1970	Base: basic lead chromate anticorrosive paint Top: alkyd resin paint
Turkey	Bosphorus	1973	Base: zinc spraying Top: phenolic resin micaceous iron oxide paint
	Fatih Sultan Mehmet	1988	Base: organic zinc rich paint Intermediate: epoxy resin paint Intermediate: epoxy resin micaceous iron oxide paint Top: paint chlorinated rubber resin paint
U.K.	Forth Road Severn Humber	1964 1966 1981	Base: zinc spraying Top: phenolic resin micaceous iron oxide paint
Japan	Kanmon	1973	Base: zinc spraying Intermediate: micaceous iron oxide paint Top: chlorinated rubber resin paint
	Innoshima	1983	Base: hi-build inorganic zinc rich paint Intermediate: hi-build epoxy resin paint Top: polyurethane resin paint
	Akashi Kaikyo	1998	Base: hi-build inorganic zinc rich paint Intermediate: hi-build epoxy resin paint Intermediate: epoxy resin paint Top: fluororesin paint

paint, which is highly rust-inhibitive due to the sacrificial anodic reaction of the zinc, with an epoxy resin intermediate coat and a polyurethane resin or fluororesin top coat. Because the superiority of fluororesin paint for long-term durability and in holding a high luster under ultraviolet rays has been confirmed in recent years, it was used for the Akashi Kaikyo Bridge [26].

18.5.4 Main Cable Corrosion Protection

Since the main cables of a suspension bridge are the most important structural members, corrosion protection is extremely important for the long-term maintenance of the bridge. The main cables are composed of galvanized steel wire about 5 mm in diameter with a void of about 20% which is longitudinally and cross-sectionally consecutive. Main cable corrosion is caused not only by water and ion invasion from outside, but also by dew resulting from the alternating dry and humid conditions inside the cable void. The standard corrosion protection system for the main cables ever since it was first worked out for the Brooklyn Bridge has been to use galvanized wire covered with a paste, wrapped with galvanized soft wires and then coated.

New approaches such as wrapping the wires with neoprene rubber or fiberglass acrylic or S-shaped deformed steel wires have also been attempted. A dehumidified air-injection system was developed and used on the Akashi Kaikyo Bridge [27]. This system includes wrapping to improve watertightness and the injection of dehumidified air into the main cables as shown in Figure 18.34. Examples of a corrosion protection system for the main cables in major suspension bridges are shown in Table 18.9.

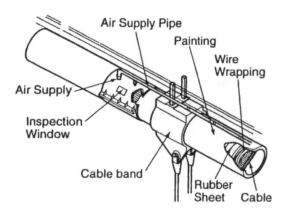

FIGURE 18.34 Dehumidified air–injection system for the main cables of the Akashi Kaikyo Bridge. (Courtesy of Honshu–Shikoku Bridge Authority.)

TABLE 18.9 Corrosion Protection Systems for Main Cable of Major Suspension Bridges

Bridge	Year of Completion	Erection Method	Cable Wire	Cable Paste	Wrapping
Brooklyn	1883	A.S.	Galvanized	Red lead paste	Galvanized wire
Williamsburg	1903	A.S.	—ᵃ	Red lead paste	Cotton duck + sheet iron coating
Golden Gate	1937	A.S.	Galvanized	Red lead paste	Galvanized wire
Chesapeake Bay II	1973	A.S.	Galvanized	—	Neoprene rubber
Verrazano Narrows	1964	A.S.	Galvanized	Red lead paste	Galvanized wire
Severn	1966	A.S.	Galvanized	Red lead paste	Galvanized wire
New Port	1969	A.S.	Galvanized	—	Glass-reinforced acrylic
Kanmon	1973	P.S.	Galvanized	Polymerized organic lead paste	Galvanized wire
Minami Bisan–Seto	1988	P.S.	Galvanized	Calcium plumbate contained polymerized organic lead paste	Galvanized wire
Hakucho	1998	P.S.	Galvanized	Aluminum triphosphate contained organic lead paste	Galvanized wire (S shape)
Akashi Kaikyo	1998	P.S.	Galvanized	—	Galvanized wire + rubber wrapping

ᵃ Coated with a raw linseed oil.

References

1. Lu, J., Large Suspension Bridges in China, *Bridge Foundation Eng.*, 7, 20, 1996 [in Japanese].
2. Navier, M., *Papport et Mémoire sur les Ponts Suspendus*, de l' Imprimerie Royale, Paris, 1823.
3. Rankine, W. J. M., *A Manual of Applied Mechanics*, 1858.
4. Ritter, W., Versteifungsfachewerke bei Bogen und Hängebrücken, *Z. bauwesen*, 1877.
5. Lévy, M., Mémoir sur le calcul des ponts suspendus rigides, *Ann. Ponts Chaussées*, 1886.
6. Melan, J., Theorie der eisernen Bogenbrücken und der Hängebrücken, *Handb. Ingenieurwissensch.*, 1888.
7. Moisseiff, L. S., The towers, cables and stiffening trusses of the bridge over the Delaware River between Philadelphia and Camden, *J. Franklin Inst.*, Oct., 1925.

8. Moisseiff, L. S. and Lienhard, F., Suspension bridge under the action of lateral forces, *Trans. ASCE,* 58, 1933.

9. Brotton, D. M., Williamson, N. M., and Millar, M., The solution of suspension bridge problems by digital computers, Part I, *Struct. Eng.,* 41, 1963.

10. Brotton, D. M., A general computer programme for the solution of suspension bridge problems, *Struct. Eng.,* 44, 1966.

11. Saafan, A. S., Theoretical analysis of suspension bridges, *Proc. ASCE,* 92, ST4, 1966.

12. Tezcan, S. S., Stiffness analysis of suspension bridges by iteration, in *Symposium on Suspension Bridges,* Lisbon, 1966.

13. Bleich, F., McCullough, C. B., Rosecrans, R., and Vincent, G. S., The Mathematical Theory of Vibration in Suspension Bridges, Department of Commerce, *Bureau of Public Roads,* Washington, D.C., 1950.

14. Birdsall, B., The suspension bridge tower cantilever problem, *Trans. ASCE,* 1942.

15. Steinman, D. B., A generalized deflection theory for suspension bridges, *Trans. ASCE,* 100, 1935.

16. Walshe, D. E. et al., A Further Aerodynamic Investigation for the Proposed Severn River Suspension Bridge, 1966.

17. Roberts, G., Severn Bridge, Institution of Civil Engineers, London, 1970.

18. Simpson, A. G., Curtis, D. J., and Choi, Y.-L., Aeroelasic aspects of the Lantau fixed crossing, Institution of Civil Engineers, London, 1981.

19. Ohashi, M., Miyata, T., Okauchi, I., Shiraishi, N., and Narita, N., Consideration for Wind Effects on a 1990 m Main Span Suspension Bridge, Pre-report 13th Int. Congress IABSE, 1988, 911.

20. Diana, G., Aeroelastic study of long span suspension bridges, the Messina Crossing, ASCE Structures Congress '93, Irvine, CA, 1993.

21. Honshu–Shikoku Bridge Authority, Wind-Resistant Design Standard for the Akashi Kaikyo Bridge, HSBA, Japan, 1990 [in Japanese].

22. Kashima, S., Yasuda, M., Kanazawa, K., and Kawaguchi, K., Earthquake Resistant Design of Akashi Kaikyo Bridge, paper presented as Third Workshop on Performance and Strengthening of Bridge Structures, Tsukuba, 1987.

23. Honshu–Shikoku Bridge Authority, Design Standard of the Main Tower for a Suspension Bridge, HSBA, Japan, 1984 [in Japanese].

24. Petersen, A. and Yamasaki, Y., Great Belt Bridge and design of its cable works, *Bridge Foundation Eng.,*1, 18, 1994 [in Japanese].

25. Abe, K. and Amano, K., Monitoring System of the Akashi Kaikyo Bridge, *Honshi Technical Report,* 86, 29, 1998 [in Japanese].

26. Honshu–Shikoku Bridge Authority, Steel Bridges Coating Standards of Honshu–Shikoku Bridge, HSBA, Japan, 1990 [in Japanese].

27. Ito, M., Saeki, S., and Tatsumi, M., Corrosion protection of bridge cables: from Japanese experiences, in *Proceedings of the International Seminar on New Technologies for Bridge Management,* IABSE, Seoul, 1996.

19

Cable-Stayed Bridges

Man-Chung Tang
T.Y. Lin International

19.1   Introduction

Since the completion of the Stromsund Bridge in Sweden in 1955, the cable-stayed bridge has evolved into the most popular bridge type for long-span bridges. The variety of forms and shapes of cable-stayed bridges intrigues even the most-demanding architects as well as common citizens. Engineers found them technically innovative and challenging. For spans up to about 1000 m, cable-stayed bridges are more economical.

The concept of a cable-stayed bridge is simple. A bridge carries mainly vertical loads acting on the girder, Figure 19.1. The stay cables provide intermediate supports for the girder so that it can span a long distance. The basic structural form of a cable-stayed bridge is a series of overlapping triangles comprising the pylon, or the tower, the cables, and the girder. All these members are under predominantly axial forces, with the cables under tension and both the pylon and the girder under compression. Axially loaded members are generally more efficient than flexural members. This contributes to the economy of a cable-stayed bridge.

At the last count, there are about 600 cable-stayed bridges in the world and the number is increasing rapidly. The span length has also increased significantly [2,7].

Some milestones: the Stromsund Bridge in Sweden, completed in 1955 with a main span of 183 m is usually recognized as the world's first major cable-stayed bridge; the Knie Bridge (320 m) and Neuenkamp Bridge (350 m) in Germany, Figure 19.2, were the longest spans in the early 1970s, until the Annacis Island–Alex Fraser Bridge (465 m) was completed in the mid 1980s. The 602-m-span Yangpu Bridge was a large step forward in 1994 but was surpassed within about half a year by the Normandie Bridge (856 m), Figure 19.3. The Tatara Bridge, with a center span of 890 m, is the world record today. Several spans in the range of 600 m are under construction. Longer spans are being planned.

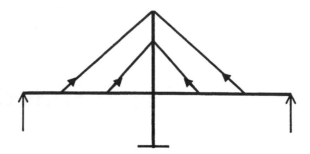

FIGURE 19.1   Concept of a cable-stayed bridge.

FIGURE 19.2   Neuenkamp Bridge.

FIGURE 19.3   Normandie Bridge.

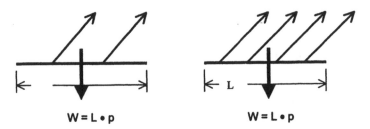

FIGURE 19.4 Cable forces in relation to load on girder.

19.2 Configuration

19.2.1 General Layout

At the early stage, the idea of a cable-stayed bridge was to use cable suspension to replace the piers as intermediate supports for the girder so that it could span a longer distance. Therefore, early cable-stayed bridges placed cables far apart from each other based on the maximum strength of the girder. This resulted in rather stiff girders that had to span the large spacing between cables, in addition to resisting the global forces.

The behavior of a cable-stayed girder can be approximately simulated by an elastically supported girder. The bending moment in the girder under a specific load can be thought of as consisting of a local component and a global component. The local bending moment between the cables is proportional to the square of the spacing. The global bending moment of an elastically supported girder is approximately [5]

$$M = a * p * \sqrt{(I/k)} \tag{19.1}$$

where a is a coefficient depending on the type of load p, I is the moment of inertia of the girder, and k is the elastic support constant derived from the cable stiffness. The global moment decreases as the stiffness of girder, I, decreases.

Considering that the function of the cables is to carry the loads on the bridge girder, which remains the same, the total quantity of cables required for a bridge is practically the same independent of the number of cables, or cable spacing, Figure 19.4. But if the cable spacing is smaller, the local bending moment of the girder between the cables is also smaller. A reduction of the local bending moment allows the girder to be more flexible. A more flexible girder attracts in turn less global moment. Consequently, a very flexible girder can be used with closely spaced cables in many modern cable-stayed bridges. The Talmadge Bridge, Savannah, Figure 19.5, is 1.45 m deep for a 335 m span, The ALRT Skytrain Bridge, Vancouver, Figure 19.6, is 1.1 m deep for a 340 m span and the design of the Portsmouth Bridge had a 84-cm-deep girder for a span of 286 m.

Because the girder is very flexible, questions concerning buckling stability occasionally arose at the beginning. However, as formulated by Tang [4], Eq. (19.2), using the energy method,

$$P(cr) = \left\{ \int EIw''^2 ds + \sum EC * Ac * Lc \right\} \Big/ \left[\int (Ps/Pc)w'^2 ds \right] \tag{19.2}$$

where E is modulus of elasticity, I is moment of inertia, A is area, L is length, w is deflection, and $(\ )'$ is derivative with respect to length s. The buckling load depends more on the stiffness of the cables than on the stiffness of the girder. Theoretically, even if the stiffness of the girder is neglected, a cable-stayed bridge can still be stable in most cases. Experience shows that even for the most flexible girder, the critical load against elastic buckling is well over 400% of the actual loads of the bridge.

FIGURE 19.5   Talmadge Bridge.

FIGURE 19.6   ALRT Skytrain Bridge.

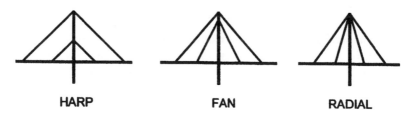

HARP   FAN   RADIAL

FIGURE 19.7   Cable arrangements.

The recently adopted design requirement that all cables be individually replaceable makes closely spaced cables more desirable. It is usually required that one cable can be detensioned, dismantled, and replaced under reduced traffic loading. The additional bending moment in the girder will not increase excessively if the cable spacing is small.

Availability of ever more powerful computers also helps. The complexity of the analysis increases as the number of cables increases. The computer offers engineers the best tool to deal with this problem.

Harp, radial, fan, Figure 19.7, or other cable configurations have all been used. However, except in very long span structures, cable configuration does not have a major effect on the behavior of the bridge.

A harp-type cable arrangement offers a very clean and delicate appearance because an array of parallel cables will always appear parallel irrespective of the viewing angle. It also allows an earlier

FIGURE 19.8   Nord Bridge.

FIGURE 19.9   Ludwighafen Bridge.

start of girder construction because the cable anchorages in the tower begin at a lower elevation. The Hoechst Bridge and the Dames Point Bridge are examples that fully utilized this advantage.

A fan-type cable arrangement can also be very attractive, especially for a single-plane cable system. Because the cable slopes are steeper, the axial force in the girder, which is an accumulation of all horizontal components of cable forces, is smaller. This feature is advantageous for longer-span bridges where compression in the girder may control the design. The Nord Bridge, Bonn, Figure 19.8, is one of the first of this type.

A radial arrangement of cables with all cables anchored at a common point at the tower is quite efficient. However, a good detail is difficult to achieve. Unless it is well treated, it may look clumsy. The Ludwighafen Bridge, Germany, Figure 19.9, is a successful example. The Yelcho Bridge, Chile, with all cables anchored in a horizontal plane in the tower top, is an excellent solution, both technically and aesthetically.

When the Stromsund Bridge was designed, long-span bridges were the domain of steel construction. Therefore, most early cable-stayed bridges were steel structures. They retained noticeable features from other types of long-span steel bridges.

In the 1960s, Morandi designed and built several relatively long span concrete cable-stayed bridges. His designs usually had few cables in a span with additional strut supports at the towers for the girder. They did not fully utilize the advantages of a cable-stayed system. The concrete cable-stayed bridge in its modern form started with the Hoechst Bridge in Germany, followed by the Brotone Bridge in France and the Dames Point Bridge in the United States, each representing a significant advance in the state of the art.

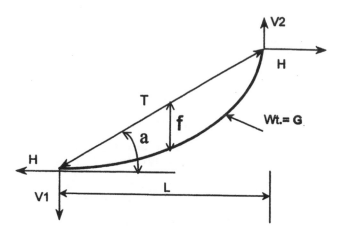

FIGURE 19.10   An inclined cable.

19.2.2   Cables

Cables are the most important elements of a cable-stayed bridge. They carry the load of the girder and transfer it to the tower and the back-stay cable anchorage.

The cables in a cable-stayed bridge are all inclined, Figure 19.10. The actual stiffness of an inclined cable varies with the inclination angle, a, the total cable weight, G, and the cable tension force, T [3]:

$$EA(\text{eff}) = EA\Big/\Big\{1 + G^2\ EA \cos^2 a\Big/\big(12\ T^3\big)\Big\} \tag{19.3}$$

where E and A are Young's modulus and the cross-sectional area of the cable. And if the cable tension T changes from $T1$ to $T2$, the equivalent cable stiffness will be

$$EA(\text{eff}) = EA\Big/\Big\{1 + G^2\ EA \cos^2 a\big(T1 + T2\big)\Big/\big(24\ T1^2\ T2^2\big)\Big\} \tag{19.4}$$

In most cases, the cables are tensioned to about 40% of their ultimate strength under permanent load condition. Under this kind of tension, the effective cable stiffness approaches the actual values, except for very long cables. However, the tension in the cables may be quite low during some construction stages so that their effectiveness must be properly considered.

A safety factor of 2.2 is usually recommended for cables. This results in an allowable stress of 45% of the guaranteed ultimate tensile strength (GUTS) under dead and live loads [9]. It is prudent to note that the allowable stress of a cable must consider many factors, the most important being the strength of the anchorage assemblage that is the weakest point in a cable with respect to capacity and fatigue behavior.

There have been significant developments in the stay cable system. Early cables were mainly lock-coil strands. At that time, the lock-coil strand was the only cable system available that could meet the more stringent requirements of cable-stayed bridges.

Over the years, many new cable systems have been successfully used. Parallel wire cables with Hi-Am sockets were first employed in 1969 on the Schumacher Bridge in Mannheim, Germany. Since then, the fabrication technique has been improved and this type cable is still one of the best cables commercially available today. A Hi-Am socket has a conical steel shell. The wires are parallel for the entire length of the cable. Each wire is anchored to a plate at the end of the socket by a button head. The space in the socket is then filled with epoxy mixed with zinc and small steel balls.

The Hi-Am parallel wire cables are prefabricated to exact length in the yard and transported to the site in coils. Because the wires are parallel and therefore all of equal length, the cable may sometimes experience difficulty in coiling. This difficulty can be overcome by twisting the cable during the coiling process. To avoid this problem altogether, the cables can be fabricated with a long lay. However, the long lay may cause a very short cable to twist during stressing.

Threadbar tendons were used for some stay cables. The first one was for the Hoechst Bridge over the Main River in Germany. The Penang Bridge and the Dames Point Bridge also have bar cables. They all have a steel pipe with cement grout as corrosion protection. Their performance has been excellent.

The most popular type of cable nowadays uses seven-wire strands. These strands, originally developed for prestressed concrete applications, offer good workability and economy. They can either be shop-fabricated or site-fabricated. In most cases, corrosion protection is provided by a high-density polyethylene pipe filled with cement grout. The technique of installation has progressed to a point where a pair of cables can be erected at the site in 1 day.

In search of better corrosion protection, especially during the construction stage before the cables have been grouted, various alternatives, such as epoxy coating, galvanization, wax and grease have all been proposed and used. Proper coating of strands must completely fill the voids between the wires with corrosion inhibitor. This requires the wires to be loosened before the coating process takes place and then retwisted into the strand configuration.

In addition to epoxy, grease, or galvanization, the strands may be individually sheathed. A sheathed galvanized strand may have wax or grease inside the sheathing. All three types of additional protection appear to be acceptable and should perform well. However, a long-term performance record is not yet available.

The most important element in a cable is the anchorage. In this respect, the Hi-Am socket has an excellent performance record. Strand cables with bonded sockets, similar to the Hi-Am socket, have also performed very well. In a recently introduced unbonded anchorage, all strands are being held in place only by wedges. Tests have confirmed that these anchorages meet the design requirements. But unbonded strand wedges are delicate structural elements and are susceptible to construction deviations. Care must be exercised in the design, fabrication, and installation if such an anchorage is to be used in a cable-stayed bridge. The advantage of an unbonded cable system is that the cable, or individual strands, can be replaced relatively easily.

Cable anchorage tests have shown that, in a bonded anchor, less than half of the cyclic stress is transferred to the wedges. The rest is dissipated through the filling and into the anchorage directly by bond. This is advantageous with respect to fatigue and overloading.

The Post Tensioning Institute's "Recommendations for Stay Cable Design and Testing," [9] was published in 1986. This is the first uniformly recognized criteria for the design of cables. In conjunction with the American Society of Civil Engineers' "Guidelines for the Design of Cable-Stayed Bridges" [10], they give engineers a much-needed base to start their design.

There have been various suggestions for using composite materials such as carbon fiber, etc. as stays and small prototypes have been built. However, actual commercial application still requires further research and development.

19.2.3   Girder

Although the Stromsund Bridge has a concrete deck, most other early cable-stayed bridges have an orthotropic deck. This is because both cable-stayed bridge and orthotropic deck were introduced to the construction industry at about the same time. Their marriage was logical. The fact that almost all long-span bridges were built by steel companies at that time made such a choice more understandable.

A properly designed and fabricated orthotropic deck is a good solution for a cable-stayed bridge. However, with increasing labor costs, the orthotropic deck becomes less commercially attractive except for very long spans.

Many concrete cable-stayed bridges have been completed. In general, there have been two major developments: cast-in-place construction and precast construction.

FIGURE 19.11   Form traveler of the Dames Point Bridge.

FIGURE 19.12   Sunshine Skyway Bridge – precast box and erection.

Cast-in-place construction of cable-stayed bridges is a further development of the free cantilever construction method of box-girder bridges. The typical construction is by means of a form traveler. The box girder is a popular shape for the girder in early structures, such as the Hoechst Bridge, the Barrios de Luna Bridge, and the Waal Bridge. But simpler cross sections have proved to be attractive: the beam and slab arrangement in the Penang Bridge, the Dames Point Bridge, and the Talmadge Bridge, or the solid slab cross section as in the Yelcho Bridge, the Portsmouth Bridge, and the Diepold-sau Bridge are both technically sound and economical. Use of the newly developed cable-supported form traveler, Figure 19.11, makes this type girder much more economical to build [8].

Precast construction can afford a slightly more complicated cross section because precasting is done in the yard. The segments, however, should all be similar to avoid adjustment in the precasting forms. The weight of the segment is limited by the transportation capability of the equipment used. Box is the preferred cross section because it is stiffer and easier to erect. The Brotonne Bridge, the Sunshine Skyway Bridge, Figure 19.12, and the Chesapeake and Delaware Canal Bridge are good examples. However, several flexible girder cable-stayed bridges have been completed successfully, notably the Pasco–Kennewick Bridge, the East Huntington Bridge, and the ALRT Skytrain Bridge.

As concrete technology advances, today's cable-stayed bridges may consider using high-strength lightweight concrete for the girder, especially in high seismic areas.

Although the steel orthotropic deck is too expensive for construction in most countries at this time, the composite deck with a concrete slab on a steel frame can be very competitive. In the Stromsund Bridge, the concrete deck was not made composite with the steel girder. Such a construction is not economical because the axial compressive force in the girder must be taken entirely by

FIGURE 19.13 Baytown–LaPorte Bridge and Yang Pu Bridge.

the steel girder. Making the deck composite with the steel girder by shear studs reduces the steel quantity of the girder significantly. The compressive stress in the concrete deck also improves the performance of the deck slab. The Hootley Bridge was the first major composite bridge designed, but the Annacis Island Bridge was completed first. The Yang Pu Bridge is the longest span today. The Baytown–LaPorte Bridge, Figure 19.13, has the largest deck area.

Precast slab panels are usually used for composite bridges. Requiring the precast panels to be stored for a period of time, say, 90 days, before erection reduces the effect of creep and shrinkage significantly. The precast panels are supported by floor beams and the edge girders during erection. The gaps between the panels are filled with nonshrinking concrete. The detail for these closure joints must be carefully executed to avoid cracking due to shrinkage and other stresses.

Most portions of the girder are under high compression, which is good for concrete members. However, tensile stresses may occur in the middle portion of the center span and at both ends of the end spans. Post-tensioning is usually used in these areas to keep the concrete under compression.

Several hybrid structures, with concrete side spans and steel main span, such as the Flehe Bridge, Germany and the Normandie Bridge, have been completed. There are two main reasons for the hybrid combination: to have heavier, shorter side spans to balance the longer main span or to build the side spans the same way as the connecting approaches. The transition, however, must be carefully detailed to avoid problems.

19.2.4 Tower

The towers are the most visible elements of a cable-stayed bridge. Therefore, aesthetic considerations in tower design is very important. Generally speaking, because of the enormous size of the structure,

FIGURE 19.14   Tower shapes — H (Talmadge), inverted Y (Flehe), diamond (Glebe), double diamond (Baytown), and special configuration (Dan Chiang).

a clean and simple configuration is preferable. The free-standing towers of the Nord Bridge and the Knie Bridge look very elegant. The H towers of the Annacis Bridge, the Talmadge Bridge, and the Nan Pu Bridge are the most logical shape structurally for a two-plane cable-stayed bridge, Figure 19.14. The A shape (as in the East Hungtington Bridge), the inverted Y (as in the Flehe Bridge), and the diamond shape (as in the Baytown–LaPorte and Yang Pu Bridges) are excellent choices for long-span cable-stayed bridges with very flexible decks. Other variations in tower shape are possible as long as they are economically feasible. Under special circumstances, the towers can also serve as tourist attractions such as the one proposed for the Dan Chiang Bridge in Taiwan.

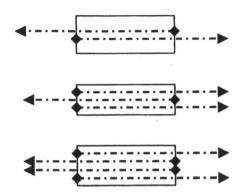

FIGURE 19.15 Crisscrossing cables at tower column.

Although early cable-stayed bridges all have steel towers, most recent constructions have concrete towers. Because the tower is a compression member, concrete is the logical choice except under special conditions such as in high earthquake areas.

Cables are anchored at the upper part of the tower. There are generally three concepts for cable anchorages at the tower: crisscrossing, dead-ended, and saddle.

Early cable-stayed bridges took their anchorage details from suspension bridges that have saddles. Those saddles were of the roller, fixed, or sector types. Roller-type saddles have a roller at the base similar to a bridge bearing. Fixed-type saddles are similar except the base is fixed instead of rolling. Sector-type saddles rotate around a pin. Each of these saddles satisfies a different set boundary condition in the structural system. Cable strands, basically lock-coil strands, were placed in the saddle trough as in a suspension bridge. To assure that the strands do not slide under unequal tension, a cover plate is usually clamped to the saddle trough to increase friction. This transverse pressure increases friction but reduces the strength of the cable. This type of saddle is very expensive and has not been used for recent cable-stayed bridges.

A different type of saddle was used in the Brotonne Bridge and the Sunshine Skyway Bridge. Here the seven-wire strands were bundled into a cable and then pulled through a steel saddle pipe which was fixed to the tower. Grouting the cable fixed the strands to the saddle pipe. However, the very high contact pressure between the strand wires must be carefully considered and dealt with in the design. Because the outer wires of a strand are helically wound around a straight, center king wire, the strands in a curved saddle rest on each other with point contacts. The contact pressure created by the radial force of the curved strands can well exceed the yield strength of the steel wires. This can reduce the fatigue strength of the cable significantly.

Crisscrossing the cables at the tower is a good idea in a technical sense. It is safe, simple, and economical. The difficulty is in the geometry. To avoid creating torsional moment in the tower column, the cables from the main span and the side span should be anchored in the tower in the same plane, Figure 19.15. This, however, is physically impossible if they crisscross each other. One solution is to use double cables so that they can pass each other in a symmetrical pattern as in the case of the Hoechst Bridge. If A-shaped or inverted-Y-shaped towers are used, the two planes of cables can also be arranged in a symmetrical pattern.

If the tower cross section is a box, the cables can be anchored at the front and back wall of the tower, Figure 19.16. Post-tensioning tendons are used to prestress the walls to transfer the anchoring forces from one end wall to the other. The tendons can be loop tendons that wrap around three side walls at a time or simple straight tendons in each side wall independently. The Talmadge, Baytown–LaPorte, and Yang Pu Bridges all employ such an anchoring detail. During the design of the Yang Pu Bridge, a full-scale model was made to confirm the performance of such a detail.

As an alternative, some bridges have the cables anchored to a steel member that connects the cables from both sides of the tower. The steel member may be a beam or a box. It must be connected

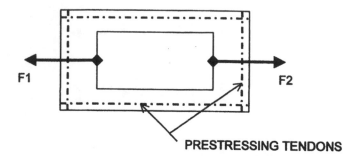

FIGURE 19.16   Tendon layout at the anchorage area.

to the concrete tower by shear studs or other means. This anchorage detail simulates the function of the saddle. However, the cables at the opposite sides are independent cables. The design must therefore consider the loading condition when only one cable exists.

19.3   Design

19.3.1   Permanent Load Condition

A cable-stayed bridge is a highly redundant, or statically indeterminate structure. In the design of such a structure, the treatment of the permanent load condition is very important. This load condition includes all structural dead load and superimposed dead load acting on the structure, all prestressing effects as well as all secondary moments and forces. It is the load condition when all permanent loads act on the structure.

Because the designer has the liberty to assign a desired value to every unknown in a statically indeterminate structure, the bending moments and forces under permanent load condition can be determined solely by the requirements of equilibrium, $\Sigma H = 0$, $\Sigma V = 0$, and $\Sigma M = 0$. The stiffness of the structure has no effect in this calculation. There are an infinite number of possible combinations of permanent load conditions for any cable-stayed bridge. The designer can select the one that is most advantageous for the design when other loads are considered.

Once the permanent load condition is established by the designer, the construction has to reproduce this final condition. Construction stage analysis, which checks the stresses and stability of the structure in every construction stage, starts from this selected final condition backwards. However, if the structure is of concrete or composite, creep and shrinkage effect must be calculated in a forward calculation starting from the beginning of the construction. In such cases, the calculation is a combination of forward and backward operations.

The construction stage analysis also provides the required camber of the structure during construction.

19.3.2   Live Load

Live-load stresses are mostly determined by evaluation of influence lines. However, the stress at a given location in a cable-stayed bridge is usually a combination of several force components. The stress, f, of a point at the bottom flange, for example, can be expressed as:

$$f = (1/A) * P + (y/I) * M + c * K \qquad (19.5)$$

where A is the cross-sectional area, I is the moment of inertia, y is the distance from the neutral axis, and c is a stress influence coefficient due to the cable force K anchored at the vicinity. P is the axial force and M is the bending moment. The above equation can be rewritten as

$$f = a1 * P + a2 * M + a3 * K \qquad (19.6)$$

where the constants $a1$, $a2$, and $a3$ depend on the effective width, location of the point, and other global and local geometric configurations. Under live load, the terms P, M, and K are individual influence lines. Thus, f is a combined influence line obtained by adding up the three terms multiplied by the corresponding constants $a1$, $a2$, and $a3$, respectively.

In lieu of the combined influence lines, some designs substitute P, M, and K with extreme values, i.e., maximum and minimum of each. Such a calculation is usually conservative but fails to present the actual picture of the stress distribution in the structure.

19.3.3 Thermal Loads

Differential temperature between various members of the structure, especially that between the cables and the rest of the bridge, must be considered in the design. Black cables tend to be heated up and cooled down much faster than the towers and the girder, thus creating a significant temperature difference. Light-colored cables, therefore, are usually preferred.

Orientation of the bridge toward the sun is another factor to consider. One face of the towers and some group of cables facing the sun may be warmed up while the other side is in the shadow, causing a temperature gradient across the tower columns and differential temperature among the cable groups.

19.3.4 Dynamic Loads

19.3.4.1 Structural Behavior

The girder of a cable-stayed bridge is usually supported at the towers and the end piers. Depending on the type of bearing or supports used, the dynamic behavior of the structure can be quite different. If very soft supports are used, the girder acts like a pendulum. Its fundamental frequency will be very low. Stiffening up the supports and bearings can increase the frequency significantly.

Seismic and aerodynamics are the two major dynamic loads to be considered in the design of cable-stayed bridges. However, they often have contradictory demands on the structure. For aerodynamic stability a stiffer structure is preferred. But for seismic design, except if the bridge is founded on very soft soil, a more flexible bridge will have less response. Some compromise between these two demands is required.

Because the way these two dynamic loads excite the structure is different, special mechanical devices can be used to assist the structure to adjust to both load conditions. Aerodynamic responses build up slowly. For this type of load the forces in the connections required to minimize the vibration buildup are relatively small. Earthquakes happen suddenly. The response will be especially sudden if the seismic motion also contains large flings. Consequently, a device that connects the girder and the tower, which can break at a certain predetermined force will help in both events. Under aerodynamic actions, it will suppress the onset of the vibrations as the connection makes the structure stiffer. Under seismic load, the connection breaks at the predetermined load and the structure becomes more flexible. This reduces the fundamental frequency of the bridge.

19.3.4.2 Seismic Design

Most cable-stayed bridges are relatively flexible with long fundamental periods in the range of 3.0 s or longer. Their seismic responses are usually not very significant in the longitudinal direction. In the transverse direction, the towers are similar to a high-rise building. Their responses are also manageable.

Experience shows that, except in extremely high seismic areas, earthquake load seldom controls the design. On the other hand, because most cable-stayed bridges are categorized as major structures, they are usually required to be designed for more severe earthquake loads than regular structures.

Various measures have been used to reduce the seismic response of cable-stayed bridges [1]. They range from a simple shock damper with a hydraulic cylinder that freezes at fast motion, to different types of friction dampers. Letting the bridge girder swing for a certain distance like a pendulum is another efficient way of reducing the seismic response. This is especially effective for out-of-phase motions. Because many cable-stayed spans are in the range of the half wavelengths of seismic motions, the out-of-phase motion can create very large reactions in the structure. A partially floating girder is often found to be very advantageous.

19.3.4.3 Aerodynamics

Aerodynamic stability of cable-stayed bridges was a major concern for many bridge engineers in the early years. This was probably because cable-stayed bridges are extremely slender. Lessons learned from aerodynamic problems in suspension bridges lead engineers to worry about cable-stayed bridges.

In reality, cable-stayed bridges, especially concrete cable-stayed bridges, have been found to be surprisingly stable aerodynamically. Although the prediction that cable-stayed bridges could not seriously vibrate under wind due to interference of the different unrelated modes that exist in such a structure was not correct, extremely few cable-stayed bridges were found to be susceptible to wind action after construction. The superior aerodynamic behavior of cable-stayed bridges is one reason for this. But the lessons learned from suspension bridge have educated many engineers so that they are aware of aerodynamic problems and can identify the preferred cross sections against wind actions. The wider deck width of most modern cable-stayed bridges also makes the structure more stable.

Several bridges did require special treatment against aerodynamic action. The Annacis Island Bridge added wind fairing to the main span; the Quincy Bridge added vertical plates to the girder in addition to horizontal fairings. Longs Creek Bridge has a tapered wind nose at each side of the girder.

During the design of the Knie Bridge in the early 1960s, a wind tunnel study was performed to search for a good solution to increase the aerodynamic stability of the bridge in case its responses were found to be unacceptable. Among various alternatives, the tapered nose was the most efficient option. This same idea has been used in many cable-stayed bridges and suspension bridges since then.

Although a cable-stayed bridge is mostly stable in its final condition, it is often vulnerable during construction stages. During the construction of the Knie Bridge, bottom bracing was added to provide the required torsional stiffness in the girder to eliminate flutter possibility. Most high-level cable-stayed bridges in high-wind areas have required wind tie-downs to stabilize the structure against buffeting.

Back in the 1960s, it was thought that a structure would not vibrate in a turbulent flow. However, it was found that by having a certain intensity and frequency content in a turbulent wind, buffeting may occur. In many cases, the buffeting responses of a cable-stayed bridge during construction can be quite severe unless specific measures are taken to stabilize the structure.

The most efficient way to stabilize the structure against buffeting is to increase its fundamental frequency by tie-downs, Figure 19.17. Most tie-downs are simple cables of seven-wire strands anchored to pile foundations, dead weights, or soil anchors such as in the Annacis and Baytown–LaPorte Bridges. Stabilizing the tower by front and back staying cables can also have the same effect as in the ALRT Fraser River and East Huntington Bridges.

Use of tie-downs can also help reduce the unbalanced bending moment in the tower during construction which is inherent in a cantilever method.

The amount of damping can have a decisive effect on the aerodynamic behavior of a cable-stayed bridge, especially on the critical flutter wind speed. Consequently, the assumption of a proper damping ratio is very important in the design. Table 19.1 shows the calculated critical flutter wind speed of the Sidney Lanier Bridge, Georgia, based on sectional model test results.

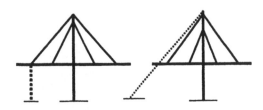

FIGURE 19.17 Temporary stabilization by tie-down cables.

TABLE 19.1 Critical Flutter Wind Speed — Sidney Lanier Bridge

Damping ratio (% of critical)	0.5%	1.0%	2.0%	3.0%	4.0%
Critical flutter wind sped, km/h	160	180	230	340	450

Practically all field measurements of damping ratio of cable-stayed bridges were performed with small amplitudes. Generally, it was found to be between 0.5 and 1.0%. Such measured values correspond to the behavior of low-amplitude votex-shedding responses, which usually happen under relatively low wind speed. However, flutter is a phenomenon represented by large amplitudes; the actual damping coefficient is much higher.

Flutter is considered an extreme natural event that may happen once every 1000 to 2000 years. There will also be no people on the bridge under such high winds. Therefore, very large amplitude oscillations with cracking of concrete and partial yielding of steel are considered acceptable. Under such conditions, the damping increases significantly. A cable-stayed bridge, being a very redundant structure that may allow many plastic hinges to form, offers decisive advantages over regular girder bridges.

A damping ratio of 5% is usually assumed in seismic analysis, which is a similar extreme natural event. A value of 2 to 4% may be conservatively assumed for a concrete cable-stayed bridge. Higher damping can also be achieved by installation of artificial dampers.

The knowledge of cable vibrations has also progressed extensively. The first stay cable vibration problem appeared in the Neuenkamp Bridge. This was a wake vibration of two parallel and horizontally located cables. The problem was new at that time. It was identified, and further vibration was suppressed by connecting the pair of cables together with a damper. This concept was used for several other subsequently built bridges.

Severe cable vibrations were observed in the Brotone Bridge. Dampers were installed and they were successful in suppressing the vibrations. The same concept was used for the Sunshine Skyway.

Rain–wind vibrations were discovered in a few bridges. This phenomenon appears only during light rain in combination with light wind. This problem was found to be caused by the change of shape of rain water mantle on the cable. Increasing the damping of cables can suppress this vibration. Tying cables together by wires, Figure 19.18, and draining the water away from the cable before it accumulates are the more common and effective methods to combat this problem. Adding dimples or spiral-wound ridges on the cable surface have also been found to be effective.

19.4 Superlong Spans

Conceptually, cable-stayed bridges can be used for very long spans. Because a cable-stayed bridge is a closed structural system, similar to a self-anchored suspension bridge, but can be built without temporary supports, it is especially advantageous in areas where the soil condition is not good and anchoring the main suspension cables becomes prohibitively expensive. Span length of over 2000 m is entirely feasible.

Three major details must be properly attended to in the design of superlong-span cable-stayed bridges: the effectiveness of very long cables, the compression in the deck, and the torsional stiffness of the girder for wind stability. When a cable is too long, it becomes ineffective. This can be resolved

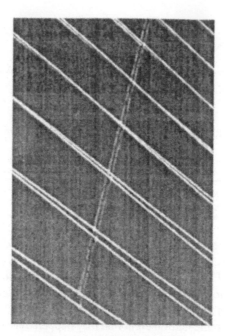

FIGURE 19.18   Tie cables against rain–wind vibrations — Dames Point Bridge.

by providing intermediate supports for the cable to reduce its sag. The compression stress in the deck increases proportionally to the span length. Depending on the type of material used for the deck girder, the maximum span length of a uniform girder is limited by its allowable stresses. This problem can best be solved by using a girder with variable cross section. By increasing the girder section gradually toward the towers, where compression is the highest, the compression stress can be reduced to an acceptable level. The torsional stiffness of the girder can be supplemented by having sufficiently wide towers so that the cables are inclined in the transverse direction.

If the deck is too narrow for the very long span, horizontal staying cables in the transverse direction at the deck level can provide stiffness in that direction.

19.5   Multispan Cable-Stayed Bridges

Most cable-stayed bridges have either three or two cable-stayed spans. The back-staying cables and the anchor pier play an important role in stabilizing the tower. When a bridge has more than three spans, the bending moment in the towers will be very large. One solution is to design the towers to carry the large bending moment [6,11]. This is usually not the most economical solution.

Several methods are available to strengthen the towers of a multispan cable-stayed bridge [6], as shown in Figure 19.19:

1. Tying the tower tops together with horizontal cables;
2. Tying the tower tops to the girder and tower intersection point at the adjacent towers;
3. Adding additional tie-down piers at span centers; and
4. Adding crossing cables at midspans.

19.6   Aesthetic Lighting

Aesthetic lighting is now part of the design of most cable-stayed bridges. Lighting enhances the beauty and visibility of the bridge at night. There are various schemes of lighting a cable-stayed bridge as shown in Figure 19.20.

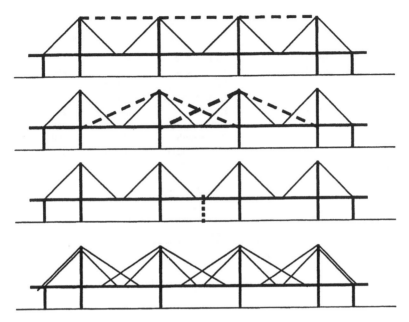

FIGURE 19.19 Multispan cable-stayed bridge system.

FIGURE 19.20 Aesthetic lighting of cable-stayed bridges.

19.7 Summary

Cable-stayed bridges are beautiful structures. Their popular appeal to engineers and nonengineers alike has been universal. In a pure technical sense this bridge type fills the gap of efficient span range between conventional girder bridges and the very long span suspension bridges.

TABLE 19.2   Milestone Cable-Stayed Bridges (as of April 1998)

Features	Bridge Name	Country	Span (m)	Year of Completion
First successfully built cable-stayed bridge	Stromsunde	Sweden	183	1955
First bridge with closely spaced cables	Bonn–Nord[a]	Germany	280	1967
All-steel open-deck plate girder	Knie[a]	Germany	320	1969
Center spine, single plane, all steel	Neuenkamp[a]	Germany	350	1971
First major concrete span, box spine girder	Hoescht	Germany	148	1972
First major precast box girder	Sunshine Skyway	U.S.	366	1986
First solid concrete flat-plate girder designed	Portsmouth	U.S.	286	Not built
First major composite girder completed	Annacis Island[a]	Canada	365	1986
First flat-plate girder completed	ALRT Skytrain	Canada	340	1988
Flexible girder built by cable-supported traveler	Dame Point	U.S.	396	1988
First major composite girder designed	Second Hoogly	India	457	1992
Longest composite girder	Yang Pu[a]	China	602	1993
Hybrid steel main span + concrete side spans	Normandy[a]	France	856	1994
Steel girder, conc. towers, longest span to date	Tatara[a]	Japan	890	1998

[a] World record span at time of completion.

Table 19.2 is a list of milestone cable-stayed bridges. It illustrates the general evolution of modern cable-stayed bridges.

References

1. Abdel-Ghaffar, A.M. Cable-stayed bridges under seismic action, in *Cable-Stayed Bridges*, Elsevier, Amsterdam, 1991.
2. Podolny, W. and Scalzi, J. B. *Construction and Design of Cable-Stayed Bridges*, John Wiley & Sons, New York, 1986.
3. Tang, M. C., Analysis of cable-stayed bridges, *J. Struct. Div. ASCE*, Aug., 1971.
4. Tang, M. C., Buckling of cable-stayed girder bridges, *J. Struct. Div. ASCE*, Sept.,1976.
5. Tang, M. C., Concrete cable-stayed bridges, presented at ACI Convention, Kansas City, Sept.,1983.
6. Tang, M. C. Multispan cable-stayed bridges, in *Proceedings, International Bridge Conference — Bridges into the 21st Century*, Hong Kong, Oct., 1995.
7. Troitsky, M. S., *Cable-Stayed Bridges*, Van Nostrand Reinhold, New York, 1988.
8. Dame Point Bridge reaches for a record, *Eng. News Rec.*, Jan. 7, 1988.
9. Recommendations for Stay Cable Design and Testing, Post-Tensioning Institute, Jan., 1986. Latest rev. 1993.
10. Guidelines for the Design of Cable-Stayed Bridges, American Society of Civil Engineers, 1992.
11. *Festschrift Ulrich Finsterwalder — 50 Jahre für Dywidag*, Dyckerhoff and Widmann, Germany, 1973.

20

Timber Bridges

Kenneth J. Fridley
Washington State University

20.1 Introduction

Wood is one of the earliest building materials, and as such often its use has been based more on tradition than on principles of engineering. However, the structural use of wood and wood-based materials has increased steadily in recent times, including a renewed interest in the use of timber as a bridge material. Supporting this renewed interest has been an evolution of our understanding of wood as a structural material and our ability to analyze and design safe, durable, and functional timber bridge structures.

An accurate and complete understanding of any material is key to its proper use in structural applications, and structural timber and other wood-based materials are no exception to this requirement. This chapter focuses on introducing the fundamental mechanical and physical properties of wood that govern its structural use in bridges. Following this introduction of basic material properties, a presentation of common timber bridge types will be made, along with a discussion of fundamental considerations for the design of timber bridges.

20.1.1 Timber as a Bridge Material

Wood has been widely used for short- and medium-span bridges. Although wood has the reputation of being a material that provides only limited service life, wood can provide long-standing and serviceable bridge structures when properly protected from moisture. For example, many covered bridges from the early 19th century still exist and are in use. Today, rather than protecting wood

---

Parts of this chapter were previously published by CRC Press in *Handbook of Structural Engineering*, W. F. Chen, Ed., 1997.

by a protective shelter as with the covered bridge of yesteryear, wood preservatives which inhibit moisture and biological attack have been used to extend the life of modern timber bridges.

As with any structural material, the use of wood must be based on a balance between its inherent advantages and disadvantages, as well as consideration of the advantages and disadvantages of other construction materials. Some of the advantages of wood as a bridge material include:

- Strength
- Light weight
- Constructibility
- Energy absorption
- Economics
- Durability, and
- Aesthetics

These advantages must be considered against the three primary disadvantages:

- Decay
- Insect attack, and
- Combustibility

Wood can withstand short-duration overloading with little or no residual effects. Wood bridges require no special equipment for construction and can be constructed in virtually any weather conditions without any negative effects. Wood is competitive with other structural materials in terms of both first costs and life-cycle costs. Wood is a naturally durable material resistant to freeze–thaw effects as well as deicing agents. Furthermore, large-size timbers provide good fire resistance as a result of natural charring. However, if inadequately protected against moisture, wood is susceptible to decay and biological attack. With proper detailing and the use of preservative treatments, the threat of decay and insects can be minimized. Finally, in many natural settings, wood bridges offer an aesthetically pleasing and unobtrusive option.

20.1.2 Past, Present, and Future of Timber Bridges

The first bridges built by humans were probably constructed with wood, and the use of wood in bridges continues today. As recently as a century ago, wood was still the dominant material used in bridge construction. Steel became an economical and popular choice for bridges in the early 1900s. Also during the early part of the 20th century, reinforced concrete became the primary bridge deck material and an economical choice for the bridge superstructure. However, important advances were made in wood fastening systems and preservative treatments, which would allow for future developments for timber bridges. Then, in the mid-20th century, glued-laminated timber (or glulams) was introduced as a viable structural material for bridges. The use of glulams grew to become the primary material for timber bridges and has continued to grow in popularity. Today, there is a renewed interest in all types of timber bridges. Approximately 8% (37,000) of the bridges listed in the National Bridge Inventory in the United States having spans greater than 6.10 m are constructed entirely of wood and 11% (51,000) use wood as one of the primary structural materials [9]. The future use of timber as a bridge material will not be restricted just to new construction. Owing to its high strength-to-weight ratio, timber is an ideal material for bridge rehabilitation of existing timber, steel, and concrete bridges.

20.2 Properties of Wood and Wood Products

It is important to understand the basic structure of wood in order to avoid many of the pitfalls relative to the misuse and/or misapplication of the material. Wood is a natural, cellular, anisotropic,

TABLE 20.1 Moisture Content (%) of Wood in Equilibrium with Temperature and Relative Humidity

Temp. (°C)	Relative Humidity (%)																			
	5	10	15	20	25	30	35	40	45	50	55	60	65	70	75	80	85	90	95	98
0	1.4	2.6	3.7	4.6	5.5	6.3	7.1	7.9	8.7	9.5	10.4	11.3	12.4	13.5	14.9	16.5	18.5	21.0	24.3	26.9
5	1.4	2.6	3.7	4.6	5.5	6.3	7.1	7.9	8.7	9.5	10.4	11.3	12.3	13.5	14.9	16.5	18.5	21.0	24.3	26.9
10	1.4	2.6	3.6	4.6	5.5	6.3	7.1	7.9	8.7	'9.5	10.3	11.2	12.3	13.4	14.8	16.4	18.4	20.9	24.3	26.9
15	1.3	2.5	3.6	4.6	5.4	6.2	7.0	7.8	8.6	9.4	10.2	11.1	12.1	13.3	14.6	16.2	18.2	20.7	24.1	26.8
20	1.3	2.5	3.5	4.5	5.4	6.2	6.9	7.7	8.5	9.2	10.1	11.0	12.0	13.1	14.4	16.0	17.9	20.5	23.9	26.6
25	1.3	2.4	3.5	4.4	5.3	6.1	6.8	7.6	8.3	9.1	9.9	10.8	11.7	12.9	14.2	15.7	17.7	20.2	23.6	26.3
30	1.2	2.3	3.4	4.3	5.1	5.9	6.7	7.4	8.1	8.9	9.7	10.5	11.5	12.6	13.9	15.4	17.3	20.8	23.3	26.0
35	1.2	2.3	3.3	4.2	5.0	5.8	6.5	7.2	7.9	8.7	9.5	10.3	11.2	12.3	13.6	15.1	17.0	20.5	22.9	25.6
40	1.1	2.2	3.2	4.1	5.0	5.7	6.4	7.1	7.8	8.6	9.3	10.1	11.1	12.2	13.4	14.9	16.8	20.3	22.7	25.4
45	1.1	2.2	3.2	4.0	4.9	5.6	6.3	7.0	7.7	8.4	9.2	10.0	11.0	12.0	13.2	14.7	16.6	20.1	22.4	25.2
50	1.1	2.1	3.0	3.9	4.7	5.4	6.1	6.8	7.5	8.2	8.9	9.7	10.6	11.7	12.9	14.4	16.2	18.6	22.0	24.7
55	1.0	2.0	2.9	3.7	4.5	5.2	5.9	6.6	7.2	7.9	8.7	9.4	10.3	11.3	12.5	14.0	15.8	18.2	21.5	24.2

Adapted from USDA, 1987 [10].

hygrothermal, and viscoelastic material, and by its natural origins contains a multitude of inclusions and other defects.* The reader is referred to basic texts that present a description of the fundamental structure and physical properties of wood as a material [e.g., Refs. 5, 6, 10].

20.2.1 Physical Properties of Wood

One physical aspect of wood that deserves attention here is the effect of moisture on the physical and mechanical properties and performance of wood. Many problems encountered with wood structures, especially bridges, can be traced to moisture. The amount of moisture present in wood is described by the moisture content (MC), which is defined by the weight of the water contained in the wood as a percentage of the weight of the oven-dry wood. As wood is dried, water is first evaporated from the cell cavities, then, as drying continues, water from the cell walls is drawn out. The point at which *free* water in the cell cavities is completely evaporated, but the cell walls are still saturated, is termed the *fiber saturation point* (FSP). The FSP is quite variable among and within species, but is on the order of 24% to 34%. The FSP is an important quantity since most physical and mechanical properties are dependent on changes in MC below the FSP, and the MC of wood in typical structural applications is below the FSP. Finally, wood releases and absorbs moisture to and from the surrounding environment. When the wood equilibrates with the environment and moisture is not transferring to or from the material, the wood is said to have reached its equilibrium moisture content (EMC). Table 20.1 provides the average EMC as a function of dry-bulb temperature and relative humidity. The *Wood Handbook* [10] provides other tables that are specific for given species or species groups and allow designers better estimates of in-service moisture contents that are required for their design calculations.

Wood shrinks and swells as its MC changes below the FSP; above the FSP, shrinkage and swelling can be neglected. Wood machined to a specified size at an MC higher than that expected in service will therefore shrink to a smaller size in use. Conversely, if the wood is machined at an MC lower than that expected in service, it will swell. Either way, shrinkage and swelling due to changes in MC must be taken into account in design. In general, the shrinkage along the grain is significantly less than that across the grain. For example, as a rule of thumb, a 1% dimensional

---

*The term *defect* may be misleading. Knots, grain characteristics (e.g., slope of grain, spiral grain, etc.), and other naturally occurring irregularities do reduce the effective strength of the member, but are accounted for in the grading process and in the assignment of design values. On the other hand, splits, checks, dimensional warping, etc. are the result of the drying process and, although they are accounted for in the grading process, they may occur after grading and may be more accurately termed *defects*.

change across the grain can be assumed for each 4% change in MC, whereas a 0.02% dimensional change in the longitudinal direction may be assumed for each 4% change in MC. More-accurate estimates of dimensional changes can be made using published values of shrinkage coefficients for various species, [10].

In addition to simple linear dimensional changes in wood, drying of wood can cause warp of various types. Bow (distortion in the weak direction), crook (distortion in the strong direction), twist (rotational distortion), and cup (cross-sectional distortion similar to bow) are common forms of warp and, when excessive, can adversely affect the structural use of the member. Finally, drying stresses (internal stress resulting from differential shrinkage) can be quite significant and can lead to checking (cracks formed along the growth rings) and splitting (cracks formed across the growth rings).

20.2.2 Mechanical Properties of Wood

The mechanical properties of wood also are functions of the MC. Above the FSP, most properties are invariant with changes in MC, but most properties are highly affected by changes in the MC below the FPS. For example, the modulus of rupture of wood increases by nearly 4% for a 1% decrease in moisture content below the FSP. The following equation is a general expression for relating any mechanical property to MC:

$$P_{MC} = P_{12} \left(\frac{P_{12}}{P_g} \right)^{(12-MC)/(FSP-MC)} \tag{20.1}$$

where P_{MC} = property of interest at any MC below the FSP, P_{12} = the property at 12% MC, and P_g = property in the green condition (at FSP).

For structural design purposes, using an equation such as (20.1) would be cumbersome. Therefore, design values are typically provided for a specific maximum MC (e.g., 19%) and adjustments are made for "wet use."

Load history can also have a significant effect on the mechanical performance of wood members. The load that causes failure is a function of the rate and duration of the load applied to the member. That is, a member can resist higher magnitude loads for shorter durations or, stated differently, the longer a load is applied, the less able is a wood member to resist that load. This response is termed *load duration effects* in wood design. Figure 20.1 illustrates this effect by plotting the time-to-failure as a function of the applied stress expressed in terms of the short-term (static) ultimate strength. There are many theoretical models proposed to represent this response, but the line shown in Figure 20.1 was developed at the U.S. Forest Products Laboratory in the early 1950s [11] and is the basis for current design "load duration" adjustment factors.

The design factors derived from the relationship illustrated in Figure 20.1 are appropriate only for stresses and not for stiffness or, more precisely, the modulus of elasticity. Related to load duration effects, the deflection of a wood member under sustained load increases over time. This response, termed *creep effects*, must be considered in design when deformation or deflections are critical from either a safety or serviceability standpoint. The main parameters that significantly affect the creep response of wood are stress level, moisture content, and temperature. In broad terms, a 50% increase in deflection after a year or two is expected in most situations, but can easily be upward of 200% given certain conditions [7]. In fact, if a member is subjected to continuous moisture cycling, a 100 to 150% increase in deflection could occur in a matter of a few weeks. Unfortunately, the creep response of wood, especially considering the effects of moisture cycling, is poorly understood and little guidance is available to the designer.

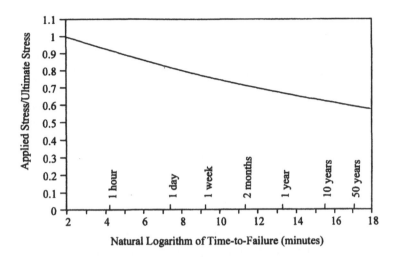

FIGURE 20.1 Load Duration behavior of wood.

Wood, being a fibrous material, is naturally resistant to fatigue effects, particularly when stressed along the grain. However, the fatigue strength of wood is negatively affected by the natural presence of inclusions and other defects. Knots and slope of grain in particular reduce fatigue resistance. Regardless of this, wood performs well in comparison with structural steel and concrete. In fact, the fatigue strength of wood has been shown to be approximately double that of most metals when evaluated at comparable stress levels relative to the ultimate strength of the material [10]. The potential for fatigue-induced failure is considered to be rather low for wood, and thus fatigue is typically not considered in timber bridge design.

20.2.3 Wood and Wood-Based Materials for Bridge Construction

The natural form of timber is the log. In fact, many primitive and "rustic" timber bridges are nothing more than one or more logs tied together. For construction purposes, however, it is simpler to use rectangular elements in bridges and other structures rather than round logs. Solid sawn lumber is cut from logs and was the mainstay of timber bridge construction for years. Solid sawn lumber comes in a variety of sizes including boards (less than 38 mm thick and 38 to 387 mm wide), dimension lumber (38 to 89 mm thick and 38 to 387 mm wide), and timbers (anything greater than 89 by 89 mm). Based on size and species, solid sawn lumber is graded by various means, including visual grading, machine-evaluated lumber (MEL), and machine stress rated (MSR), and engineering design values are assigned.

In the mid-1900s glulam timber began to receive significant use in bridges. Glulams are simply large sections formed by laminating dimension lumber together. Sections as large as 1.5 m deep are feasible with glulams. Today, while solid sawn lumber is still used extensively, the changing resource base and shift to plantation-grown trees has limited the size and quality of the raw material. Therefore, it is becoming increasingly difficult to obtain high-quality, large-dimension timbers for construction. This change in raw material, along with a demand for stronger and more cost-effective material, initiated the development of alternative products that can replace solid lumber such as glulams.

Other engineered products such as wood composite I-joists and structural composite lumber (SCL) also resulted from this evolution. SCL includes such products as laminated veneer lumber (LVL) and parallel strand lumber (PSL). These products have steadily gained popularity and now are receiving widespread use in building construction, and they are beginning to find their way into bridge construction as well. The future may see expanded use of these and other engineered wood composites.

20.2.4 Preservation and Protection

As mentioned previously, one of the major advances in the 20th century allowing for continued and expanded use of timber as a bridge material is pressure treatment. Two basic types of wood preservatives are used: oil-type preservatives and waterborne preservatives. Oil-type preservatives include creosote, pentachlorophenol (or "penta"), and copper naphthenate. Creosote can be considered the first effective wood preservative and has a long history of satisfactory performance. Creosote also offers protection against checking and splitting caused by changes in MC. While creosote is a natural by-product from coal tar, penta is a synthetic pesticide. Penta is an effective preservative treatment; however, it is not effective against marine borers and is not used in marine environments. Penta is a "restricted-use" chemical, but wood treated with penta is not restricted. Copper naphthenate has received recent attention as a preservative treatment, primarily because it is considered an environmentally safe chemical while still giving satisfactory protection against biological attack. Its primary drawback is its high cost relative to other treatments. All these treatments generally leave the surface of the treated member with an oily and unfinishable surface. Furthermore, the member may "bleed" or leach preservative unless appropriate measures are taken.

Most timber bridge applications utilize oil-type preservatives for structural elements such as beams, decks, piles, etc. They offer excellent protection against decay and biological attack, are noncorrosive, and are relatively durable. Oil-type preservatives are not, however, recommended for bridge elements that may have frequent or repeated contact by humans or animals since they can cause skin irritations.

Waterborne preservatives have the advantage of leaving the surface of the treated material clean and, after drying, able to be painted or stained. They also do not cause skin irritations and, therefore, can be used where repeated human and/or animal contact is expected. Waterborne preservatives use formulations of inorganic arsenic compounds in a water solution. They do, however, leave the material with a light green, gray, or brownish color. But again, the surface can be later painted or stained. A wide variety of waterborne preservatives are available, but the most common include chromated copper arsenate (CCA), ammoniacal copper arsenate (ACA), and ammoniacal copper zinc arsenate (ACZA). Leaching of these chemicals is not a problem with these formulations since they each are strongly bound to the wood. CCA is commonly used to treat southern pine, ponderosa pine, and red pine, all of which are relatively accepting of treatment. ACA and ACZA are used with species that are more difficult to treat, such as Douglas fir and larch. One potential drawback to CCA and ACA is a tendency to be corrosive to galvanized hardware. The extent to which this is a problem is a function of the wood species, the specific preservative formulation, and service conditions. However, such corrosion seems not to be an issue for hot-dipped galvanized hardware typical in bridge applications.

Waterborne preservatives are used for timber bridges in applications where repeated or frequent contact with humans or animals is expected. Such examples include handrails and decks for pedestrian bridges. Additionally, waterborne preservatives are often used in marine applications where marine borer hazards are high.

Any time a material is altered due to chemical treatment its microlevel structure may be affected, thus affecting its mechanical properties. Oil-type preservatives do not react with the cellular structure of the wood and, therefore, have little to no effect on the mechanical properties of the material. Waterborne preservatives do react, however, with the cell material, thus they can affect properties. Although this is an area of ongoing research, indications are that the only apparent effect of waterborne preservatives is to increase load duration effects, especially when heavy treatment is used for saltwater applications. Currently, no adjustments are recommended for design values of preservative treated wood vs. untreated materials.

In addition to preservative treatment, fire-retardant chemical treatment is also possible to inhibit combustion of the material. These chemicals react with the cellular structure in wood and can cause significant reductions in the mechanical properties of the material, including strength. Generally, fire retardants are not used in bridge applications. However, if fire-retardant-treated material is used, the designer should consult with the material producer or treater to obtain appropriate design values.

20.3 Types of Timber Bridges

Timber bridges come in a variety of forms, many having evolved from tradition. Most timber bridges designed today, however, are the results of fairly recent developments and advances in the processing and treating of structural wood. The typical timber bridge is a single- or two-span structure. Single-span timber bridges are typically constructed with beams and a transverse deck or a slab-type longitudinal deck. Two-span timber bridges are often beam with transverse decks. These and other common timber bridge types are presented in this section.

20.3.1 Superstructures

As with any bridge, the structural makeup can be divided into three basic components: the superstructure, the deck, and the substructure. Timber bridge superstructures can be further classified into six basic types: beam superstructures, longitudinal deck (or slab) superstructures, trussed superstructures, trestles, suspension bridges, and glulam arches.

Beam Superstructures

The most basic form of a timber beam bridge is a log bridge. It is simply a bridge wherein logs are laid alternately tip-to-butt and bound together. A transverse deck is then laid over the log beams. Obviously, spans of this type of bridge are limited to the size of logs available, but spans of 6 to 18 m are reasonable. The service life of a log bridge is typically 10 to 20 years.

The sawn lumber beam bridge is another simple form. Typically, made of closely spaced 100 to 200-mm-wide by 300 to 450-mm-deep beams, sawn lumber beams are usually used for clear spans up to 9 m. With the appropriate use of preservative treatments, sawn lumber bridges have average service lives of approximately 40 years. A new alternative to sawn lumber is structural composite lumber (SCL) bridges. Primarily, laminated veneer lumber (LVL) has been used in replacement of solid sawn lumber in bridges. LVL can be effectively treated and can offer long service as well.

Glulam timber beam bridges are perhaps the most prevalent forms of timber bridges today. A typical glulam bridge configuration is illustrated in Figure 20.2. This popularity is primarily due to the large variety of member sizes offered by glulams. Commonly used for clear spans ranging from 6 to 24 m, glulam beam bridges have been used for clear spans up to 45 m. Transportation restrictions rather than material limitations limit the length of beams, and, therefore, bridges. Since glulam timber can be satisfactorily treated with preservatives, they offer a durable and long-lasting structural element. When designed such that field cutting, drilling, and boring are avoided, glulam bridges can provide a service life of at least 50 years.

Longitudinal Deck Superstructures

Longitudinal deck (or slab) superstructures are typically either glulam or nail-laminated timber placed longitudinally to span between supports. A relatively new concept in longitudinal deck systems is the stress-laminated timber bridge, which is similar to the previous two forms except that continuity in the system is developed through the use of high-strength steel tension rods. In any case, the wide faces of the laminations are oriented vertically rather than horizontally as in a typical glulam beam. Figure 20.3 illustrates two types of glulam longitudinal decks: noninterconnected and interconnected. Since glulam timbers have depths typically less than the width of a bridge, two or more segments must be used. When continuity is needed, shear dowels must be used to provide interconnection between slabs. When continuity is not required, construction is simplified. Figure 20.4 illustrates a typical stress-laminated section.

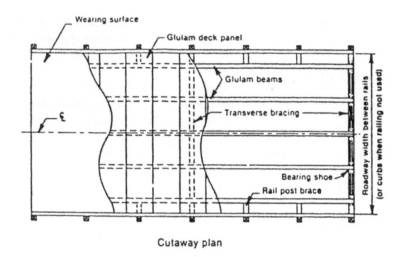

Cutaway plan

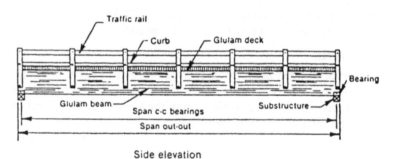

Side elevation

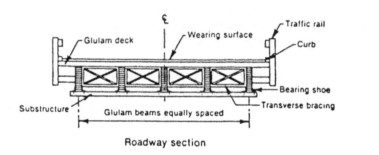

Roadway section

FIGURE 20.2 Glulam beam bridge with transverse deck. (*Source:* Ritter, M.A., EM7700-8, USDA Forest Service, Washington, D.C., 1990.)

Longitudinal deck systems are relatively simple and offer a relatively low profile, making them an excellent choice when vertical clearance is a consideration. Longitudinal decks are economical choices for clear spans up to approximately 10 m. Since the material can be effectively treated, the average service life of a longitudinal timber deck superstructure is at least 50 years. However, proper maintenance is required to assure an adequate level of prestress is maintained in stress-laminated systems.

Trussed Superstructures

Timber trusses were used extensively for bridges in the first half of the 20th century. Many different truss configurations were used including king post, multiple king posts, Pratt, Howe, lattice, long, and bowstring trusses, to name a few. Clear spans of up to 75 m were possible. However, their

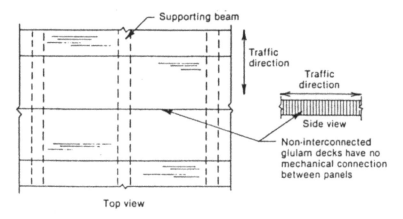

Non-interconnected glulam deck

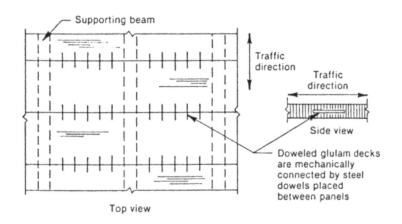

Doweled glulam deck

FIGURE 20.3 Glulam longitudinal decks. (*Source:* Ritter, M.A., EM7700-8, USDA Forest Service, Washington, D.C., 1990.)

use has declined due primarily to high fabrication, erection, and maintenance costs. When timber trusses are used today, it is typically driven more by aesthetics than by structural performance or economics.

Trestles

Another form of timber bridge which saw its peak usage in the first half of the 20th century was the trestle. A trestle is a series of short-span timber superstructures supported on a series of closely spaced timber bents. During the railroad expansion during the early to mid 1900s, timber trestles were a popular choice. However, their use has all but ceased because of high fabrication, erection, and maintenance costs.

Suspension Bridges

A timber suspension bridge is simply a timber deck structure supported by steel cables. Timber towers, in turn, support the steel suspension cables. Although there are examples of vehicular timber suspension bridges, the more common use of this form of timber bridge is as a pedestrian bridge. They are typically used for relatively long clear spans, upward of 150 m. Since treated wood can be used throughout, 50-year service lives are expected.

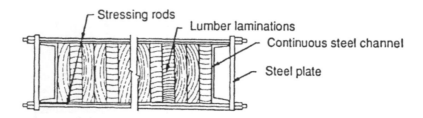

External rod configuration
(rods placed above and below the lumber laminations)

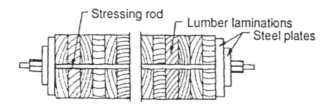

Internal rod configuration
(rods placed through the lumber laminations)

FIGURE 20.4 Stress laminated bridge. (*Source:* Ritter, M.A., EM7700-8, USDA Forest Service, Washington, D.C., 1990.)

Glued Laminated Arches

One of the most picturesque forms of timber bridges is perhaps the glulam arch. Constructed from segmented circular or parabolic glulam arches, either two- or three-hinge arches are used. The glulam arch bridge can have clear spans in excess of 60 m, and since glulam timber can be effectively treated, service lives of at least 50 years are well within reason. Although the relative first and life-cycle costs of arch bridges have become high, they are still a popular choice when aesthetics is an issue.

20.3.2 Timber Decks

The deck serves two primary purposes: (1) it is the part of the bridge structure that forms the roadway, and (2) it distributes the vehicular loads to the supporting elements of the superstructure. Four basic types of timber decks are sawn lumber planks, nailed laminated decks, glulam decks, and composite timber–concrete decks. The selection of a deck type depends mainly on the level of load demand.

Lumber Planks

The lumber plank deck is perhaps the simplest deck type. It is basically sawn lumber, typically 75 to 150 mm thick and 250 to 300 mm wide, placed flatwise and attached to the supporting beams with large spikes. Generally, the planks are laid transverse to the beams and traffic flow, but can be placed longitudinally on cross beams as well. Lumber planks are only used for low-volume bridges. They are also of little use when protection of the supporting members is desired since water freely travels between adjacent planks. Additionally, when a wearing surface such as asphalt is desired, lumber planks are not recommended since deflections between adjacent planks will result in cracking and deterioration of the wearing surface.

Nailed Laminated and Glulam Decks

Nailed laminated and glulam decks are essentially as described previously for longitudinal deck (or slab) superstructures. Nailed laminated systems are typically 38-mm-thick by 89- to 285-mm-deep lumber placed side by side and nailed or spiked together along its length. The entire deck is nailed together to act as a composite section and oriented such that the lumber is laid transverse to the bridge span across the main supporting beams, which are spaced from 0.6 to 1.8 m. Once a quite popular deck system, its use has declined considerably in favor of glulam decks.

A glulam deck is a series of laminated panels, typically 130- to 220-mm thick by 0.9 to 1.5 m wide. The laminations of the glulam panel are oriented with their wide face vertically. Glulam decks can be used with the panels in the transverse or longitudinal direction. They tend to be stronger and stiffer than nailed laminated systems and offer greater protection from moisture to the supporting members. Finally, although doweled glulam panels (see Figure 20.3) cost more to fabricate, they offer the greatest amount of continuity. With this continuity, thinner decks can be used, and improved performance of the wearing surface is achieved due to reduced cracking and deterioration.

Composite Timber–Concrete Decks

The two basic types of composite timber–concrete deck systems are the T-section and the slab (see Figure 20.5). The T-section is simply a timber stem, typically a glulam, with a concrete flange that also serves as the bridge deck. Shear dowels are plates that are driven into the top of the timber stem and develop the needed shear transfer. For a conventional single-span bridge, the concrete is proportioned such that it takes all the compression force while the timber resists the tension. Composite T-sections have seen some use in recent years; however, high fabrication costs have limited their use.

Composite timber–concrete slabs were used considerably during the second quarter of the 20th century, but receive little use today. They are constructed with alternating depths of lumber typically nailed laminated with a concrete slab poured directly on top of the timber slab. With a simple single span, the concrete again carries the compressive flexural stresses while the timber carries the flexural stresses. Shear dowels or plates are driven into the timber slab to provide the required shear transfer between the concrete and the timber.

20.3.3 Substructures

The substructure supports the bridge superstructure. Loads transferred from the superstructures to the substructures are, in turn, transmitted to the supporting soil or rock. Specific types of substructures that can be used are dependent on a number of variables, including bridge loads, soil and site conditions, etc. Although a timber bridge superstructure can be adapted to virtually any type of substructure regardless of material, the following presentation is focused on timber substructures, specifically timber abutments and bents.

Abutments

Abutments serve the dual purpose of supporting the bridge superstructure and the embankment. The simplest form of a timber abutment is a log, sawn lumber, or glulam placed directly on the embankment as a spread footing. However, this form is not satisfactory for any structurally demanding situation. A more common timber abutment is the timber pile abutment. Timber piles are driven to provide the proper level of load-carrying capacity through either end bearing or friction. A backwall and wing walls are commonly added using solid sawn lumber to retain the embankment. A continuous cap beam is connected to the top of the piles on which the bridge superstructure is supported. A timber post abutment can be considered a hybrid between the spread footing and pile abutment. Timber posts are supported by a spread footing, and a backwall and wing walls are added to retain the embankment. Pile abutments are required when soil conditions do not provide adequate support for a spread footing or when uplift is a design concern.

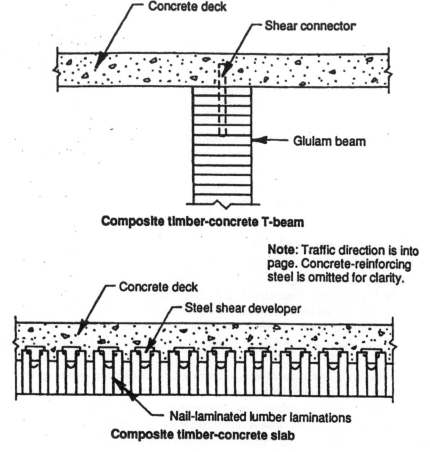

FIGURE 20.5 Composite timber–concrete decks. (*Source*: Ritter, M.A., EM7700-8, USDA Forest Service, Washington, D.C., 1990.)

Bents

Bents are support systems used for multispan bridges between the abutments. Essentially, timber bents are formed from a set of timber piles with lumber cross bracing. However, when the height of the bent exceeds that available for a pile, frame bents are used. Frame bents were quite common in the early days of the railroad, but, due to high cost of fabrication and maintenance, they are not used often for new bridges.

20.4 Basic Design Concepts

In this section, the basic design considerations and concepts for timber bridges are presented. The discussion should be considered an overview of the design process for timber bridges, not a replacement for specifications or standards.

20.4.1 Specifications and Standards

The design of timber bridge systems has evolved over time from what was tradition and essentially a "master-builder" approach. Design manuals and specifications are available for use by engineers involved with or interested in timber bridge design. These include *Timber Bridges: Design, Construction, Inspection, and Maintenance* [8], AASHTO *LRFD Bridge Design Specifications* [1], and AASHTO

Standard Specifications for Highway Bridges [2]. The wood industry, through the American Forest and Paper Association (AF&PA), published design values for solid sawn lumber and glulam timber for both allowable stress design [4] and load and resistance factor design [3] formats. Rather than presenting those aspects of bridge design common to all bridge types, the focus of the following presentation will be on those aspects specific to timber bridge design. Since bridge design is often governed by AASHTO, focus will be on AASHTO specifications. However, AF&PA is the association overseeing the engineering design of wood, much like ACI is for concrete and AISC is for steel, and AF&PA-recommended design procedures will also be presented.

20.4.2 Design Values

Design values for wood are provided in a number of sources, including AF&PA specifications and AASHTO specifications. Although the design values published by these sources are based on the same procedures per ASTM standards, specific values differ due to assumptions made for end-use conditions. The designer must take care to use the appropriate design values with their intended design specification(s). For example, the design should not use AF&PA design values directly in AASHTO design procedures since AF&PA and AASHTO make different end-use assumptions.

AF&PA "Reference" Design Values

The AF&PA *Manual for Engineered Wood Construction: Load and Resistance Factor Design* [3] provides nominal design values for visually and mechanically graded lumber, glulam timber, and connections. These values include reference bending strength, F_b; reference tensile strength parallel to the grain, F_t; reference shear strength parallel to the grain, F_v; reference compressive strength parallel and perpendicular to the grain, F_c and $F_{c\perp}$, respectively; reference bearing strength parallel to the grain, F_g; and reference modulus of elasticity, E. These are appropriate for use with the LRFD provisions.

Similarly, the Supplement to the NDS® provides tables of design values for visually graded and machine stress rated lumber, and glulam timber for use in allowable stress design (ASD). The basic quantities are the same as with the LRFD, but are in the form of allowable stresses and are appropriate for use with the ASD provisions of the NDS. Additionally, the NDS provides tabulated allowable design values for many types of mechanical connections.

One main difference between the ASD and LRFD design values, other than the ASD prescribing allowable stresses and the LRFD prescribing nominal strengths, is the treatment of duration of load effects. Allowable stresses (except compression perpendicular to the grain) are tabulated in the NDS and elsewhere for an assumed 10-year load duration in recognition of the duration of load effect discussed previously. The allowable compressive stress perpendicular to the grain is not adjusted since a deformation definition of failure is used for this mode rather than fracture as in all other modes; thus the adjustment has been assumed unnecessary. Similarly, the modulus of elasticity is not adjusted to a 10-year duration since the adjustment is defined for strength, not stiffness. For the LRFD, short-term (i.e., 20 min) nominal strengths are tabulated for all strength values. In the LRFD, design strengths are reduced for longer-duration design loads based on the load combination being considered. Conversely, in the NDS, allowable stresses are increased for shorter load durations and decreased only for permanent (i.e., greater than 10 year) loading.

AASHTO-LRFD "Base" Design Values

AASHTO-LRFD publishes its own design values which are different from those of the AF&PA LRFD. AASHTO publishes base bending strength, F_{bo}; base tensile strength parallel to the grain, F_{to}; base shear strength parallel to the grain, F_{vo}; base compressive strength parallel and perpendicular to the grain, F_{co} and $F_{c\perp}$, respectively; and base modulus of elasticity, E_0. While the NDS publishes design values based on an assumed 10-year load duration and the AF&PA LRFD assumes a short-term (20-min) load duration, AASHTO publishes design values based on an assumed 2-month duration.

TABLE 20.2 Factors to Convert NDS-ASD Values to AASTHO-LRFD Values

Material	Property					
	F_b	F_v	F_c	$F_{c\perp}$	F_t	E
Dimension lumber	2.35	3.05	1.90	1.75	2.95	0.90
Beams and stringers, posts and timbers	2.80	3.15	2.40	1.75	2.95	1.00
Glulam	2.20	2.75	1.90	1.35	2.35	0.83

Unfortunately, the AASHTO published design values are not as comprehensive (with respect to species, grades, sizes, as well as specific properties) as thsoe of AF&PA. The AASHTO-LRFD does, however, provide for adjustments from AF&PA-published reference design values so they can be used in AASHTO specifications. For design values not provided in the AASHTO-LRFD, conversion factors are provided from NDS allowable stresses to AASHTO-LRFD base strengths. Table 20.2 provides these adjustments for solid sawn and glulam timbers. The designer is cautioned that these conversion factors are from the NDS allowable stresses, *not* the AF&PA-LRFD strength values.

20.4.3 Adjustment of Design Values

In addition to the providing *reference* or *base* design values, the AF&PA-LRFD, the NDS, and the AASHTO-LRFD specifications provide adjustment factors to determine final *adjusted* design values. Factors to be considered include load duration (termed *time effect* in the LRFD), wet service, temperature, stability, size, volume, repetitive use, curvature, orientation (form), and bearing area. Each of these factors will be discussed further; however, it is important to note that not all factors are applicable to all design values, nor are all factors included in all the design specifications. The designer must take care to apply the appropriate factors properly.

AF&PA Adjustment Factors

LRFD reference strengths and ASD allowable stresses are based on the following specified reference conditions: (1) dry use in which the maximum EMC does not exceed 19% for solid wood and 16% for glued wood products; (2) continuous temperatures up to 32°C, occasional temperatures up to 65°C (or briefly exceeding 93°C for structural-use panels); (3) untreated (except for poles and piles); (4) new material, not reused or recycled material; and (5) single members without load sharing or composite action. To adjust the reference design value for other conditions, adjustment factors are provided which are applied to the published reference design value:

$$R' = R \cdot C_1 \cdot C_2 \cdots C_n \qquad (20.2)$$

where R' = adjusted design value (resistance), R = reference design value, and $C_1, C_2, \ldots C_n$ = applicable adjustment factors. Adjustment factors, for the most part, are common between LRFD and ASD. Many factors are functions of the type, grade, and/or species of material while other factors are common to all species and grades. For solid sawn lumber, glulam timber, piles, and connections, adjustment factors are provided in the AF&PA LRFD manual and the NDS. For both LRFD and ASD, numerous factors need to be considered, including wet service, temperature, preservative treatment, fire-retardant treatment, composite action, load sharing (repetitive use), size, beam stability, column stability, bearing area, form (i.e., shape), time effect (load duration), etc. Many of these factors will be discussed as they pertain to specific designs; however, some of the factors are unique for specific applications and will not be discussed further. The four factors that are applied to all design properties are the wet service factor, C_M; temperature factor, C_t; preservative treatment factor, C_{pt}; and fire-retardant treatment factor, C_{rt}. Individual treaters provide the two treatment factors, but the wet service and temperature factors are provided in the AF&PA LRFD Manual. For example, when considering the design of solid sawn lumber members, the adjustment

TABLE 20.3　AF&PA LRFD Wet Service Adjustment Factors, C_M

Thickness	Size Adjusted[a] F_b		F_t	Size Adjusted[a] F_c		F_v	$F_{c\perp}$	E, E_{05}
	≤20 MPa	>20 MPa		≤12.4 MPa	>12.4 MPa			
≤90 mm	1.00	0.85	1.00	1.00	0.80	0.97	0.67	0.90
>90 mm	1.00	1.00	1.00	0.91	0.91	1.00	0.67	1.00

[a] Reference value adjusted for size only.

TABLE 20.4　AF&PA-LRFD Temperature Adjustment Factors, C_t

Sustained Temperature (°C)	Dry Use		Wet Use	
	E, E_{05}	All Other Prop.	E, E_{05}	All Other Prop.
32 < T ≤ 48	0.9	0.8	0.9	0.7
48 < T ≤ 65	0.9	0.7	0.9	0.5

values given in Table 20.3 for wet service, which is defined as the maximum EMC exceeding 19%, and Table 20.4 for temperature, which is applicable when continuous temperatures exceed 32°C, are applicable to all design values. Often with bridges, since they are essentially exposed structures, the MC will be expected to exceed 19%. Similarly, temperature may be a concern, but not as commonly as MC.

Since, as discussed, LRFD and ASD handle time (duration of load) effects so differently and since duration of load effects are somewhat unique to wood design, it is appropriate to elaborate on it here. Whether using ASD or LRFD, a wood structure is designed to resist all appropriate load combinations – unfactored combinations for ASD and factored combinations for LRFD. The time effects (LRFD) and load duration (ASD) factors are meant to recognize the fact that the failure of wood is governed by a creep–rupture mechanism; that is, a wood member may fail at a load less than its short-term strength if that load is held for an extended period of time. In the LRFD, the time effect factor, λ, is based on the load combination being considered. In ASD, the load duration factor, C_D, is given in terms of the assumed cumulative duration of the design load.

AASHTO-LRFD Adjustment Factors

AASHTO-LRFD base design values are based on the following specified reference conditions: (1) wet use in which the maximum EMC exceeds 19% for solid wood and 16% for glued wood products (this is opposite from the dry use assumed by AF&PA, since typical bridge use implies wet use); (2) continuous temperatures up to 32°C, occasional temperatures up to 65°C; (3) untreated (except for poles and piles); (4) new material, not reused or recycled material; and (5) single members without load sharing or composite action. AASHTO has fewer adjustments available for the designer to consider, primarily but not entirely due to the specific application. To adjust the base design value for other conditions, AASHTO-LRFD provides the following adjustment equation:

$$F = F_0 \cdot C_F \cdot C_M \cdot C_D \tag{20.3}$$

where F = adjusted design value (resistance), F_0 = base design value, C_F = size adjustment factor, C_M = moisture content adjustment factor, C_D = deck adjustment factor, and C_S = stability adjustment factor.

The size factor is applicable only to bending and is essentially the same as that used by AF&PA for solid sawn lumber and the same as the volume effect factor used by AF&PA for glulam timber. For solid sawn lumber and vertically laminated lumber, the size factor is defined as

$$C_F = \left(\frac{300}{d}\right)^{1/9} \leq 1.0 \tag{20.4}$$

TABLE 20.5 AASHTO-LRFD Moisture Content Adjustment Factors, C_M, for Glulam

		Property			
F_b	F_v	F_c	F_{cp}	F_t	E
1.25	1.15	1.35	1.90	1.25	1.20

where d = width (mm). The equation implies if lumber less than or equal to 300 mm in width is used, no adjustment is made. If, however, a width greater than 300 mm is used, a reduction in the published base bending design value is required. For horizontally glulam timber, the "size" factor may more appropriately be termed a *volume* factor (per the AF&PA). The size factor for glulam is given as

$$C_F = \left[\left(\frac{300}{d} \right) \left(\frac{130}{b} \right) \left(\frac{6400}{L} \right) \right]^a \leq 1.0 \qquad (20.5)$$

where d = width (mm), b = thickness (mm), L = span (mm), and a = 0.05 for southern pine and 0.10 for all other species glulam. As with the previous size adjustment, if the dimensions of the glulam exceed 130 by 300 by 6400, then a reduction in the bending strength is required.

Unlike the size factor, the moisture factor is applicable to all published design values, not just bending strength. The moisture adjustment factor, C_M, is again similar to that provided by AF&PA; however, it is embedded in the published base design values. Unless otherwise noted, C_M should be assumed as unity. The only exception is when glulams are used and the moisture content is expected to be less than 16%. An increase in the design values is then allowed per Table 20.5. A similar increase is not allowed for lumber used at moisture contents less than 19% per AASHTO. This is a conservative approach in comparison with that of AF&PA.

The deck adjustment factor, C_D, is again specific for the bending resistance, F_b, of 50- to 100-mm-wide lumber used in stress-laminated and mechanically (nail or spike) laminated deck systems. For stress-laminated decks, the bending strength can be increased by a factor of C_D = 1.30 for select structural grade lumber, and C_D = 1.5 for No. 1 and No. 2 grade. For mechanically laminated decks, the bending strength of all grades can be increased by a factor of C_D = 1.15.

Since, as discussed, AF&PA LRFD and ASD handle time (duration of load) effects so differently and since duration of load effects are somewhat unique to wood design, it is appropriate to elaborate on it here and understand how time effects are accounted for by AASHTO-LRFD. Implicit in the AASHTO-LRFD Specification, $\lambda = 0.8$ is assumed for vehicle live loads. The published base design values are reduced by a factor of 0.80 to account for time effects. For strength load combination IV, however, a reduction of 75% is required. This load combination is for dead load only. The rationale behind this reduction is found in the AF&PA-LRFD time effects factors. For live-load-governed load combinations, AF&PA requires $\lambda = 0.8$; and for dead load only, $\lambda = 0.6$ is used. The ratio of the dead-load time effect factor to the live-load time effect factor is 0.6/0.8 = 0.75.

20.4.4 Beam Design

The focus of the remaining discussion will be on the design provisions specified in the AASHTO-LRFD for wood members. The design of wood beams follows traditional beam theory. The flexural strength of a beam is generally the primary concern in a beam design, but consideration of other factors such as horizontal shear, bearing, and deflection are also crucial for a successful design.

Moment Capacity

In terms of moment, the AASHTO-LRFD design factored resistance, M_r, is given by

$$M_r = \phi_b M_n = \phi_b F_b S C_s \qquad (20.6)$$

where ϕ_b = resistance factor for bending = 0.85, M_n = nominal adjusted moment resistance, F_b = adjusted bending strength, S = section modulus, and C_s = beam stability factor.

The beam stability factor, C_s, is only used when considering strong axis bending since a beam oriented about its weak axis is not susceptible to lateral instability. Additionally, the beam stability factor need not exceed the value of the size effects factor. The beam stability factor is taken as 1.0 for members with continuous lateral bracing; otherwise C_s is calculated from

$$C_s = \frac{1+A}{1.9} - \sqrt{\frac{(1+A)^2}{3.61} - \frac{A}{0.95}} \leq C_F \qquad (20.7)$$

where

$$A = \frac{0.438\,EB^2}{L_e dF_b} \quad \text{for visually graded solid sawn lumber} \qquad (20.8)$$

and

$$A = \frac{0.609\,Eb^2}{L_e dF_b} \quad \text{for mechanically graded lumber and glulams} \qquad (20.9)$$

where E = modulus of elasticity, b = net thickness, d = net width, L_e = effective length, and F_b = adjusted bending strength. The effective length, L_e, accounts for both the lateral motion and torsional phenomena and is given in the AASHTO-LRFD specification for specific unbraced lengths, L_u, defined as the distance between points of lateral and rotations support. For $L_u/d < 7$, the effective unbraced length, $L_e = 2.06 L_u$; for $7 \leq L_u/d \leq 14.3$, $L_e = 1.63 L_u + 3d$; and for $l_u/d > 14.3$, $L_e = 1.84 L_u$.

While the basic adjustment factor for beam stability is quite similar between AASHTO and AF&PA, the consideration of beam stability and size effects combined differs significantly from the approach used by AF&PA. For solid sawn lumber, AF&PA requires both the size factor and the beam stability factor apply. For glulams, AF&PA prescribes the lesser of the volume factor or the stability factor be used. AASHTO compared with AF&PA is potentially nonconservative with respect to lumber elements and conservative with respect to glulam elements.

Shear Capacity

Similar to bending, the basic design equation for the factored shear resistance, V_r is given by

$$V_r = \phi_v V_n = \phi_v \frac{F_v bd}{1.5} \qquad (20.10)$$

where ϕ_v = resistance factor for shear = 0.75, V_n = nominal adjusted shear resistance, F_v = adjusted shear strength, and b and d = thickness and width, respectively. Obviously, the last expression in Eq. (20.10) assumes a rectagular section, the nominal shear resistance could be determined from the relationship

$$V_n = \frac{F_v Ib}{Q} \qquad (20.11)$$

where I = moment of inertia and Q = statical moment of an area about the neutral axis.

In timber bridges, notches are often made at the support to allow for vertical clearances and tolerances as illustrated in Figure 20.6; however, stress concentrations resulting from these notches

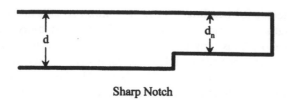

Sharp Notch

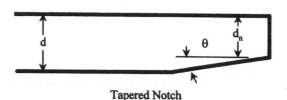

Tapered Notch

FIGURE 20.6 Notched beam: (a) sharp notch; (b) tapered notch.

significantly affect the shear resistance of the section. AASHTO-LRFD does not address this condition, but AF&PA does provide the designer with some guidance. At sections where the depth is reduced due to the presence of a notch, the shear resistance of the notched section is determined from

$$V' = \left(\frac{2}{3} F_v' b d_n\right)\left(\frac{d_n}{d}\right) \tag{20.12}$$

where $d =$ depth of the unnotched section and $d_n =$ depth of the member after the notch. When the notch is made such that it is actually a gradual tapered cut at an angle θ from the longitudinal axis of the beam, the stress concentrations resulting from the notch are reduced and the above equation becomes

$$V' = \left(\frac{2}{3} F_v' b d_n\right)\left(1 - \frac{(d - d_n)\sin\theta}{d}\right) \tag{20.13}$$

Similar to notches, connections too can produce significant stress concentrations resulting in reduced shear capacity. Where a connection produces at least one half the member shear force on either side of the connection, the shear resistance is determined by

$$V' = \left(\frac{2}{3} F_v' b d_e\right)\left(\frac{d_e}{d}\right) \tag{20.14}$$

where $d_e =$ effective depth of the section at the connection which is defined as the depth of the member less the distance from the unloaded edge (or nearest unloaded edge if both edges are unloaded) to the center of the nearest fastener for dowel-type fasteners (e.g., bolts).

Bearing Capacity

The last aspect of beam design to be covered in this section is bearing at the supports. The governing design equation for factored bearing capacity perpendicular to the grain, $P_{r\perp}$, is

$$P_{r\perp} = \phi_c P_{n\perp} = \phi_c F_{c\perp} A_b C_b \tag{20.15}$$

where ϕ_c = resistance factor for compression = 0.90, P_{np} = nominal adjusted compression resistance perpendicular to the grain, F_{cp} = adjusted compression strength perpendicular to the grain, A_b = bearing area, and C_b = bearing factor.

The bearing area factor, C_b, allows an increase in the compression strength when the bearing length along the grain, l_b, is no more than 150 mm along the length of the member, is at least 75 mm from the end of the member, and is not in a region of high flexural stress. The bearing factor C_b is given by AF&PA as

$$C_b = (l_b + 9.5)/l_b \tag{20.16}$$

where l_b is in mm. This equation is the basis for the adjustment factors presented in the AASHTO-LRFD. For example, if a bearing length of 50 mm is used, the bearing strength can be increased by a factor of $(50 + 9.5)/50 = 1.19$.

20.4.6 Axially Loaded Members

The design of axially loaded members is quite similar to that of beams. Tension, compression, and combined axial and bending are addressed in AASHTO-LRFD.

Tension Capacity

The governing design equation for factored tension capacity parallel to the grain, P_{rt}, is

$$P_{rt} = \phi_t P_{nt} = \phi_t F_t A_n \tag{20.17}$$

where ϕ_t = resistance factor for tension = 0.80, P_{nt} = nominal adjusted tension resistance parallel to the grain, F_t = adjusted tension strength, and A_n = smallest net area of the component.

Compression Capacity

In terms of compression parallel to the grain, the AASHTO-LRFD design factored resistance, P_{rc} is given by

$$P_{rc} = \phi_c P_{nc} = \phi_c F_c A C_p \tag{20.18}$$

where ϕ_c = resistance factor for bending = 0.85, and P_{nc} = nominal adjusted compression resistance, F_c = adjusted compression strength, A = cross-sectional area, and C_p = column stability factor.

The column stability factor, C_p, accounts for the tendency of a column to buckle. The factor is taken as 1.0 for members with continuous lateral bracing; otherwise, C_p is calculated from one of the following expressions, depending on the material:

For sawn lumber:

$$C_p = \frac{1+B}{1.6} - \sqrt{\frac{(1+B)^2}{2.56} - \frac{B}{0.80}} \tag{20.19}$$

For round timber piles:

$$C_p = \frac{1+B}{1.7} - \sqrt{\frac{(1+B)^2}{2.89} - \frac{B}{0.85}} \tag{20.20}$$

For mechanically graded lumber and glued laminated timber:

$$C_p = \frac{1+B}{1.8} - \sqrt{\frac{(1+B)^2}{3.24} - \frac{B}{0.9}} \tag{20.21}$$

where

$$B = \frac{4.32Ed^2}{L_e^2 F_b} \quad \text{for visually graded solid sawn lumber} \tag{20.22}$$

and

$$B = \frac{60.2Ed^2}{L_e^2 F_b} \quad \text{for mechanically graded lumber and glulams} \tag{20.23}$$

where E = modulus of elasticity, d = net width (about which buckling may occur), L_e = effective length = effective length factor times the unsupported length = KL_u, and F_b = adjusted bending strength.

Combined Tension and Bending

AASHTO uses a linear interaction for tension and bending:

$$\frac{P_u}{P_{rt}} + \frac{M_u}{M_r} \leq 1.0 \tag{20.24}$$

where P_u and M_u = factored tension and moment loads on the member, respectively, and P_{rt} and M_r are the factored resistances as defined previously.

Combined Compression and Bending

AASHTO uses a slightly different interaction for compression and bending than tension and bending:

$$\left(\frac{P_u}{P_{rc}}\right)^2 + \frac{M_u}{M_r} \leq 1.0 \tag{20.25}$$

where P_u and M_u = factored compression and moment loads on the member, respectively, and P_{rc} and M_r are the factored resistances as defined previously. The squared term on the compression term was developed from experimental observations and is also used in the AF&PA LRFD. However, AF&PA includes secondary moments in the determination of M_u, which AASHTO neglects. AF&PA also includes biaxial bending in its interaction equations.

20.4.7 Connections

The final design consideration to be discussed in this section is that of connections. AASHTO-LRFD does not specifically address connections, so the designer is referred to the AF&PA LRFD. Decks must be attached to the supporting beams and beams to abutments such that vertical, longitudinal, and transverse loads are resisted. Additionally, the connections must be easily installed in the field. The typical timber bridge connection is a dowel-type connection directly between two wood components, or with a steel bracket.

The design of fasteners and connections for wood has undergone significant changes in recent years. Typical fastener and connection details for wood include nails, staples, screws, lag screws, dowels, and bolts. Additionally, split rings, shear plates, truss plate connectors, joist hangers, and many other types of connectors are available to the designer. The general LRFD design checking equation for connections is given as follows:

$$Z_u \leq \lambda \phi_z Z'$$

(20.26)

where Z_u = connection force due to factored loads, λ = applicable time effect factor, ϕ_z = resistance factor for connections = 0.65, and Z' = connection resistance adjusted by the appropriate adjustment factors.

It should be noted that, for connections, the moisture adjustment is based on both in-service condition and on conditions at the time of fabrication; that is, if a connection is fabricated in the wet condition but is to be used in service under a dry condition, the wet condition should be used for design purposes due to potential drying stresses which may occur. It should be noted that C_M does not account for corrosion of metal components in a connection. Other adjustments specific to connection type (e.g., end grain factor, C_{eg}; group action factor, C_g; geometry factor, C_Δ; penetration depth factor, C_d; toe-nail factor, C_{tn}; etc.) will be discussed with their specific use. It should also be noted that when failure of a connection is controlled by a nonwood element (e.g., fracture of a bolt), then the time-effects factor is taken as unity since time effects are specific to wood and not applicable to nonwood components.

In both LRFD and ASD, tables of reference resistances (LRFD) and allowable loads (ASD) are available which significantly reduce the tedious calculations required for a simple connection design. In this section, the basic design equations and calculation procedures are presented, but design tables are not provided herein.

The design of general dowel-type connections (i.e., nails, spikes, screws, bolts, etc.) for lateral loading are currently based on possible yield modes. Based on these possible yield modes, lateral resistances are determined for the various dowel-type connections. Specific equations are presented in the following sections for nails and spikes, screws, bolts, and lag screws. In general, however, the dowel bearing strength, F_e, is required to determine the lateral resistance of a dowel-type connection. Obviously, this property is a function of the orientation of the applied load to the grain, and values of F_e are available for parallel to the grain, $F_{e\parallel}$, and perpendicular to the grain, $F_{e\perp}$. The dowel bearing strength or other angles to the grain, $F_{e\theta}$, is determined by

$$F_{e\theta} = \frac{F_{e\parallel} F_{e\perp}}{F_{e\parallel} \sin^2 \theta + F_{e\perp} \cos^2 \theta}$$

(20.27)

where θ = angle of load with respect to a direction parallel to the grain.

Nails, Spikes, and Screws

Nails, spikes, and screws are perhaps the most commonly used fastener in wood construction. Nails are generally used when loads are light such as in the construction of diaphragms and shear walls; however, they are susceptible to working loose under vibration or withdrawal loads. Common wire nails and spikes are quite similar, except that spikes have larger diameters than nails. Both a 12d (i.e., 12-penny) nail and spike are 88.9 mm in length; however, a 12d nail has a diameter of 3.76 mm while a spike has a diameter of 4.88 mm. Many types of nails have been developed to provide better withdrawal resistance, such as deformed shank and coated nails. Nonetheless, nails and spikes should be designed to carry laterally applied load and not withdrawal. Screws behave in a similar manner to nails and spikes, but also provide some withdrawal resistance.

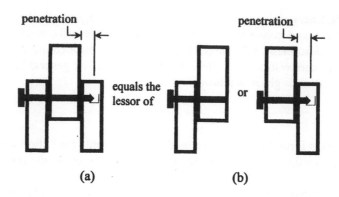

FIGURE 20.7 Double-shear connection: (a) complete connection; (b) left and right shear planes.

Lateral Resistance

The reference lateral resistance of a single nail or spike in single shear is taken as the least value determined by the four governing modes:

$$\text{I}_s: \quad Z = \frac{3.3\, D t_s F_{es}}{K_D} \tag{20.28}$$

$$\text{III}_m: Z = \frac{3.3\, k_1 D p F_{em}}{K_D (1 + 2 R_e)} \tag{20.29}$$

$$\text{III}_s: \quad Z = \frac{3.3\, k_2 D t_s F_{em}}{K_D (2 + R_e)} \tag{20.30}$$

$$\text{IV}: \quad Z = \frac{3.3 D^2}{K_D} \sqrt{\frac{2 F_{em} F_{yb}}{3(1 + R_e)}} \tag{20.31}$$

where $D =$ shank diameter; $t_s =$ thickness of the side member; $F_{es} =$ dowel bearing strength of the side member; $p =$ shank penetration into member (see Figure 20.7); $R_e =$ ratio of dowel bearing strength of the main member to that of the side member $= F_{em}/F_{es}$; $F_{yb} =$ bending yield strength of the dowel fastener (i.e., nail or spike in this case); $K_D =$ factor related to the shank diameter as follows: $K_D = 2.2$ for $D \le 4.3$ mm, $K_D = 0.38D + 0.56$ for 4.3 mm $< D \le 6.4$ mm, and $K_D = 3.0$ for $D > 6.4$ mm; and k_1 and $k_2 =$ factors related to material properties and connection geometry as follows:

$$k_1 = -1 + \sqrt{2(1 + R_e) + \frac{2 F_{yb}(1 = 2R_e) D^2}{3 F_{em} p^2}} \tag{20.32}$$

$$k_2 = -1 + \sqrt{\frac{2(1 + R_e)}{R_e} + \frac{2 F_{yb}(1 + 2R_e) D^2}{3 F_{em} t_s^2}} \tag{20.33}$$

Similarly, the reference lateral resistance of a single wood screw in single shear is taken as the least value determined by the three governing modes:

$$\mathbf{I_s:} \quad Z = \frac{3.3\,Dt_s F_{es}}{K_D} \tag{20.34}$$

$$\mathbf{III_s:} \quad Z = \frac{3.3\,k_3 Dt_s F_{em}}{K_D(2+R_e)} \tag{20.35}$$

$$\mathbf{IV:} \quad Z = \frac{3.3\,D^2}{K_D}\sqrt{\frac{1.75\,F_{em}F_{yb}}{3(1+R_e)}} \tag{20.36}$$

where K_D is defined for wood screws as it was for nails and spikes, and $k_3 =$ a factor related to material properties and connection geometry as follows:

$$k_3 = -1 + \sqrt{\frac{2(1+R_e)}{R_e} + \frac{F_{yb}(2+R_e)D^2}{2F_{em}t_s^2}} \tag{20.37}$$

For nail, spike, or wood screw connections with steel side plates, the above equations for yield mode $\mathbf{I_s}$ is not appropriate. Rather, the resistance for that mode should be computed as the bearing resistance of the fastener on the steel side plate. When double shear connections are designed (Figure 20.7a), the reference lateral resistance is taken as twice the resistance of the weaker single shear representation of the left and right shear planes (Figure 20.7b).

For multiple nail, spike, or wood screw connections, the least resistance, as determined from Eqs. (20.28) through (20.31) for nails and spikes or Eqs. (20.34) through (20.36) for wood screws, is simply multiplied by the number of fasteners, n_f, in the connection detail. When multiple fasteners are used, the minimum spacing between fasteners in a row is $10D$ for wood side plates and $7D$ for steel side plates, and the minimum spacing between rows of fasteners is $5D$. Whether a single or a multiple nail, spike, or wood screw connection is used, the minimum distance from the end of a member to the nearest fastener is $15D$ with wood side plates and $10D$ with steel side plates for tension members, and $10D$ with wood side plates and $5D$ with steel side plates for compression members. Additionally, the minimum distance from the edge of a member to the nearest fastener is $5D$ for an unloaded edge and $10D$ for a loaded edge.

The reference lateral resistance must be multiplied by all the appropriate adjustment factors. It is necessary to consider penetration depth, C_d, and end grain, C_{eg}, for nails, spikes, and wood screws. For nails and spikes, the minimum penetration allowed is $6D$, while for wood screws the minimum is $4D$. The penetration depth factor, $C_d = p/12D$, is applied to nails and spikes when the penetration depth is greater than the minimum, but less than $12D$. Nails and spikes with a penetration depth greater than $12D$ assume a $C_d = 1.0$. The penetration depth factor, $C_d = p/7D$, is applied to wood screws when the penetration depth is greater than the minimum, but less than $7D$. Wood screws with a penetration depth greater than $7D$ assume a $C_d = 1.0$. Whenever a nail, spike, or wood screw is driven into the end grain of a member, the end grain factor, $C_{eg} = 0.68$, is applied to the reference resistance. Finally, in addition to C_d and C_{eg}, a toe-nail factor, $C_{tn} = 0.83$, is applied to nails and spikes for "toe-nail" connections. A toe-nail is typically driven at an angle of approximately $30°$ to the member.

Axial Resistance

For connections loaded axially, tension is of primary concern and is governed by either fastener capacity (e.g., yielding of the nail) or fastener withdrawal. The tensile resistance of the fastener (i.e., nail, spike, or screw) is determined using accepted metal design procedure. The reference withdrawal resistance for nails and spikes with undeformed shanks in the side grain of the member is given by

$$Z_w = 31.6\,DG^{2.5}pn_f \tag{20.38}$$

where Z_w = reference withdrawal resistance in newtons and G = specific gravity of the wood. For nails and spikes with deformed shanks, design values are determined from tests and supplied by fastener manufactures, or Eq. (20.38) can be used conservatively with D = least shank diameter. For wood screws in the side grain,

$$Z_w = 65.3\ DG^2\ pn_f \qquad (20.39)$$

A minimum wood screw depth of penetration of at least 25 mm or one half the nominal length of the screw is required for Eq. (20.39) to be applicable. No withdrawal resistance is assumed for nails, spikes, or wood screws used in end grain applications.

The end grain adjustment factor, C_{eg}, and the toe-nail adjustment factor, C_{tn}, as defined for lateral resistance, are applicable to the withdrawal resistances. The penetration factor is not applicable, however, to withdrawal resistances.

Combined Load Resistance

The adequacy of nail, spike, and wood screw connections under combined axial tension and lateral loading is checked using the following interaction equation:

$$\frac{Z_u \cos \alpha}{\lambda \phi_z Z'} + \frac{Z_u \sin \alpha}{\lambda \phi_z Z'_w} \le 1.0 \qquad (20.40)$$

where α = angle between the applied load and the wood surface (i.e., 0° = lateral load and 90° = withdrawal/tension).

Bolts, Lag Screws, and Dowels

Bolts, lag screws, and dowels are commonly used to connect larger-dimension members where larger connection capacities are required. The provisions presented here are valid for bolts, lag screws, and dowels with diameters in the range of 6.3 mm $\le D \le$ 25.4 mm.

Lateral Resistance

The reference lateral resistance of a bolt or dowel in single shear is taken as the least value determined by the six governing modes:

$$\mathbf{I_m}: \quad Z = \frac{0.83\ Dt_m F_{em}}{K_\theta} \qquad (20.41)$$

$$\mathbf{I_s}: \quad Z = \frac{0.83\ Dt_s F_{es}}{K_\theta} \qquad (20.42)$$

$$\mathbf{II}: \quad Z = \frac{0.93 k_1 D F_{es}}{K_\theta} \qquad (20.43)$$

$$\mathbf{III_m}: \quad Z = \frac{1.04\ k_2 Dt_m F_{em}}{K_\theta (1 + 2R_e)} \qquad (20.44)$$

$$\mathbf{III_s}: \quad Z = \frac{1.04\ k_3 Dt_s F_{em}}{K_\theta (2 + R_e)} \qquad (20.45)$$

$$\mathbf{IV}: \quad Z = \frac{1.04\ D^2}{K_\theta} \frac{2 F_{em} F_{yb}}{3(1 + R_e)} \qquad (20.46)$$

where D = shank diameter; t_m and t_s = thickness of the main and side member, respectively; F_{em} = F_{es} = dowel bearing strength of the main and side member, respectively; R_e = ratio of dowel bearing strength of the main member to that of the side member = F_{em}/F_{es}; F_{yb} = bending yield strength of the dowel fastener (i.e., nail or spike in this case); K_θ = factor related to the angle between the load and the main axis (parallel to the grain) of the member = $1 + 0.25(\theta/90)$; and k_1, k_2, and k_3 = factors related to material properties and connection geometry as follows:

$$k_1 = \frac{\sqrt{R_e + 2R_e^2\left(1 + R_t + R_t^2\right) + R_t^2 R_e^3} - R_e\left(1 + R_t\right)}{1 + R_e} \tag{20.47}$$

$$k_2 = -1 + \sqrt{2\left(1 + R_e\right) + \frac{2F_{yb}\left(1 + 2R_e\right)D^2}{3F_{em}t_m^2}} \tag{20.48}$$

$$k_3 = -1 + \sqrt{\frac{2\left(1 + R_e\right)}{R_e} + \frac{2F_{yb}\left(1 + 2R_e\right)D^2}{3F_{em}t_s^2}} \tag{20.49}$$

where R_t = ratio of the thickness of the main member to that of the side member = t_m/t_s.

The reference lateral resistance of a bolt or dowel in double shear is taken as the least value determined by the four governing modes:

$$\mathbf{I_m}: \quad Z = \frac{0.83\, Dt_m F_{em}}{K_\theta} \tag{20.50}$$

$$\mathbf{I_s}: \quad Z = \frac{1.66\, Dt_s F_{es}}{K_\theta} \tag{20.51}$$

$$\mathbf{III_s}: \quad Z = \frac{2.08\, k_3 Dt_s F_{em}}{K_\theta\left(2 + R_e\right)} \tag{20.52}$$

$$\mathbf{IV}: \quad Z = \frac{2.08\, D^2}{K_\theta}\sqrt{\frac{2\, F_{em} F_{yb}}{3\left(1 + R_e\right)}} \tag{20.53}$$

where k_3 is defined by Eq. (20.49)

Similarly, the reference lateral resistance of a single lag screw in single shear is taken as the least value determined by the three governing modes:

$$\mathbf{I_s}: \quad Z = \frac{0.83\, Dt_s F_{es}}{K_\theta} \tag{20.54}$$

$$\mathbf{III_s}: \quad Z = \frac{1.19\, k_r Dt_s F_{em}}{K_\theta\left(1 + R_e\right)} \tag{20.55}$$

$$\mathbf{IV}: \quad Z = \frac{1.11\, D^2}{K_\theta}\sqrt{\frac{1.75\, F_{em} F_{yb}}{3\left(1 + R_e\right)}} \tag{20.56}$$

where k_4 = a factor related to material properties and connection geometry as follows:

$$k_4 = -1 + \sqrt{\frac{2(1+R_e)}{R_e} + \frac{F_{yb}(2+R_e)D^2}{2\,F_{em}t_s^2}} \tag{20.57}$$

When double shear lag screw connections are designed, the reference lateral resistance is taken as twice the resistance of the weaker single shear representation of the left and right shear planes as was described for nail and wood screw connections.

Wood members are often connected to nonwood members with bolt and lag screw connections (e.g., wood to concrete, masonry, or steel). For connections with concrete or masonry main members, the dowel bear strength, F_{em}, for the concrete or masonry can be assumed the same as the wood side members with an effective thickness of twice the thickness of the wood side member. For connections with steel side plates, the equations for yield modes I_s and I_m are not appropriate. Rather, the resistance for that mode should be computed as the bearing resistance of the fastener on the steel side plate.

For multiple bolt, lag screw, and dowel connections, the least resistance is simply multiplied by the number of fasteners, n_f, in the connection detail. When multiple fasteners are used, the minimum spacings, edge distances, and end distances are dependent on the direction of loading. When loading is primarily parallel to the grain, the minimum spacing between fasteners in a row (parallel to the grain) is $4D$, and the minimum spacing between rows (perpendicular to the grain) of fasteners is $1.5D$ but not greater than 127 mm.* The minimum edge distance is dependent on l_m = length of the fastener in the main member for spacing in the main member or total fastener length in the side members for side member spacing relative to the diameter of the fastener. For shorter fasteners ($l_m/D \le 6$), the minimum edge distance is $1.5D$, while for longer fasteners ($l_m/D > 6$), the minimum edge distance is the greater of $5D$ or one half the spacing between rows (perpendicular to the grain). The minimum end distance is $7D$ for tension members and $4D$ for compression members. When loading is primarily perpendicular to the grain, the minimum spacing within a row (perpendicular to the grain) is typically limited by the attached member but not to exceed 127 mm,* and the minimum spacing between rows (parallel to the grain) is dependent on l_m. For shorter fastener lengths ($l_m/D \le 2$), the spacing between rows is limited to $2D$; for medium fastener lengths ($2 < l_m/D < 6$), the spacing between rows is limited to $(5l_m + 10D)/8$; and for longer fastener lengths ($l_m/D \ge 6$), the spacing is limited to $5D$; but never should the spacing exceed than 127 mm.* The minimum edge distance is $4D$ for loaded edges and $1.5D$ for unloaded edges. Finally, the minimum end distance for members loaded primarily perpendicular to the grain is $4D$.

The reference lateral resistance must be multiplied by all appropriate adjustment factors. It is necessary to consider group action, C_g, and geometry, C_Δ for bolts, lag screws, and dowels. In addition, penetration depth, C_d, and end grain, C_{eg}, need to be considered for lag screws. The group action factor accounts for load distribution between bolts, lag screw, or dowels when one or more rows of fasteners are used and is defined by

$$C_g = \frac{1}{n_f} \sum_{i=1}^{n_r} a_i \tag{20.58}$$

where n_f = number of fasteners in the connection, n_r = number of rows in the connection, and a_i = effective number of fasteners in row i due to load distribution in a row and is defined by

$$a_i = \left(\frac{1+R_{EA}}{1-m}\right)\left[\frac{m(1-m^{2n_i})}{\left(1+R_{EA}m^{n_i}\right)(1+m)-1+m^{2n_i}}\right] \tag{20.59}$$

*The limit of 127 mm can be violated if allowances are made for dimensional changes of the wood.

where

$$m = u - \sqrt{u^2 - 1} \qquad (20.60\text{a})$$

$$u = 1 + \gamma \frac{s}{2} \left(\frac{1}{(EA)_m} + \frac{1}{(EA)_s} \right) \qquad (20.60\text{b})$$

and where γ = load/slip modulus for a single fastener; s = spacing of fasteners within a row; $(EA)_m$ and $(EA)_s$ = axial stiffness of the main and side member, respectively; R_{EA} = ratio of the smaller of $(EA)_m$ and $(EA)_s$ to the larger of $(EA)_m$ and $(EA)_s$. The load/slip modulus, γ, is either determined from testing or assumed as $\gamma = 0.246D^{1.5}$ kN/mm for bolts, lag screws, or dowels in wood-to-wood connections or $\gamma = 0.369D^{1.5}$ kN/mm for bolts, lag screws, or dowels in wood-to-steel connections.

The geometry factor, C_Δ, is used to adjust for connections in which either end distances and/or spacing within a row does not meet the limitations outlined previously. Defining a = actual minimum end distance, a_{min} = minimum end distance as specified previously, s = actual spacing of fasteners within a row, and s_{min} = minimum spacing as specified previously, the lesser of the following geometry factors are used to reduce the adjusted resistance of the connection:

1. End distance: for, $a \geq a_{min}$, $C_\Delta = 1.0$
                          for $a_{min}/2 \leq a < a_{min}$, $C_\Delta = a/a_{min}$
2. Spacing: for, $s \geq s_{min}$, $C_\Delta = 1.0$
                          for $3D \leq s < s_{min}$, $C_\Delta = s/s_{min}$

In addition to group action and geometry, the penetration depth factor, C_d, and end grain factor, C_{eg}, are applicable to lag screws (not bolts and dowels). The penetration of a lag screw, including the shank and thread less the threaded tip, is required to be at least $4D$. For penetrations of at least $4D$ but not more than $8D$, the connection resistance is multiplied by $C_d = p/8D$, where p = depth of penetration. For penetrations of at least $8D$, $C_d = 1.0$. The end grain factor, C_{eg}, is applied when a lag screw is driven in the end grain of a member and is given as $C_{eg} = 0.67$.

Axial Resistance
Again, the tensile resistance of the fastener (i.e., bolt, lag screw, or dowel) is determined using accepted metal design procedure. Withdrawal resistance is only appropriate for lag screws since bolts and dowels are "through-member" fasteners. For the purposes of lag screw withdrawal, the penetration depth, p, is assumed as the threaded length of the screw less the tip length, and the minimum penetration depth for withdrawal is the lesser of 25 mm or one half the threaded length. The reference withdrawal resistance of a lag screw connection is then given by

$$Z_w = 92.6 \, D^{0.75} G^{1.5} pn_f \qquad (20.61)$$

where Z_w = reference withdrawal resistance in newtons and G = specific gravity of the wood.

The end grain adjustment factor, C_{eg}, is applicable to the withdrawal resistance of lag screws and is defined as $C_{eg} = 0.75$.

Combined Load Resistance
The resistance of a bolt, dowel, or lag screw connection to combined axial and lateral load is given by

$$Z'_\alpha = \frac{Z'Z'_w}{Z' \sin^2 \alpha + Z'_w \cos^2 \alpha} \qquad (20.62)$$

where Z'_α = adjusted resistance at an angle and α = angle between the applied load and the wood surface (i.e., $0°$ = lateral load and $90°$ = withdrawal/tension).

References

1. American Association of State Highway and Transportation Officials (AASHTO), *AASHTO LRFD Bridge Design Specifications*, AASHTO, Washington, D.C., 1994.
2. American Association of State Highway and Transportation Officials (AASHTO), *Standard Specifications for Highway Bridges*, 16th ed. AASHTO, Washington, D.C., 1996.
3. American Forest and Paper Association (AF&PA), *Manual of Wood Construction: Load and Resistance Factor Design*, AF&PA, Washington, D.C., 1996.
4. American Forest and Paper Association (AF&PA), *National Design Specification for Wood Construction* and *Supplement*, AF&PA, Washington, D.C., 1997.
5. Bodig, J. and Jayne, B., *Mechanics of Wood and Wood Composites*, Van Nostrand Reinhold, New York, 1982.
6. Freas, A.D., Wood properties, in *Wood Engineering and Construction Handbook*, 2nd ed., K.F. Faherty and T.G. Williamson, Eds., McGraw Hill, New York, 1995.
7. Fridley, K.J., Designing for creep in wood structures, *For. Prod. J.*, 42(3):23–28, 1992.
8. Ritter, M.A., Timber Bridges: Design, Construction, Inspection, and Maintenance, EM 7700-8, USDA, Forest Service, Washington, D.C., 1990.
9. Ritter, M.A. and Ebeling, D.W., Miscellaneous wood structures, in *Wood Engineering and Construction Handbook*, 2nd ed., K.F. Faherty and T.G. Williamson, Eds. McGraw-Hill, New York, 1995.
10. U.S. Department of Agriculture (USDA), Wood Handbook: Wood as an Engineering Material, Agriculture Handbook 72, Forest Products Laboratory, USDA, Madison, WI, 1987.
11. Wood, L.W., Relation of Strength of Wood to Duration of Load, USDA Forest Service Report No. 1916, Forest Products Laboratory, USDA, Madison, WI, 1951.

Further Reading

1. American Institute of Timber Construction (AITC), *Glulam Bridge Systems*, AITC, Englewood, CO, 1988.
2. American Institute of Timber Construction (AITC), AITC 117-93, *Design Standard Specifications for Structural Glued Laminated Timber of Softwood Species*, 1993. AITC, Englewood, CO,1993.
3. American Institute of Timber Construction (AITC), *Timber Construction Manual*, Wiley Inter-Science, New York, 1994.
4. Western Wood Products Association (WWPA), *Western Wood Use Book*, 4th ed. WWPA, Portland, OR, 1996.
5. Wipf, T.J., Klaiber, F.W., and Sanders, W.W., Load Distribution Criteria for Glued Laminated Longitudinal Timber Deck Highway Bridges, Transportation Research Record 1053, Transportation Research Board, Washington, D.C.

21
Movable Bridges

Michael J Abrahams
Parsons Brinckerhoff Quade &
Douglas, Inc.

21.1 Introduction

Movable bridges have been an integral part of the U.S. transportation system, their development being in concert with that of (1) the development of the railroads and (2) the development of our highway system. While sometimes referred to as draw bridges, movable bridges have proved to be an economical solution to the problem of how to carry a rail line or highway across an active waterway. It is not surprising to learn that movable bridges are found most commonly in states that have low coastal zones such as California, Florida, Louisiana, and New Jersey, or a large number of inland waterways such as Michigan, Illinois, and Wisconsin.

Jurisdiction for movable bridges currently lies with the U.S. Coast Guard. In most instances, marine craft have priority, and the movable span must open to marine traffic upon demand. This precedence is reflected in the terms closed and open, used to describe the position of the movable span(s). A "closed" movable bridge has closed the waterway to marine traffic, while an "open" bridge has opened the waterway to marine traffic. Highway bridges are typically designed to remain in the closed position and only to be opened when required by marine traffic. However, movable railroad bridges can be designed to remain in either the open or closed position, depending on how frequently they are used by train traffic. The difference is important as different wind and seismic load design conditions are used to design for a bridge that is usually open vs. one that is usually closed.

The first specification for the design of movable bridges was published by the American Railway Engineering Association (AREA) in its 1922 *Manual of Railway Engineering* [1]. Until 1938 this specification was used to design both movable highway and railroad bridges, when the American Association of State Highway Officials published its *Standard Specifications for Movable Highway Bridges* [2]. Both specifications are very similar, but have remained separate. Today, movable railroad bridges are designed in accordance with the AREA Manual, Chapter 15, Part 6 [3], and movable

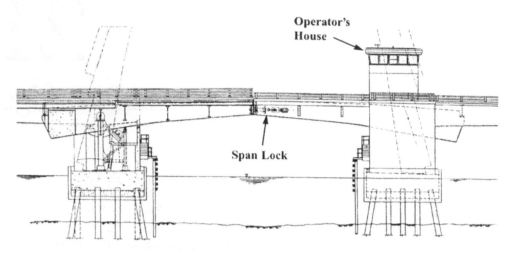

FIGURE 21.1   South Slough (Charleston) Bridge, Coos County, Oregon.

highway bridges are designed in accordance with the American Association of State Highway and Transportation Officials (AASHTO) *Standard Specifications for Movable Highway Bridges* [4]. These specifications primarily cover the mechanical and electrical aspects of a movable bridge; the structural design of the bridge is covered in other parts of the AREA Manual for railroad bridges or the AASHTO *Standard Specifications for Highway Bridges* [5].

21.2   Types of Movable Bridges

The three major categories of movable bridges are swing, bascule, and vertical lift. This list is not exclusive and there are other types, such as jackknife, reticulated, retracting, and floating that are not common and will not be described here. However, the reader should be aware that movable bridges can be crafted to suit specific site needs and are not restricted to the types discussed below.

21.2.1   Bascule Bridges

Bascule bridges are related to medieval drawbridges that protected castles and are familiar illustrations in schoolbooks. The function is the same; the bascule span leaf (or leaves if there are two) rotates from the horizontal (closed) position to the vertical (open) position to allow use of the waterway below. Figure 21.1 illustrates a typical double-leaf deck-girder bascule bridge, the South Slough (Charleston) Bridge, Coos County, Oregon, which spans 126 ft (38.4 m) between trunnions.

This highway bridge includes a number of features. It is a trunnion type as the bascule span rotates about a trunnion. The counterweight, which is at the back end of the leaf and serves to balance the leaf about the trunnion, is placed outside of the pier so that it is exposed. This is advantageous in that it minimizes the width of the pier. Also note that the tail or back end of the leaf reacts against the flanking span to stop the span and to resist uplift when there is traffic (live load) on the span. There is a lock bar mechanism between the two leaves that transfers live-load shear between the leaves as the live load moves from one leaf to the other. The locks (also called center locks to distinguish them from end locks that are provided at the tail end of some bascule bridges) transfer shear only and allow rotation, expansion, and contraction to take place between the leaves. The bridge shown in Figure 21.1 is operated mechanically, with drive machinery in each pier to raise and lower the leaves.

Another feature to note is the operator's house, also referred to as the control house. It is situated so that the operator has a clear view both up and down the roadway and waterway, which is required when the leaves are both raised and lowered. The lower levels of the operator's house typically house

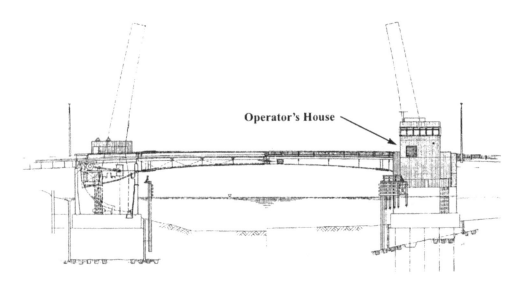

FIGURE 21.2 3rd Street Bridge, Wilmington, Delaware.

the electrical switchgear, emergency generator, bathroom, workshop, and storage space. This bridge has a free-standing fender system that is intended to guide shipping through the channel while protecting the pier from impact. Although not directly related to the bascule bridge, the use of precast footing form and tremie fill shown in the figure can be an excellent solution to constructing the pier as it minimizes the pier depth and avoids excavation at the bottom of the waterway.

Figure 21.2, the 3rd Street Bridge, Wilmington, Delaware, shows a through-girder double-leaf bascule span illustrating other typical bascule span design features. It has a center-to-center trunnion distance of 188 ft (57.3 m). For this bridge the tail or back end of the leaf, including the counterweight, is totally enclosed in the pier and the live-load reaction is located at the front wall of the pier. In addition, this bascule is shown with a mechanical drive. A larger pier is required to protect the enclosed counterweight. The advantage of an enclosed pier is that it allows the counterweight to swing below the waterline within the confines of the bascule pier pit. And, as can be seen, the bascule pier is constructed within a cofferdam. For this bridge there was not enough depth to place a full tremie seal so underwater tie-downs were used to tie the seal to the rock below. Also note the architectural detailing of the cast-in-place concrete substructure, was achieved using form liners. This was done because the bridge is located in a park and needed to be compatible with a parklike setting.

Figure 21.3, the Pelham Bay Bridge, New York, illustrates a single-leaf Scherzer rolling lift bridge or bascule railroad bridge, typical of many movable railroad bridges. The design was developed and patterned by William Scherzer in 1893 and is both simple and widely used. This is a through truss with a span of 81 ft 7 in. (24.9 m). Railroad bascule bridges are always single span, which is required by the AREA Manual, as the heavy live loads associated with heavy rail preclude a joint at midspan. This problem does not occur with light rail (trolley) live loads and combined highway/trolley double-leaf bascule bridges were frequently used in the 1920s. Also, railroad bascule bridges such as the one shown are usually through-truss spans, again due to the heavy rail live loads. This bridge has an overhead counterweight, a typical feature of Scherzer-type bridges. This allows the bridge to be placed relatively close to the water and permits a very simple pier. The track is supported by a steel girder and two simple open piers. As illustrated, the leaf rolls back on a track rather than pivoting about a trunnion. The advantage of this feature is that it is not restricted by the capacity of the trunnion shafts and it minimizes the distance between the front face of the pier and the navigation channel. As the span rotates, it rolls back away from the channel. The drive machinery is located on the moving leaf and typically uses a mechanically or hydraulically driven rack and

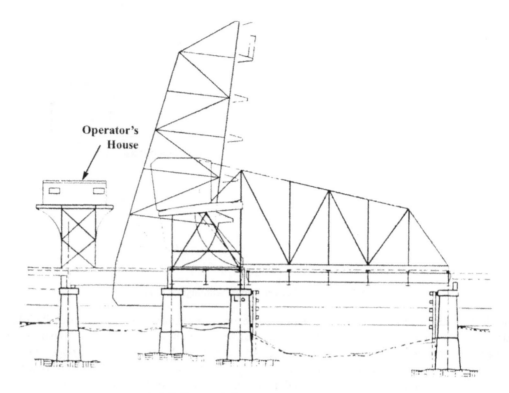

FIGURE 21.3   Pelham Bay Bridge, New York.

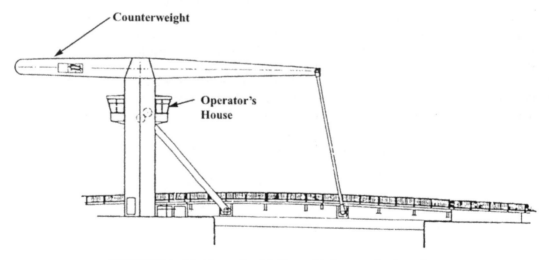

FIGURE 21.4   Manchester Road Bridge at the Canary Wharf, London.

pinion to move the span. The machinery must thus be able to operate as it rotates, and for hydraulic machinery this means the reservoir needs to be detailed accordingly. However, this is not always the case and designs have been developed that actuate the span with external, horizontally mounted hydraulic cylinders. The pier needs to be designed to accommodate the large moving load of the bascule leaf as it rolls back. Conversely, the reaction from the leaf in a trunnion-type span is concentrated in one location, simplifying the design of the pier. More-complicated bascule bridges with overhead counterweight designs have been developed where the counterweight is supported by a scissors-type frame and by trunnions that pivot.

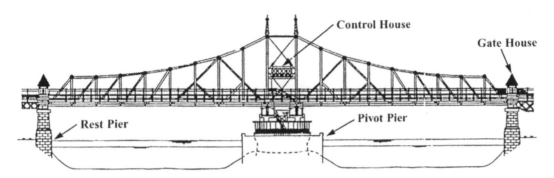

FIGURE 21.5 Macombs Dam Bridge over the Harlem River, New York City.

Figure 21.4, the Manchester Road Bridge at the Canary Wharf, London, illustrates a modern interpretation of an overhead counterweight bascule bridge. It has a span of 109 ft (33.2 m). In addition to being attractive, it is a very practical design with all of the structure above the roadway level, allowing the profile to be set as close to the waterline as desired. The design is not new and is found in many small hand-operated bridges in Holland, perhaps the most famous of which appears in Van Gogh's 1888 painting *The Langlois Bridge.*

21.2.2 Swing Spans

Swing spans were widely used by the railroads. However, they only allowed a limited opening and the center pivot pier was often viewed as a significant impediment to navigation. The pivot pier could also require an elaborate, difficult-to-maintain, and expensive fender system. As a result, swing spans are infrequently used for movable spans. However, they can be a cost-effective solution, particularly for a double-swing span, and should be considered when evaluating options for a new movable bridge.

Figure 21.5 shows a typical through-truss swing span, the Macombs Dam Bridge over the Harlem River in New York City, constructed in 1895. This span is 415 ft (126.5 m) long. The large pivot pier in the middle of the channel illustrates the navigation issue with this design. The piers at either end of the swing span are referred to as the rest piers. By using a through truss, the depth of structure (the distance between the profile grade line and the underside of the structure) is minimized — thus minimizing the height and length of the approaches. The turning mechanism is located at the pivot pier and the entire dead load of the swing span is supported on the pivot pier. As the two arms of the swing span are equal, they are balanced. This bridge is operated with a mechanical drive that utilizes a rack-and-pinion system. There are live-load end lifts at the ends of the swing span that are engaged when the span is closed in order to allow the movable span to act as a two-span continuous bridge under live load. The end lifts, as the name suggests, lift the ends of the swing span, which are free cantilevers when the span operates. The operator's house is typically located on the swing span within the truss but above the roadway, as this location provides good visibility. On older bridges one may also find tenders' houses located at the ends of the swing span. These were for gate tenders who would stop traffic, manually close the traffic gates, and hold horses if necessary. The tenders have been replaced with automatic traffic signals and gates but, on this bridge, their houses remain.

Figure 21.6, the Potato Slough Bridge, San Joaquin County, California, illustrates a good example of a modern highway swing span. This bridge has a 310-ft (94.5-m) swing span that uses simple composite deck, steel girder construction. It is very economical on a square foot basis compared with a bascule or vertical lift bridge, due to its simplicity and lack of a large counterweight. One way of looking at this is that on a bascule or vertical lift bridge a large amount of structure is composed of the counterweight and its supports. These elements do not contribute to effective load-carrying area. The swing span back span, on the other hand, not only acts as a counterweight

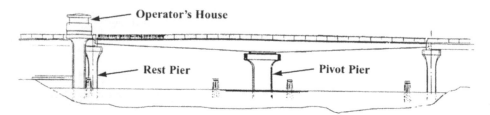

FIGURE 21.6 Potato Slough Bridge, San Joaquin County, California.

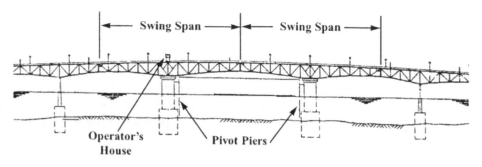

FIGURE 21.7 Coleman Bridge over the York River, Virginia.

but also carries traffic making for a more cost-effective solution. One disadvantage of the deck girder design is that it does not minimize the depth of construction, as does a through-truss or through-girder design. On this bridge the swing span is symmetrical and thus balanced. Nevertheless, some small counterweights may be required to correct any transverse imbalance. The operator's house is located in an adjacent independent structure, again in an area that provides good visibility upstream, downstream, and along the roadway. The pivot pier can accommodate switchgear and a generator. The roadway joints at the ends of the span are on a radius. These could also be detailed as beveled joints, provided that the span only needs to swing in one direction. However, some designers believe it is preferable to design a swing bridge to swing in either direction to allow the bridge to be opened away from oncoming marine traffic and to minimize damage if the structure is struck and needs to swing free.

Figure 21.7 illustrates the double-swing span Coleman Bridge across the York River in Virginia. The two swing spans are each 500 ft (152.4 m) long and provide a 420-ft (128.0-m) wide navigation channel, wide enough to accommodate the range of U.S. Navy vessels that traverse the opening. The bridge is a double-swing deck truss. At this site the river banks are relatively high, so the depth of structure was not a significant issue. Because the bridge is located adjacent to a national park, the low profile of a deck truss was a major advantage. The bridge uses hydraulic motors to drive the span, driving through a rack-and-pinion system similar to that used in large slewing excavators. Unlike the single-swing bridges above, there are lock bars at all three movable span joints. These are driven when the span is in the closed position and function in the same manner as lock bars between the leaves of a double-leaf bascule. There are wedges at each pivot pier to support the live load. As shown, the operator's house is located above one of the swing spans. The control equipment is located inside the operator's house and the generator and switchgear is located on the swing spans below deck. This bridge superstructure was replaced in 1996 and uses a lightweight concrete deck. The piers were constructed in 1952 when the bridge was first built using concrete-filled steel shell caissons that were placed by dredging through open wells.

Figure 21.8, the Tchefuncte River Bridge, Madisonville, Louisiana illustrates a bobtail swing, which is used where only a small channel is required. The structure is a through girder (this minimizes the depth of construction) with a main span of 160 ft (48.8 m). The 80-ft (24.4 m) long

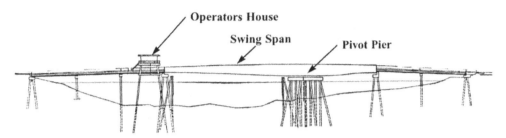

FIGURE 21.8   Tchefuncte River Bridge, Madisonville, Louisiana.

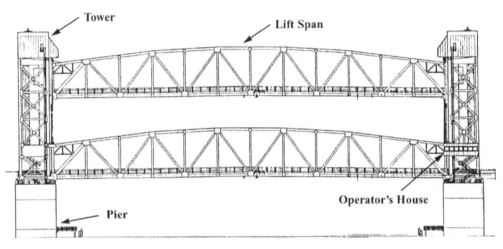

FIGURE 21.9   James River Bridge, Virginia.

bobtail end contains a concrete counterweight that balances the weight of the longer front span. This type of design is particularly well suited where the profile is near the waterline. A relatively simple foundation can support the swing span and no structure is required above deck level. This bridge is operated using hydraulic cylinders on the pivot pier. Girder swing spans tend to be flexible and need a wedge or end lift system that can lift the ends of the span and provide a live-load support when it is in the closed position.

21.2.3   Vertical Lift Bridges

Vertical lift bridges, the last of the three major types of movable bridges, are most suitable for longer spans, particularly for railroad bridges.

Figure 21.9 shows a through-truss highway lift span — the James River Bridge in Virginia — which has a span of 415 ft (126.5 m). The maximum span for this type of design to date is approximately 550 ft (167.7 m) long. The weight of the lift span is balanced by counterweights, one in each tower. Wire ropes that pass over sheaves in the towers are attached to the lift span at one end and the counterweight at the other. A secondary counterweight system is often required to balance the weight of the wire ropes as the span moves up and down and the weight of the wire ropes shifts from one side of the sheaves to the other.

Two types of drive systems are commonly employed, tower drive and span drive. A span drive places the drive machinery in the center of the lift span and, through drive shafts, operates a winch and hauling rope system to raise and lower the span. A tower drive — as the name implies — uses drive machinery in each tower to operate the span. The advantage of the span drive is that it ensures that the two ends lift together, whereas a tower drive requires coordinating the movement

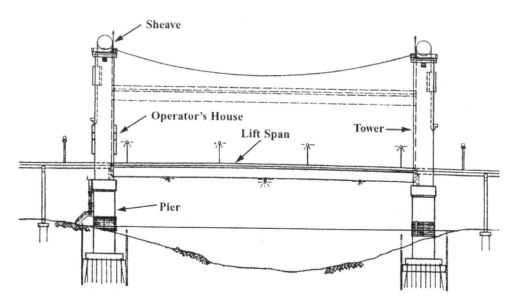

FIGURE 21.10   Danziger Bridge, New Orleans, Louisiana.

at each end. The disadvantages of the span drive are that it tends to be ugly and the lift span, ropes, sheaves, and counterweights must carry the additional weight of the operating machinery. Consequently, tower drives are favored on new bridges.

The machinery drive can be either mechanical or hydraulic. Guide wheels guide the span as it moves along the tower legs, and they must be detailed so as to allow expansion and contraction at one end of the lift span to accommodate changes in temperature. Span locks are used at each end of the lift span to ensure that it does not drift up when in the down (closed) position. If the bridge is normally in the open position, an additional set of span locks needs to be provided. As shown, the operator's house is located on one of the towers. For this bridge, the house partially wraps around the tower to provide good visibility of both the waterway and roadway.

Figure 21.10, the Danziger Bridge, New Orleans, is a vertical lift bridge that uses an orthotropic deck with steel box girders for the lift span and welded steel boxes for the tower. The lift span is 320 ft (97.6 m) long. While the depth of construction is greater than that of an equivalent through truss, the appearance is cleaner, the load to lift should be less and the height of the towers lower than that of an equivalent through truss. The foundations for both of these vertical lift bridges used deep cofferdam construction, which may be advantageous for longer spans because the mass and rigidity of such a foundation should be better able to resist the forces from collision with a large ship.

21.3   Structural Design

21.3.1   Design Criteria

In the closed positions movable bridges are designed for the same design conditions as fixed bridges. However, a movable bridge must also be designed for the following conditions. The load combinations described below are from the AASHTO Specifications [4], and are based on allowable stresses. Similar provisions apply to railroad bridges.

1. *Impact Loads*: Dead load plus 20%. This is applied to structural parts in which the member stress varies with the movement of the span. It is not combined with live-load stresses. For structural parts with stresses caused by machinery or forces applied for moving or stopping the span, 100% impact is used. For end floorbeams, live load plus 100% impact is used.

2. *Wind Loads:*
   a. Movable Span Closed:
      i. Structure to be designed as a fixed span.
   b. Movable Span Open:
      ii. When the movable span is normally left in the closed position, the structure is designed for 30 pounds per square foot (psf) (1.436 kPa) wind load on the structure, combined with dead load, and 20% of dead load to allow for impact, at 1.25 times the allowable unit stresses. For swing bridges, the design is also checked for 30 psf (1.436 kPa) wind load on one arm and 20 psf (0.958 kPa) wind load on the other arm.
      iii. When the movable span is normally left in the open position, the structure is designed for 50 psf (2.394 kPa) wind load on the structure, combined with dead load, at 1.33 times allowable unit stresses. For swing bridges the design is also checked for 50 psf (2.394 kPa) wind load on one arm and 35 psf (1.676 kPa) wind load on the other arm, applied simultaneously).
3. *Ice/Snow Loads:* These are typically not considered in structural design but must be considered in designing the operating machinery.
4. *Bascule Bridges:* The stresses in the main and counterweight trusses or girders are checked for the following load cases:
   a. Case I   Dead load: Bridge open in any position
   b. Case II   Dead load: Bridge closed
   c. Case III   Dead load. Bridge closed with counterweights independently supported
   d. Case IV   Live load plus impact: Bridge closed with live loads thereon
5. *Swing Bridges:* The main trusses or girders are checked for the following load cases:
   a. Case I   Dead load: Bridge open, or closed with end wedges (lifts) not driven
   b. Case II   Dead load: Bridge closed, with its wedges lifted to give positive end reaction, equal to the reaction due to temperature plus 1.5 times the maximum negative reaction of the live load and impact, or the force required to lift the span 1 in. (25 mm) whichever is the greater
   c. Case III   Live load plus impact: Bridge closed, with one arm loaded and considered as a simple span, but with end wedges (lifts) not driven
   d. Case IV   Live load plus impact: Bridge closed and considered as a continuous structure
6. *Vertical Lift Bridges:* The main trusses or girders and towers are checked for the following load cases:
   a. Case I   Dead load: Bridge open
   b. Case II   Dead load: Bridge closed
   c. Case III   Dead load with bridge closed and counterweights independently supported (it should be noted that vertical lift bridges need to include provisions to support the counterweights independently)
   d. Case IV.   Bridge closed with live loads thereon

All of the above applies to the structural design of the moving span and its supports. For design of the operating machinery, there are other load cases contained in the AREA Manual [3] and the AASHTO Specifications [4].

21.3.2 Bridge Balance

Almost all movable bridges are counterweighted so that the machinery that moves the span only needs to overcome inertia, friction, wind, ice, and imbalance. It is prudent to design bridges with a healthy allowance for imbalance as the as-built conditions are never perfect, particularly over time. Recently, at least one bascule bridge and several lift spans have been designed without counterweights, relying instead on the force of the hydraulic machinery to move the span. While this saves the cost of the counterweight and reduces the design dead loads, one needs to compare carefully the reduced construction costs against the present value of the added machinery costs and future annual electric utility demand and service costs (utility rates are based not only on how much energy is consumed but also on how much it costs the utility to be able to supply the energy on demand).

Counterweights are designed to allow for adjustment of the bridge balance, recognizing that during its lifetime, the weight and weight distribution of the bridge can change. The typical reasons for these changes are deck replacement, paint, repairs, or new span locks, among others. Typically, contract drawings show the configuration, estimated concrete volume, and location of the counterweights, but require that the contractor be responsible for balancing the span. This is reasonable as the designer does not know the final weight of the elements to be used, such as the size of the splice plates, the lock bar machinery, concrete unit weight, and other variables. Balance checks can be made during construction or retrofit using detailed calculations accounting for every item that contributes to the weight of the moving span. These calculations need to account for the location of the weight in reference to the horizontal and vertical global axes of the span and, for an asymmetrical span such as a swing span, the transverse axis. For bascule and vertical lift bridges, current practice is to attach strain gauges to the machinery drive shafts and measure the strain in the shafts as the span is actuated through a full cycle, thereby accurately determining the balance. Strain gauge balancing was developed for trunnion bascule type bridges [6,7]. The method has been extended to rolling lift bascule bridges as well as vertical lift bridges.

21.3.3 Counterweights

Figure 21.11 illustrates the typical counterweight configuration for a vertical lift bridge. Both the AREA Manual [3] and AASHTO Specifications [4] require that a pocket be provided in the counterweight for adjustment. The pockets are then partially filled with smaller counterweight blocks, which can be moved by hand to adjust the balance of the bridge. Counterweights are typically made up of a concrete surrounding a steel frame or a reinforced steel box that is filled with normal-weight concrete. Heavyweight concrete can be used to minimize the size of the counterweight. Punchings from bolt holes can be mixed in with concrete to increase its density or concrete can be made using heavyweight aggregate, although this is seldom done due to cost considerations. However, there is at least one vertical lift bridge where cast-iron counterweights were used because the counterweights needed to be as small as possible as they were concealed in the towers. If there is not enough space left for added blocks or if there are no longer any blocks available, counterweight adjustments can always be made by adding steel plates, shapes, or rails.

Figure 21.12 shows the results of a balance check of a rolling lift bridge. In this case, the bridge had been in operation for many years and the owner wanted to replace the timber ties with newer, heavier ties. As shown, the imbalance varied with the position of the span and in the open position the center of gravity was behind the center of rotation. It would be preferable to have all the imbalance on the span side, and to reduce the imbalance. One needs to be careful as an increased imbalance can have a chain reaction and cause an increase in the drive machinery and bridge power requirements.

In general, it is good practice to balance a span so that it is slightly toe heavy for a bascule bridge and slightly span heavy for a vertical lift bridge, the idea being that the span will tend to stay closed under its own weight and will not bounce under live load (although once the span locks are engaged the span cannot rise). The amount of imbalance needs to be included in designing the bridge-operating machinery, so that it can tolerate the imbalance in combination with all the other machinery design loads.

21.3.4 Movable Bridge Decks

An important part of the design of movable bridges is to limit the moving dead load which affects the size of the counterweight, the overall size of the main structural members, and, to a lesser extent,

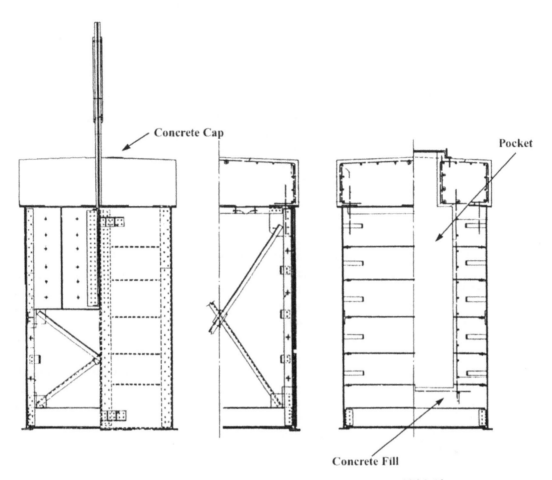

FIGURE 21.11   Typical counterweight configuration for a vertical lift bridge.

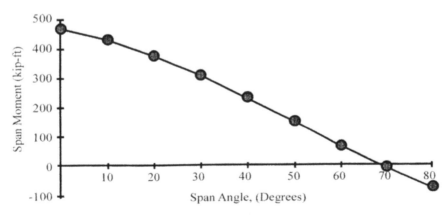

FIGURE 21.12   Results of balance check of a rolling lift bridge.

the machinery depending upon the type of movable bridge. For movable railroad bridges this is typically not a problem, as movable span decks are designed with open decks (timber ties on stringers) and the design live load is such a large part of the overall design load that the type of deck is not an issue. For highway bridges, however, the type of deck needs to be carefully selected to provide a minimum weight while providing an acceptable riding surface. Early movable spans used timber decks, but they are relatively heavy and have poor traction and wear. Timber was replaced by open steel grid, which at 20 to 25 psf (98 to 122 kg/m²) was a good solution that is both lightweight and long wearing. In addition, the open grid reduced the exposed wind area, particularly for bascule bridges in the open position. However, with higher driving speeds, changes in tires and greater congestion, steel grid deck has become the source of accidents, particularly when wet or icy. Now most new movable bridge decks are designed with some type of solid surface. Depending on the bridge, this can be a steel grid partially filled with concrete or epoxy, an orthotopic deck, lightweight concrete, or the Exodermic system. Aluminum and composite decks are also now being developed and may prove to be a good solution. While orthotropic decks would seem to be a good solution, as the deck can be used as part of the overall structural system, they have not yet seen widespread use in new designs. The reader is referred to Chapters 14 and 24 for information on bridge decks.

21.3.5 Vessel Collision

Movable bridges are typically designed with the minimum allowable channel. As a result, vessel collision is an important aspect as there may be a somewhat higher probability of ship collision than with a fixed bridge with a larger span. There are two factors that are unique to movable bridges with regard to fender (and vessel collision) design. The first is that if a large vessel is transiting the crossing, the bridge will be in the open position and traffic will be halted away from the main span. As a result, the potential consequences of a collision are less than they would be with a fixed bridge ship collision. On the other hand, a movable bridge is potentially more vulnerable to misalignment or extensive damage than a fixed bridge. This is because not only are the spans supported by machinery, but movable spans by their very nature lack the continuity of a fixed bridge. There is no code to govern these issues, but they need to be considered in the design of a movable bridge. The configuration of the piers is an important aspect of this consideration. The reader is referred to Chapter 60, Vessel Collision Analysis and Design.

21.3.6 Seismic Design

The seismic design of movable bridges is also a special issue because they represent a large mass, which may include a large counterweight, supported on machinery that is not intended to behave in a ductile manner. In addition, the movable span is not joined to the other portions of the structure thus allowing it to respond in a somewhat independent fashion. The AREA Manual [3], Chapter 9 covers the seismic design of railroad bridges. However, these guidelines specifically exclude movable bridges. For movable highway bridges, the AASHTO *Standard Specification for Movable Highway Bridges* [4] requires that movable bridges that are normally in the closed position shall be designed for one half the seismic force in the open position. The interpretation of this provision is left up to the designer. The reader is referred to Part IV for an additional discussion of seismic investigation.

21.4 Bridge Machinery

Currently, bridge machinery is designed with either a mechanical or hydraulic drive for the main drive and usually a mechanical drive for the auxiliary machinery items such as span locks and wedges. This is true for all types of movable bridges and the choice of mechanical vs. hydraulic drive is usually based on a combination of owner preference and cost — although other factors may also be considered. Mechanical drives are typically simple configurations based on machinery design principles that were developed long before movable bridges, although now drives use modern enclosed speed reducers and bearings. Overall, these systems have performed very well with sometimes limited maintenance. More recently, hydraulic machinery has been introduced in movable bridge design and it has proved to be an effective solution, as the hydraulics can be closely matched

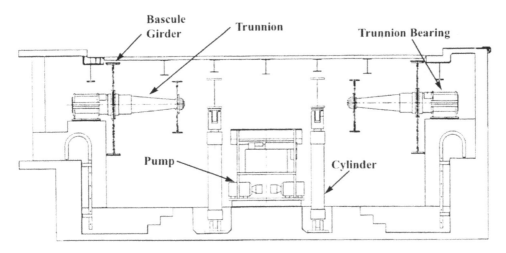

FIGURE 21.13   Section through a bascule pier showing girder trunnions and hydraulic cylinders.

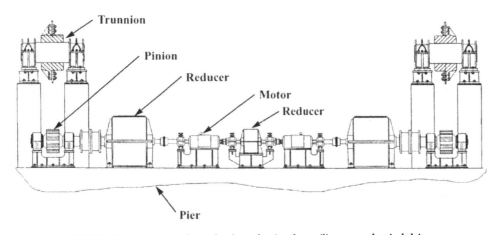

FIGURE 21.14   Section through a bascule pier that utilizes a mechanical drive.

to the power demands, which require good speed control over a wide range of power requirements. Also, there are many firms that furnish hydraulic machinery. However, the systems also require a more-specialized knowledge and maintenance practice than was traditionally the case with mechanical drives.

Figure 21.13 shows a section through a bascule pier illustrating the layout of the bascule girder trunnions (about which the bascule girders rotate) as well as the hydraulic cylinders used to operate the span. Typical design practice is to provide multiple cylinders so that one or more can be removed for maintenance while the span remains in operation. The cylinder end mounts incorporate spherical bearings to accommodate any misalignments. Note that the hydraulic power pack, consisting of a reservoir, motors, pumps, and control valves, is located between the cylinders. Typically redundant motors and pumps are used and the valves can be hand operated if the control system fails. As movable bridges are located in waterways, the use of biodegradable hydraulic fluids is favored in case of a leak or spill.

Figure 21.14 shows a similar section through a bascule pier that utilizes a mechanical drive. What is not shown is the rack attached to the bascule girder. Note the different arrangement here of the trunnions, with bearings on either side of the girders. The central reducer contains a differential, similar to the differential in a vehicle, that serves to equalize the torque in these two drive shafts. As shown, there are two drive motors and typically the span will be designed to operate with only

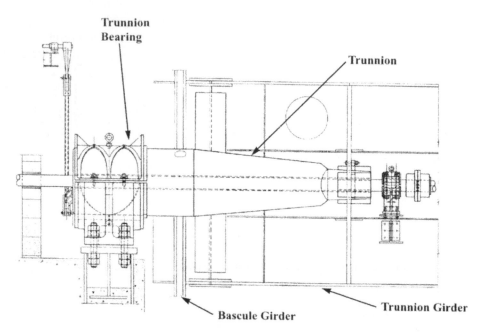

FIGURE 21.15   Trunnion and trunnion bearing.

one motor in operation either as a normal or emergency condition. Also note the extensive use of welded steel frames to support the machinery. It is important that they be stress relieved after assembly but prior to machining and that they be carefully detailed to avoid reentrant corners that could, in time, be a source of cracks.

Figure 21.15 is an illustration of a trunnion and trunnion bearing. The trunnions are fabricated from forged steel and, in this case, are supported on one end by a trunnion bearing and on the other by a trunnion girder that spans between the bascule girders. In this figure a sleeve-type trunnion bearing is shown. The use of sleeve bearings in this type of arrangement is not favored by some designers because of concern with uneven stress on the lining due to deformation of the trunnions and trunnion girder, particularly as the span rotates. Alternative solutions include high-capacity spherical roller bearings and large spherical plain bearings. The crank arrangement shown on the left side of the figure is associated with a position indicator.

Figure 21.16 shows a typical arrangement of the treads for a rolling lift bascule.

Figure 21.17 is a typical drive mechanism for a vertical lift bridge, with a tower drive. The drive is somewhat similar to that used for a bascule bridge except that the pinion drives the rack attached to a sheave rather than a rack attached to a bascule girder. Although a mechanical drive is shown, a similar arrangement could be accomplished with hydraulic motors.

Figure 21.18 is a typical welded sheave used for a vertical lift bridge. As shown, there are 16 rope grooves so this would be associated with a large vertical lift bridge. Typically there are four sheaves for a vertical lift bridge, one at each corner of the lift span. The trunnion bearing is not shown but would be similar to that shown in a bascule bridge trunnion. While sleeve bearings are commonly used, spherical type bearings are also considered to allow for trunnion flexure.

Figure 21.19 shows a span lock typically used between the leaves of a two-leaf bascule bridge. In this case a manufactured unit is illustrated. It incorporates a motor, brake, reducer, and lock bars. Alternative arrangements with a standard reducer are also used, although for this type of an installation the compactness and limited weight favor a one piece unit. It is important that provisions be included for replacement of the wearing surfaces in the lock bar sockets and realignment as they receive considerable wear.

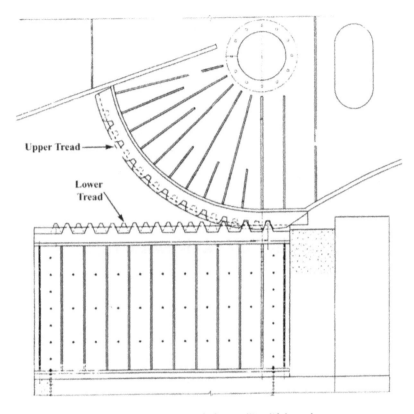

FIGURE 21.16 Treads for a rolling lift bascule.

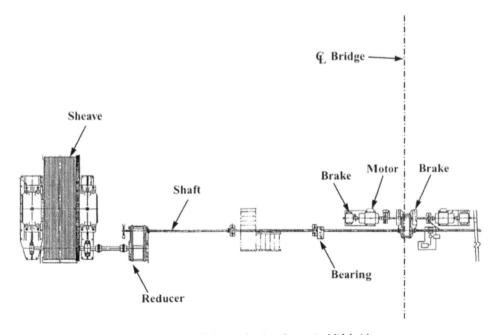

FIGURE 21.17 Drive mechanism for vertical lift bridge.

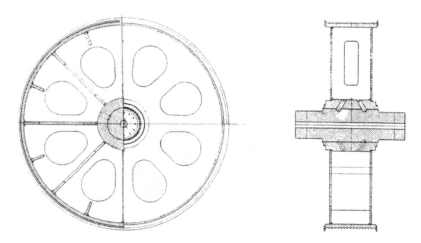

FIGURE 21.18   Welded sheave for a vertical lift bridge.

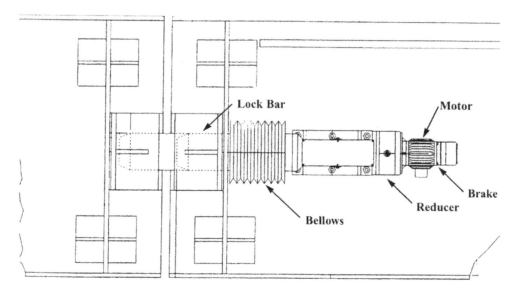

FIGURE 21.19   Span lock between leaves of a two-leaf bascule bridge.

Figures 21.20 and 21.21 show a pivot bearing, balance wheel, and live-load wedge arrangement typically used for a center pivot swing span. For highway bridges AASHTO states, "Swing bridges shall preferably be the center bearing type." No such preference is indicated by AREA. The center pivot, which contains a bronze bearing disk, carries the dead load of the swing span. The balance wheels are only intended to accommodate unbalanced wind loads when the span moves so that they are adjusted to be just touching the roller track. The wedges are designed to carry the bridge live load and are retracted prior to swinging the span.

Figure 21.22 shows a rim-bearing-type swing span arrangement. Note that it is much more complicated than the center pivot arrangement shown above. The rollers must be designed to carry dead, live, and impact loads and, unlike the intermittent rollers used for a center pivot bridge, need to be placed in a continuous fashion all around the rim. The purpose of the center pivot is to keep the rollers centered, and for some bridges to carry a portion of the dead and live load. Figure 21.23 shows an end lift device used for a swing span.

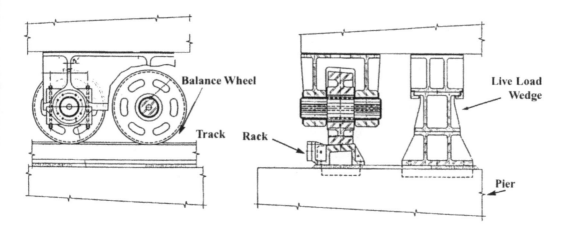

Typical Balance Wheels

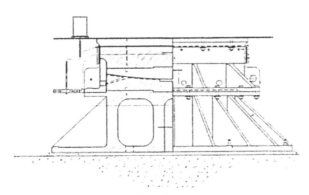

Pivot Bearing

FIGURE 21.20 Balance wheels and pivot bearing for a center pivot swing span.

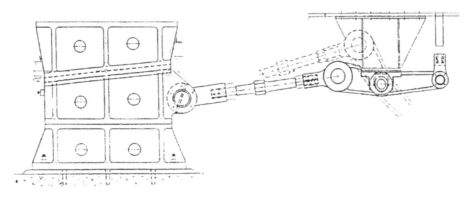

FIGURE 21.21 Live-load wedge arrangement for center pivot swing span.

Figure 21.24 shows a mechanical drive arrangement for a swing span, and similar arrangements can be adapted to both pivot and rim-bearing bridges. A common problem with this arrangement is the pinion attachment to the structural supports as very high forces can be induced in braking the swing span when stopping and these supports tend to be a maintenance problem. Figure 21.25

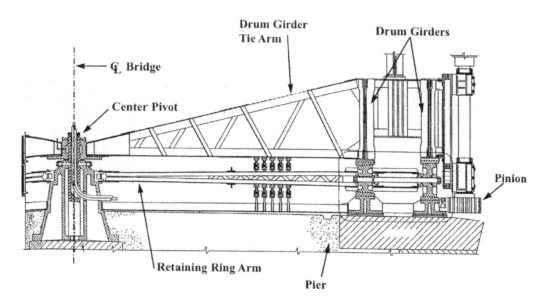

FIGURE 21.22 Rim–bearing swing span arrangement.

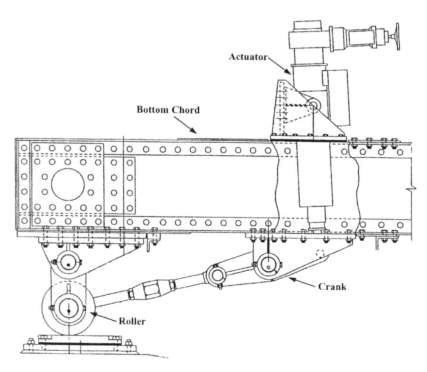

FIGURE 21.23 End lift device for a swing span.

illustrates one of four hydraulic drives from the Coleman Bridge. This drive has an eccentric ring mount so that the pinion/rack backlash can be adjusted.

Figure 21.26 shows a hydraulic drive for a swing span using hydraulic cylinders.

Figure 21.27 shows a typical air buffer. These are provided at the ends of the movable span. With modern control systems, particularly with hydraulics, buffers may not be required to assist in seating. For many years these were custom-fabricated but, if required, one can now utilize off-the-shelf commercial air or hydraulic buffers, as is shown here.

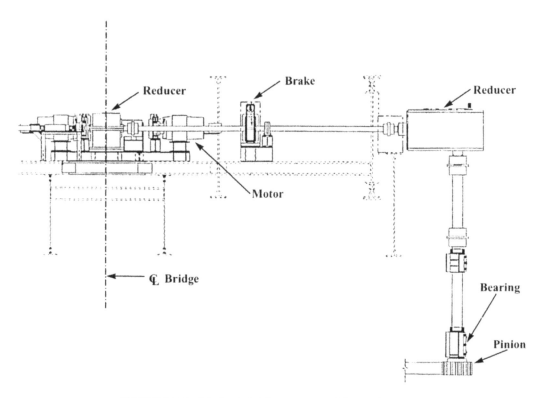

FIGURE 21.24 Mechanical drive arrangement for a swing span.

21.5 Bridge Operation, Power, and Controls

21.5.1 Bridge Operations

Movable bridges are designed to be operated following a set protocol, and this protocol is incorporated into the control system as a series of permissive interlocks. The normal sequence of operation is as follows:

Vessel signals for an opening, usually through a marine radio but it can be through a horn. For a highway bridge the operator sounds a horn, activates the traffic signals, halting traffic, lowers the roadway gates, then lowers the barrier gates. For a rail bridge the operator needs to get a permissive signal from the train dispatcher.

After the barrier gates are lowered, a permissive signal allows the operator to withdraw the locks and/or wedges and lifts and, once that is completed, to open the span. The vessel then can proceed through the opening. To close the bridge, the steps are reversed.

The controls are operated from a control desk and Figure 21.28 shows a typical control desk layout. Note that the control desk includes a position indicator to demonstrate the movable span(s) position as well as an array of push buttons to control the operation. A general objective in designing such a desk is to have the position of the buttons mimic the sequence of operations. Typically, the buttons are lit to indicate their status.

21.5.2 Bridge Power

In the early years, when the streets of most cities had electric trolleys, movable bridges were operated on the 500 VDC trolley power. As the trolleys were removed, rectifiers were installed on the bridges to transform the utility company AC voltage to DC voltage. Many of the historical movable bridges

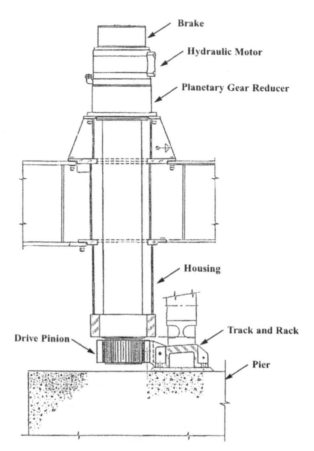

FIGURE 21.25   One of four hydraulic drives from the Coleman Bridge, York River, Virginia.

that are still operating on their original DC motors and drum switch/relay speed controls have these rectifiers. Most of the movable bridges that have been rehabilitated in recent years, but still retain the original DC motors, now have silicon controlled rectifier (SCR) controllers that use AC voltage input and produce a variable DC voltage output directly to the motor.

The most common service voltage for movable bridges is 480 Vac, 3 phase, although in some locations, the service is 240 or 208 Vac, 3 phase. Economics and the utility company policies are the primary determinant factors in what voltage is used. Electrical power, simplified, equals volts times amperes. Thus, for a given horsepower, the motor current at 480 V is one half of that at 240 V. The economics are obvious when one considers the motor frame size, motor controllers, electrical switching equipment, and conductors are all physically smaller for 480 V than for 240 V. However, some utility companies do not normally provide 480 V and they are not willing to maintain a single 480 V service without passing along substantial costs to the bridge owner. If these additional service costs exceed the savings of using 480-V motors and controls, 240-V service becomes more attractive.

The choice of one voltage over the other has no bearing on the cost of power for a movable bridge. Power is power and the rate per kilowatt-hour is the same regardless of voltage. A service cost factor that is sometimes overlooked is the demand charge that utility companies impose on very large intermittent loads. These charges are to offset the utility company cost of reserving power generation and transmission capability to serve the demands of a facility that is normally not online. These charges are based on the peak load, measured at the meter, over a period of, typically, 15 min. The charges are amortized over the year following the last highest reading and added to the billing for the actual amount of electrical power used. In the case of a bridge that has a very high power demand, even if it is opened only once or twice a year, the annual electrical costs are very

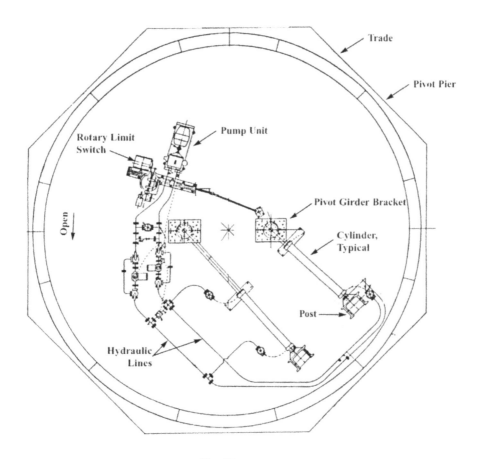

Plan View

FIGURE 21.26 Hydraulic drive for a swing span using hydraulic cylinders.

high because the owner has to pay for the demand capacity whether it is used or not. Referring to the earlier discussion on counterweights, it is very important that the design of the bridge is such that it is as energy efficient as possible.

Both AREA [3] and AASHTO [4] require a movable bridge to have an emergency means of operation should primary power be lost. Most bridges are designed with a backup engine driven generator and operate the bridge on the normal electrical motor drives. For safety and reliability, diesel engines are preferred by most bridge owners. Hand operation can be provided as backup for auxiliary devices such as locks, gates, and wedges.

However, there are many different types of backup systems, such as the following:

1. Internal combustion engines or air motors on emergency machinery that can be engaged when needed;
2. Smaller emergency electrical motors on emergency machinery to reduce the size of the emergency generator;
3. A receptacle for a portable emergency generator to reduce the capital investment for emergency power for several bridges, as well as other municipality–owned facilities.

21.5.3 Bridge Control

The predominant control system in use in newly constructed or rehabilitated movable bridges is the programmable logic controller (PLC). This is a computer-based system that has been adapted

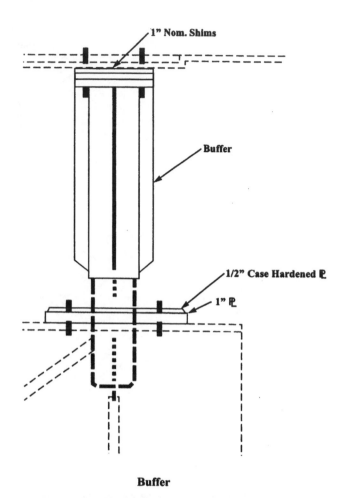

Buffer

FIGURE 21.27 Typical air buffer.

from other industrial-type applications. The PLC offers the ability to automate the operation of a bridge completely. However, most agencies have used the PLC as a replacement for a relay-based system to reduce the cost of initial construction and to reduce the space required for the control system. Other common applications for the PLC include generation of alarm messages to help reduce time in troubleshooting and maintenance of the systems.

As an example of their widespread use, the New Jersey Department of Transportation has PLCs on all of its bridges and has a proactive training program for its operations and technical staff. However, not all states are using PLCs, as the Florida and Washington State Departments of Transportation are now returning to the relay-based systems because they do not have the technical staff to maintain the PLC.

A more recent development is the use of PLCs for remote operation. For example, the city of Milwaukee, Wisconsin has several bridges that are controlled remotely by means of computer modem links and closed-circuit TV. This reduces the staff to one tender per three bridges. The potential liability of this type of system needs to be carefully evaluated as the bridge operator may not be able to observe adequately all parts of the bridge when operating the span.

Environmental regulations have made the installation permits for submarine cables difficult to obtain. PLC and radio modems have been used in several states to replace the control wiring that would otherwise be in a submarine cable.

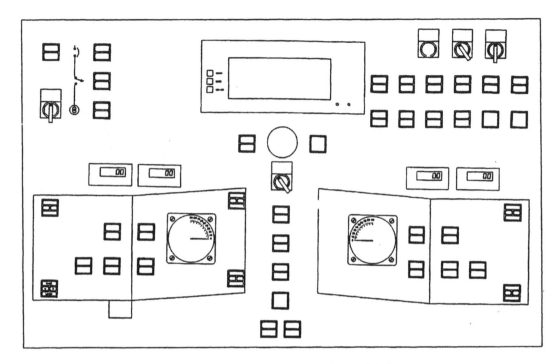

FIGURE 21.28   Layout of a typical control desk.

The selection of a drive system is performance oriented. Reliability and cost are key issues. The most common drives for movable bridges over that past 80 years have been DC and wound rotor AC motors with relays and drum switches. These two technologies remain the most common today although there have been many advances in DC and AC motor controls and the old systems are being rapidly replaced with solid-state drives.

The modern DC drives on movable bridges are digitally controlled, fully regenerative, four-quadrant, SCR motor controllers. In more general terms, this is a solid-state drive that provides infinitely variable speed and torque control in both forward and reverse directions. They have microprocessor programming that provides precise adjustment of operating parameters, and once a system is set up, it rarely needs to be adjusted. Programmable parameters include acceleration, deceleration, preset speeds, response rate, current/torque limit, braking torque, and sequence logic. This type of drive has been proved to provide excellent speed and torque control for bridge-operating conditions.

The wound rotor motor drive technology has also moved into digital control. The new SCR variable voltage controllers are in essence crane control systems. While they are not quite as sophisticated as the DC drives, they have similar speed and torque control capabilities. Most of the movable bridge applications have been retrofitted using the existing motors.

Adjustable frequency controllers (AFC) control speed by varying the frequency of the AC voltage and current to a squirrel cage induction motor. This type of drive has been used on movable bridges with some success but it is not well suited for this type of application. There are two primary reasons. First, this type of drive was designed for the control of pumps and fans, not high-inertia loads. Second, at low speeds, it does not provide sufficient braking torque to maintain control of an overhauling load. This is a significant concern when seating a span with an ice and snow load.

The first flux vector-controled AFC has been in use on a movable bridge for approximately 6 years now. It is a somewhat sophisticated drive system that controls magnetic flux to create slip artificially and thus control torque at any speed including full-rated motor torque at zero speed.

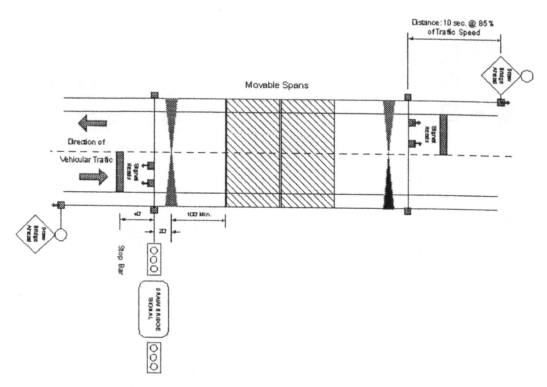

FIGURE 21.29   Typical layout of movable bridge signals and gates.

The drive controller uses input from a digital shaft encoder to locate the motor rotor position and then calculates how much voltage and current to provide to each motor lead. The drive is capable of 100% rated torque at zero speed which gives it excellent motion control at low speeds.

21.6 Traffic Control

Rail traffic control for movable railroad bridges involves interlocking the railroad signal system with the bridge-operating controls. For a movable bridge that is on a rail line that has third rail or catenary power, the interlocking must include the traction power system. In principle, the interlocking needs to be designed so that the railroad signals indicate that the track is closed and the power is deenergized prior to operating the span. However, the particulars of how this is accomplished depends upon the railroad in question and will not be addressed here. For a movable highway bridge, highway traffic control is governed by the AASHTO Movable Highway Bridge Specifications [4], as well as the Manual for Uniform Traffic Control Devices (MUTCD) [8]. Each owner may impose additional requirements but the Manual is typically used in the United States. As a minimum this will include a DRAWBRIDGE AHEAD warning sign, traffic signal, warning (or roadway) gates, and usually resistance (or barrier) gates. One possible arrangement is shown in Figure 21.29 for a two-leaf bascule bridge, note that there are no resistance gates. AASHTO [4] requires that a resistance gate (positive barrier) be placed prior to a movable span opening except where the span itself, such as a bascule leaf, blocks the opening.

For marine traffic, navigation lighting must follow the requirements of the Bridge Permit as approved by the U.S. Coast Guard. The permit typically follows the Coast Guard requirements as found in the U.S. Code of Federal Regulations 33, Part 118, Bridge Lighting and Other Systems [9]. These regulations identify specific types and arrangements for navigation lights depending upon the type of movable bridge.

References

1. American Railway Engineering Association, *Manual of Railway Engineering*, 1922.
2. American Association of State Highway Officials, *Standard Specifications for Movable Highway Bridges*, 1938.
3. *AREA Manual for Railway Engineering*, Chapter 15, Steel Structures, Part 6, Movable Bridges, 1997.
4. AASHTO, *Standard Specifications for Movable Highway Bridges*, 1988.
5. AASHTO, *Standard Specifications for Highway Bridges*, 16th Ed., 1996.
6. Ecale, H., G. Brown, and P. Kocsis, Chicago type bascule balancing: a new technique, Technical Notes, *ASCE J. Struct. Div.*, 103(ST11), 2269–2272, November 1977.
7. Lu, Malvern, Jenkins, Allred and Biwas, Balancing of trunnion type bascule bridges, *ASCE J. Struct. Div.*, V. 108, (ST10), October 1982.
8. Federal Highway Administration, *Manual on Uniform Traffic Control Devices*.
9. U.S. Code of Federal Regulations 33, Part 118, Bridge Lighting and Other Systems.

22

Floating Bridges

M. Myint Lwin
*Washington State Department
of Transportation*

22.1 Introduction

Floating bridges are cost-effective solutions for crossing large bodies of water with unusual depth and very soft bottom where conventional piers are impractical. For a site where the water is 2 to 5 km wide, 30 to 60 m deep and there is a very soft bottom extending another 30 to 60 m, a floating bridge is estimated to cost three to five times less than a long-span fixed bridge, tube, or tunnel.

A modern floating bridge may be constructed of wood, concrete, steel, or a combination of materials, depending on the design requirements. A 124-m-long floating movable wood pontoon railroad bridge was built in 1874 across the Mississippi River in Wisconsin in the United States. It was rebuilt several times before it was abandoned. A 98-m-long wood floating bridge is still in service in Brookfield, Vermont. The present Brookfield Floating Bridge is the seventh replacement structure, and was built by the Vermont Agency of Transportation in 1978. The first 2018-m-long Lake Washington Floating Bridge in Seattle [1,2], Washington, was built of concrete and opened to traffic in 1940 (Figure 22.1). Since then, three more concrete floating bridges have been built [2,3]. These concrete floating bridges form major transportation links in the state and interstate highway systems in Washington State. The Kelowna Floating Bridge on Lake Okanagan in British Columbia, Canada [4], was built of concrete and opened to traffic in 1958. It is 640 m long and carries two lanes of traffic. The 1246-m-long Salhus Floating Bridge and the 845-m-long Bergsoysund Floating Bridge in Norway (Figure 22.2) were constructed of concrete pontoons and steel superstructures [5]. They were opened to traffic in the early 1990s.

FIGURE 22.1   First Lake Washington floating bridge.

Washington State's experience has shown that reinforced and prestressed concrete floating bridges are cost-effective, durable, and low in maintenance as permanent transportation facilities. Concrete is highly corrosion resistant in a marine environment when properly designed, detailed, and constructed. Concrete is a good dampening material for vibration and noise, and is also far less affected by fire and heat than wood, steel, or other construction materials.

22.2 Basic Concept

The concept of a floating bridge takes advantage of the natural law of buoyancy of water to support the dead and live loads. There is no need for conventional piers or foundations. However, an anchoring or structural system is needed to maintain transverse and longitudinal alignments of the bridge.

A floating bridge is basically a beam on an elastic foundation and supports. Vertical loads are resisted by buoyancy. Transverse and longitudinal loads are resisted by a system of mooring lines or structural elements.

The function of a floating bridge is to carry vehicles, trains, bicycles, and pedestrians across an obstacle — a body of water. Inasmuch as a floating bridge crosses an obstacle, it creates an obstacle for marine traffic. Navigational openings must be provided for the passage of pleasure boats, smaller water crafts, and large vessels. These openings may be provided at the ends of the bridge. However, large vessels may impose demands for excessive horizontal and vertical clearances. In such cases, movable spans will have to be provided to allow the passage of large vessels. The Hood Canal Floating Bridge in Washington State has a pair of movable spans capable of providing a total of 183 m of horizontal clearance (Figure 22.3). Opening of the movable spans for marine traffic will cause interruption to vehicular traffic. Each interruption may be as long as 20 to 30 min. If the frequency of openings is excessive, the concept of a floating bridge may not be appropriate for the site. Careful consideration should be given to the long-term competing needs of vehicular traffic and marine traffic before the concept of a highway floating bridge is adopted.

FIGURE 22.2   The Bergsoysund floating bridge.

FIGURE 22.3   Movable spans for large vessels.

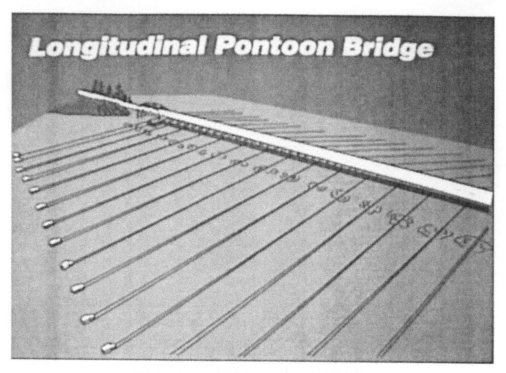

FIGURE 22.4   Continuous pontoon-type structure.

22.3   Types

22.3.1   Floating Structure

Floating bridges have been built since time immemorial. Ancient floating bridges were generally built for military operations [6]. All of these bridges took the form of small vessels placed side by side with wooden planks used as a roadway. Subsequently, designers added openings for the passage of small boats, movable spans for the passage of large ships, variable flotation to adjust for change in elevations, and so on.

Modern floating bridges generally consist of concrete pontoons with or without an elevated superstructure of concrete or steel. The pontoons may be reinforced concrete or prestressed concrete posttensioned in one or more directions. They can be classified into two types, namely, the continuous pontoon type and the separate pontoon type. Openings for the passage of small boats and movable spans for large vessels can be incorporated into each of the two types of modern floating bridges.

A continuous pontoon floating bridge consists of individual pontoons joined together to form a continuous structure (Figure 22.4). The size of each individual pontoon is based on the design requirements, the construction facilities, and the constraints imposed by the transportation route. The top of the pontoons may be used as a roadway or a superstructure may be built on top of the pontoons. All the present floating bridges in Washington State are of the continuous pontoon floating bridge type.

A separate pontoon floating bridge consists of individual pontoons placed transversely to the structure and spanned by a superstructure of steel or concrete (Figure 22.5). The superstructure must be of sufficient strength and stiffness to maintain the relative position of the separated pontoons. The two floating bridges in Norway are of the separate pontoon floating bridge type.

Both types of floating structures are technically feasible and relatively straightforward to analyze. They can be safely designed to withstand gravity loads, wind and wave forces, and extreme events,

FIGURE 22.5   Separate pontoon-type structure.

such as vessel collisions and major storms. They perform well as highway structures with a high-quality roadway surface for safe driving in most weather conditions. They are uniquely attractive and have low impact on the environment. They are very cost-effective bridge types for water crossing where the water is deep (say, over 30 m) and wide (say, over 900 m), but the currents must not be very swift (say, over 6 knots), the winds not too strong (say, average wind speed over 160 km/h), and the waves not too high (say, significant wave height over 3 m).

22.3.2 Anchoring Systems

A floating structure may be held in place in many ways — by a system of piling, caissons, mooring lines and anchors, fixed guide structures, or other special designs. The most common anchoring system consists of a system of mooring lines and anchors. This system is used in all the existing floating bridges in Washington State. The mooring lines are galvanized structural strands meeting ASTM A586. Different types of anchors may be used, depending on the water depth and soil condition. Four types of anchors are used in anchoring the floating bridges on Lake Washington, Seattle.

Type A anchors (Figure 22.6) are designed for placement in deep water and very soft soil. They are constructed of reinforced concrete fitted with pipes for water jetting. The anchors weigh from 60 to 86 tons each. They are lowered to the bottom of the lake and the water jets are turned on allowing the anchors to sink into the soft lake bottom to embed the anchors fully. Anchor capacity is developed through passive soil pressure.

Type B anchors (Figure 22.7) are pile anchors designed for use in hard bottom and in water depth less than 27 m. A Type B anchor consists of two steel H-piles driven in tandem to a specified depth. The piles are tied together to increase capacity.

Type C anchors (Figure 22.8) are gravity-type anchors, constructed of reinforced concrete in the shape of a box with an open top. They are designed for placement in deep water where the soil is too hard for jetting. The boxes are lowered into position and then filled with gravel to the specified weight.

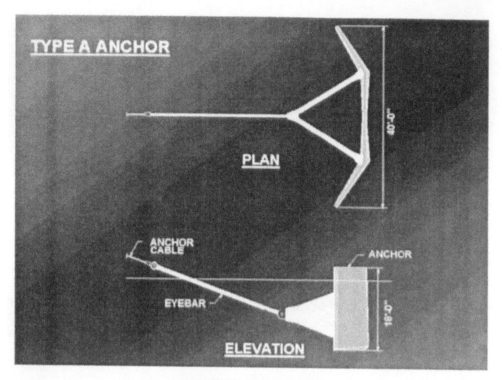

FIGURE 22.6   Type A anchor.

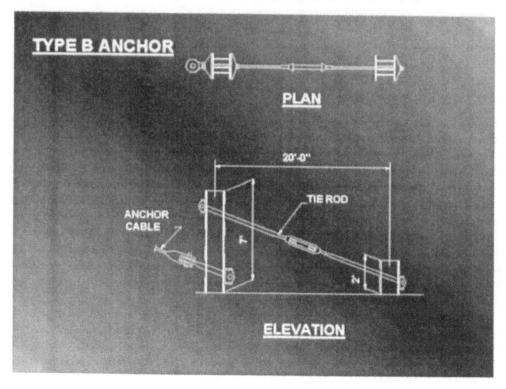

FIGURE 22.7   Type B anchor.

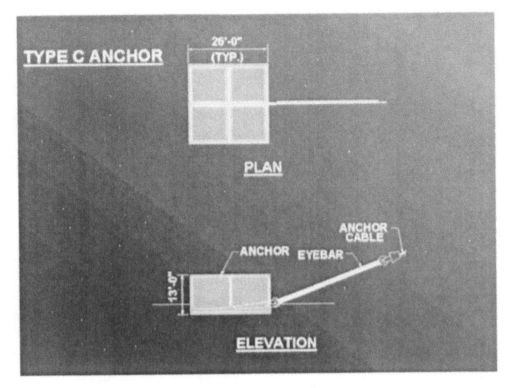

FIGURE 22.8 Type C anchor.

Type D (Figure 22.9) anchors are also gravity-type anchors like the Type C anchors. They consist of solid reinforced concrete slabs, each weighing about 270 tons. They are designed for placement in shallow and deep water where the soil is too hard for water jetting. The first slab is lowered into position and then followed by subsequent slabs. The number of slabs is determined by the anchor capacity required. Type D anchors are the choice over Type B and Type C anchors, because of the simplicity in design, ease in casting, and speed in placement.

22.4 Design Criteria

The design of a floating bridge follows the same good engineering practices as for land-based concrete or steel bridges. The design and construction provisions stipulated in the AASHTO *Standard Specifications for Highway Bridges* [7] or the *LRFD Bridge Design Specifications* [8] are applicable and should be adhered to as much as feasible. However, due to the fact that a floating bridge is floating on fresh or marine waters, the design criteria must address some special conditions inherent in floating structures. The performance of a floating bridge is highly sensitive to environmental conditions and forces, such as winds, waves, currents, and corrosive elements. The objectives of the design criteria are to assure that the floating bridge will

- Have a long service life of 75 to 100 years with low life-cycle cost;
- Meet functional, economical, and practical requirements;
- Perform reliably and be comfortable to ride on under normal service conditions;
- Sustain damage from accidental loads and extreme storms without sinking;
- Safeguard against flooding and progressive failure.

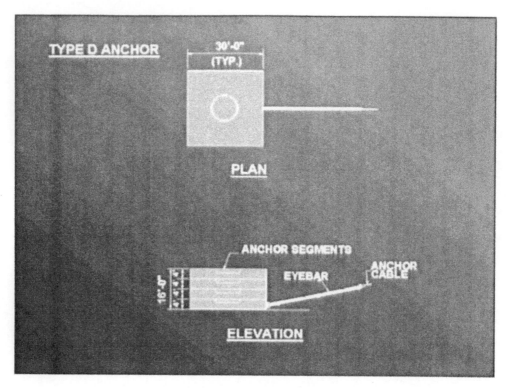

FIGURE 22.9   Type D anchor.

22.4.1   Loads and Load Combinations

The structure should be proportioned in accordance with the loads and load combinations for service load design and load factor design outlined in the AASHTO *Standard Specifications for Highway Bridges* or the AASHTO *LRFD Bridge Design Specifications*, except the floating portion of the structure shall recognize other environmental loads and forces and modify the loads and load combinations accordingly.

Winds and waves are the major environmental loads, while currents, hydrostatic pressures, and temperatures also have effects on the final design. Depending on the site conditions, other loadings, such as tidal variations, marine growth, ice, drift, etc., may need to be considered.

22.4.2   Winds and Waves

Winds and waves exert significant forces on a floating structure (Figure 22.10). Yet these environment loads are the most difficult to predict. Generally, there is a lack of long-term climatological data for a bridge site. A long record of observations of wind data is desirable for developing more accurate design wind speeds and wave characteristics. It is advisable to install instruments on potential bridge sites to collect climatological data as early as possible.

Wind blowing over water generates a sea state that induces horizontal, vertical, and torsional loads on the floating bridge. These loads are a function of wind speed, wind direction, wind duration, fetch length, channel configuration, and depth. Consideration must be given to the normal and extreme storm wind and wave conditions for the site. The normal storm conditions are defined as the storm conditions that have a mean recurrence interval of 1 year, which is the maximum storm that is likely to occur once a year. The extreme storm conditions are defined as the storm conditions that have a mean recurrence interval of 100 years, which is the maximum storm that is likely to occur once in 100 years. These wind and wave forces may be denoted by

FIGURE 22.10   Wind and wave forces (figure computer generated).

> WN = Normal Wind on Structure — 1-year Storm
> NW = Normal Wave on Structure — 1-year Storm
> WS = Extreme Storm Wind on Structure — 100-year Storm
> SW = Extreme Storm Wave on Structure — 100-year Storm

The following modifications are recommended for the AASHTO Load Combinations where wind loads are included:

1. Substitute WS + SW for W, and WN + NW for 0.3W.
2. Use one half the temperature loads in combination with WS and SW.
3. Omit L, I, WL, and LF loads when WS and SW are used in the design.

A 20-year wind storm condition is normally used to make operational decisions for closing the bridge to traffic to ensure safety and comfort to the traveling public. This is especially important when there is excessive motion and water spray over the roadway. When there is a movable span in the floating bridge for providing navigation openings, a 20-year wind storm is also used to open the movable span to relieve the pressures on the structure.

Following is an example of a set of wind and wave design data:

Return Interval	Wind Speed (1-min average)	Significant Wave Height	Period
1-year wind storm	76 km/h	0.85 m	3.23 s
20-year wind storm	124 km/h	1.55 m	4.22 s
100-year wind storm	148 km/h	1.95 m	4.65 s

22.4.3 Potential Damage

A floating bridge must have adequate capacity to safely sustain potential damages (DM) resulting from small vessel collision, debris or log impact, flooding, and loss of a mooring cable or component. Considering only one damage condition and location at any one time, the pontoon structure must be designed for at least the following:

1. *Collision:* Apply a 45-kN horizontal collision load as a service load to the pontoon exterior walls. Apply a 130-kN horizontal collision load as a factored load to the pontoon exterior wall. The load may be assumed to be applied to an area no greater than 0.3×0.3 m.
2. *Flooding:*
   - Flooding of any two adjacent exterior cells along the length of the structure
   - Flooding of all cells across the width of the pontoon
   - Flooding of all the end cells of an isolated pontoon during towing
   - Flooding of the outboard end cells of a partially assembled structure
3. *Loss of a mooring cable or component.*
4. *Complete separation of the floating bridge by a transverse or diagonal fracture.* This condition should apply to the factored load combinations only.

The above potential DM loadings should be combined with the AASHTO Groups VI, VIII, and IX combinations for Service Load Design and Load Factor Design. If the AASHTO LRFD Bridge Design Specifications are used in the design, the loads from Items 2 to 4 may be considered as extreme events.

Every floating structure is unique and specific requirements must be established accordingly. Maritime damage criteria and practices, such as those for ships and passenger vessels, should be reviewed and applied where applicable in developing damage criteria for a floating structure. However, a floating bridge behaves quite differently than a vessel, in that structural restraint is much more dominant than hydrostatic restraint. The trim, list, and sinkage of the flooded structure are relatively small. With major damage, structural capacity is reached before large deformations occurred or were observed. This is an important fact to note when comparing with stability criteria for ships.

22.4.4 Control Progressive Failure

While water provides buoyancy to keep a floating bridge afloat, water leaking into the interior of a floating bridge can cause progressive failure, eventually sinking the bridge. Time is of the essence when responding to damage or flooding. Maintenance personnel must respond to damage of a floating bridge quickly, especially when water begins to leak into the structure. An electronic cell monitoring system with water sensors to detect water entry and provide early warning to the maintenance personnel should be installed to assure timely emergency response. A bilge piping system should also be installed in the bridge for pumping out water.

It is important to control progressive failure in a floating bridge caused by flooding resulting from structural damage. Damage to the floating bridge could occur from a wind storm, a collision by a boat, severing of mooring lines, or other unforeseen accidents. The interior of the pontoons should be divided into small watertight compartments or cells (Figure 22.11) to confine flooding to only a small portion of the bridge. Access openings in the exterior or interior wall or bulkheads shall be outfitted with watertight doors.

Water sensors may be installed in each watertight compartment for early detection and early warning of water entry. A bilge piping system may be installed in the compartments for pumping out water when necessary. In such cases, pumping ports and quick disconnect couplings should be provided for pumping from a boat or vehicle equipped with pumps.

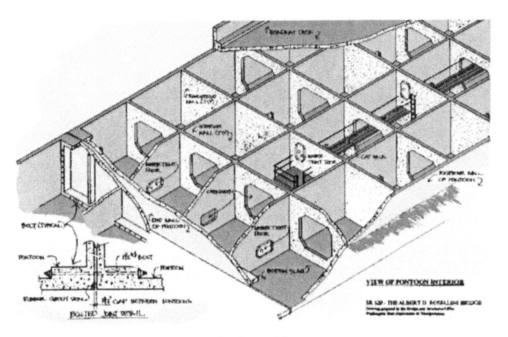

FIGURE 22.11   Small watertight compartments.

22.4.5   Design of Concrete Members

The design of reinforced concrete members should be based on behavior at service load conditions as per AASHTO Standard Specifications, or the service limit state as per AASHTO LRFD Bridge Design Specifications; except sections where reinforcement is to resist sustained hydrostatic forces the allowable stress in the reinforcing steel should not exceed 97 Mpa to limit crack width to a maximum of 0.10mm.

Prestressed members should be designed under the applicable service load and load factor provisions in the AASHTO Standard Specifications or the limit states provisions in the AASHTO LRFD Bridge Design Specifications, except the allowable concrete tensile stress under final conditions in the precompressed tensile zone should be limited to zero.

The ultimate flexural strength of the overall pontoon section should be computed for a maximum crack width of 0.25 mm and should not be less than the loads from the factored load combinations or 1.3 times the cracking moment, M_{cr}.

In a moderate climate, the following temperature differentials between the various portions of a floating bridge may be used:

1. Between the exposed portion and the submerged portion of a pontoon: ±19°C. The exposed portion may be considered as the top slab, and the remaining part of the pontoon as submerged. If the top slab is shaded by an elevated structure, the differential temperature may be reduced to ±14°C

2. Between the top slab and the elevated structure of the pontoon: ±14°C

The effects of creep and shrinkage should be considered while the pontoons are in the dry only. Creep and shrinkage may be taken as zero once the pontoons are launched. The time-dependent effects of creep and shrinkage may be estimated in accordance with the AASHTO Specifications. A final differential shrinkage coefficient of 0.0002 should be considered between the lower portion of the pontoon and the top slab of the pontoon.

High-performance concrete containing fly ash and silica fume is most suitable for floating bridges [9,10]. The concrete is very dense, impermeable to water, highly resistant to abrasion, and relatively

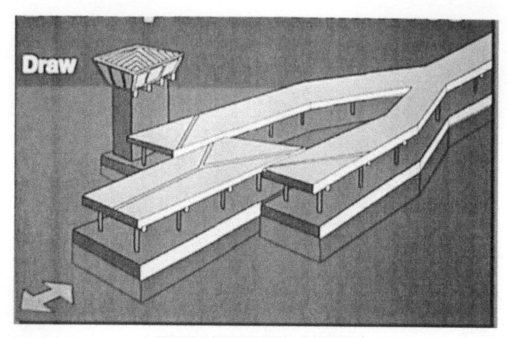

FIGURE 22.12 Draw-type movable span.

crack free. High-performance concrete also has high strength, low creep, and low shrinkage. Concrete mixes may be customized for the project.

Recommended minimum concrete cover of reinforcing steel:

	Fresh Water	Salt Water
Top of roadway slab	65 mm	65 mm
Exterior surfaces of pontoons and barrier	38 mm	50 mm
All other surfaces	25 mm	38 mm

22.4.6 Anchoring System

An anchoring system should be installed in the floating bridge to maintain transverse and horizontal alignment. The anchoring system should be designed to have adequate capacity to resist transverse and longitudinal forces from winds, waves, and current.

Adequate factors of safety or load and resistance factors consistent with the type of anchoring should be included in the design of the components of the system.

22.4.7 Movable Span

A floating bridge creates an obstruction to marine traffic. Movable spans may need to be provided for the passage of large vessels. The width of opening that must be provided depends on the size and type of vessels navigating through the opening. Movable spans of up to a total opening of 190 m have been used.

Two types of movable spans are used in Washington State — the draw-type and the lift/draw type. In the draw type movable span, the draw pontoons retract into a "lagoon" formed by flanking pontoons (Figure 22.12). Vehicles must maneuver around curves at the "bulge" where the "lagoon" is formed. In the lift/draw type of movable span, part of the roadway will be raised for the draw pontoons to retract underneath it (Figure 22.13). As far as traffic safety and flow are concerned, the lift/draw-type movable span is superior over the draw type. Traffic moves efficiently on a straight alignment with no curves to contend with.

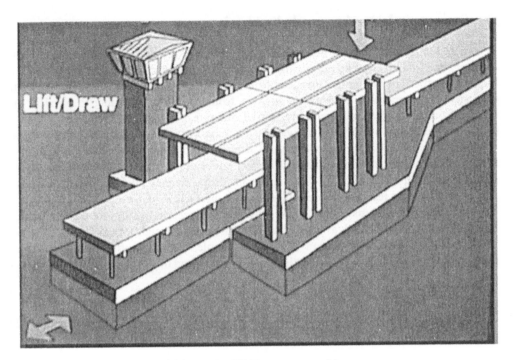

FIGURE 22.13   Lift/Draw-type movable span.

Movable spans may be operated mechanically or hydraulically. The design of the movable spans should be in accordance with the latest AASHTO *Standard Specifications for Movable Highway Bridges* [11].

22.4.8 Deflection and Motion

Floating bridges should be designed so that they are comfortable to ride on during normal storm (1-year storm) conditions and also to avoid undesirable structural effects during extreme storm (100-year storm) conditions. Deflection and motion criteria have been used to meet these objectives. The following deflection and motion limits for normal storm (1-year storm) conditions may be used as guidelines:

Loading Condition	Type of Deflection or Motion	Maximum Deflection	Maximum Motion
Vehicular load	Vertical	$L/800$	
Winds — static	Lateral (drift)	0.3 m	
	Rotation (heel)	0.5°	
Waves — dynamic	Vertical (heave)	±0.3 m	0.5 m/s²
	Lateral (sway)	±0.3 m	0.5 m/s²
	Rotation (roll)	±0.5°	0.05 rad/s²

The objective of the motion limits are to assure that the people will not experience discomfort walking or driving across the bridge during a normal storm. The motion limit for rotation (roll) under the dynamic action of waves should be used with care when the roadway is elevated a significant distance above the water surface. The available literature contains many suggested motion criteria for comfort based on human perceptions [12]. A more-detailed study on motion criteria may be warranted for unusual circumstances.

22.5 Structural Design and Analysis

The design and analysis of floating bridges have gone through several stages of progressive development since the first highway floating bridge was designed and built in the late 1930s across Lake Washington, Seattle. The design has advanced from empirical methods to realistic approach, from the equivalent static approach to dynamic analysis, from computer modeling to physical model testing, and from reinforced concrete to prestressed concrete.

The most difficult part of early designs was the prediction of winds and waves, and the response due to wind–wave–structure interaction. Climatological data were very limited. The wind–wave–structure interaction was not well understood. Current state of the knowledge in atmospheric sciences, computer science, marine engineering, finite-element analysis, structural engineering, and physical model testing provides more accurate prediction of wind and wave climatology, more realistic dynamic analysis of wind–wave–structure interactions, and more reliable designs.

Designing for static loads, such as dead and live loads, is very straightforward using the classical theory on beam on elastic foundation [13]. For example, the maximum shear, moment, and deflection due to a concentrated load, P, acting away from the end of a continuous floating structure are given by

$$V_{max} = \frac{P}{2} \tag{22.1}$$

$$M_{max} = \frac{P}{4\lambda} \tag{22.2}$$

$$y_{max} = \frac{P\lambda}{2k} \tag{22.3}$$

where k = modulus of foundation

$$\lambda = 4\sqrt{\frac{k}{4EI}}$$

Designing for the response of the structure to winds and waves is more complex, because of the random nature of these environment loads. To determine the dynamic response of the bridge to wind generated waves realistically, a dynamic analysis is necessary.

22.5.1 Preliminary Design

The design starts with selecting the type, size, and location of the floating bridge (Figure 22.14). Assuming a concrete box section of cellular construction with dimensions as shown in Figure 22.15, the first step is to determine the freeboard required. The height of the freeboard is selected to avoid water spray on the roadway deck from normal storms. The draft can then be determined as necessary to provide the selected freeboard.

The freeboard and draft of the floating structure should be calculated based on the weight of concrete, weight of reinforcing steel, weight of appurtenances, weight of marine growth as appropriate, and vertical component of anchor cable force. The weight of the constructed pontoon is generally heavier than the computed weight, because of form bulging and other construction tolerances. Based on the experience in Washington State, the weight increase varies from 3 to 5% of the theoretical weight. This increase should be included in the draft calculation. Additionally, floating pontoons experience loss in freeboard in the long term. The main reason is due to weight

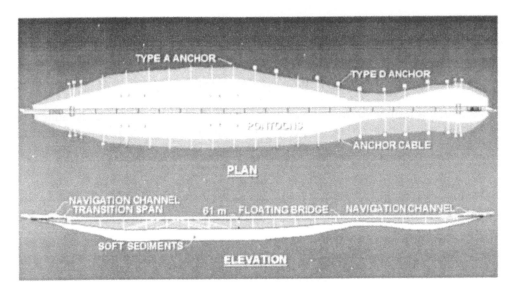

FIGURE 22.14 General layout.

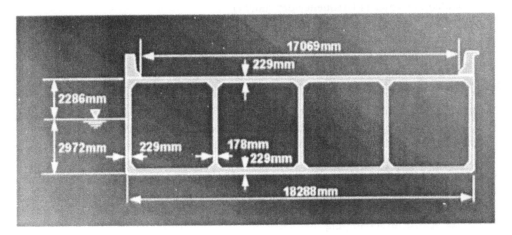

FIGURE 22.15 Typical cross section.

added as a result of modifications in the structural and mechanical elements throughout the service life of the structure. It is prudent to make allowance for this in the design. This can be done by allowing 150 to 230 mm of extra freeboard in the new bridge.

The thicknesses for the walls and slabs should be selected for local and global strength and constructibility. The wall thickness should be the minimum needed to provide adequate concrete cover to the reinforcing steel and adequate space for post-tensioning ducts when used. There must also be adequate room for depositing and consolidating concrete. The objective should be to keep the weight of the structure to a minimum, which is essential for cost-effective design of a floating bridge.

The exterior walls of the pontoons should be designed for wave plus hydrostatic pressures and the collision loads. The bottom slab should be designed for wave plus hydrostatic pressures. The interior walls should be designed for hydrostatic pressures due to flooding of a cell to full height of the wall. The roadway slab should be designed for the live load plus impact in the usual way.

The preliminary design gives the overall cross-sectional dimensions and member thicknesses required to meet local demands and construction requirements. The global responses of the floating bridge will be predicted by dynamic analysis.

22.5.2 Dynamic Analysis

The basic approach to dynamic analysis is to solve the equation of motion:

$$M\ddot{X} + C\ddot{X} + KX = F(t) \tag{22.4}$$

This equation is familiar to structural engineering in solving most structural dynamic problems of land-based structures. However, in predicting the dynamic response of a floating bridge, the effects of water–structure interaction must be accounted for in the analysis. As a floating bridge responds to the incident waves, the motions (heave, sway, and roll) of the bridge produce hydrodynamic effects generally characterized in terms of added mass and damping coefficients. These hydrodynamic coefficients are frequency dependent. The equation of motion for a floating structure takes on the general form:

$$[M + A]\ddot{X} + [C_1 + C_2]\dot{X} + [K + k]X = F(t) \tag{22.5}$$

where
$X, \dot{X}, \ddot{X}$ = generalized displacement, velocity and acceleration at each degree of freedom
M = mass–inertia matrix of the structure
A = added mass matrix (frequency dependent)
C_1 = structural damping coefficient
C_2 = hydrodynamic damping coefficient (frequency dependent)
K = structural stiffness matrix (elastic properties, including effects of mooring lines when used)
k = hydrostatic stiffness (hydrostatic restoring forces)
$F(t)$ = forces acting on the structure

A substantial amount of experimental data has been obtained for the hydrodynamic coefficients for ships and barges [14,15]. Based on these experimental data, numerical methods and computer programs have been developed for computing hydrodynamic coefficients of commonly used cross-sectional shapes, such as the rectangular shape. For structural configurations for which no or limited data exist, physical model testing will be necessary to determine the basic sectional added mass, damping, and wave excitation loads.

Structural damping is an important source of damping in the structure. It significantly affects the responses. A structural damping coefficient of 2 to 5% of critical damping is generally assumed for the analysis. It is recommended that a better assessment of the damping coefficient be made to better represent the materials and structural system used in the final design.

The significant wave height, period, and central heading angle may be predicted using the public domain program NARFET developed by the U.S. Army Coastal Engineering Research Center [16,17]. This program accounts for the effective fetch to a location on the floating structure by a set of radial fetch lines to the point of interest. The Joint North Sea Wave Project (JONSWAP) spectrum is commonly used to represent the frequency distribution of the wave energy predicted by the program NARFET. This spectrum is considered to represent fetch-limited site conditions very well. A spreading function is used to distribute the energy over a range of angles of departure from the major storm heading to the total energy [18]. The spreading function takes the form of an even cosine function, $\cos^{2n}\theta$, where θ is the angle of the incident wave with respect to the central heading angle. Usually $2n$ is 2 or greater. $2n = 2$ is generally used for ocean structures where the structures are relatively small with respect to the open sea. In the case of a floating bridge, the bridge length is very large in comparison to the body of water, resulting in very little energy distributed away from the central heading angle. A larger number of $2n$ will have to be used for a floating bridge. The larger the number of $2n$ the more focused the wave direction near the central heading angle. A 2*n* value of 12 to 16 have been used in analyzing the floating bridges in Washington State. The value of 2*n* should be selected with care to reflect properly the site condition and the wind and wave directions.

The equation of motion may be solved by the time-domain (deterministic) analysis or the frequency-domain (probabilistic) analysis. The time-domain approach involves solving differential equations when the coefficients are constants. The equations become very complex when the coefficients are frequency dependent. This method is tedious and time-consuming. The frequency-domain approach is very efficient in handling constant and frequency-dependent coefficients. The equations are algebraic equations. However, time-dependent coefficients are not admissible and nonlinearities will have to be linearized by approximation. For the dynamic analysis of floating bridges subjected to the random nature of environmental forces and the frequency-dependent hydrodynamic coefficients, frequency-domain analysis involves only simple and fast calculations.

22.5.3 Frequency-Domain Analysis

The frequency-domain dynamic analysis is based on the principles of naval architecture and the strip theory developed for use in predicting the response of ships to sea loads [19,20]. The essence of this approach is the assumption that the flow at one section through the structure does not affect the flow at any other section. Additional assumptions are (1) the motions are relatively small, (2) the fluid is incompressible and invisid, and (3) the flow is irrotational. By using the strip theory, the problem of wave–structure interaction can be solved by applying the equation of motion in the frequency domain [21,22]. By Fourier transform, the equation of motion may be expressed in terms of frequencies, ω, as follows:

$$\left\{-\omega^2[M+A]+i\omega[C_1+C_2]+[K+k]\right\}=\left\{F(\omega)\right\} \tag{22.6}$$

This equation may be solved as a set of algebraic equations at each frequency and the responses determined. The maximum bending moments, shears, torsion, deflections, and rotations can then be predicted using spectral analysis and probability distribution [23,24]. The basic steps involved in a frequency-domain analysis are

1. Compute the physical properties of the bridge — geometry of the bridge elements, section properties, connections between bridge elements, mass–inertia, linearized spring constants, structural damping, etc.
2. Compute hydrodynamic coefficients — frequency-dependent added mass and frequency-dependent damping.
3. Compute hydrostatic stiffnesses.
4. Calculate wind, wave, and current loads, and other loading terms.
5. Build a finite-element computer model of the bridge [25] as a collection of nodes, beam elements, and spring elements. The nodes form the joints connecting the beam elements and the spring elements, and each node has six degrees of freedom.
6. Solve the equation of motion in the frequency domain to obtain frequency responses, the magnitudes of which are referred to as response amplitude operators (RAOs).
7. Perform spectral analysis, using the RAOs and the input sea spectrum, to obtain the root mean square (RMS) of responses.
8. Perform probability analysis to obtain the maximum values of the responses with the desired probability of being exceeded.
9. Combine the maximum responses with other loadings, such as wind, current, etc., for final design.

22.6 Fabrication and Construction

Concrete pontoons are generally used for building major floating bridges. The fabrication and construction of the concrete pontoons must follow the best current practices in structural and marine engineering in concrete design, fabrication, and construction, with added emphasis on high-quality concrete and watertightness. Quality control should be the responsibility of the fabricators/contractors. Final quality assurance and acceptance should be the responsibility of the owners. In addition to these traditional divisions of responsibilities, the construction of a floating bridge necessitates a strong "partnership" arrangement to work together, contracting agencies and contractors, to provide full cooperation and joint training, share knowledge and expertise, share responsibility, and to help each other succeed in building a quality floating bridge. The contractors should have experience in marine construction and engage the services of naval architects or marine engineers to develop plans for monitoring construction activities and identifying flood risks, and prepare contingency plans for mitigating the risks.

Knowledge is power and safety. The construction personnel, including inspectors from the contracting agencies and the contractors, should be trained on the background of the contract requirements and the actions necessary to implement the requirements fully. Their understanding and commitment are necessary for complete and full compliance with contract requirements that bear on personal and bridge safety.

Construction of floating bridges is well established. Many concrete floating bridges have been built successfully using cast-in-place, precast, or a combination of cast-in-place and precast methods. Construction techniques are well developed and reported in the literature. Owners of floating bridges have construction specifications and other documents and guidelines for the design and construction of such structures.

Floating bridges may be constructed in the dry in graving docks or on slipways built specifically for the purpose. However, construction on a slipway requires more extensive preparation, design, and caution. The geometry and strength of the slipway must be consistent with the demand of the construction and launching requirements. Construction in a graving dock utilizes techniques commonly used in land-based structures. Major floating bridges around the world have been constructed in graving docks (Figure 22.16).

Because of the size of a floating bridge, the bridge is generally built in segments or pontoons compatible with the graving dock dimensions and draft restrictions. The segments or pontoons are floated and towed (Figure 22.17) to an outfitting dock where they are joined and completed in larger sections before towing to the bridge site, where the final assembly is made (Figure 22.18).

It is important to explore the availability of construction facilities and decide on a feasible facility for the project. These actions should be carried out prior to or concurrent with the design of a floating bridge to optimize the design and economy. Some key data that may be collected at this time are

- Length, width, and draft restrictions of the graving dock;
- Draft and width restrictions of the waterways leading to the bridge site;
- Wind, wave, and current conditions during tow to and installation at the job site.

The designers will use the information to design and detail the structural plans and construction specifications accordingly.

22.7 Construction Cost

The construction costs for floating bridges vary significantly from project to project. There are many variables that affect the construction costs. The following construction costs in U.S. dollars for the floating bridges in Washington State are given to provide a general idea of the costs of building

FIGURE 22.16 Construction in a graving dock.

FIGURE 22.17 Towing pontoon.

FIGURE 22.18   Final assembly.

concrete floating bridges in the past. These are original bid costs for the floating portions of the bridges. They have not been adjusted for inflation and do not include the costs for the approaches.

Name of Bridge	Length, m	Width of Pontoon, m	Lanes of Traffic	Cost, U.S. $ million
Original Lacey V. Murrow Bridge[a]	2018	17.9	4	3.25 (1938)
Evergreen Point Bridge[a]	2310	18.3	4	10.97 (1960)
Original Hood Canal Bridge[a]	1972	15.2	2	17.67 (1961)
Homer Hadley Bridge	1771	22.9	5	64.89 (1985)
New Lacey V. Murrow Bridge	2018	18.3	3	73.78 (1991)

[a] These bridges have movable spans which increased construction costs.

22.8   Inspection, Maintenance, and Operation

A floating bridge represents a major investment of resources and a commitment to efficiency and safety to the users of the structure and the waterway. To assure trouble-free and safe performance of the bridge, especially one with a movable span, an inspection, maintenance, and operation manual (Manual) should be prepared for the bridge. The main purpose of the Manual is to provide guidelines and procedures for regular inspection, maintenance, and operation of the bridge to extend the service life of the structure. Another aspect of the Manual is to define clearly the responsibilities of the personnel assigned to inspect, maintain, and operate the bridge. The Manual must address the specific needs and unique structural, mechanical, hydraulic, electrical, and safety features of the bridge.

The development of the Manual should begin at the time the design plans and construction specifications are prepared. This will assure that the necessary inputs are given to the designers to help with preparation of the Manual later. The construction specifications should require the contractors to submit documents, such as catalog cuts, schematics, electrical diagrams, etc., that will be included in the Manual. The Manual should be completed soon after the construction is completed and the bridge is placed into service. The Manual is a dynamic document. Lessons learned and modifications made during the service life of the structure should be incorporated into the Manual on a regular basis.

A training program should be developed for the supervisors and experienced co-workers to impart knowledge to new or inexperienced workers. The training program should be given regularly and aimed at nurturing a positive environment where workers help workers understand and diligently apply and update the guidelines and procedures of the Manual.

22.9 Closing Remarks

Floating bridges are cost-effective alternatives for crossing large lakes with unusual depth and soft bottom, spanning across picturesque fjords, and connecting beautiful islands. For conditions like Lake Washington in Seattle where the lake is over 1610 m wide, 61 m deep, and another 61 m of soft bottom, a floating bridge is estimated to cost three to five times less than a long-span fixed bridge, tube, or tunnel.

The bridge engineering community has sound theoretical knowledge, technical expertise, and practical skills to build floating bridges to enhance the social and economic activities of the people. However, it takes time to plan, study, design, and build floating bridges to form major transportation links in a local or national highway system. There are environmental, social, and economic issues to address. In the State of Washington, the first floating bridge was conceived in 1920, but was not built and opened to traffic until 1940. After over 30 years of planning, studies, and overcoming environmental, social, and economic issues, Norway finally has the country's first two floating bridges opened to traffic in 1992 and 1994. It is never too early to start the planning process and feasibility studies once interest and a potential site for a floating bridge is identified.

Every floating bridge is unique and has its own set of technical, environmental, social, and economic issues to address during preliminary and final engineering:

- *Winds and waves*: Predicting accurately the characteristics of winds and waves has been a difficult part of floating bridge design. Generally there is inadequate data. It is advisable to install instruments in potential bridge sites to collect climatological data as early as possible. Research in the area of wind–wave–structure interactions will assure safe and cost-effective structures.
- *Earthquake*: Floating bridges are not directly affected by ground shakings from earthquakes.
- *Tsunami and seiches*: These may be of particular significant in building floating bridges at sites susceptible to these events. The dynamic response of floating bridges to tsunami and seiches must be studied and addressed in the design where deemed necessary.
- *Corrosion*: Materials and details must be carefully selected to reduce corrosion problems to assure long service life with low maintenance.
- *Progressive failure*: Floating bridges must be designed against progressive failures by dividing the interiors of pontoons into small watertight compartments, by installing instruments for detecting water entry, and by providing means to discharge the water when necessary.
- *Riding comfort and convenience*: Floating bridges must be comfortable to ride on during minor storms. They must have adequate stiffness and stability. They must not be closed to vehicular traffic frequently for storms or marine traffic.
- *Public acceptance*: Public acceptance is a key part of modern civil and structural engineering. The public must be educated regarding the environmental, social, and economic impacts of

FIGURE 22.19   Floating Bridges on Lake Washington.

proposed projects. Reaching out to the public through community meetings, public hearings, news releases, tours, exhibits, etc. during the early phase of project development is important to gain support and assure success. Many major public projects have been delayed for years and years because of lack of interaction and understanding.

• *Design criteria*: The design criteria must be carefully developed to meet site-specific requirements and focused on design excellence and cost-effectiveness to provide long-term performance and durability. Design excellence and economy come from timely planning, proper selection of materials for durability and strength, and paying attention to design details, constructibility, and maintainability. The design team should include professionals with knowledge and experience in engineering, inspection, fabrication, construction, maintenance, and operation of floating bridges or marine structures.

• *Construction plan*: It is essential to have a good set of construction plans developed jointly by the contracting agency and the contractor to clearly address qualifications, materials control, quality control, quality assurance, acceptance criteria, post-tensioning techniques, repair techniques, launching and towing requirements, weather conditions, flood control and surveillance.

Well-engineered and maintained floating bridges are efficient, safe, durable, and comfortable to ride on. They form important links in major transportation systems in different parts of the world (Figure 22.19).

References

1. Andrew, C. E., The Lake Washington pontoon bridge, *Civil Eng.*, 9(12), 1939.
2. Lwin, M. M., Floating bridges — solution to a difficult terrain, in *Proceedings of the Conference on Transportation Facilities through Difficult Terrain*, Wu, J. T. H. and Barrett, R. K., Eds., A.A. Balkema, Rotterdam, 1993.
3. Nichols, C. C., Construction and performance of hood canal floating bridge, in *Proceedings of Symposium on Concrete Construction in Aqueous Environment*, ACI Publication SP-8, Detroit, MI, 1962.
4. Pegusch, W., The Kelowna floating bridge, in *The Engineering Journey*, the Engineering Institute of Canada, Canada, 1957.
5. Landet, E., Planning and construction of floating bridges in Norway, Proceedings of International Workshop on Floating Structures in Coastal Zone, Port and Harbour Research Institute, Japan, 1994.
6. Gloyd, C. S., Concrete floating bridges, *Concrete Int.*, 10(7), 1988.
7. AASHTO, *Standard Specifications for Highway Bridges*, 16th ed., AASHTO, Washington, D.C., 1996.
8. AASHTO, *LRFD Bridge Design Specifications*, AASHTO, Washington, D.C., 1994.
9. Lwin, M. M., Bruesch, A. W., and Evans, C. F., High performance concrete for a floating bridge, in *Proceedings of the Fourth International Bridge Engineering Conference*, Vol. 1, Federal Highway Administration, Washington, D.C., 1995.
10. Lwin, M. M., Use of high performance concrete in highway bridges in Washington State, in *Proceedings International Symposium on High Performance Concrete*, Prestressed Concrete Institute and Federal Highway Administration, New Orleans, 1997.
11. AASHTO, *Standard Specifications for Movable Highway Bridges*, AASHTO, Washington, D.C., 1988.
12. Bachman, H. and Amman, W., Vibrations in Structure Induced by Men and Machines, Structural Engineering Document No. 3e, International Association for Bridge and Structural Engineering, Zurich, Switzerland, 1987.
13. Hetenyi, M., *Beams on Elastic Foundation*, Ann Arbor, MI, 1979.
14. Frank, W., Oscillation of cylinders in or below the surface of deep fluids, Report No. 2375, Naval Ships Research and Development Center, 1967.
15. Garrison, C. J., Interaction of oblique waves with an infinite cylinder, *Appl. Ocean Res.*, 6(1), 1984.
16. *Shore Protection Manual*, Vol. 1, Coastal Engineering Research Center, Department of the Army, 1984.
17. Program NARFET, Waterways Experiment Station, Corps of Engineers, Coastal Engineering Research Center, Vicksburg, MI.
18. Mitsuyasu, H., Observations of the directional spectrum of ocean waves using a clover leaf buoy, *J. Phys. Oceanogr.*, Vol. 5, 1975.
19. Comstock, J., *Principles of Naval Architecture*, Society of Naval Architects and Marine Engineers, New York, 1975.
20. Salvesen, N., Tuck, E. O., and Faltinsen, O., Ship motions and sea loads, *Trans. Soc. Naval Architects Marine Eng.*, 78, 1970.
21. Engel, D. J. and Nachlinger, R. R., Frequency domain analysis of dynamic response of floating bridge to waves, in *Proceedings of Ocean Structural Dynamics Symposium*, Oregon State University, 1982.
22. Hutchison, B. L., Impulse response techniques for floating bridges and breakwaters subject to short-crested seas, *Marine Technol.*, 21(3), 1984.
23. Marks, W., *The Application of Spectral Analysis and Statistics to Seakeeping*, T&R Bulletin No. 1-24, Society of Naval Architects and Marine Engineers, New York, 1963.
24. Ochi, M. K., On prediction of extreme values, *J. Ship Res.*, 17(1), 1973.
25. Gray, D. L. and Hutchison, B. L., A resolution study for computer modeling of floating bridges, in *Proceedings of Ocean Structural Dynamics Symposium*, Oregon State University, 1986.

23

Railroad Bridges

Donald F. Sorgenfrei
Modjeski and Masters, Inc.

W. N. Marianos, Jr.
Modjeski and Masters, Inc.

23.1 Introduction

23.1.1 Railroad Network

The U.S. railroad network consists predominantly of privately owned freight railroad systems classified according to operating revenue, the government-owned National Railroad Passenger Corporation (Amtrak), and numerous transit systems owned by local agencies and municipalities.

Since the deregulation of the railroad industry brought about by the 1980 Staggers Act, there have been numerous railway system mergers. By 1997 there remained 10 Class I (major) Railroads, 32 Regional Railroads, and 511 Local Railroads operating over approximately 150,000 track miles. The 10 Class I Railroads comprise only 2% of the number of railroads in the United States but account for 73% of the trackage and 91% of freight revenue.

By far the present leading freight commodity is coal, which accounts for 25% of all the carloads. Other leading commodities in descending order by carloads are chemicals and allied products, farm products, motor vehicles and equipment, food and sundry products, and nonmetallic minerals. Freight equipment has drastically changed over the years in container type, size and wheelbase, and carrying capacity. The most predominant freight car is the hopper car used with an open top for coal loading and the covered hopper car used for chemicals and farm products. In more recent years special cars have been developed for the transportation of trailers, box containers, and automobiles. The

---

It should be noted that much of this material was developed for the American Railway Engineering and Maintenance of Way Association (AREMA) Structures Loading Seminar. This material is used with the permission of AREMA.

average freight car capacity (total number of freight cars in service divided by the aggregate capacity of those cars) has risen approximately 10 tons each decade with the tonnage ironically matching the decades, i.e., 1950s — 50 tons, 1960s — 60 tons, and so on. As the turn of the century approaches, various rail lines are capable of handling 286,000 and 315,000-lb carloads, often in dedicated units.

In 1929 there were 56,936 steam locomotives in service. By the early 1960s they were nearly totally replaced by diesel electric units. The number of diesel electric units has gradually decreased as available locomotive horsepower has increased. The earlier freight trains were commonly mixed freight of generally light railcars, powered by heavy steam locomotives. In more recent years that has given way to heavy railcars, unit trains of common commodity (coal, grain, containers, etc.) with powerful locomotives. Newer locomotives generally have six axles, weigh 420,000 lbs, and can generate up to 8000 Hp.

These changes in freight hauling have resulted in concerns for railroad bridges, many of which were not designed for these modern loadings. The heavy, steam locomotive with steam impact governed in design considerations. Present bridge designs are still based on the steam locomotive wheel configuration with diesel impact, but fatigue cycles from the heavy carloads are of major importance.

The railroad industry records annual route tonnage referred to as "million gross tons" (MGT). An experienced railroader can fairly well predict conditions and maintenance needs for a route based on knowing the MGT for that route. It is common for Class I Railroads to have routes of 30 to 50 MGT with some coal routes in the range of 150 MGT.

Passenger trains are akin to earlier freight trains, with one or more locomotives (electric or diesel) followed by relatively light cars. Likewise, transit cars are relatively light.

23.1.2 Basic Differences between Railroad and Highway Bridges

A number of differences exist between railroad and highway bridges:

1. The ratio of live load to dead load is much higher for a railroad bridge than for a similarly sized highway structure. This can lead to serviceability issues such as fatigue and deflection control governing designs rather than strength.
2. The design impact load on railroad bridges is higher than on highway structures.
3. Simple-span structures are preferred over continuous structures for railroad bridges. Many of the factors that make continuous spans attractive for highway structures are not as advantageous for railroad use. Continuous spans are also more difficult to replace in emergencies than simple spans.
4. Interruptions in service are typically much more critical for railroads than for highway agencies. Therefore, constructibility and maintainability without interruption to traffic are crucial for railroad bridges.
5. Since the bridge supports the track structure, the combination of track and bridge movement cannot exceed the tolerances in track standards. Interaction between the track and bridge should be considered in design and detailing.
6. Seismic performance of highway and railroad bridges can vary significantly. Railroad bridges have performed well during seismic events.
7. Railroad bridge owners typically expect a longer service life from their structures than highway bridge owners expect from theirs.

23.1.3 *Manual for Railway Engineering*, AREMA

The base document for railroad bridge design, construction, and inspection is the American Railway Engineering Maintenance of Way Association (AREMA) *Manual for Railway Engineering* (*Manual*) [1].

Early railroads developed independent specifications governing the design loadings, allowable strains, quality of material, fabrication, and construction of their own bridges. There was a proliferation of specifications written by individual railroads, suppliers, and engineers. One of the earliest general specifications is titled *Specification for Iron Railway Bridges and Viaducts*, by Clarke, Reeves and Company (Phoenix Bridge Company). By 1899 private railroads joined efforts in forming AREMA. Many portions of those original individual railroad specifications were incorporated into the first manual titled *Manual of Recommended Practice for Railway Engineering and Maintenance of Way* published in 1905. In 1911 the Association dropped "Maintenance of Way" from its name and became the American Railway Engineering Association (AREA); however, in 1997 the name reverted back to the original name with the consolidation of several railroad associations.

The *Manual* is not deemed a specification but rather a recommended practice. Certain provisions naturally are standards by necessity for the interchange of rail traffic, such as track gauge, track geometrics, clearances, basic bridge loading, and locations for applying loadings. Individual railroads may, and often do, impose more stringent design requirements or provisions due to differing conditions peculiar to that railroad or region of the country, but basically all railroads subscribe to the provisions of the *Manual.*

Although the *Manual* is a multivolume document, bridge engineering provisions are grouped in the *Structural Volume* and subdivided into applicable chapters by primary bridge material and special topics, as listed:

Chapter 7	Timber Structures
Chapter 8	Concrete Structures and Foundations
Chapter 9	Seismic Design for Railway Structures
Chapter 10	Structures Maintenance & Construction (New)
Chapter 15	Steel Structures
Chapter 19	Bridge Bearings
Chapter 29	Waterproofing

The primary structural chapters each address bridge loading (dead load, live load, impact, wind, seismic, etc.) design, materials, fabrication, construction, maintenance/inspection, and capacity rating. There is uniformity among the chapters in the configuration of the basic live load, which is based on the Cooper E-series steam locomotive. The present live-load configuration is two locomotives with tenders followed by a uniform live load as shown in Fig. 23.1. There is not uniformity in the chapters in the location and magnitude of many other loads due to differences in the types of bridges built with different materials and differences in material behavior. Also it is recognized that each chapter has been developed and maintained by separate committee groups of railroad industry engineers, private consulting engineers, and suppliers. These committees readily draw from railroad industry experiences and research, and from work published by other associations such as AASHTO, AISC, ACI, AWS, APWA, etc.

23.2 Railroad Bridge Philosophy

Railroad routes are well established and the construction of new railroad routes is not common; thus, the majority of railroad bridges built or rehabilitated are on existing routes and on existing right-of-way. Simply stated, the railroad industry first extends the life of existing bridges as long as economically justified. It is not uncommon for a railroad to evaluate an 80- or 90-year-old bridge, estimate its remaining life, and then rehabilitate it sufficiently to extend its life for some economical period of time.

Bridge replacement generally is determined as a result of a lack of load-carrying capacity, restrictive clearance, or deteriorated physical condition. If bridge replacement is necessary, then simplicity, cost, future maintenance, and ease of construction without significant rail traffic disruptions typically govern the design. Types of bridges chosen are most often based on the capability of a railroad to do its own construction work. Low-maintenance structures, such as ballasted deck prestressed concrete box-girder spans with concrete caps and piles, are preferred by some railroads. Others may prefer weathering steel elements.

In a review of the existing railroad industry bridge inventory, the majority of bridges by far are simple-span structures over streams and roadways. Complex bridges are generally associated with crossing major waterways or other significant topographical features. Signature bridges are rarely constructed by railroads. The enormity of train live loads generally preclude the use of double-leaf bascule bridges and suspension and cable-stayed bridges due to bridge deflection and shear load transfer, respectively. Railroads, where possible, avoid designing skewed or curved bridges, which also have inherent deflection problems.

When planning the replacement of smaller bridges, railroads first determine if the bridge can be eliminated using culverts. A hydrographic review of the site will determine if the bridge opening needs to be either increased or can be decreased.

The *Manual* provides complete details for common timber structures and for concrete box-girder spans. Many of the larger railroads develop common standards, which provide complete detailed plans for the construction of bridges. These plans include piling, pile bents, abutments and wing walls, spans (timber, concrete, and steel), and other elements in sufficient detail for construction by in-house forces or by contract. Only site-specific details such as permits, survey data, and soil conditions are needed to augment these plans.

Timber trestles are most often replaced by other materials rather than in kind. However, it is often necessary to renew portions of timber structures to extend the life of a bridge for budgetary reasons. Replacing pile bents with framed bents to eliminate the need to drive piles or the adding of a timber stringer to a chord to increase capacity is common. The replacement of timber trestles is commonly done by driving either concrete or steel piling through the existing trestle, at twice the present timber span length and offset from the existing bents. This is done between train movements. Either precast or cast-in-place caps are installed atop the piling beneath the existing timber deck. During a track outage period, the existing track and timber deck is removed and new spans (concrete box girders or rolled steel beams) are placed. In this type of bridge renewal, key factors are use of prefabricated bridge elements light enough to be lifted by railroad track mounted equipment (piles, caps, and spans), speed of installation of bridge elements between train movements, bridge elements that can be installed in remote site locations without outside support, and overall simplicity in performing the work.

The railroad industry has a large number of 150 to 200 ft span pin-connected steel trusses, many with worn joints, restrictive clearances, and low carrying capacity, for which rehabilitation cannot be economically justified. Depending on site specifics, a common replacement scenario may be to install an intermediate pier or bent and replace the span with two girder spans. Railroad forces have perfected the technique of laterally rolling out old spans and rolling in new prefabricated spans between train movements.

Railroads frequently will relocate existing bridge spans to other sites in lieu of constructing new spans, if economically feasible. This primarily applies to beam spans and plate girder spans up to 100 ft in length.

In general, railroads prefer to construct new bridges online rather than relocating or doglegging to an adjacent alignment. Where site conditions do not allow ready access for direct span replacement, a site bypass, or runaround, called a "shoofly" is constructed which provides a temporary bridge while the permanent bridge is constructed.

The design and construction of larger and complex bridges is done on an individual basis.

23.3 Railroad Bridge Types

Railroad bridges are nearly always simple-span structures. Listed below in groupings by span length are the more common types of bridges and materials used by the railroad industry for those span lengths.

Short spans	to 16 ft	Timber stringers
		Concrete slabs
		Rolled steel beams
	to 32 ft	Conventional and prestressed concrete box girders and beams
		Rolled steel beams
	to 50 ft	Prestressed concrete box girders and beams
		Rolled steel beams, deck and through girders
Medium spans, 80 to 125 ft		Prestressed concrete beams
		Deck and through plate girders
Long spans		Deck and through trusses (simple, cantilever, and arches)

Suspension bridges are not used by freight railroads due to excessive deflection.

23.4 Bridge Deck

23.4.1 General

The engineer experienced in highway bridge design may not think of the typical railroad bridge as having a deck. However, it is essential to have a support system for the rails. Railroad bridges typically are designed as either open deck or ballast deck structures. Some bridges, particularly in transit applications, use direct fixation of the rails to the supporting structure.

23.4.2 Open Deck

Open deck bridges have ties supported directly on load-carrying elements of the structure (such as stringers or girders). The dead loads for open deck structures can be significantly less than for ballast deck structures. Open decks, however, transfer more of the dynamic effects of live load into the bridge than ballast decks. In addition, the bridge ties required are both longer and larger in cross section than the standard track ties. This adds to their expense. Bridge tie availability has declined, and their supply may be a problem, particularly in denser grades of structured timber.

TABLE 23.1 Weight of Rails, Inside Guard Rails, Ties, Guard Timbers, and Fastenings for Typical Open Deck (Walkway not included)

Item	Weight (plf of track)
Rail (136 RE):	
(136 lb/lin. yd × 2 rails/track × 1 lin. yd/3 lin. ft)	91
Inside guard rails:	
(115 lb/lin. yd × 2 rails/track × 1 lin. yd/3 lin. ft)	77
Ties (10 in. × 10 10 ft bridge ties):	
(10 in. × 10 in. × 10 ft × 1 ft²/144 in.² × 60 lb/ft³ × 1 tie/14 in. × 12 in./1 ft)	357
Guard Timbers (4 × 8 in.):	
(4 in. × 8 in. × 1 ft × 1 ft²/144 in.³ × 60 lb/1 ft³ × 2 guard timbers/ft)	27
Tie Plates (7¾ × 14¾ in. for rail with 6 in. base):	
24.32 lb/plate × 1 tie/14 in. × 12 in./ft × 2 plates/tie)	42
Spikes (⅝ × ⅝ in. × 6 in. reinforced throat)	
(0.828 lb/spike × 18 spikes/tie × 1 tie/14 in. × 12 in./1 ft)	13
Miscellaneous Fastenings (hook bolts and lag bolts):	
(Approx. 2.25 lb/hook bolt + 1.25 lb/lag screw × 2 bolts/tie × 1 tie/14 in. × 12 in./ft)	6
Total weight	613

TABLE 23.2 Weight of Typical Ballast Deck

Item	Weight (plf of track)
Rail (136 RE):	
(136 lb/lin. yd. × 2 rails/track × 1 lin. yd/3 lin. ft)	91
Inside Guard Rails:	
(115 lb/lin. yd × 2 rails/track × 1 lin. yd/3 lin. ft)	77
Ties (neglect, since included in ballast weight)	—
Guard Timbers (4 × 8 in.):	
(4 in. × 8 in. × 1 ft × 1 ft²/144 in.² × 60 lb/1 ft³ × 2 guard timbers/ft)	27
Tie Plates (7¾ × 14¾ in. for rail with 6 inc. base):	
(24.32 lb/plate × 1 tie/19.5 in. × 12 in./ft × 2 plates/tie)	30
Spikes (⅝ × ⅝ × 6 in. reinforced throat)	
(0.828 lb/spike × 18 spikes/tie × 1 tie/19.5 in. × 12 in./1 ft)	9
Ballast (assume 12 in. additional over time)	
(Approx. 120 lb/ft³ × 27 in. depth/12 in./1 ft × 16 ft)	4320
Waterproofing:	
(Approx. 150 lb/ft³ × 0.75 in. depth/12 in./1 ft × 20 ft)	188
Total weight:	4742

23.4.3 Ballast Deck

Ballast deck bridges have the track structure supported on ballast, which is carried by the structural elements of the bridge. Typically, the track structure (rails, tie plates, and ties) is similar to track constructed on grade. Ballast deck structures offer advantages in ride and maintenance requirements. Unlike open decks, the track alignment on ballast deck spans can typically be maintained using standard track maintenance equipment. If all other factors are equal, most railroads currently prefer ballast decks for new structures.

In ballast deck designs, an allowance for at least 6 in. of additional ballast is prudent. Specific requirements for additional ballast capacity may be provided by the railroad. In addition, the required depth of ballast below the tie should be verified with the affected railroad. Typical values for this range from 8 to 12 in. or more. The tie length used will have an effect on the distribution of live-load effects into the structure. Ballast decks are also typically waterproofed. The weight of waterproofing should be included in the dead load. Provisions for selection, design, and installation of waterproofing are included in Chapter 29 of the AREMA *Manual.*

23.4.4 Direct Fixation

Direct fixation structures have rails supported on plates anchored directly to the bridge deck or superstructure. Direct fixation decks are much less common than either open decks or ballast decks and are rare in freight railroad service. While direct fixation decks eliminate the dead load of ties and ballast, and can reduce total structure height, they transfer more dynamic load effects into the bridge. Direct fixation components need to be carefully selected and detailed.

23.4.5 Deck Details

Walkways are frequently provided on railroad bridge decks. They may be on one or both sides of the track. Railroads have their own policies and details for walkway placement and construction.

Railroad bridge decks on curved track should allow for superelevation. With ballast decks, this can be accomplished by adjusting ballast depths. With open decks, it can require the use of beveled ties or building the superelevation into the superstructure.

Continuous welded rail (CWR) is frequently installed on bridges. This can affect the thermal movement characteristics of the structure. Check with the affected railroad for its policy on anchorage of CWR on structures. Long-span structures may require the use of rail expansion joints.

23.5 Design Criteria

23.5.1 Geometric Considerations

Railroad bridges have a variety of geometric requirements. The AREMA *Manual* has clearance diagrams showing the space required for passage of modern rail traffic. It should be noted that lateral clearance requirements are increased for structures carrying curved track. Track spacing on multiple-track structures should be determined by the affected railroad. Safety concerns are leading to increased track-spacing requirements.

If possible, skewed bridges should be avoided. Skewed structures, however, may be required by site conditions. A support must be provided for the ties perpendicular to the track at the end of the structure. This is difficult on open deck structures. An approach slab below the ballast may be used on skewed ballast deck bridges.

23.5.2 Proportioning

Typical depth-to-span length ratios for steel railroad bridges are around 1:12. Guidelines for girder spacing are given in Chapter 15 of the *Manual.*

23.5.3 Bridge Design Loads

23.5.3.1 Dead Load

Dead load consists of the weight of the structure itself, the track it supports, and any attachments it may carry. Dead loads act due to gravity and are permanently applied to the structure. Unit weights for calculation of dead loads are given in AREMA Chapters 7, 8, and 15. The table in Chapter 15 is reproduced below:

Unit Weights for Dead Load Stresses

Type	Pounds per Cubic Foot
Steel	490
Concrete	150
Sand, gravel, and ballast	120
Asphalt-mastic and bituminous macadam	150
Granite	170
Paving bricks	150
Timber	60

Dead load is applied at the location it occurs in the structure, typically as either a concentrated or distributed load.

The *Manual* states that track rails, inside guard rails, and rail fastenings shall be assumed to weigh 200 pounds per linear foot (plf) of track. The 60 pound per cubic foot weight given for timber should be satisfactory for typical ties. Exotic woods may be heavier. Concrete ties are sometimes used, and their heavier weight should be taken into account if their use is anticipated.

In preliminary design of open deck structures, a deck weight of 550 to 650 plf of track can be assumed. This should be checked with the weight of the specific deck system used for final design. Example calculations for track and deck weight for open deck and ballast deck structures are included in this chapter.

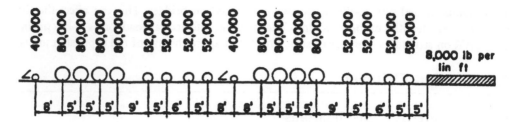

FIGURE 23.1   Cooper E80 live load.

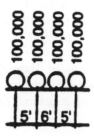

FIGURE 23.2   Alternate live load.

Railroad bridges frequently carry walkways and signal and communication cables and may be used by utilities. Provisions (both in dead load and physical location) may need to be made for these additional items. Some structures may even carry ornamental or decorative items.

23.5.3.2   Live Load

Historically, freight railroads have used the Cooper E load configuration as a live-load model. The Cooper E80 load is currently the most common design live load. The E80 load model is shown in Figure 23.1. The 80 in E80 refers to the 80 kip weight of the locomotive drive axles. An E60 load has the same axle locations, but all loads are factored by 60/80. Some railroads are designing new structures to carry E90 or E100 loads.

The Cooper live-load model does not match the axle loads and spacings of locomotives currently in service. It did not even reflect all locomotives at the turn of the 20th century, when it was introduced by Theodore Cooper, an early railroad bridge engineer. However, it has remained in use throughout the past century. One of the reasons for its longevity is the wide variety of rail rolling stock that has been and is currently in service. The load effects of this equipment on given spans must be compared, as discussed in Section 23.6. The Cooper live-load model gives a universal system with which all other load configurations can be compared. Engineering personnel of each railroad can calculate how the load effects of each piece of equipment compare to the Cooper loading.

The designated steel bridge design live load also includes an "Alternate E80" load, consisting of four 100-kip axles. This is shown in Figure 23.2. This load controls over the regular Cooper load on shorter spans.

A table of maximum load effects over various span lengths is included in Chapter 15, Part 1 of the AREMA *Manual.*

23.5.3.3   Impact

Impact is the dynamic amplification of the live-load effects on the bridge caused by the movement of the train across the span. Formulas for calculation of impact are included in Chapters 8 and 15 of the AREMA manual. The design impact values are based on an assumed train speed of 60 mph. It should be noted that the steel design procedure allows reduction of the calculated impact for ballast deck structures. Different values for impact from steam and diesel locomotives are used. The steam impact values are significantly higher than diesel impact over most span lengths.

Impact is not applied to timber structures, since the capacity of timber under transient loads is significantly higher than its capacity under sustained loads. Allowable stresses for timber design are based on the sustained loads.

23.5.3.4 Centrifugal Force

Centrifugal force is the force a train moving along a curve exerts on a constraining object (track and supporting structure) which acts away from the center of rotation. Formulas or tables for calculation of centrifugal force are included in Chapters 7, 8, and 15 of the AREMA manual. The train speed required for the force calculation should be obtained from the railroad.

Although the centrifugal action is applied as a horizontal force, it can produce overturning moment due to its point of application above the track. Both the horizontal force and resulting moment must be considered in design or evaluation of a structure.

The horizontal force tends to displace the structure laterally:

- For steel structures (deck girders, for example), it loads laterals and cross frames.
- For concrete structures (box girders, for example), the superstructure is typically stiff enough in the transverse direction that the horizontal force is not significant for the superstructure.

For all bridge types, the bearings and substructure must be able to resist the centrifugal horizontal force.

The overturning moment tends to increase the live-load force in members on the outside of the curve and reduce the force on inside members. However, interior members are not designed with less capacity than exterior members. Substructures must be designed to resist the centrifugal overturning moment. This will increase forces toward the outside of the curve in foundation elements. The centrifugal force is applied at the location of the axles along the structure, 6 ft above the top of rail, at a point perpendicular to the center of a line connecting the rail tops. The effect of track superelevation may compensate somewhat for centrifugal force. The plan view location of the curved track on the bridge (since railroad bridge spans are typically straight, laid out along the curve chords) can also be significant. Rather than applying the centrifugal force at each axle location, some railroads simply increase the calculated live-load force by the centrifugal force percentage, factor in the effect of the force location above the top of rail, and use the resulting value for design.

23.5.3.5 Lateral Loads from Equipment

This item includes all lateral loads applied to the structure due to train passage, other than centrifugal force. The magnitude and application point of these loads varies among Chapters 7, 8, and 15. For timber, a load of 20 kips is applied horizontally at the top of rail. For steel, a load of one quarter of the heaviest axle of the specified live load is applied at the base of rail. In both cases, the lateral load is a moving concentrated load that can be applied at any point along the span in either horizontal direction. It should be noted that lateral loads from equipment are not included in design of concrete bridges. However, if concrete girders are supported on steel or timber substructures, lateral loads should be applied to the substructures.

Lateral loads from equipment are applied to lateral bracing members, flanges of longitudinal girders or stringers without a bracing system, and to chords of truss spans. Experience has shown that very high lateral forces can be applied to structures due to lurching of certain types of cars. Wheel hunting is another phenomenon that applies lateral force to the track and structure. Damaged rolling stock can also create large lateral forces.

It should be noted that there is not an extensive research background supporting the lateral forces given in the AREMA *Manual.* However, the lateral loads in the *Manual* have historically worked well when combined with wind loads to produce adequate lateral resistance in structures.

23.5.3.6 Longitudinal Force from Live Load

Longitudinal forces are typically produced from starting or stopping trains (acceleration or deceleration) on the bridge. They can be applied in either longitudinal direction. These forces are transmitted through the rails and distributed into the supporting structure.

Chapters 7, 8, and 15 all take the longitudinal force due to braking to be 15% of the vertical live load, without impact. The chapters differ slightly in their consideration of the acceleration (traction) aspect of the force. Chapter 7 uses 25% of the drive axle loads for traction, while Chapters 8 and 15 use 25% of the axles of the regular Cooper E80 train configuration. In each chapter, the braking and traction forces are compared, and the larger value used in design. Chapters 7, 8, and 15 differ in the point of application of the longitudinal force. Chapter 7 applies it 6 ft above the top of rail. Chapters 8 and 15 apply the braking force at 8 ft above the top of rail and the traction force 3 ft above the top of rail.

All three chapters recognize that some of the longitudinal force is carried through the rails off the structure. (The extent of this transfer depends on factors such as rail continuity, rail anchorage, and the connection of the bridge deck to the span.) Where a large portion of the longitudinal force is carried to the abutments or embankment, Chapter 7 allows neglecting longitudinal force in the design of piles, posts, and bracing of bents. Chapters 8 and 15 allow taking the applied longitudinal force as half of what was initially calculated on short (<200 feet) ballast deck bridges with short spans (<50 feet), if the continuity of members or frictional resistance will direct some of the longitudinal force to the abutments.

Chapters 8 and 15 also state that the longitudinal load is to be applied to one track only, and can be distributed to bridge components based on their relative stiffness and the types of bearings. For multiple-track structures, it may be prudent to include longitudinal force on more than one track, depending on the bridge location and train operation at the site.

Longitudinal force is particularly significant in long structures, such as viaducts, trestles, or major bridges. Large bridges may have internal traction or braking trusses to carry longitudinal forces to the bearings. Viaducts frequently have braced tower bents at intervals to resist longitudinal force.

The American Association of Railroads (AAR) is currently conducting research on the longitudinal forces in bridges induced by the new high-adhesion locomotives now coming into service. In addition, the introduction of new mechanical systems such as the load-empty brake and electronically controlled brakes are affecting the longitudinal forces introduced into the track. Transit equipment can have high acceleration and deceleration rates, which can lead to high longitudinal forces on transit structures.

23.5.3.7 Wind Loading

Wind loading is the force on the structure due to wind action on the bridge and train. Chapters 7, 8, and 15 deal with wind on the structure slightly differently:

1. *Timber:* Use 30 psf as a moving horizontal load acting in any direction.
2. *Concrete:* Use 45 psf as a horizontal load perpendicular to the track centerline.
3. *Steel:* As a moving horizontal load:
   a. Use 30 psf on loaded bridge.
   b. Use 50 psf on unloaded bridge.

The application areas of the wind on structure vary as well:

1. *Timber:* For trestles, the affected area is 1.5 times the vertical projection of the floor system. For trusses, the affected area is the full vertical projection of the spans, plus any portion of the leeward trusses not shielded by the floor system. For trestles and tower substructures, the affected area is the vertical projections of the components (bracing, posts, and piles).
2. *Steel:* Similar to timber, except that for girder spans 1.5 times the vertical projection of the span is used.
3. *Concrete:* Wind load is applied to the vertical projection of the structure. Note that 45 psf = 1.5 (30 psf).

For all materials, the wind on the train is taken as 300 plf, applied 8 ft above the top of rail.

The 30-psf wind force on a loaded structure and 50-psf force on an unloaded structure used in Chapter 15 reflect assumptions on train operations. It was assumed that the maximum wind velocity under which train operations would be attempted would produce a force of 30 psf. Hurricane winds, under which train operations would not be attempted, would produce a wind force of 50 psf.

For stability of spans and towers against overturning due to wind on a loaded bridge, the live load is reduced to 1200 plf, without impact being applied. This value represents an unloaded, stopped train on the bridge.

It should be noted that Chapter 15 has a minimum wind load on loaded bridges of 200 plf on the loaded chord or flange and 150 plf on the unloaded chord or flange.

Virtually every bridge component can be affected by wind. However, wind is typically most significant in design of

1. Lateral bracing and cross frames
2. Lateral bending in flanges
3. Vertical bending in girders and trusses due to overturning
4. Tower piles or columns
5. Foundations

23.5.3.8 Stream Flow, Ice, and Buoyancy

These loads are experienced by a portion of the structure (usually a pier) because of its location in a body of water. These topics are only specifically addressed in Chapter 8, because they apply almost entirely to bridge substructures, which typically consist of concrete.

Buoyancy, stream flow, and ice pressure are to be applied to any portion of the structure that can be exposed to them. This typically includes piers and other elements of the substructure. Buoyancy can be readily calculated for immersed portions of the structure.

While the AREMA *Manual* does not address design forces for stream flow and ice pressure, other design criteria, such as the AASHTO *LRFD Bridge Design Specification* does include procedures for calculating them. The designer can use these sources for guidance until specific forces are included by AREMA.

Spans may be floated off piers due to buoyancy, stream flow, and ice pressure. Loaded ballast cars are sometimes parked on bridges during floods or ice buildup to resist this. Drift or debris accumulation adjacent to bridges can be a significant problem, reducing the flow area through the bridge and effectively increasing the area exposed to force from stream flow.

Two other factors concerning waterways must be considered. The first is vessel collision (or, more correctly) allision with piers. Pier protection is covered in Part 23, Spans over Navigable Streams, of Chapter 8. These requirements should be addressed when designing a bridge across a navigable waterway. The second factor to be considered is scour. Scour is a leading cause of bridge failure. The AASHTO *LRFD Bridge Design Specification* contains scour analysis and protection guidelines. Hydraulic studies to determine required bridge openings should be performed when designing new structures or when hydrologic conditions upstream of a bridge change.

23.5.3.9 Volume Changes

Volume changes in structures can be caused by thermal expansion or contraction or by properties of the structural materials, such as creep or shrinkage. Volume changes in themselves, if unrestrained, have relatively little effect on the forces on the structure. Restrained volume changes, however, can produce significant forces in the structure. The challenge to the designer is to provide a means to relieve volume changes or to provide for the forces developed by restrained changes.

Chapter 7 does not specifically state thermal expansion movement requirements. Due to the nature of the material and type of timber structures in use, it is unlikely that thermal stresses will be significant in timber design. Chapter 15 requires an allowance of 1 in. of length change due to temperature per every 100 ft of span length in steel structures. Chapter 8 provides the following table for design temperature rise and fall values for concrete bridges:

Climate	Temperature Rise	Temperature Fall
Moderate	30°F	40°F
Cold	35°F	45°F

It should be noted that the tabulated values refer to the temperature of the bridge concrete. A specific railroad may have different requirements for thermal movement.

Expansion bearings are the main design feature typically used to accommodate volume changes. Common bearing types include:

1. Sliding steel plates
2. Rocker bearings
3. Roller bearings (cylindrical and segmental)
4. Elastomeric bearing pads

Provision should be made for span length change due to live load. For spans longer than 300 ft, provision must be made for expansion and contraction of the bridge floor system within the trusses.

For concrete structures, provisions need to be made for concrete shrinkage and creep. Specific guidelines are given in Chapter 8, Parts 2 and 17 for these properties. It is important to remember that creep and shrinkage are highly variable phenomena, and allowance should be made for higher-than-expected values. It also should be noted the AREMA *Manual* requires 0.25 in²/ft minimum of reinforcing steel in exposed concrete surfaces.

Chapter 8 also requires designing for longitudinal force due to friction or shear resistance at expansion bearings. This is in recognition of the fact that most expansion bearings have some internal resistance to movement. This resistance applies force to the structure as the bridge expands and contracts. The AREMA *Manual* contains procedures for calculating the shear force transmitted through bearing pads. Loads transmitted through fixed or expansion bearings should be included in substructure design.

Bearings must also be able to resist wind and other lateral forces applied to the structure. Chapter 19 of the AREMA *Manual for Railway Engineering* covers bridge bearings. It is included in the 1997 *Manual*, and should be applied for bearing design and detailing.

It should be noted that movement of bridge bearings affects the tolerances of the track supported by the bridge. This calls for careful selection of bearings for track with tight tolerances (such as high-speed lines). Maintenance requirements are also important when selecting bearings, since unintended fixity due to freezing of bearings can cause significant structural damage.

23.5.3.10 Seismic Loads

Seismic design for railroads is covered in Chapter 9 of the *Manual*. The philosophical background of Chapter 9 recognizes that railroad bridges have historically performed well in seismic events. This is due to the following factors:

1. The track structure serves as an effective restraint (and damping agent) against bridge movement.
2. Railroad bridges are typically simple in their design and construction.
3. Trains operate in a controlled environment, which makes types of damage permissible for railroad bridges that might not be acceptable for structures in general use by the public.

Item 3 above is related to the post-seismic event operation guidelines given in Chapter 9. These guidelines give limits on train operations following an earthquake. The limits vary according to earthquake magnitude and distance from the epicenter. For example, following an earthquake of magnitude 6.0 or above, all trains within a 100-mile radius of the epicenter must stop until the track and bridges in the area have been inspected and cleared for use. (Note that specific railroad policies may vary.)

Three levels of ground motion are defined in Chapter 9:

- Level 1 — Motion that has a reasonable probability of being exceeded during the life of the bridge.
- Level 2 — Motion that has a low probability of being exceeded during the life of the bridge.
- Level 3 — Motion for a rare, intense earthquake.

Three performance limit states are given for seismic design of railroad bridges. The serviceability limit state requires that the structure remain elastic during Level 1 ground motion. Only moderate damage and no permanent deformations are acceptable. The ultimate limit state requires that the structure suffer only readily detectable and repairable damage during Level 2 ground motion. The survivability limit state requires that the bridge not collapse during Level 3 ground motion. Extensive damage may be allowed. For some structures, the railroad may elect to allow for irreparable damage, and plan to replace the bridges following a Level 3 event.

An in-depth discussion of seismic analysis and design is beyond the scope of this section. Guidelines are given in Chapter 9 of the manual. Base acceleration coefficient maps for various return periods are included in the chapter. It should be noted that no seismic analysis is necessary for locations where a base acceleration of 0.1 g or less is expected with a 475-year return period. For most locations in North America, therefore, a seismic analysis would not be needed.

Section 1.4 of Chapter 9 addresses seismic design. Important structures (discussed in its Section 1.3.3) should be designed to resist higher seismic loads than nonimportant structures.

Even if no specific seismic analysis and design is required for a structure, it is good practice to detail structures for seismic resistance if they are in potentially active areas. Specific concerns are addressed in Chapter 9. Provision of adequate bearing areas and designing for ductility are examples of inexpensive seismic detailing.

23.5.4 Load Combinations

A variety of loads can be applied to a structure at the same time. For example, a bridge may experience dead load, live load, impact, centrifugal force, wind, and stream flow simultaneously. The AREMA *Manual* chapters on structure design recognize that it is unlikely that the maximum values of all loads will be applied concurrently to a structure. Load combination methods are given to develop maximum credible design forces on the structure.

Chapter 7, in Section 2.5.5.5, Combined Stresses, states: "For stresses produced by longitudinal force, wind or other lateral forces, or by a combination of these forces with dead and live loads and centrifugal force, the allowable working stresses may be increased 50%, provided the resulting sections are not less than those required for dead and live loads and centrifugal force."

Chapter 15, in Section 1.3.14.3, Allowable Stresses for Combinations of Loads or Wind Loads Only, states:

a. Members subject to stresses resulting from dead load, live load, impact load and centrifugal load shall be designed so that the maximum stresses do not exceed the basic allowable stresses of Section 1.4, of Basic Allowable Stresses, and the stress range does not exceed the allowable fatigue stress range of Article 1.3.13.
b. The basic allowable stresses of Section 1.4, Basic Allowable Stresses, shall be used in the proportioning of members subject to stresses resulting from wind loads only, as specified in Article 1.3.8.

    c. Members, except floorbeam hangers, which are subject to stresses resulting from lateral loads, other than centrifugal load, and/or longitudinal loads, may be proportioned for stresses 25% greater than those permitted by paragraph a, but the section of the member shall not be less than that required to meet the provisions of paragraph a or paragraph b alone.

    d. Increase in allowable stress permitted by paragraph c shall not be applied to allowable stress in high strength bolts.

Chapter 8, in Part 4 on Pile Foundations, defines primary and secondary loads. Primary loads include dead load, live load, centrifugal force, earth pressure, buoyancy, and negative skin friction. Secondary (or occasional) loads include wind and other lateral forces, ice and stream flow, longitudinal forces, and seismic forces. Section 4.2.2.b allows a 25% increase in allowable loads when designing for a combination of primary and secondary loads, as long as the design satisfies the primary load case at the allowable load.

These three load combination methods are based on service load design. Chapter 8, in Part 2, Reinforced Concrete Design, addresses both service load and load factor design

Chapter 8, Section 2.2.4 gives several limitations on the load combination tables. For example, load factor design is not applicable to foundation design or for checking structural stability. In addition, load factors should be increased or allowable stresses adjusted if the predictability of loads is different than anticipated in the chapter.

For stability of towers, use the 1200 plf vertical live load as described in the Wind Loading section.

As a general rule, the section determined by a load combination should never be smaller than the section required for dead load, live load, impact, and centrifugal force. It is important to use the appropriate load combination method for each material and component in the bridge design. Combination methods from different sections and chapters should not be mixed.

23.5.5 Serviceability Considerations

23.5.5.1 Fatigue

Fatigue resistance is a critical concern in the design of steel structures. It is also a factor, although of less significance, in the design of concrete bridges. A fatigue design procedure, based on allowable stresses, impact values, number of cycles per train passage, fracture criticality of the member, and type of details, is applied to steel bridges. Fatigue can be the controlling design case for many new steel bridges.

23.5.5.2 Deflection

Live load-deflection control is a significant serviceability criterion. Track standards limit the amount of deflection in track under train passage. The deflection of the bridge under the live load accumulates with the deflection of the track structure itself. This total deflection can exceed the allowable limits if the bridge is not sufficiently stiff. The stiffness of the structure can also affect its performance and longevity. Less stiff structures may be more prone to lateral displacement under load and out-of-plane distortions. Specific deflection criteria are given in Chapter 15 for steel bridges. Criteria for concrete structures are given in Chapter 8 using span-to-depth ratios.

Long-term deflections should also be checked for concrete structures under the sustained dead load to determine if any adverse effects may occur due to cracking or creep.

23.5.5.3 Others

Other serviceability criteria apply to concrete structures. Reinforced concrete must be checked for crack control. Allowable stress limits are given for various service conditions for prestressed concrete members.

23.6 Capacity Rating

23.6.1 General

Rating is the process of determining the safe capacity of existing structures. Specific guidelines for bridge rating are given in Chapters 7, 8, and 15 of the AREMA *Manual*. Ratings are typically performed on both as-built and as-inspected bridge conditions. The information for the as-built condition can be taken from the bridge as-built drawings. However, it is important to check the current condition of the structure. This is done by performing an inspection of the bridge and adjusting the as-built rating to include the effects of any deterioration, damage, or modifications to the structure since its construction. Material property testing of bridge components may be very useful in the capacity rating of an older structure.

Structure ratings are normally presented as the Cooper E value live load that the bridge can safely support. The controlling rating is the lowest E value for the structure (based on a specific force effect on a critical member or section). For example, a structure rating may be given as E74, based on bending moment at the termination of a flange cover plate.

As discussed in the Live Load section, there are a wide variety of axle spacings and loadings for railroad equipment. Each piece of equipment can be rated to determine the maximum force effects it produces for a given span length. The equipment rating is given in terms of the Cooper load that would produce the equivalent force effect on the same span length. Note that this equivalent force effect value will probably be different for shear and moment on each span length.

In addition to capacity ratings, fatigue ratings can be performed on structures to estimate their remaining fatigue life. These are typically only calculated for steel structures. Guidelines for this can be found in the commentary section of Chapter 15.

23.6.2 Normal Rating

The normal rating of the structure is the load level which can be carried by the bridge for an indefinite time period. This indefinite time period can be defined as its expected service life. The allowable stresses used for normal rating are the same as the allowable stresses used in design. The impact effect calculation, however, is modified from the design equation. Reduction of the impact value to reflect the actual speed of trains crossing the structure (rather than the 60 mph speed assumed in the design impact) is allowed. Formulas for the impact reduction are included in the rating sections of the AREMA *Manual* chapters.

23.6.3 Maximum Rating

The maximum rating of the structure is the maximum load level which can be carried by the bridge at infrequent intervals. This rating is used to check if extraheavy loads can cross the structure. Allowable stresses for maximum rating are increased over the design allowable values.

The impact reduction for speed can be applied as for a normal rating. In addition, "slow orders" or speed restrictions can be placed on the extraheavy load when crossing the bridge. This can allow further reduction of the impact value, thus increasing the maximum rating of the structure. (Note that this maximum rating value would apply only at the specified speed.)

References

AREMA, *Manual for Railway Engineering*, AREMA, Landover, MD, 1997.
American Association of Railroads, *Railroad Facts*, 1997 Edition, Washington, D.C., 1997.
Waddell, J.A.L., *Bridge Engineering*, 1916.

24

Bridge Decks and Approach Slabs

Michael D. Keever
*California Department
of Transportation*

John H. Fujimoto
*California Department
of Transportation*

24.1 Introduction

This chapter discusses bridge decks and structure approach slabs, the structural riding surface that typically is the responsibility of bridge design engineers when developing contract plans and details. Decks and approach slabs are usually not specially designed for each bridge project, but are instead taken from tables and standard plans either developed and provided by the owner or approved for use by the owner. However, as the load and resistance factor design (LRFD) is adopted by various agencies, standards that were developed previously must be reviewed and revised to comply with these new standards.

Not only do decks and approach slabs provide the riding surface for vehicular traffic, but they also serve several structural purposes. The bridge deck distributes the vehicular wheel loads to the girders, which are the primary load-carrying members on a bridge superstructure. The deck is often composite with the main girders and, with reinforcement distributed in the effective regions of the deck, serves to impart flexural strength and torsional rigidity to the bridge. The structure approach slab is a transitional structure between the bridge, which has relatively little settlement, and the roadway approach, which is subject to varying levels of approach settlement, sometimes significant. The approach slab serves as a bridge between the roadway and the primary bridge and is intended to reduce the annoying and sometimes unsafe "bump" that is often felt when approaching and leaving a bridge.

24.2 Bridge Decks

There are several different types of bridge decks, with the most common being cast-in-place reinforced concrete [1]. Other alternative deck types include precast deck panels, prestressed cast-in-place decks, post-tensioned concrete panels, filled and unfilled steel grid, steel orthotropic decks, and timber. These less common types may be used when considering deck rebar corrosion, deck replacement, traffic, maintenance, bridge weight, aesthetics, and life cycle costs, among other reasons. This chapter will emphasize cast-in-place reinforced concrete decks, including a design example. The alternative deck types will be discussed in less detail later in the chapter.

24.2.1 Cast-in-Place Reinforced Concrete

The extensive use of cast-in-place concrete bridge decks is due to several reasons including cost, acceptable skid resistance, and commonly available materials and contractors to do the work. Despite these advantages, this type of deck is not without disadvantages.

The most serious drawback is the tendency of the deck rebar to corrode. Deicing salts used in regions that must contend with snow and ice have problems associated with corrosion of the rebar in the deck. In these areas, cracks caused by corrosion are aggravated by the results of freeze–thaw action. Damage due to rebar corrosion often results in the cost and inconvenience to the traveling public of replacing the deck. In an effort to minimize this problem, concrete cover can be increased over the rebar, sealants can be placed on the deck, and epoxy-coated or galvanized rebar can be used in the top mat of deck steel. However, these solutions do not prevent the deck from cracking, which initiates the damage.

A common means of reducing deck rebar corrosion is the use of coated rebar. One drawback to this type of rebar is that it is often difficult to protect epoxy-coated and galvanized rebar during construction. Small nicks to the rebar, common when using normal construction methods, may be repaired in the field if they are detected. However, it is difficult to make repairs that are as good as the original coatings.

Despite corrosion problems associated with a cast-in-place reinforced bridge deck, it continues to be the most common type of deck built, and therefore a design example will be included in Section 24.2.1.3.

24.2.1.1 Traditional Design Method

Traditionally, cast-in-place bridge decks are designed assuming that the bridge deck is a continuous beam spanning across the girders, which are assumed to be unyielding supports. Although it is known that the girders do indeed deflect, it greatly simplifies the analysis to assume they do not. By using this method the maximum moments are determined and the deck is designed. This design method, in which the deck is designed as a series of strips transverse to the girders, is referred to as the "approximate strip design method." This method has been refined over time and has now been adapted to LRFD in the 1994 AASHTO-LRFD Specifications [2]. All references to this code will be shown in brackets.

24.2.1.2 Empirical Design Method

More recently, an alternative method of isotropic bridge deck design has been developed for cast-in-place concrete bridge decks in which it is assumed instead that the deck resists the loads using an arching effect between the girders. The 1994 AASHTO-LRFD Specification includes an empirical design method for decks using isotropic reinforcement based on these arch design principles [9.7.2]. Under this method no analysis is required. Instead, four layers of reinforcement are placed with no differentiation between the transverse and longitudinal direction. However, to use this method the conditions outlined under [9.7.2.4] must be satisfied. These conditions include requirements for the effective length between girders, the depth of the slab, the length of the deck overhang, and the specified concrete strength.

The deck overhang itself is not designed using this method. Instead cantilever overhangs are designed using the approximate strip method described above and used in the design example below.

24.2.1.3 Design Example

Given
Consider a cast-in-place conventionally reinforced bridge deck (Figure 24.1). The superstructure has six girders spaced at 2050 mm. The deck width is 11,890 mm wide and the overhang beyond

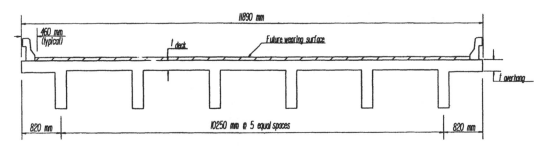

FIGURE 24.1 Typical section.

the exterior girder is 820 mm from the girder centerline. The unit weight of concrete is assumed to be 23.5 kN/m³.

Find

Based on the 1994 AASHTO-LRFD Specifications, use the Approximate Strip Design Method for Decks [4.6.2.1] to determine the deck thickness, design moments, and the detailing requirements necessary to design the bridge deck reinforcement.

Solution:

1. **Determine the Deck Thickness** [Table 2.5.2.6.3-1] [9.7.1.1]

$$t_{deck} = (S + 3000)/30 = (2050 + 3000)/30 = 168.3 \text{ mm}$$

where $S =$ the girder spacing. The minimum required deck thickness, excluding provisions for grinding, grooving, and sacrificial surface is $t_{deck} = 175$ mm. $\Leftarrow$ controls

The deck overhang is often a different thickness than the deck thickness. This may be for aesthetic, structural, or other reasons. For this example, assume the deck overhang is a constant thickness of $t_{overhang} = 250$ mm.

2. **Determine Unfactored Dead Loads**

For simplicity the deck will be designed as a 1-m-wide one-way slab. Therefore, all loads will be determined on a per meter width.

Slab: $q_{DS} = (23.5 \text{ kN/m}^3)(0.175 \text{ m})(1 \text{ m}) = 4.11 \times 10^{-3} \text{ kN/m}$
Overhang: $q_{DO} = (23.5 \text{ kN/m}^3)(0.250 \text{ m})(1 \text{ m}) = 5.88 \times 10^{-3} \text{ kN/m}$
Barrier rail: $P_{DB} = 5.8$ kN per 1 m width
Wearing surface: $q_{DW} = (1.70 \text{ kN/m}^2)(1 \text{ m}) = 1.70 \text{ kN/m}$

3. **Determine Unfactored Live Loads** [4.6.2.1] [3.6.1.3.3] [3.6.1.2.2]
   a. *Wheel load:*

   Truck axle load = 145 kN/axle.

   The axle load of 145 kN is distributed equally such that each wheel load is 72.5 kN. These 72.5-kN wheel loads are moved within each lane, with an edge spacing within the lane of 0.6 m, except at the deck overhang where the edge spacing is 0.3 m as specified in [Figure 3.6.1.2.2-1].
   b. *Calculate the number of live load lanes* [3.6.1.1.1]:
   Assume for this example the barrier rail width is 460 mm, which implies that the clear distance between the face of rail $w = 11,890$ mm $- 2(460$ mm$) = 10,970$ mm. Therefore, the number of lanes is $N = (10,970)/3600 = 3.05$, i.e., 3 using just the integer portion of the solution as required.

TABLE 24.1   Wheel Load Layout

Alternative	Wheel Load Layout (Span/Distance from left end of span)
M_{+ve} Alternative 1 (one lane)	Span 2/820 mm
	Span 3/570 mm
M_{+ve} Alternative 2 (one lane)	Span 2/1030 mm
	Span 3/780 mm
M_{+ve} Alternative 3 (two lanes)	Span 2/820 mm
	Span 3/570 mm
	Span 4/820 mm
	Span 5/570 mm
M_{-ve} Alternative 1 (one lane)	Span 2/1150 mm
	Span 3/900 mm

c. *Determine the wheel load distribution* [4.6.2.1.6] [Table 4.6.2.1.3-1] [4.6.2.1.3]:
For cast-in-place concrete decks the strip width, w_{strip}, is
Overhangs:

$$w_{strip} = 1140 + 0.833X$$

where $X =$ the distance from the location of the load to the centerline of the support. If it is assumed that the wheel load is pushed as close to the face of the barrier as permitted by the Code, i.e., 300 mm:

$$X = 820 \text{ mm} - 460 \text{ mm} - 300 \text{ mm} = 60 \text{ mm}$$

Therefore, for the overhang $w_{strip} = 1140 + 0.833(60) = 1190$ mm.
Interior slab:

Positive moment (M_{+ve}): $w_{strip} = 660 + 0.55S = 660 + 0.55(2050) = 1788$ mm

Negative moment (M_{-ve}): $w_{strip} = 1220 + 0.25S = 1220 + 0.25(2050) = 1733$ mm

where $S =$ the girder spacing of 2050 mm.
d. *Determine the live loads on a 1-m strip*:
The unfactored wheel loads placed on a 1-m strip are

Overhang: 72.5 kN/1.190 m = 60.924 kN/m

Positive moment: 72.5 kN/1.788 m = 40.548 kN/m

Negative moment: 72.5 kN/1.733 m = 41.835 kN/m

4. **Determine the Wheel Load Location to Maximize the Live-Load Moments**
   Three alternative wheel load layouts will be considered to determine the maximum positive moment for live load. Three alternatives are investigated to illustrate the method used to determine the maximum moments. Only the controlling location of the wheel loads will be used to determine the maximum factored negative moment. In Table 24.1 , deck spans are numbered from left to right, with the left cantilever being span 1. Distances are measured from the leftmost girder of the span. Wheel loads are placed at the locations listed in the table.

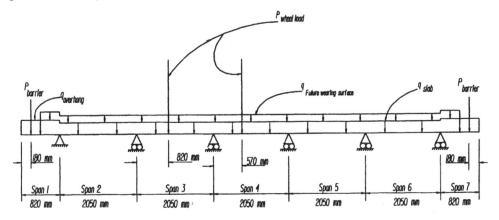

FIGURE 24.2 Loads: M_{+ve} Alternative 1 (one lane).

5. Calculate Unfactored Moments

Apply the unfactored loads determined in the steps above to a continuous 1-m-wide beam spanning across the girders (Figure 24.2). Based on these loads, the unfactored moments for the dead load and the live load alternatives are listed in Table 24.2. The bridge is symmetrical about the end of span 3; thus loads are only given for the left half of the bridge. The locations investigated to determine the controlling moments are shown in bold type.

6. Determine the Load Factors [1.3.2.1] [3.4.1]

$$Q = \eta \Sigma \gamma_i q_i$$

where
Q = factored load
η = load modifier
γ = load factor
q = unfactored loads

a. *Load modifier:* [1.3.3] [1.3.4] [1.3.5]:

$$\eta = \eta_D \eta_R \eta_i > 0.95$$

where
$\eta_D = 0.95$
$\eta_R = 0.95$
$\eta_i = 0.95$

Therefore, η is the maximum of $\eta = (0.95)(0.95)(0.95) = 0.86$ or $\eta = 0.95 \Leftarrow$ controls.

b. *Load factor:* [Table 3.4.1-1] [Table 3.4.1-2] [3.4.1] [3.3.2]:
γ_{DL}:

Maximum Load Factor	Minimum Load Factor	
$\gamma_{DCmax} = 1.25$	$\gamma_{DCmin} = 0.90$	$\Leftarrow$ slab and barrier rail
$\gamma_{DWmax} = 1.50$	$\gamma_{DWmin} = 0.65$	$\Leftarrow$ future wearing surface

γ_{LL} (Strength-1 Load Combination)

$$\gamma_{LL} = 1.75$$

$$\gamma_{IM} = 1.75$$

TABLE 24.2   Unfactored Moments

Location		Dead-Load Moment (kN–m)			Live-Load Moment (kN–m)			
		M_{DD}	M_{DB}	M_{DW}	$M_{+ve\,LL}$ Alt 1	$M_{+ve\,LL}$ Alt 2	$M_{+ve\,LL}$ Alt3	$M_{-ve\,LL}$ Alt 1
Span 1	Left	0	0	0	0	0	0	0
	.1 pt	−.02	0	0	0	0	0	0
	.2 pt	−.08	0	0	0	0	0	0
	.3 pt	−.18	−.38	0	0	0	0	0
	.4 pt	−.32	−.86	0	0	0	0	0
	.5 pt	−.49	−1.33	0	0	0	0	0
	.6 pt	−.71	−1.81	0	0	0	0	0
	.7 pt	−.97	−2.29	−.01	0	0	0	0
	.8 pt	−1.26	−2.76	−.03	0	0	0	0
	.9 pt	−1.60	−3.24	−.07	0	0	0	0
	Right	−1.98	−3.71	−.11	0	0	0	0
Span 2	Left	−1.98	−3.71	−.11	0	0	0	0
	.1 pt	−1.13	−3.24	.15	3.62	2.64	3.75	2.22
	.2 pt	−.46	−2.77	.34	7.24	5.29	7.51	4.43
	.3 pt	.04	−2.31	.46	10.87	7.93	11.26	6.65
	.4 pt	**.37**	**−1.84**	**.50**	**14.49**	**10.57**	**15.01**	**8.87**
	.5 pt	.52	−1.37	.48	9.80	13.22	10.45	11.08
	.6 pt	.50	−.90	.38	5.11	7.75	5.89	9.95
	.7 pt	.31	−.43	.21	.42	2.08	1.33	3.60
	.8 pt	−.05	.04	−.03	−4.28	−3.59	−3.23	−2.76
	.9 pt	−.59	.51	−.34	−8.97	−9.26	−7.79	−9.12
	Right	−1.30	.98	−.72	−13.66	−14.93	−12.35	−15.48
Span 3	Left	−1.30	.98	−.72	−13.66	−14.93	−12.35	−15.48
	.1 pt	−.54	.86	−.39	−6.50	−8.61	−5.84	−9.52
	.2 pt	.05	.74	−.12	.67	−2.29	.67	−3.56
	.3 pt	.46	.63	.07	6.00	4.03	5.35	2.40
	.4 pt	.71	.51	.20	4.85	8.72	3.54	8.36
	.5 pt	.78	.39	.25	3.70	6.73	1.74	9.09
	.6 pt	.67	.27	.23	2.55	4.73	−.07	6.47
	.7 pt	.40	.16	.13	1.40	2.74	−1.87	3.86
	.8 pt	−.05	.04	−.03	.25	.74	−3.68	1.24
	.9 pt	−.67	−.08	−.26	−.91	−1.25	−5.48	−1.38
	Right	−1.47	−.20	−.57	−2.06	−3.25	−7.29	−3.99
Span 4	Left	−1.47	−.20	−.57	−2.06	−3.25	−7.29	−3.99

M_{DD}, M_{DB}, M_{DW} represent the moments due to dead loads: deck(including slab and overhang), barrier rail, and future wearing surface moments respectively.

$M_{+ve\,LL}$ and $M_{-ve\,LL}$ represent the positive and negatiave live-load moments, respectively.

c. *Multiple presence factor* [Table 3.6.1.1.2-1]:

$$m_{1lane} = 1.20; \quad m_{2lanes} = 1.00; \quad m_{3lanes} = 0.85$$

d. *Dynamic load allowance* [3.6.1.2.2] [3.6.2]

$$IM = 0.33$$

7. Calculate the Factored Moments

$$M_u = \eta[\gamma_{DC}(M_{DD}) + \gamma_{DC}(M_{DB}) + \gamma_{DW}(M_{DW}) + (m)(1+IM)(\gamma_{LL})(M_{LL})]$$

TABLE 24.3 AASHTO LRFD Bridge Design Specifications

Type of Deck	Direction of Primary Strip Relative to Traffic	Width of Primary Strip (mm)
Concrete		
• Cast-in-place	Overhang	$1140 + 0.833X$
	Either Parallel or Perpendicular	$+M$: $660 + 0.55S$
		$-M$: $1220 + 0.25S$
• Cast-in-place with stay-in-place concrete formwork	Either Parallel or Perpendicular	$+M$: $660 + 0.55S$
		$-M$: $1220 + 0.25S$
• Precast, post-tensioned	Either Parallel or Perpendicular	$+M$: $660 + 0.55S$
		$-M$: $1220 + 0.25S$
Steel		
• Open grid	Main bars	$0.007P + 4.0S_b$
• Filled or partially filled grid	Main bars	Article 4.6.2.1.8 applies
• Unfilled, composite grids	Main bars	Article 9.8.2.4 applies
Wood		
• Prefabricated glulam		
• Non-interconnected	Parallel	$2.0h + 760$
	Perpendicular	$2.0h + 1020$
• Interconnected	Parallel	$2280 + 0.07L$
	Perpendicular	$4.0h + 760$
• Stress-laminated	Parallel	$0.066S + 2740$
	Perpendicular	$0.84S + 610$
• Spike-laminated	Parallel	$2.0h + 760$
• Continuous decks or	Perpendicular	$2.0h + 1020$
interconnected panels	Parallel	$2.0h + 760$
• Noninterconnected panels	Perpendicular	$2.0h + 1020$
• Planks		Plank width

Note: 1996 Interim Revisions, Table 4.6.2.1.3-1 — Equivalent Strips.

Positive Moment:
a. M_{+ve} Alternative 1 with one lane of live load (0.4 point of span 2):

$$M_u = 0.95[(1.25)(0.37) + (0.90)(-1.84) + (1.50)(0.50) + (1.20)(1.33)(1.75)(14.49)]$$
$$= 38.03 \text{ kNm} \Longleftarrow \text{controls positive moment}$$

b. M_{+ve} Alternative 2 with one lane of live load (0.5 point of span 2):

$$M_u = 0.95[(1.25)(0.52) + (0.90)(-1.37) + (1.50)(0.48) + (1.20)(1.33)(1.75)(13.22)]$$
$$= 35.20 \text{ kNm}$$

c. M_{+ve} Alternative 3 with two lanes of live load (0.4 point of span 2):

$$M_u = 0.95[(1.25)(0.37) + (0.90)(-1.84) + (1.50)(0.50) + (1.00)(1.33)(1.75)(15.01)]$$
$$= 32.77 \text{ kNm}$$

Negative Moment:
M_{-ve} Alternative 1 with one lane of live load (0.0 point of span 3):

$$M_u = 0.95[(1.25)(-1.30) + (0.90)(0.98) + (1.50)(-0.72) + (1.20)(1.33)(1.75)(-15.48)]$$
$$= -42.81 \text{ kNm} \Longleftarrow \text{controls negative moment}$$

As specified in [4.6.2.1.1] the entire width of the deck shall be designed for these maximum moments.

In reviewing the magnitude of the dead loads in comparison to the live loads it becomes apparent that the total combined dead load plus live-load moment is clearly dominated by the live-load moments. Therefore, performing complex analysis to determine exact dead-load moments is not justified. Using elementary beam formulas or other approximate methods is probably sufficient in most cases.

8. **Determine the Slab Reinforcement Detailing Requirements**
   a. *Determine the top deck reinforcement cover* [Table 5.12.3-1]:
   The top deck requires a minimum cover of 50 mm over the top mat reinforcement, unless environmental conditions at the site require additional cover. This cover does not include additional concrete placed on the deck for sacrificial purposes, grooving, or grinding. The clearance between the bottom mat reinforcement and the bottom of the deck slab is 25 mm, up to a No. 35 bar.
   b. *Determine deck reinforcement spacing requirements* [5.10.3.2]:

   $$s_{max} = 1.5(175 \text{ mm}) = 262 \text{ mm} \Leftarrow \text{controls}$$

   or $\qquad s_{max} = 450$ mm

   The minimum spacing of reinforcement is determined by [5.10.3.1] and is dependent on the bar size chosen and aggregate size.
   c. *Determine distribution reinforcement requirements* [9.7.2.3] [9.7.3.2]:
   Reinforcement is needed in the bottom of the slab in the direction of the girders in order to distribute the deck loads to the primary deck slab reinforcement, which is oriented transversely to traffic. The effective span length (S) is dependent on the girder type, which was not specified for this example in order to make the solution general. However, with the girder spacing of 2050 mm used in this example, the maximum value of 67% in the formula $(3840)/(\sqrt{S}) \leq 67\%$ would control. This value is a percentage of the primary slab reinforcement that is to be used for distribution reinforcement in the bottom of the slab and is placed parallel to the main girders.
   d. *Determine the minimum top slab reinforcement parallel to the girders* [5.10.8.2]:
   The top slab reinforcement shall be a minimum as required for shrinkage and temperature of $(0.75)(A_g)/f_y$. The top slab reinforcement may be controlled by the negative moment reinforcement needs of the main girders which would likely be greater than the shrinkage and temperature reinforcement requirements.

9. **Design**
   The entire width of the deck should be designed for the maximum positive and negative moments as specified in [4.6.2.1.1]. The positive and negative reinforcement is designed like a typical concrete beam. Concrete design is covered in many civil engineering texts and it is not the intent of this chapter to cover this topic. See Chapter 9 of this text for a discussion of concrete design methods.

24.2.2 Precast Concrete Bridge Decks

This type of bridge deck (Figure 24.3) has the advantage of not requiring significant curing and setup time prior to being loaded with traffic loads as is required for a cast-in-place deck. Therefore, this type of deck is often used for deck replacement [1]. Work can be done overnight or during off-peak traffic times when traffic can be temporarily detoured around the bridge or when a reduced number of traffic lanes can be provided and the deck is replaced in longitudinal sections while traffic continues in adjacent lanes.

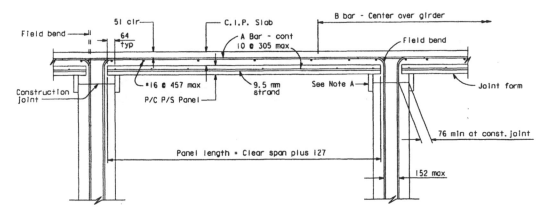

FIGURE 24.3 Precast concrete bridge deck. (From California Department of Transportation, Bridge Standard Detail Sheets, Sacramento, 1993. With permission.)

Precast decks can either serve as the final deck riding surface or a cast-in-place surface can be added on top. A cast-in-place surface uses the precast panels as the deck formwork, which would be placed between the girders. In adding a cast-in-place concrete surface, the problems associated with filling and maintaining the joints between the panels are reduced, and it assists in making the bridge deck composite with the girders. However, this method is at odds with the desire to open the deck to traffic as soon as possible. This led to the development of methods that do not require an additional final surface. If a final concrete surface is not placed on top, the joints between each panel must be successfully filled to avoid leakage and avoid future maintenance problems. This is typically done with expansive grouts or special epoxy crack sealers.

Precast panels may be prestressed, which reduces the depth of the precast panel between the main longitudinal girders or provides for increased spacing between the main girders for a given deck thickness. Perhaps more importantly, prestressing reduces the cracking in the deck. This is especially important for bridges exposed to aggressive environments.

Future widenings of decks using transverse deck prestressing is more difficult than a deck with conventional reinforcement. While a prestressed deck is likely to require less maintenance than a conventionally reinforced deck, repairs that may be required will be more difficult than for a conventionally reinforced deck [1].

24.2.3 Steel Grid Bridge Decks

Steel grids (Figure 24.4) can either be constructed off site as individual panels, constructed at the job site, or can even be assembled on the site of individual components. These grids are then usually welded or mechanically fastened to the supporting components. A significant advantage of open grid decks is their light weight. This can be especially important for existing bridges where the girders would require strengthening if extra dead load were added to the structure, and for movable bridges where dead load is minimized in order to limit loads on the mechanical systems.

However, open grids can result in poor riding quality and a loud whine as traffic crosses them. In addition, rainfall and possible spills fall directly through the deck and cannot be captured or controlled. This can lead to corrosion of components below the bridge and in environmental problems below the bridge because spills can fall directly into waterways. In addition to this, careful detailing is required to avoid fatigue problems associated with steel grid systems [1]. Interlocking grid systems that do not require welding may eliminate stresses that cause these types of failures. Older open steel grid systems have had problems with skid resistance, but newer systems are available that meet today's standards. However, open grids can wear over time, reducing the skid resistance.

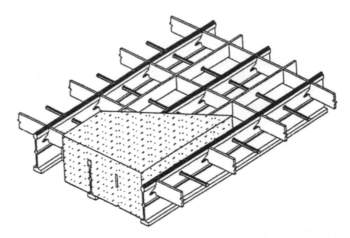

FIGURE 24.4   Steel grid bridge deck. (From American Grid, *Weldless Bridge Deck Systems*. With permission.)

Some of these problems can be solved by filling, or partially filling, the grid with concrete. The grid is then made composite with the concrete and acts as rebar in tension would in a conventionally reinforced-concrete beam. Although this type of deck is heavier than an open grid, it is still lighter than a conventional concrete bridge deck. Typically, an overfill of the grid is specified, which is assumed also to act compositely with the grid. This overfill provides added cover to minimize cracking, reduce corrosion of the steel grid, and improve ridability.

A variation of the filled and unfilled grid type uses an unfilled grid with shear studs. This makes it composite with a reinforced concrete slab which is set on top of the grid in an attempt to combine the advantages of a concrete and a steel grid deck.

24.2.4   Timber Bridge Decks

This was a common deck type prior to the advent of the automobile (Figure 24.5). Because of durability problems this type of deck is rarely used today, except on very low volume bridges, often in rural areas [1]. Timber bridge decks may be constructed using glulam timber panel decks and stress-laminated decks that are post-tensioned together. An asphalt wearing surface may be placed on top of the deck in an attempt to increase the durability of this type of bridge deck.

24.2.5   Steel Orthotropic Bridge Decks

An orthotropic steel deck is a deck plate acting as the flange section and is stiffened with longitudinal ribs and transverse floor beams. A wearing surface is added to act compositely with the deck plate. This subject is covered in greater depth in Chapter 14 of this text.

24.3   Approach Slabs

The structure approach slab provides motorists a smooth transition between the paved highway surface and the riding surface of a bridge. The most common construction consists of reinforced portland cement concrete (PCC). The need for an approach slab is generated by the differential vertical displacement that often occurs between the generally unyielding bridge structure and adjacent fill approaches.

The settlement of the adjacent fill can be gradual, over lengthy periods of time, or sudden, such as when ground motion from an earthquake "liquefies" unconsolidated material in the presence of groundwater. During such a catastrophic event, the role of the approach slab becomes paramount in enabling the passage of emergency vehicles immediately after the earthquake.

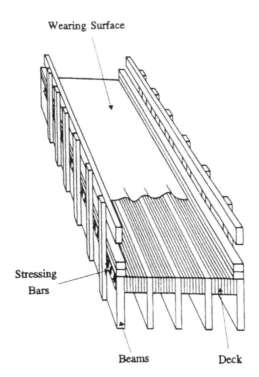

FIGURE 24.5 Timber bridge deck. (From Davalos, J. F., Wolcott, M. P., Dickson, B., and Brokaw J., *Quality Assurance and Inspection Manual for Timber Bridges,* 1992. With permission.)

The design and maintenance of an approach slab is greatly dependent on numerous factors which can affect the amount and rate of settlement that occurs. Careful attention to address these problematic features properly will lead to a serviceable, low-maintenance structure.

24.3.1 Structural Design Considerations

The reinforced-concrete approach slab is designed as a simply supported element, spanning from where it rests on the bridge abutment to the end that meets the roadway. The length of the slab is generally standardized to provide for the majority of applications. Currently, in California, the most commonly used approach slab is one which provides a 9-m transition from roadway to bridge.

The slab itself is usually about 300 mm thick and is reinforced with two-way mats of reinforcing steel both top and bottom. Although the slab is designed to support any passing live load adequately, equal importance is given to preventing the slab from cracking and "breaking up" under the constant cycles of loading by traffic, causing costly maintenance repairs and poor ridability for the motoring public. Reinforcement ratio indexes for the longitudinal bottom reinforcing bars of 0.0110, 0.0022 for the bottom transverse bars, and 0.0031 for the longitudinal top reinforcing bars have historically provided satisfactory performance for the 9-m-long slab. A minimum amount of transverse top reinforcing steel satisfies temperature and shrinkage requirements.

24.3.2 Settlement Problems

Settlement of the adjacent fill abutting a bridge is related to the geologic properties of the fill material and the native soil that underlies it. Considerations should be given to include composition and compaction of the fill material, settlement of the native soil due to the overburden imposed on it, and drainage conditions of the approaches.

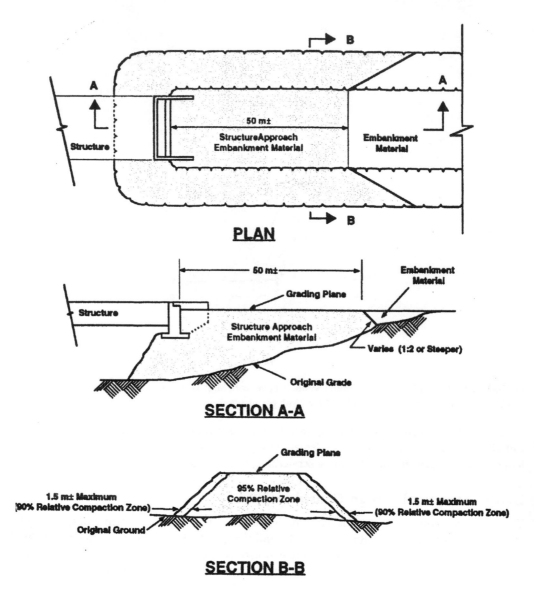

FIGURE 24.6 Limits of structure approach embankment material. (From California Department of Transportation, *Highway Design Manual,* Sacramento, 1995. With permission.)

The selection of fill material that creates the approaches to a structure is critical. Volume changes of the fill material result from rearrangement of soil particles, loss of moisture, gain of moisture, or frost and ice action. The fill material should not consist of soil types that are subject to influence under those conditions, such as cohesionless soils, which tend to settle under the vibration of traffic, or highly plastic soils, which are difficult to work with and highly compressible [3]. Approach fills should be constructed with selected material of slightly cohesive granular soil and extend well beyond the limits of abutments.

Adequate compaction of the fill material will limit the amount of future settlement by removing the potential volume changes of the soil. A minimum relative compaction value of 95% provides a reasonable lower limit that will minimize detrimental magnitudes of settlement of the fill. In the state of California, current practice is to require this type of "structure approach embankment material" 50 m from the back of the abutment (Figure 24.6).

Special attention is required in the areas immediately adjacent to the abutment or wingwalls, where confined spaces may limit the accessibility of large compaction equipment. Typically, small, hand-operated pieces of equipment are resorted to in order to compact the fill in these locations. It is imperative that proper procedures, which should be carefully addressed in the specifications, are adhered to and diligently monitored for quality assurance. This will limit the potential for differential settlement between the fill adjacent to the bridge structure, which is difficult to compact, and the fill away from the structure, which can be compacted by large equipment.

Subsidence of the underlying native soil is another major contribution to the poor performance of a bridge approach. With the additional mass of the overburden in the form of the approach fill, the native soil may not have sufficient strength to support the additional weight and will settle. Several options are possible to alleviate this problem.

The use of a surcharge to preload the approach site and preconsolidate compressible native material can be an effective solution to limit future, postconstruction settlement. The effectiveness of utilizing a surcharge is dependent on the amount of time available for this operation. Naturally, the more time that can elapse before construction of the approaches, the better. Care must be exercised that the amount and rate of applied surcharge does not exceed the shear strength of the native soil. This typically results in the necessity to load the site incrementally [3].

The use of drains in soft and compressible, slow-draining soil can accelerate its consolidation by basically dewatering the site and allowing primary consolidation to occur. This works well in thick, homogeneous layers of clay. Examples of vertical drains are sand drains and wick drains, which allow the migration of water within the soil to the surface where it can be collected and removed.

At sites where relatively shallow thicknesses of unsuitable, compressible soil may exist, a simple solution is to remove this layer of material. The replacement material would then be imported rock or suitable, well-compacted material.

A final alternative may be the use of lightweight fill material to limit the overburden on the native soil. Appropriate materials suitable for this purpose can consist of furnace slag, expanded shale, coal waste refuse, lightweight or cellular concrete, polystyrene foam, and other materials having small unit weights [3].

Proper drainage of the approach site, particularly behind and between the bridge abutment and wingwalls is essential to minimize surface and subsurface erosion and to reduce lateral hydrostatic pressure against these structural elements. An effective drainage system within the approach fill will consist of a medium located immediately adjacent to the abutment and wingwalls that will allow for the migration of water to bottom drains that will move the water away from the area. This can consist of pervious material such as gravel or a geocomposite drain and filter fabric, which is a man-made waffle-like material that channels water vertically downward. The placement of pervious material concurrently with the approach backfill is a difficult and more tedious process than the use of the geocomposite drain. The geocomposite drain system is the preferred method and is applied as shown in Figure 24.7.

24.3.3 Additional Considerations

Some related considerations that appear to have no effect on the performance and serviceability of approach slabs are worth mentioning. These include location of approaches relative to the structure, fill height, and average daily traffic (ADT).

The location of the approach, either leading to or away from the structure, seems to make negligible difference in the performance of the approach slab. One might think that the approach leading away from the structure would experience greater impact from vehicles as they come off the structure, resulting in greater settlement and damage to the approach slab. Studies show, however, essentially no difference between slabs at either end of the structures [5].

The height of approach fill also appears to have little effect on the long-term approach performance. It would seem reasonable to conclude that taller fill heights would exhibit the highest continued

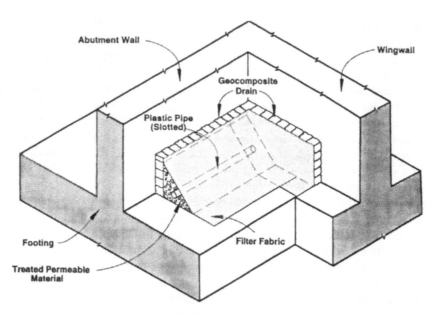

FIGURE 24.7   Abutment drainage details. (From California Department of Transportation, Highway Design Manual, Sacramento, 1995. With permission.)

settlement, but studies again show essentially uniform ground subsidence over all fill heights. A possible explanation is that higher fills are more likely to have been preconditioned in the form of allowing for a settlement period or surcharge of the fill prior to construction of the approach slab [5].

Finally, the volume of average daily traffic similarly exhibits negligible impact to the PCC structure approaches. AC approach data were inconclusive. It appears that the number of cycles of loading has no bearing on the long-term condition of the approach slabs. Maintenance records indicate essentially equal amounts of patching required for varying ADT levels [5].

24.4 Summary

There are several different types of bridge decks available to the design engineer. It is the responsibility of the engineer to determine the most appropriate type of bridge deck for a given site and situation. This can range from a new, short-span bridge with limited traffic to rehabilitation of a long-span bridge over water to be replaced at night because of significant traffic demands.

For most new bridges, cast-in-place concrete bridge decks are chosen as the most appropriate deck type. Typically, these types of decks are designed as a transverse beam supported by the main longitudinal girders. However, just as alternative types of bridge decks are available and are gaining greater acceptance, a new empirical method of designing cast-in-place concrete bridge decks that considers the arching effect between the girders has been introduced. While cast-in-place concrete decks designed as transverse beams have been the standard for decades, bridge deck type and design is continuing to evolve.

The major considerations for approach slabs are settlement of the approach fill, settlement of the underlying native soil, adequate drainage of the fill area behind the abutments and between the wingwalls, and adequate reinforcement; all must be addressed when designing a structure approach slab. Careful selection of suitable fill material and proper consolidation of the fill will limit post-construction settlement. Settlement of the native underlying soil can be eliminated through the practice of applying a surcharge, the use of preconstruction drains in clay soils, the removal of unsuitable compressible soil layers, and, finally, the use of lightweight fills. Proper drainage adjacent to the structure abutments and under the approach slabs will prevent subsurface erosion which would eventually undermine the underlying subbase of the approach slab.

References

1. Bettigole, N. H., and Robison, R., *Bridge Decks*, American Society of Civil Engineers, Danvers, 1997, 3, 40–62.
2. *AASHTO LRFD Bridge Design Specifications*, American Society of State Highway and Transportation Officials, Washington, D.C., 1994.
3. Wolde-Tinsae, A. M., Aggour, M. S., and Chini, S. A., Structural and Soil Provisions for Approaches to Bridges, Report FHWA/MD-89/04, Maryland Department of Transportation, Baltimore, July 1987, 1–48.
4. California Department of Transportation Highway Design Manual, 5th ed., State of California Department of Transportation, Sacramento, 1995, 600-46–600-55.
5. Stewart, C. F., Highway Structure Approaches, Report FHWA/CA/SD-85-05, California Department of Transportation, Sacramento, 1985.
6. Bridge Deck Construction Manual, State of California Department of Transportation, Sacramento, 1991.
7. Barker, R. M. and Puckett, J. A., *Design of Highway Bridges*, John Wiley & Sons, New York, 1997, 504–562.
8. *AASHTO LRFD Bridge Design Specifications*, 1996 Interim Revisions, American Society of State Highway and Transportation Officials, Washington, D.C., 1996.

25

Expansion Joints

Ralph J. Dornsife
Washington State Department of Transportation

25.1 Introduction

Expansion joint systems are integral, yet often overlooked, components designed to accommodate cyclic movements. Properly functioning bridge expansion joint systems accommodate these movements without imposing significant secondary stresses on the superstructure. Sealed expansion joint systems provide barriers preventing runoff water and deicing chemicals from passing through the joint onto bearing and substructure elements below the bridge deck. Water and deicing chemicals have a detrimental impact on overall structural performance by accelerating degradation of bridge deck, bearing, and substructure elements. In extreme cases, this degradation has resulted in premature, catastrophic structural failure. In fulfilling their functions, expansion joints must provide a reasonably smooth ride for motorists.

Perhaps because expansion joints are generally designed and installed last, they are often relegated to peripheral status by designers, builders, and inspectors. As a result of their geometric configuration and the presence of multiple-axle vehicles, expansion joint elements are generally subjected to a significantly larger number of loadings than other structural members. Impact, a consequence of bridge discontinuity inherent at a joint, exacerbates loading. Unfortunately, specific expansion joint systems are often selected based upon their initial cost with minimal consideration for long-term performance, durability, and maintainability. Consequently, a plethora of bridge maintenance problems plague them.

In striving to improve existing and develop new expansion joint systems, manufacturers present engineers with a multitudinous array of options. In selecting a particular system, the designer must carefully assess specific requirements. Magnitude and direction of movement, type of structure, traffic volumes, climatic conditions, skew angles, initial and life cycle costs, and past performance of alternative systems must all be considered. For classification in the ensuing discussion, expansion joint systems will be grouped into three broad categories depending upon the total movement range accommodated. Small movement range joints encompass all systems capable of accommodating total motion ranges of up to about 45 mm. Medium movement range joints include systems accommodating total motion ranges between about 45 mm and about 130 mm. Large movement range joints accommodate total motion ranges in excess of about 130 mm. These delineated ranges are somewhat arbitrary in that some systems can accommodate movement ranges overlapping these broad categories.

25.2 General Design Criteria

Expansion joints must accommodate movements produced by concrete shrinkage and creep, post-tensioning shortening, thermal variations, dead and live loads, wind and seismic loads, and structure settlements. Concrete shrinkage, post-tensioning shortening, and thermal variations are generally taken into account explicitly in design calculations. Because of uncertainties in predicting, and the increased costs associated with accommodating large displacements, seismic movements are usually not explicitly included in calculations.

Expansion joints should be designed to accommodate all shrinkage occurring after their instal-lation. For unrestrained concrete, ultimate shrinkage strain after installation, β, may be estimated as 0.0002 [1]. More-detailed estimations can be used which include the effect of ambient relative humidity and volume-to-surface ratios [2]. Shrinkage shortening of the bridge deck, Δ_{shrink}, in mm, is calculated as

$$\Delta_{shrink} = (\beta) \cdot (\mu) \cdot (L_{trib}) \cdot (1000 \text{ mm/m}) \tag{25.1}$$

where

L_{trib} = tributary length of structure subject to shrinkage; m
β = ultimate shrinkage strain after expansion joint installation; estimated as 0.0002 in lieu of more-refined calculations
μ = factor accounting for restraining effect imposed by structural elements installed before slab is cast [1]
   = 0.0 for steel girders, 0.5 for precast prestressed concrete girders, 0.8 for concrete box girders and T-beams, 1.0 for flat slabs

Thermal displacements are calculated using the maximum and minimum anticipated bridge deck temperatures. These extreme values are functions of the geographic location of the structure and the bridge type. Thermal movement, in mm, is calculated as

$$\Delta_{temp} = (\alpha) \cdot (L_{trib}) \cdot (\delta T) \cdot (1000 \text{ mm/m}) \tag{25.2}$$

where

α = coefficient of thermal expansion; 0.000011 m/m/°C for concrete and 0.000012 m/m/°C for steel
L_{trib} = tributary length of structure subject to thermal variation; m
δT = temperature variation; °C

Any other predictable movements following expansion joint installation, such as concrete post-tensioning shortening and creep, should also be included in the design calculations.

25.3 Jointless Bridges

Bridge designers have used superstructure continuity in an effort to avoid some of the maintenance problems associated with expansion joints [3]. This evolution from simple-span construction was facil-itated by the development of the moment distribution procedure published by Hardy Cross [4] in 1930.

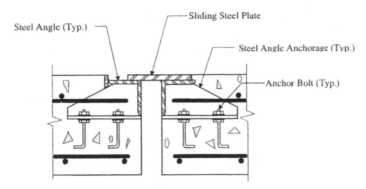

FIGURE 25.1   Sliding plate joint (cross section).

In recent years, some transportation agencies have extended this strategy by developing jointless bridge designs. Jointless bridges are characterized by continuous spans built integrally with their abutments. In many instances, approach slabs are tied to the superstructure slab or to the abutments. The resulting designs are termed *integral* or *semi-integral* depending upon the degree of continuity developed among superstructure, substructure, and approach slab elements. Design methods and details for jointless bridges vary considerably [3,5]. Many transportation agencies have empirically established maximum lengths for jointless bridges [5].

Jointless bridges should not be considered a panacea for addressing expansion jointmaintenance problems. As superstructure movements are restrained in jointless bridges, secondary stresses are induced in superstructure and substructure elements. Stresses may also be induced in approach slabs. If inadequately addressed during design, these stresses can damage structural elements and adjacent asphalt pavements. Damaged structural elements, slabs, and pavements are accompanied by increased probability of moisture infiltration, further exacerbating deterioration. Most jointless bridges have been built relatively recently [5]. Their long-term performance and durability will determine how extensively the jointless bridge concept is applied to future construction.

25.4   Small Movement Range Joints

Many different systems exist for accommodating movement ranges under about 45 mm. These include, but are not limited to, steel sliding plates, elastomeric compression seals, preformed closed cell foam, epoxy-bonded elastomeric glands, asphaltic plug joints, bolt-down elastomeric panels, and poured sealants. In this section, several of these systems will be discussed with an emphasis on design procedures and past performance.

25.4.1   Sliding Plate Joints

Steel sliding plates, shown in Figure 25.1, have been used extensively in the past for expansion joints in both concrete and timber bridge decks. Two overlapping steel plates are attached to the bridge deck, one on each side of the expansion joint opening. They are generally installed so that the top surfaces of the plates are flush with the top of the bridge deck. The plates are generally bolted to timber deck panels or embedded with steel anchorages into a concrete deck. Steel plate widths are sized to accommodate anticipated total movements. Plate thicknesses are determined by structural requirements.

Standard steel sliding plates do not generally provide an effective seal against intrusion of water and deicing chemicals into the joint and onto substructure elements. As a result of plate corrosion and debris collection, the steel sliding plates often bind up, impeding free movement of the superstructure. Repeated impact and weathering tend to loosen or break anchorages to the bridge deck.

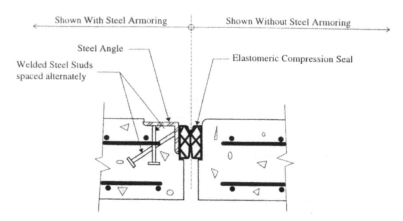

FIGURE 25.2   Compression seal joint (cross section).

Consequently, sliding plate systems are rarely specified for new bridge construction today. Nevertheless, sliding plate systems still exist on many older structures. These systems can be replaced with newer systems providing increased resistance against water and debris infiltration. In situations where the integrity of the deck anchorage has not been compromised, sliding plates can be retrofitted with poured sealants or elastomeric strip seals.

25.4.2   Compression Seal Joints

Compression seals, shown in Figure 25.2, are continuous elastomeric sections, typically with extruded internal web systems, installed within an expansion joint gap to seal the joint effectively against water and debris infiltration. Compression seals are held in place by mobilizing friction against adjacent vertical joint faces. Hence, design philosophy requires that they be sized and installed to be always in a state of compression. Compression seals may be installed against smooth concrete faces or against steel armoring. When installed directly against concrete, polymer concrete nosing material is often used to provide added impact resistance. Combination lubricant/adhesive is typically used to install the seal in its compressed state.

Because compression seals are held in place by friction, their performance is extremely dependent upon the close correlation of constructed joint width and design joint width. If the joint opening is constructed too wide, friction force will be insufficient to prevent the compression seal from slipping out of the joint at wider expansion gap widths. Relaxation of the elastomer and debris accumulation atop the seal contribute to seal slippage. To minimize slippage and maximize compression seal performance, a joint may be formed narrower than the design width, then sawcut immediately prior to compression seal installation. The sawcut width is calculated based upon ambient bridge deck temperature and the degree of slab shrinkage which has already occurred. As an alternative to sawcutting, block outs can be formed on each side of the joint during bridge deck casting. Prior to compression seal installation, concrete is cast into the block outs, often with steel armoring, to form an expansion gap width compatible with ambient conditions.

In design calculations, the maximum and minimum compressed widths of the seal are generally set at 85 and 40% of the uncompressed width [1]. These widths are measured perpendicular to the axis of the joint. It is also generally assumed that the width of the seal at about 20°C is 60% of its uncompressed width. For skewed joints, bridge deck movement must be separated into components perpendicular to and parallel to the joint axis. Shear displacement of the compression seal should be limited to a specified percentage of its uncompressed width, usually set at about 22% [1]. Additionally, the expansion gap width should be set so that the compression seal can be installed over a reasonably wide range of construction temperatures. Manufacturers' catalogues generally specify the minimum expansion gap widths into which specific size compression seals can be installed. The expansion gap width should be specified on the contract drawings as a function of the bridge deck temperature.

Design relationships can be stated as follows:

$$\Delta_{temp\text{-}normal} = \Delta_{temp} \cdot \cos\theta \qquad \text{[thermal movement normal to joint]} \qquad (25.3)$$

$$\Delta_{temp\text{-}parallel} = \Delta_{temp} \cdot \sin\theta \qquad \text{[thermal movement parallel to joint]} \qquad (25.4)$$

$$\Delta_{shrink\text{-}normal} = \Delta_{shrink} \cdot \cos\theta \qquad \text{[shrinkage movement normal to joint]} \qquad (25.5)$$

$$\Delta_{shrink\text{-}parallel} = \Delta_{shrink} \cdot \sin\theta \qquad \text{[shrinkage movement parallel to joint]} \qquad (25.6)$$

$$W_{min} = W_{install} - [(T_{max} - T_{install})/(T_{max} - T_{min})]\,\Delta_{temp\text{-}normal} > 0.40W \qquad (25.7)$$

$$W_{max} = W_{install} + [(T_{install} - T_{min})/(T_{max} - T_{min})]\,\Delta_{temp\text{-}normal} + \Delta_{shrink\text{-}normal} < 0.85W \qquad (25.8)$$

where
θ = skew angle of expansion joint, measured with respect to a line perpendicular to the bridge longitudinal axis; degrees
W = uncompressed width of compression seal; mm
$W_{install}$ = expansion gap width at installation
$T_{install}$ = bridge deck temperature at time of installation; °C
W_{min}, W_{max} = minimum and maximum expansion gap widths; mm
T_{min}, T_{max} = minimum and maximum bridge deck temperatures; °C

Multiplying Eq. (25.7) by −1.0, adding to Eq. (25.8), and rearranging yields:

$$W > (\Delta_{temp\text{-}normal} + \Delta_{shrink\text{-}normal})/0.45 \qquad (25.9)$$

Similarly,

$$W > (\Delta_{temp\text{-}parallel} + \Delta_{shrink\text{-}parallel})/0.22 \qquad (25.10)$$

Now, assuming $W_{install} = 0.6\,W$,

$$W_{max} = 0.6W + [(T_{install} - T_{min})/(T_{max} - T_{min})]\,\Delta_{temp\text{-}normal} + \Delta_{shrink\text{-}normal} < 0.85W \qquad (25.11)$$

which, upon rearranging, yields:

$$W > 4\,[(T_{install} - T_{min})/(T_{max} - T_{min}) \cdot (\Delta_{temp\text{-}normal}) + \Delta_{shrink\text{-}normal}] \qquad (25.12)$$

Equations (25.9), (25.10), and (25.12) are used to calculate the required compression seal size. Next, expansion gap widths at various construction temperatures can be evaluated.

25.4.3 Asphaltic Plug Joints

Asphaltic plug joints comprise liquid polymer binder and graded aggregates compacted in pre-formed block outs as shown in Figure 25.3. The compacted composite material is referred to as polymer modified asphalt (PMA). These joints have been used to accommodate movement ranges up to 50 mm. This expansion joint system was developed in Europe and can be adapted for use with concrete or asphalt bridge deck surfaces. The PMA is installed continuously within a block out centered over the expansion joint opening with the top of the PMA flush with the roadway surface. A steel plate retains the PMA at the bottom of the block out during installation. The polymer

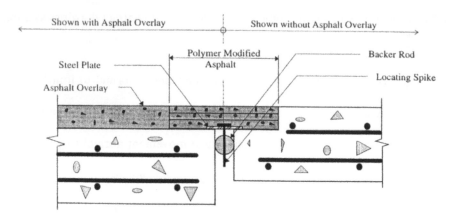

FIGURE 25.3 Asphaltic plug joint (cross section).

binder material is generally installed in heated form. Aggregate gradation, binder properties, and construction quality are critical to asphaltic plug joint performance.

The asphaltic plug joint is designed to provide a smooth, seamless roadway surface. It is relatively easy to repair, is not as susceptible to snowplow damage as other expansion joint systems, and can be cold-milled and/or built up for roadway resurfacing. The performance of asphaltic plug joints in the United States has been somewhat erratic [6]. The material properties of PMA vary with temperature. Asphaltic plug joints have demonstrated a proclivity to soften and creep at warmer temperatures, exhibiting wheel rutting and eventual migration of PMA out of the block outs. In very cold temperatures, the PMA can become brittle and crack at the plug joint-to-pavement interface, making the joint susceptible to water infiltration. Ongoing research is investigating these issues and developing comprehensive design guidelines, material specifications, and installation procedures to improve their performance [6].

As with all expansion joint systems, designers must understand the limitations of asphaltic plug joints. These joints were not designed for, and should not be used to, accommodate differential vertical displacements, as may occur at longitudinal joints. Because of PMA creep susceptibility, asphaltic plug joints should not be used where the roadway is subject to significant traffic acceleration and braking. Examples include freeway off-ramps and roadway sections in the vicinity of traffic signals. Asphaltic plug joints have also performed poorly in highly skewed applications and in applications subjected to large rotations. Maintaining the minimum block-out depth specified by the manufacturer is particularly critical to successful performance. In spite of these limitations, asphaltic plug joints do offer advantages not inherent in other expansion joint systems.

25.4.4 Poured Sealant Joints

Durable low-modulus sealants, poured cold to provide watertight expansion joint seals as shown in Figure 25.4, have been used in new construction and in rehabilitation projects. Properties and application procedures vary between products. Most silicone sealants possess good elastic performance over a wide range of temperatures while demonstrating high levels of resistance to ultraviolet and ozone degradation. Rapid-curing sealants are ideal candidates for rehabilitation in situations where significant traffic disruption from extended traffic lane closure is unacceptable. Other desirable properties include self-leveling and self-bonding capabilities. Installation procedures vary among different products, with some products requiring specialized equipment for mixing individual components. Designers must assess the design and construction requirements, weighing desirable properties against material costs for alternative sealants.

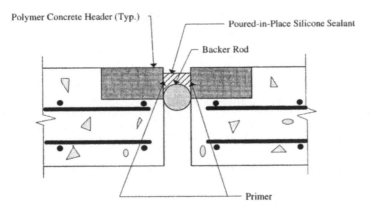

FIGURE 25.4 Poured sealant joint (cross section).

Most sealants can be installed against either concrete or steel. Particularly in rehabilitation projects, it is extremely critical that the concrete or steel substrates be thoroughly cleaned before the sealant is placed. Some manufacturers require application of specific primers onto substrate surfaces prior to sealant placement to enhance bonding. Debonding of sealant from substrate concrete or steel, compromising the integrity of the watertight seal, has previously plagued poured sealant joints. The latest products are relatively new, but have demonstrated good short-term performance and versatility of use in bridge rehabilitation. Their long-term durability will determine the extent of their future application.

Poured sealant joints should be designed based upon manufacturers' recommendations. Maximum and minimum working widths of the poured sealant joint are generally recommended as a percentage of the sealant joint width at installation. A minimum recess is typically required between the top of the roadway surface and the top of the sealant. This recess is critical in preventing tires from contacting and debonding the sealant from its substrate material.

25.4.5 Design Example 1

Given
A reinforced-concrete box-girder bridge has an overall length of 70 m. A compression seal expansion joint at each abutment will accommodate half of the total bridge movement. These expansion joints are skewed 20°. Bridge deck temperatures are expected to range between –15°C and 40°C during the life of the structure.

Find
Compression seal sizes and construction gap widths at 5, 20, and 30°C.

Solution
Step 1: Calculate temperature and shrinkage movement.

$$\text{Temperature: } \Delta_{temp} = (\tfrac{1}{2})(0.000011 \text{ m/m/°C})(55°C)(70 \text{ m})(1000 \text{ mm/m}) = 21 \text{ mm}$$

$$\text{Shrinkage: } \Delta_{shrink} = (\tfrac{1}{2})(0.0002 \text{ m/m})(0.8)(70 \text{ m})(1000 \text{ mm/m}) \qquad = \underline{6 \text{ mm}}$$

$$\text{Total deck movement at the joint:} \qquad 27 \text{ mm}$$

$$\Delta_{temp\text{-}normal} + \Delta_{shrink\text{-}normal} = (27 \text{ mm})(\cos 20°) = 25 \text{ mm}$$

$$\Delta_{temp\text{-}parallel} + \Delta_{shrink\text{-}parallel} = (27 \text{ mm})(\sin 20°) = 9.2 \text{ mm}$$

Step 2: Determine compression seal width required from Eqs. (25.9), (25.10), and (25.12).

$W > 25$ mm/0.45 = 56 mm

$W > 9.2$ mm/0.22 = 42 mm

$W > 4 \cdot [(20°C + 15°C)/(40°C + 15°C) \cdot (21$ mm$) + 6$ mm$] \cdot \cos 20° = 73$ mm

<u>Use 75 mm compression seal</u>.

Step 3: Evaluate construction gap widths for various temperatures for a 75 mm compression seal.

Construction width at 20°C = $0.6 \cdot (75$ mm$) = 45$ mm

Construction width at 5°C = 45 mm + $[(20°C - 5°C)/(40°C + 15°C)] \cdot (21$ mm$) \cdot (\cos 20°) = 50$ mm

Construction width at 30°C = 45 mm − $[(30°C - 20°C)/(40°C + 15°C)] \cdot (21$ mm$) \cdot (\cos 20°) = 41$ mm

Conclusion
<u>Use a 75-mm compression seal</u>. Construction gap widths for installation temperatures of 5, 20, and 30°C are 50, 45, and 41 mm, respectively.

25.5 Medium Movement Range Joints

Medium movement range expansion joints accommodate movement ranges from about 45 mm to about 130 mm and include sliding plate systems, bolt-down panel joints (elastomeric expansion dams), strip seals, and steel finger joints. Sliding plate systems were previously discussed under small motion range joints.

25.5.1 Bolt-Down Panel Joints

Bolt-down panel joints, also referred to as elastomeric expansion dams, consist of monolithically molded elastomeric panels reinforced with steel plates as shown in Figure 25.5. They are bolted into block outs formed in the concrete bridge deck on each side of an expansion joint gap. Manufacturers fabricate bolt-down panels in varying widths roughly proportional to the total allowable movement range. Expansion is accompanied by uniform stress and strain across the width of the panel joint between anchor bolt rows. Unfortunately, the bolts and nuts connecting bolt-down panels to bridge decks are prone to loosening and breaking under high-speed traffic. The resulting loose panels and hardware in the roadway present hazards to vehicular traffic, particularly motorcycles. Consequently, to mitigate liability, some transportation agencies avoid using bolt-down panel joints.

25.5.2 Strip Seal Joints

An elastomeric strip seal expansion joint system, shown in Figure 25.6, consists of a preformed elastomeric gland mechanically locked into metallic edge rails embedded into concrete on each side of an expansion joint gap. Movement is accommodated by unfolding of the elastomeric gland. Steel studs or reinforcing bars are generally welded to the edge rails to facilitate bonding with the concrete in formed block outs. In some instances the edge rails are bolted in place. Edge rails also furnish armoring for the adjacent bridge deck concrete. Properly installed strip seals have demonstrated relatively good performance. Damaged or worn glands can be replaced with minimal traffic disruptions.

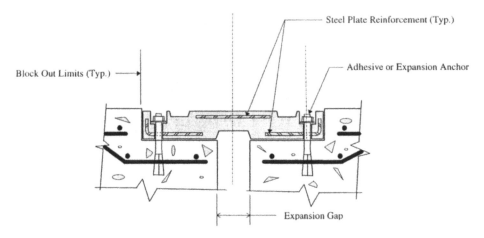

FIGURE 25.5 Bolt-down panel joint (cross section).

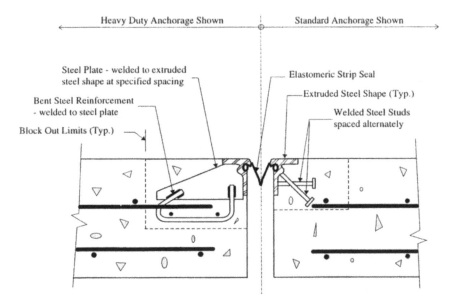

FIGURE 25.6 Elastomeric strip seal joint (cross section).

The elastomeric glands exhibit a proclivity for accumulating debris. In some instances, this debris can resist joint movement and result in premature gland failure.

25.5.3 Steel Finger Joints

Steel finger joints, shown in Figure 25.7, have been used to accommodate medium and large movement ranges. These joints are generally fabricated from steel plate and are installed in cantilever or prop cantilever configurations. The steel fingers must be designed to support traffic loads with sufficient stiffness to preclude excessive vibration. In addition to longitudinal movement, they must also accommodate any rotation or differential vertical deflection across the joint. To minimize the potential for damage from snowplow blade impact, steel fingers may be fabricated with a slight downward taper toward the joint centerline. Generally, steel finger joints do not provide a seal against water intrusion to substructure elements. Elastomeric or metallic troughs can be installed beneath the steel finger joint assembly to catch and redirect water and debris runoff. However, unless regularly maintained, these troughs clog and become ineffective [3].

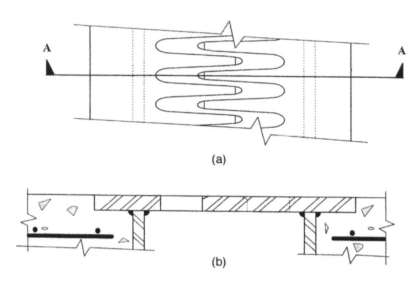

(a)

(b)

FIGURE 25.7   Steel finger joint. (a) Plan view; (b) section A–A.

25.5.4   Design Example 2

Given

A steel-plate girder bridge has a total length of 180 m. It is symmetrical and has a strip seal expansion joint at each end. These expansion joints are skewed 15°. Bridge deck temperatures are expected to range between −35°C and 50°C during the life of the structure. Assume an approximate installation temperature of 20°C.

Find

Type A and Type B strip seal sizes and construction gap widths at 5, 20, and 30°C. Type A strip seals have a 15 mm gap at full closure. Type B strip seals are able to fully close, leaving no gap.

Solution

Step 1: Calculate temperature and shrinkage movement.

Temperature: Δ_{temp} = (½)(0.000012 m/m/°C)(85°C)(180 m)(1000 mm/m) = 92 mm

Shrinkage: Δ_{shrink} = 0.0 (no shrinkage, μ = 0.0 for steel bridge)   _____

Total deck movement at the joint: 92 mm

$\Delta_{temp\text{-}normal\text{-}closing}$ = (50°C − 20°C)/(50°C + 35°C)(92 mm)(cos 15°) =   31 mm

$\Delta_{temp\text{-}normal\text{-}opening}$ = (20°C + 35°C)/(50°C + 35°C)(92 mm)(cos 15°) =   58 mm

Step 2: Determine strip seal size required. Assume a minimum construction gap width of 40 mm at 20°C.

Type A: Construction gap width of 40 mm at 20°C will not accommodate 31 mm closing and still allow a 15 mm gap at full closure. Therefore, minimum construction gap width at 20°C must be 31 mm + 15 mm = 46 mm.

Size required = 46 mm + 58 mm = 104 mm → <u>Use 100 mm strip seal</u>

Type B: Construction width of 40 mm at 20°C is adequate.

> Size required = 40 mm + 58 mm = 98 mm → <u>Use 100 mm strip seal</u>

Step 3: Evaluate construction gap widths for various temperatures for a 100 mm strip seal.
Type A: Required construction gap width at 20°C = 15 mm + 31 mm = 46 mm

> Construction gap width at 5°C = 46 mm + (20°C – 5°C)/(20°C + 35°C)(58 mm)
> = 62 mm

> Construction gap width at 30°C = 46 mm – (30°C – 20°C)/(50°C – 20°C)(31 mm)
> = 36 mm

Type B: Construction width of 40 mm at 20°C is adequate.

> Construction gap width at 5°F = 40 mm + (20°C – 5°C)/(20°C + 35°C)(58 mm)
> = 56 mm

> Construction gap width at 30°F = 40 mm – (30°C – 20°C)/(50°C – 20°C)(31 mm)
> = 30 mm

Conclusion
<u>Use a 100 mm strip seal</u>. Construction gap widths for Type A strip seals at installation temperatures of 5, 20, and 30°C are 62, 46, and 36 mm, respectively. Construction gap widths for Type B strip seals at installation temperatures of 5, 20, and 30°C are 56, 40, and 30 mm, respectively.

25.6 Large Movement Range Joints

Large movement range joints accommodate more than 130 mm of total movement and include bolt-down panel joints (elastomeric expansion dams), steel finger joints, and modular expansion joints. Bolt-down panel and steel finger joints were previously discussed as medium movement range joints.

25.6.1 Modular Bridge Expansion Joints

Modular bridge expansion joints (MBEJ), shown in Figure 25.8, are complex, expensive, structural systems designed to provide watertight wheel load transfer across wide expansion joint openings. These systems were developed in Europe and introduced in the United States in the 1960s [7]. They are generally shipped to the construction site for installation in a completely assembled configuration. MBEJs comprise a series of center beams supported atop support bars. The center beams are oriented parallel to the joint axis while the support bars span parallel to the primary direction of movement. MBEJs can be classified as either single-support bar systems or multiple-support bar systems. In multiple-support bar systems, each center beam is supported by a separate support bar at each support location. Figure 25.8 depicts a multiple-support bar system. In the more complex single-support bar system, one support bar supports all center beams at each support location. This design concept requires that each center beam be free to translate along the longitudinal axis of the support bar as the joint opens and closes. This is accomplished by attaching steel yokes to the underside of the center beams. The support bar passes through the openings in the yokes. Elastomeric springs between the underside of each center beam and the top of the support bar and between the bottom of the support bar and the bottom of the yoke support each center beam and permit it to translate along the longitudinal axis of the support bar.

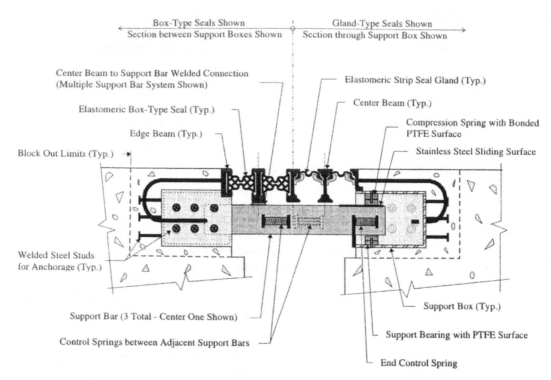

FIGURE 25.8 Modular bridge expansion joint (multiple support bar system), cross section.

The support bars are, in turn, supported on sliding bearings mounted within support boxes. Polytetrafluorethylene (PTFE)-to-stainless-steel interfaces between elastomeric support bearings and support bars facilitate unimpeded translation of the support bars as the expansion gap varies. Control springs between adjacent support bars and between support bars and support boxes of multiple-support bar MBEJs are designed to maintain equal distances between center beams as the expansion gap varies. The support boxes are embedded in bridge deck concrete on each side of the expansion joint. Elastomeric strip seals or elastomeric box-type seals attach to adjacent center beams, providing resistance to water and debris intrusion.

The highly repetitive nature of axle loads predisposes MBEJ components and connections to high fatigue susceptibility, particularly at connections of center beam to support bar. Bolted connections have, generally, performed poorly. Welded connections are preferred, but must be carefully designed, fatigue-tested, fabricated, and inspected to assure satisfactory performance and durability. Field-welded center beam splices are also highly fatigue susceptible, requiring careful detailing, welding, and inspection. A lack of understanding of the dynamic response of these systems, connection detail complexity, and the competitive nature of the marketplace have exacerbated fatigue susceptibility. Fortunately, current research is developing fatigue-resistant structural design specifications in addition to focusing on developing minimum performance standards, performance and acceptance test methods, and installation guidelines for MBEJs [7,8].

Calculated total movements establish MBEJ size. Often, an allowance is made to provide a nominal factor of safety on the calculated movements. Currently available systems permit 75 mm of movement per strip seal element; hence, the total movement rating provided will be a multiple of 75 mm. To minimize impact and wear on bearing elements, the maximum gap between adjacent center beams is limited, typically to about 90 mm [9]. To facilitate installation within concrete block outs, contract drawings should specify the face-to-face distance of edge beams as a function of temperature at the time of installation.

Design relationships can be expressed as:

$$n = MR/mr \qquad (25.13)$$

$$G_{min} = (n - 1) \cdot (w) + (n) \cdot (g) \qquad (25.14)$$

$$G_{max} = G_{min} + MR \qquad (25.15)$$

where

MR   = total movement rating of the MBEJ system; mm
mr   = movement rating per strip seal element; mm
n = number of seals
$n - 1$ = number of center beams
w = width of each center beam; mm
g = minimum gap per strip seal element at full closure; mm
G_{min} = minimum face-to-face distance of edge beams; mm
G_{max} = maximum face-to-face distance of edge beams; mm

Structural design of MBEJs is generally performed by the manufacturer. Project specifications should require that the manufacturer submit structural calculations, detailed fabrication drawings, and applicable fatigue tests for approval. All elements and connections must be designed and detailed to resist fatigue stresses imposed by repetitive vertical and horizontal wheel loadings. Additionally, MBEJs should be detailed to provide access for inspection and periodic maintenance, including replacement of seals, control springs, and bearing components.

25.6.2 Design Example 3

Given
Two cast-in-place post-tensioned concrete box-girder bridge frames meet at an intermediate pier where they are free to translate longitudinally. Skew angle is 0° and bridge deck ambient temperatures range from −15 to 50°C. A MBEJ will be installed 60 days after post-tensioning operations have been completed. Specified creep is 150% of elastic shortening. Assume that 50% of shrinkage has already occurred at installation time. The following longitudinal movements were calculated for each of the two frames:

	Frame A	Frame B
Shrinkage	30 mm	15 mm
Elastic shortening	36 mm	20 mm
Creep (1.5 × elastic shortening)	54 mm	30 mm
Temperature fall (20 to −15°C)	76 mm	38 mm
Temperature rise (20 to 50°C)	66 mm	33 mm

Find
MBEJ size required to accommodate the total calculated movements and the installation gaps measured face to face of edge beams, "$G_{install}$," at 5, 20, and 30°C.

Solution
Step 1: Determine MBEJ size.

Total opening movement (Frame A) = (0.5)(30 mm) + 54 mm + 76 mm = 145 mm
Total opening movement (Frame B) = (0.5)(15 mm) + 30 mm + 38 mm = 76 mm
Total opening movement (both frames) = 145 mm + 76 mm = 221 mm
Total closing movement (both frames) = 66 mm + 33 mm = 99 mm

Determine size of modular joint, including a 15% allowance:

1.15(221 mm + 99 mm) = 368 mm → <u>Use 375 mm movement rating MBEJ</u>.

Step 2: Evaluate installation gaps measured face to face of edge beams at 5, 20, and 30°C.

MR = 375 mm (MBEJ movement range)
mr = 75 mm (maximum movement rating per strip seal element)
n = 375 mm/75 mm = 5 strip seal elements
$n - 1$ = 4 center beams
w = 65 mm (center beam top flange width)
g = 0 mm
G_{min} = (4)(65 mm) + (4)(0 mm) = 260 mm
G_{max} = 260 mm + 375 mm = 635 mm
G_{20C} = G_{min} + Total closing movement from temperature rise
= 260 mm + 1.15(99 mm) = 374 mm → <u>Use 375 mm</u>.
G_{5C} = 375 mm + [(20°C – 5°C)/(20°C + 15°C)] · (76 mm + 38 mm) = 424 mm
G_{30C} = 375 mm – [(30°C – 20°C)/(50°C – 20°C)] · (66 mm + 33 mm) = 342 mm

Check spacing between center beams at minimum temperature:

$$G_{-15C} = 375 \text{ mm} + 221 \text{ mm} = 596 \text{ mm}$$

Maximum spacing = [596 mm – (4) · (65 mm)]/5 = 67 mm < 90 mm   OK

Check spacing between center beams at 20°C for seal replacement:

Spacing = [375 mm – 4(65 mm)]/5 = 23 mm < 40 mm

Therefore, center beams must be mechanically jacked in order to replace strip seal elements.

Conclusion
Use a MBEJ with a 375 mm movement rating. Installation gaps measured face to face of edge beams at installation temperatures of 5, 20, and 30°C are 424, 375, and 342 mm, respectively.

25.7 Construction and Maintenance

In conjunction with appropriate design procedures, the long-term performance and durability of expansion joint systems require the synergistic application of high-quality fabrication, competent construction practices, assiduous inspection, and routine maintenance. Expansion joint components and connections experience severe loading under harsh environmental conditions. An adequately designed system must be properly manufactured, installed, and maintained to assure adequate performance under these conditions. The importance of quality control must be emphasized. Contract drawings and specifications must explicitly state design, material, fabrication, installation, and quality control requirements. Structural calculations and detailed fabrication drawings should be submitted to the bridge designer for careful review and approval prior to fabrication.

Experience and research will continue to improve expansion joint system technology [10,11]. It is vitally important that design engineers keep abreast of new technological developments. Interdisciplinary and interagency communication facilitates exchange of important information. Maintenance personnel can furnish valuable feedback to designers for implementation in future designs. Designers can provide valuable guidance to maintenance personnel with the goal of increasing service life. Manufacturers furnish designers and maintenance crews with guidelines and limitations for successfully designing and maintaining their products. In turn, designers and maintenance personnel provide feedback to manufacturers on the performance of their products and how they might be improved. Communication among disciplines is key to improving long-term performance and durability.

References

1. Washington State Department of Transportation, Miscellaneous design, in Bridge Design Manual, Washington State Department of Transportation, Olympia, 1997, chap. 8.
2. American Association of State Highway and Transportation Officials, Concrete structures, in *AASHTO LRFD Bridge Design Specifications*, AASHTO Washington, D.C., 1994, sect. 5, 5–14.
3. Burke, M. P., Jr., Bridge Deck Joints, National Cooperative Highway Research Program Synthesis of Highway Practice Report 141, Transportation Research Board, National Research Council, Washington, D.C., 1984.
4. Cross, H., Analysis of continuous frames by distributing fixed-end moments, *Proc. Am. Soc. Civil Eng.*, May, 919, 1930.
5. Steiger, D. J., Field evaluation and study of jointless bridges, in *Third World Congress on Joint Sealing and Bearing Systems for Concrete Structures*, Stoyle, J. E., Ed., American Concrete Institute, Farmingham Hills, MI, 1991, 227.
6. Bramel, B. K., Puckett, J. A., Ksaibati, K., and Dolan, C. W., "Asphalt plug joint usage and perceptions in the United States, draft copy of a paper prepared for the annual meeting of the Transportation Research Board, August 1996.
7. Kaczinski, M. R., Dexter, R. J., and Connor, R. J., Fatigue design and testing of modular bridge expansion joints, in *Fourth World Congress on Joint Sealants and Bearing Systems for Concrete Structures*, Atkinson, B., Ed., American Concrete Institute, Farmingham, Hill, MI, 1996, 97.
8. Dexter, R. J., Connor, R.J., and Kaczinski, M.R., Fatigue Design of Modular Bridge Expansion Joints, National Cooperative Highway Research Program Report 402 , Transportation Research Board, National Research Council, Washington, D.C., April, 1997.
9. Van Lund, J. A., Bridge deck joints in Washington State, in *Third World Congress on Joint Sealing and Bearing Systems for Concrete Structures*, Stoyle, J. E., Ed., American Concrete Institute, Farmingham Hills, MI, 1991, 371.
10. Stoyle, J. E., Ed., *Third World Congress on Joint Sealing and Bearing Systems for Concrete Structures*, American Concrete Institute, Farmingham Hills, MI, 1991.
11. Atkinson, B., Ed., *Fourth World Congress on Joint Sealants and Bearing Systems for Concrete Structures*, American Concrete Institute, Farmingham Hills, MI, 1996.

Index

Milton Keynes UK
Ingram Content Group UK Ltd.
UKHW051901071024
449327UK00025B/2051